Soil Health and Nutrition Management

Soil Health and Nutrition Management

_____

Edited by

Naveen Chandra Joshi
Amity Institute of Microbial Technology, Amity University, India

Thomas Leustek
School of Environmental and Biological Sciences, Rutgers

Prashant Kumar Singh
Department of Biotechnology, Mizoram Central University, India

CABI

CABI is a trading name of CAB International

CABI
Nosworthy Way
Wallingford
Oxfordshire OX10 8DE
UK

Tel: +44 (0)1491 832111
E-mail: info@cabi.org
Website: www.cabi.org

CABI
200 Portland Street
Boston
MA 02114
USA

Tel: +1 (617)682-9015
E-mail: cabi-nao@cabi.org

A catalogue record for this book is available from the British Library, London, UK.

ISBN-13: 9781800624573 (hardback)
            9781800624580 (ePDF)
            9781800624597 (ePub)

DOI: 10.1079/9781800624597.0000

Commissioning Editor: Rebecca Stubbs
Editorial Assistant: Helen Elliott
Production Editor: Rosie Hayden

Typeset by Straive Pondicherry, India

Contents

Contributors

Sanghamitra Adak, Divis ion of Plant Biology, Bose Institute, Kankurgachi, Kolkata, West Bengal, India

Sakshi Arora, Amity Institute of Microbial Technology, Amity University, Noida, Uttar Pradesh, India. Email: arora.17sakshi@gmail.com

Manika Bhatia, TERI School of Advanced Studies, New Delhi, India

Anchal Chaudhary, Department of Microbiology, Guru Nanak Dev University, Amritsar, Punjab, India

Divya Chaudhary, Amity Institute of Microbial Technology, Amity University, Noida, Uttar Pradesh, India

Sonal Chaudhary, Amity Institute of Microbial Technology, Amity University, Noida, Uttar Pradesh, India

Kamlesh Choure, Department of Biotechnology, AKS University, Sherganj, Satna, Madhya Pradesh, India. Email: kamlesh.chaure@gmail.com

Priyanka Das, Division of Plant Biology, Bose Institute, Kankurgachi, Kolkata, West Bengal, India. Email: prink.bot@gmail.com

Priyasha Das, Amity Institute of Microbial Technology, Amity University, Noida, Uttar Pradesh, India

Charu Khosla Gupta, Department of Botany, Acharya Narendra Dev College, University of Delhi, Delhia, India. Email: charukhoslagupta@andc.du.ac.in

Arti Hansda, Department of Life Science, School of Science, GSFC University, Vadodara, Gujarat, India

Rojalin Hota, Department of Soil Science and Agricultural Chemistry, MITS Institute of Professional Studies, Rayagada, Odisha, India

Mukul Joshi, Department of Biological Sciences, Birla Institute of Technology and Science (BITS), Pilani Campus, Vidya Vihar, Pilani, Rajasthan, India. Email: mukul.joshi@pilani.bits-pilani.ac.in

Naveen Chandra Joshi, Amity Institute of Microbial Technology, Amity University, Noida, Uttar Pradesh, India. Email: naveen.joshi.jnu@gmail.com

Chandra Kant, Dharma Samaj College, Aligarh, Uttar Pradesh, India

Linthoi Khomdram, Hansraj College, University of Delhi, North Campus, New Delhi, India

Nikena Khwairakpam, Microbial Biotechnology Research Laboratory, Department of Biochemistry, Manipur University, Canchipur, Manipur, India

Chanchal Kumar, Department of Forensic Science, Guru Ghasidas Vishwavidyalaya, Bilaspur, Chhattisgarh, India

K. **Santosh Kumar**, Department of Chemistry, School of Science, GSFC University, Vadodara, Gujarat, India

Maneesh Kumar, Department of Biotechnology, Magadh University, Bodh Gaya, Bihar, India. Email: kumar.maneesh11@gmail.com

Rahul Kumar, Department of Soil Science and Agricultural Chemistry, Jharkhand RAI University, Ranchi, Odisha, India

Arti Kumari, Department of Biotechnology, Patna Women's College, Patna, Bihar, India

Priyanka Kumari, Department of Biotechnology, Gautam Buddha University, Greater Noida, Uttar Pradesh, India

Shalini Kaushik Love, DESM (Department of Education Science and Mathematics), Regional Institute of Education, Bhubaneswar, Odisha, India. Email: shalinikaushiklove@gmail.com

Maheshree Maibam, Microbial Biotechnology Research Laboratory, Department of Biochemistry, Manipur University, Canchipur, Manipur, India

Arun Lahiri Majumder, Division of Plant Biology, Bose Institute, Kankurgachi, Kolkata, West Bengal, India

Sushant Malhotra, Department of Biological Sciences, Birla Institute of Technology and Science (BITS), Pilani Campus, Vidya Vihar, Pilani, Rajasthan, India

Yash Mangla, Department of Botany, Kirori Mal College, University of Delhi, Delhi, India

Arti Mishra, Department of Botany, Hansraj College, University of Delhi, India; Umeå Plant Science Center, Department of Plant Physiology, Umeå University, Umeå, Sweden. Email: artimishrahrc@gmail.com

Swati Mohapatra-Department of Life Sciencey, School of Science, GSFC University, Vadodara, Gujarat, India. Email: swatimohapatraiitr@gmail.com

Jiya Navshree, South Asian University, Rajpur Road, Maidan Garhi, New Delhi, India

Debananda S. Ningthoujam, Microbial Biotechnology Research Laboratory, Department of Biochemistry, Manipur University, Canchipur, Manipur, India. Email: debananda.ningthoujam@gmail.com

Amanda Nongthombam, Microbial Biotechnology Research Laboratory, Department of Biochemistry, Manipur University, Canchipur, Manipur, India

Knight Nthebere, Department of Soil Science and Agricultural Chemistry, Jayashankar Telangana State Agricultural University, Telangana, Hyderabad, India

Chandrakant Pant, Department of Biological Sciences, Birla Institute of Technology and Science (BITS), Pilani Campus, Vidya Vihar, Pilani, Rajasthan, India

Roshni Patel, Department of Life Science, School of Science, GSFC University, Vadodara, Gujarat, India

Shalini Porwal, Amity Institute of Microbial Technology, Amity University, Noida, Uttar Pradesh, India. Email: sporwal@amity.edu

Seema Pradhan, BRIC-Institute of Life Sciences, Bhubaneswar, Odisha, India. Email: seema@ils.res.in

Piyush Kant Rai, Department of Biotechnology, AKS University, Sherganj, Satna,Madhya Pradesh, India

Usha Sabharwal, Department of Life Sciences, Parul Institue of Applied Sciences, Parul University, Waghodia, Vadodara, Gujarat, India

Lipun Sahoo, Department of Botany, Guru Ghasidas Vishwavidyalaya, Bilaspur, Chhattisgarh, India

Deviprasad Samantaray, Department of Microbiology, Orissa University of Agriculture and Technology, Bhubaneswar, India

Hansa Sehgal, Department of Biological Sciences, Birla Institute of Technology and Science (BITS), Pilani Campus, Vidya Vihar, Pilani, Rajasthan, India

Era Sharma, Department of Biological Sciences, Birla Institute of Technology and Science (BITS), Pilani Campus, Vidya Vihar, Pilani, Rajasthan, India

Neha Sharma, Department of Microbiology, Faculty of Allied Health Sciences, Shree Guru Gobind Singh Tricentenary University, Budhera, Gurugram, Haryana, India. Email: neha1_fahs@sgtuniversity.org

Nikhil Sharma, Department of Microbiology, Guru Nanak Dev University, Amritsar, Punjab, India

Prashansa Sharma, Department of Plant Molecular Biology, University of Delhi, South Campus, New Delhi, India

Smriti Shukla, Amity Institute of Environmental Toxicology, Safety and Management, Amity University, Noida, Uttar Pradesh, India

Alka Singh, Department of Chemistry, Feroze Gandhi College, Raebareli, Uttar Pradesh, India

Anupriya Singh, Delhi University, New Delhi, India

Ayushi Singh, Amity Institute of Microbial Technology, Amity University, Noida, Uttar Pradesh, India

Deepjyoti Singh, North Carolina State University, Raleigh, NC, USA

Prashant Kumar Singh, Department of Biotechnology, Mizoram Central University, Mizoram, India

Sakshi Singh, Amity Institute of Microbial Technology, Amity University, Noida, Uttar Pradesh, India

Shweta Singh, Dwarika Prasad Girls Inter College, Prayagraj, Uttar Pradesh, India

Vijayata Singh, Icahn School of Medicine at Mount Sinai, New York, USA. Email: vijayata.cb@gmail.com

Yurembam Rojiv Singh, Microbial Biotechnology Research Laboratory, Department of Biochemistry, Manipur University, Canchipur, Manipur, India

Bharati Swain, Department of Botany, Guru Ghasidas Vishwavidyalaya, Bilaspur, Chhattisgarh, India

Shantirani Thokchom, Microbial Biotechnology Research Laboratory, Department of Biochemistry, Manipur University, Canchipur, Manipur, India

Vivekanand Tiwari, Institute of Plant Sciences, Agricultural Research Organization, Volcani, Israel

Jaagriti Tyagi, Institute of Microbial Technology, Amity University, Noida, Uttar Pradesh, India; Biotechnology Department, Dr. MPS Group of Institutions, Agra, Uttar Pradesh, India. Email: jaagriti.tyagi13@gmail.com; jtyagi@amity.edu

Subodh Verma, Central European Institute of Technology (CEITEC), Brno, Czech Republic

Kanchan Vishwakarma, Amity Institute of Microbial Technology, Amity University, Noida, Uttar Pradesh, India; Swedish University of Agricultural Science Umeå Sweden, Department of Forest Ecology and Management and Umeå Plant Science Centre, Umeå, Sweden.

Nikita Wadhwa, Hansraj College, University of Delhi, North Campus, New Delhi, India

Deepanker Yadav, Department of Botany, Guru Ghasidas Vishwavidyalaya, Bilaspur, Chhattisgarh, India. Email: deepankerbhu@gmail.com

Manikyala Bhargava Narasimha Yadav, Department of Soil Science and Agricultural Chemistry, University of Agricultural Sciences, Dharwad, Karnataka, India

1 Plant Nutrient Requirements and Nutrient Homeostasis in Plants

Neha Sharma[1], Anchal Chaudhary[2], Nikhil Sharma[2], Smriti Shukla[3], Naveen Chandra Joshi[4]*, Kanchan Vishwakarma[4,5], Prashant Kumar Singh[6] and Arti Mishra[7,8]*

[1]*Department of Microbiology, Faculty of Allied Health Sciences, Shree Guru Gobind Singh Tricentenary University, Budhera, Gurugram, Haryana, India;* [2]*Department of Microbiology, Guru Nanak Dev University, Amritsar, Punjab, India;* [3]*Amity Institute of Environmental Toxicology, Safety and Management, Amity University, Noida, Uttar Pradesh, India;* [4]*Amity Institute of Microbial Technology, Amity University, Noida, Uttar Pradesh, India;* [5]*Swedish University of Agricultural Science Umeå Sweden, Department of Forest Ecology and Management and Umeå Plant Science Centre, Umeå, Sweden;* [6]*Department of Biotechnology, Mizoram Central University, Mizoram, India;* [7]*Department of Botany, Hansraj College, University of Delhi, India;* [8]*Umeå Plant Science Center, Department of Plant Physiology, Umeå University, Umeå, Sweden*

Abstract

Plants, being sessile organisms, rely on a delicate balance of essential nutrients for their growth, development and overall health. This chapter provides a comprehensive overview of plant nutrient requirements and the intricate mechanisms governing plant nutrient homeostasis. Understanding the dynamic interplay between essential elements is crucial for optimizing agricultural practices, enhancing crop yields and mitigating environmental impacts. The first section of the chapter delves into the fundamental macronutrients and micronutrients essential for plant growth, including nitrogen, phosphorus, potassium, calcium, magnesium, sulfur, iron, zinc, manganese, copper, molybdenum and boron (N, P, K, Ca, Mg, S, Fe, Zn, Mn, Cu, Mo and B, respectively). Each nutrient's specific roles, functions, and sources are discussed, highlighting the significance of a balanced nutrient supply for plant metabolic processes. The second part of the chapter explores the mechanisms through which plants maintain nutrient homeostasis, ensuring optimal growth and development while coping with fluctuations in nutrient availability. This involves intricate processes such as nutrient uptake, transport and distribution within plant tissues. The role of ion channels, transporters and signalling pathways in orchestrating nutrient homeostasis is elucidated, emphasizing the adaptability of plants to diverse environmental conditions. Furthermore, this chapter highlights the importance of symbiotic relationships, such as mycorrhizal associations, in enhancing nutrient acquisition and promoting plant resilience. The impact of abiotic stressors, including nutrient deficiencies or excesses, on plant physiology and biochemical pathways is also discussed, shedding light on the adaptive responses employed by plants to maintain equilibrium.

Keywords: Abiotic stress, plant nutrients, plant–microbe interaction, signalling molecules, nutrient homeostasis

---

*Corresponding authors: artimishrahrc@gmail.com, naveen.joshi.jnu@gmail.com

© CAB International 2025. *Soil Health and Nutrition Management*
(eds N.C. Joshi, T. Leustek and P.K. Singh)
DOI: 10.1079/9781800624597.0001

1.1   Introduction

Studying plant nutrient requirements and the complex mechanisms regulating nutrient homeostasis is pivotal for advancing agricultural practices, maximizing crop yields and minimizing environmental consequences. Plants rely on a diverse array of essential elements, including macronutrients (such as nitrogen (N), phosphorus (P) and potassium (K)) and micronutrients (such as zinc (Zn), iron (Fe) and manganese (Mn)), to support various physiological processes crucial for plant growth (Tripathi *et al.*, 2015). All nutrients are essential to the physiological processes that allow plants to develop properly; deficiencies in these nutrients result in specific diseases (Rao, 2009). A plant needs some nutrients in more significant amounts than others, yet all nutrients are necessary to complete its life cycle. The main functions of major nutrients in plant physiology and development have been extensively researched and recorded (Fageria *et al.*, 2008). It is possible to look at the roles of advantageous components in plant feeding in more detail. A complex substrate, soil serves as a reservoir for water and nutrients necessary for plant development. Plants develop a vast root system to absorb nutrients from the soil (Drinkwater and Snapp, 2007). However, all these nutrients' availability in soil can vary based on a wide range of variables. Nutrients go from the soil to the roots via different transport methods, such as mass flow, diffusion and root interception. Plants either actively absorb the nutrients using energy or passively absorb them without energy (Karthika *et al.*, 2018). The dynamic interplay between these nutrients necessitates a delicate balance as deviations from optimal concentrations can lead to nutrient deficiencies or toxicities, adversely impacting plant health. Understanding the intricacies of nutrient uptake, transport and distribution within plant tissues is crucial for developing strategies that ensure a balanced and sustainable nutrient supply (Ahmed *et al.*, 2024). The link between soil nutrients and plant development is handled by plant nutrition.

Fundamental mechanisms involved in nutrient homeostasis include specialized transporters and channels that facilitate the movement of ions across cell membranes (Jain and Zoncu, 2022). Signal transduction pathways play an important role in sensing and responding to changes in nutrient availability, allowing plants to adapt to varying environmental conditions. Moreover, the establishment of symbiotic relationships, such as mycorrhizal associations, enhances nutrient absorption efficiency and contributes to the overall resilience of plants. The implications of nutrient management extend beyond immediate crop health, with profound effects on ecosystem sustainability and environmental health. Imbalances in nutrient application can result in nutrient runoff, leading to water pollution and ecosystem disturbances (Bisht and Chauhan, 2020). The agricultural sector can mitigate adverse environmental impacts and contribute to global food security by unravelling the complexities of nutrient interactions and adopting precision nutrient management strategies. In summary, the ongoing exploration of plant nutrient requirements and the intricate mechanisms governing nutrient homeostasis is essential for developing sustainable and efficient agricultural practices. This knowledge serves as a foundation for addressing global challenges related to food production, environmental conservation and the overall well-being of ecosystems.

1.2   Essential Nutrients for Plants

Any rooting media, such as soil, provides plants with various mineral nutrients necessary for their development and metabolism. The plant absorbs the mineral nutrients as ions, which are then either stored within the cell sap or integrated into the structure of the plant. For usage in plant metabolism, the nutrients must pass through the root hair cells' plasma membrane (Mitra *et al.*, 2021).

Only light, water and around 20 other essential nutrients are required for plants to meet their metabolic needs. An ingredient is deemed vital if it satisfies these three requirements:

1. Without the element, a plant cannot complete its life cycle.
2. No other element can carry out the element's role.
3. The element directly contributes to plant nourishment.

1.2.1 Types of essential plant nutrients

Essential minerals, which contain both macro- and micronutrients, provide plants with the nourishment they need to grow and develop properly. In essence, plants need 13 different types of minerals in varied quantities. Since nitrogen is a structural component of many coenzymes, including pyrimidine, purines, porphyrins, DNA and RNA, all these kinds of chemical compounds are known to serve various functions. However, soil activity may decrease if the resulting levels of nitrogen are relatively high (Waldrop and Zak, 2006).

1.2.2 Nutrient classification

Generally, two categories of nutrients exist – macronutrients and micronutrients.

Macronutrients

These are referred to as major nutrients because plants absorb them in greater amounts (Suryawanshi et al., 2020). Structural elements are significant nutrients that the water and air absorb. They are hydrogen, oxygen and carbon. Additional categories for macronutrients include primary and secondary nutrients (Fig. 1.1).

Micronutrients

These are nutrients absorbed by plants in slight concentrations. There are various types of micronutrients such as Fe, copper (Cu), Zn,

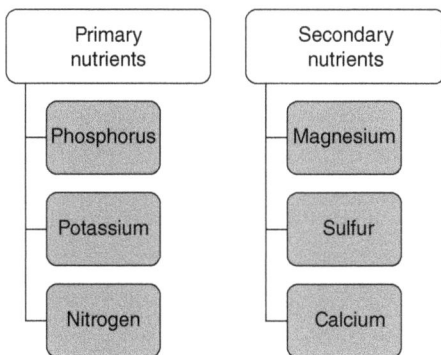

Fig. 1.1. Primary and secondary macronutrients.

boron (B), Mn, molybdenum (Mo) and chlorine (Cl) (Johnson and Mirza, 2020). Bukovac and Whitter classified nutrients into mobile, partially mobile and immobile elements (Fig. 1.2).

1.2.3 Macronutrients and micronutrients

Plants absorb the other elements through the soil. The nutrients N, P, K, calcium (Ca), magnesium (Mg) and sulfur (S) are classified as macronutrients since their concentrations in plant dry matter exceed 1–150 mg, while the nutrients Fe, Zn, Mn, Cu, B, Mo, and Cl are categorized as micronutrients since their concentrations in plant dry matter range from 0.1 to 100 mg. Their necessity in and of itself does not, however, lessen their importance for the development and metabolism of plants. Mineral nutrition elements are vital components of cell structures and metabolites, and are involved in energy transfer activities, turgor-related processes, cell osmotic relations, enzyme-catalyzed reactions and plant reproduction. The effective performance of these tasks determines the productivity of plants (Pandey, 2018). Plant macronutrients are crucial for ensuring the proper growth and development of plants because they serve various important functions involving structural components and redox-sensitive substances (Nadeem et al., 2018). Phosphorus is recognized to be a crucial part of the membranes of plants and is an important component of DNA, RNA and adenosine triphosphate.

Similarly, calcium is an essential component of living things that is especially required as calcium ions. The development and proliferation of plants depend on calcium, which also activates enzymes involved in salt balance and water flow inside plant cells, activating K^+, which regulates the opening and shutting of tiny holes called stomata. Similar to magnesium, which is essential for photosynthesis in plants and is a critical metal element in chlorophyll, too little of it results in the degradation of chlorophyll and the yellowing of leaves, or chlorosis. However, sufficient amounts of magnesium keep plants healthy. Sulfur is the most beneficial plant element for almost all living things because it performs a variety of dynamic tasks that are essential to the proper development, growth

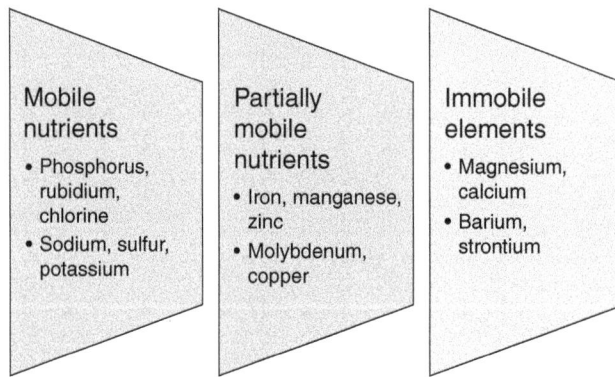

Fig. 1.2. Bukovac and Whitter classification of micronutrients.

and persistence of plant life. Sulfur is therefore regarded as a necessary nutrient for all crops to maximize output (Tripathi *et al.*, 2014).

Certain essential compounds that plants require in small amounts are called micronutrients. The output and quality of agricultural goods rise in tandem with the concentration of micronutrients, protecting both human health and animal welfare using enrichment plant materials in feed. Particularly, each vital plant element fulfils its nutritional function in a balanced ratio required for several species' optimal development and growth (Nadeem *et al.*, 2018). Manganese's divalent ions are transformed into trivalent and tetravalent ions that are essential for various oxidation–reduction mechanisms like the photosynthesis electron transport chain. Manganese also could activate various enzymes that function in the metabolism of carbohydrates, phosphorous events, citric acid cycle, carboxylation activities and oxidation reactions. Two significant enzymes are the protein manganese enzyme of photosystem II and superoxide dismutase. Of these, superoxide dismutase is present in chloroplasts, which comprise 5% of the entire mitochondrial mass, in an amount of >90%. Zinc is taken up by soil as divalent cations from the soil solution, especially in the case of calcareous soil with higher pH levels. Zinc is either bound to organic acid inside the xylem or transferred to divalent cations by chemical changes. In contrast, zinc forms an organic

complex with low molecular weight and comparatively greater amounts in phloem sap (Tavakoli *et al.*, 2014).

1.2.4   Role of macro- and micronutrients

Macro- and micronutrients have several purposes. As important components of macromolecules and coenzymes, they take part in several metabolic activities in plant cells, including permeability of the cell membrane, maintenance of the osmotic level of cell sap, electron transport mechanisms, buffering action and enzymatic activity (Table 1.1).

1.3   Nutrient Uptake and Transport

Nutrient uptake and transport mainly consist of root absorption and vascular transport. Researchers have studied how plant roots absorb mineral elements and then distribute those elements throughout the plant (Puig and Penarrubia, 2009). Plant roots take up mineral components from the soil solution. They travel through symplastic (intracellular) and/or apoplastic (extracellular) channels in the root to reach the stele, where they are put into the xylem for delivery to the shoot. Every cytotoxic action must be transferred in a chelated state via the apoplast or the symplast. Certain plants retain several mineral components in their roots.

Table 1.1. Role of macro- and micronutrients.

Nutrients	Role
Nitrogen	Since nitrogen is a crucial component of protein and is necessary for a plant's abundant growth, it is well known to play a part in the structural growth and physiological processes of plants. Nitrogen is, therefore, a very valuable plant nutrient supplied to the soil through nitrogenous fertilizers to facilitate the synthesis of various components of living cells. Metabolism of plant, protein and nucleic acid, formation of carbohydrates, chlorophyll content and photosynthesis occur in the presence of nitrogen. Vegetative growth is increased by excessive application of nitrogen. This leads to the prevention of the growth of foods because of overcrowding (Suryawanshi *et al.*, 2020)
Phosphorus	This element is crucial for the growth of several components in young plants, including sugars, nucleic acids and ATP for energy generation. This nutrient is seen in the crops' maturation, blooming, fruiting, germination, ripening and grain quality improvement. This nutrient involves biochemical processes such as the metabolism of proteins, lipids and carbohydrates. Phosphorus is a crucial nutrient in forming a plant's roots (Dhok, 2020)
Potassium	This is necessary to produce proteins, sugar movement, enzymes and coenzymes' enzymatic activity, and photosynthesis. Strengthens the crop's root system and increases its resilience to disease. Plant lodging is avoided. Tubers grow, water is seen to be regulated, starch is created and carbohydrates are formed (Johnson and Mirza, 2020)
Calcium	This constituent of plant cell walls is necessary for creating the wall. It keeps the cell wall's permeability intact. It is necessary for both cell division and elongation. Mitochondria display calcium accumulation during respiration. Protein content rises when calcium is used. Carbohydrate translocation promotes nitrogen intake and aids in nitrogen metabolism. It also affects seed formation, root growth, chromosomal integrity and other processes (Suryawanshi *et al.*, 2020; Baker *et al.*, 2006). When administered exogenously, it improves fruit firmness and storage life (Al Eryani-Raqeeb *et al.,* 2009). To enhance the quality of the fruit, foliar spraying or soil treatment is applied directly to the fruit and leaves (White and Broadley, 2003). It occurs in the fruit's surface fissures, stomata, cuticle and root ends (Saure, 2005)
Magnesium	This is necessary for the synthesis of ATP, which is necessary for energy generation. It is the primary component of chlorophyll since plants account for 15–25% of all magnesium. The cell exhibits enzyme activity and peptide chain synthesis. Magnesium is responsible for the leaf's dark green colour. Phosphorus is absorbed, and proteins, lipids and carbohydrates are formed. Magnesium is a mobile component. Translocation from older to younger leaves is constant (Dhok, 2020)
Sulfur	This is crucial for the production of chlorophyll and proteins, as well as for the function of amino acids and proteins. Increased nodule and seed production increases crop quality. Using sulfur, coenzyme A and vitamin B are created (Suryawanshi *et al.*, 2020)
Chlorine	The element is crucial for leaf turgor and is also responsible for photosynthetic processes in plants (Engel *et al.*, 2001)
Molybdenum	This is vital to leguminous plants because it maintains plant enzymatic activity and fixes nitrogen. When molybdenum functions as a catalyst, nitrogen transformation takes place and nitrogen is fixed (Johnson and Mirza, 2020)
Boron	This is necessary for the generation of cell wall and reproductive tissue. Boron is necessary for the synthesis of starch, the transit of sugar, the construction and growth of cells, hormone production, fat metabolism, photosynthesis and salt absorption. Boron acts as an enzyme activator (Suryawanshi *et al.*, 2020)
Iron	This is necessary for the plant's respiratory system and photosynthetic processes. Metabolic processes create porphyrin and manufacture enzymes such as cytochrome oxidase, peroxidase and catalase (Dhok, 2020)

Continued

Table 1.1. Continued.

Nutrients	Role
Zinc	Plant hormone synthesis and internode elongation rely on this process. Symptoms first appear at the centre of the leaf, exhibiting intermediate mobility. Auxin production and noticeable enzymatic activity are observed (Suryawanshi *et al.*, 2020)
Copper	This is necessary for respiration, synthesis of proteins and formation of chlorophyll. When applied in excessive quantities, it is referred to as plant poison. Oxidation, reduction and metabolism occur in roots
Manganese	The chloroplast is the most vulnerable cell organelle to a manganese shortage. Manganese activates the enzyme responsible for assimilating nitrogen, producing chlorophyll and regulating oxidation and reduction (Suryawanshi *et al.*, 2020)
Nickel	This is necessary for seed germination. Nickel is a metal component of urease, and is necessary to transform urea into ammonium (Dhok, 2020)

Some examples are Mo, Ca, sodium (Na), cadmium (Cd) and aluminium (Al). Certain cell types in transpiring leaf tissues absorb mineral elements from the apoplast after being transported there by the xylem. The phloem transports mineral elements to tissues lacking xylem or not transpiring, and transports them back in the plant's recirculation system. While Fe, Zn, Cu, Mo and iodine (I) are less mobile and Mn and Ca are virtually stationary in the phloem of most plants, K, Na, Mg, Cd, N, P, S, selenium (Se) and Cl are transported rapidly. The mobility of phloem differs throughout species (Brown *et al.*, 2002). Mineral elements with poor phloem mobilities are found in low quantities in fruits, seeds and tubers, and they accumulate in tissues with elevated transpiration rates (Karley and White, 2009).

Regardless of their chemical structure, all mineral elements are included in the ionome, whether necessary or not for life. The main areas of current research interest are measuring the ionome of various plant species, ecotypes and induced mutants and providing a genetic explanation for variations in ionome within and between plant species. However, environmental and developmental variables also significantly impact the ionome (Broadley *et al.*, 2010). The ionome of the plant and its tissues is determined by the intake of mineral substances and their distribution throughout the plant. For this reason, these activities are essential to the plant's mineral nutrition and to raising the levels of mineral elements in edible tissues for human consumption.

Nutrient uptake in plants is a complex process involving various mechanisms that ensure the absorption of essential elements from the soil into the plant roots. This process is vital for the plant's growth, development and overall health. A detailed overview of the mechanism of nutrient uptake in plants follows.

1.3.1   Root hair development

Root hairs are extensions of the root epidermal cells that grow tips and are crucial for both plant–soil interactions and nutrient uptake (George *et al.*, 2021). The main physiological, environmental and genetic variables that control the development and differentiation of root hairs in angiosperms are covered in the cited article. The root epidermis of various species has a variety of patterns for the arrangement of root hair cells. An intercellular gene regulation network in *Arabidopsis* (*A. thaliana* L.) produces a striped pattern of hair and non-hair files through feedback loops and protein transfer between neighbouring cells (George *et al.*, 2021; Pitsili, 2021). An elongation phase and an initiation phase, which involve bulge development and site selection, can broadly categorize root hair growth. Transcription factors, guanosine triphosphates and cell wall remodelling enzymes control the beginning phase (Stéger and Palmgren, 2022). Root hairs grow by tip growth, a kind of polarized cell proliferation limited to the rising apex, during the elongation phase. In *Arabidopsis*, root hair formation occurs in the root epidermis; hair and non-hair cells are arranged in alternating files according to the patterning mechanism in *Arabidopsis* (Gayomba and Muday, 2020). Once the fate of the hair cell has been determined, differentiation takes place and root hairs protrude from the cell surface.

In general, there are two phases of root hair differentiation: the initiation phase, which includes bulge development and site selection, and the elongation phase, which is characterized by root hair quickly elongating by tip growth.

1.3.2 Mycorrhizal associations

Relationships between plants and symbiotic fungi, or mycorrhizas, are symbiotic in plant communities. The roots of plants are connected by intricate networks of mycorrhizal hyphae, which also influence seedling establishment, nutrient flow, competitive interactions between and among plant species and, ultimately, all elements of the ecology and coexistence of plant communities (Tedersoo *et al.*, 2020). The phylum *Mucoromycota* and subphylum *Glomeromycotina* are home to arbuscular mycorrhizal fungus (AMF). All woody plants, including gymnosperm and angiosperm, composed of certain non-flowering and blooming families, are covered in AMF colonization. Soil fungi with rapid mineral and water absorption across a large surface area form a complex network of hyphae (Ahad and Ferdous, 2019).

Moreover, the formation of arbuscules – highly branching organs – occurs in the cortical cells of the roots, allowing fungi to exchange resources with the plant in both directions. About 80% of terrestrial plants have this association in their roots because fungi boost leaf photosynthesis, increase the efficiency of plants to absorb water, provide phosphorus and other mineral nutrients and increase the hydraulic conductivity of plant roots (Bhantana *et al.*, 2021; Wahab *et al.*, 2023). These advantageous consequences give plants abiotic stress tolerance, which enables them to function in challenging environmental circumstances (Fig. 1.3).

Plant transfer and uptake of nutrients are clearly the purposes of AMF mutual association. AMF improves almost all plants' absorption of nutrients, particularly P. Under low P and N conditions, AMF enhances plant growth and development

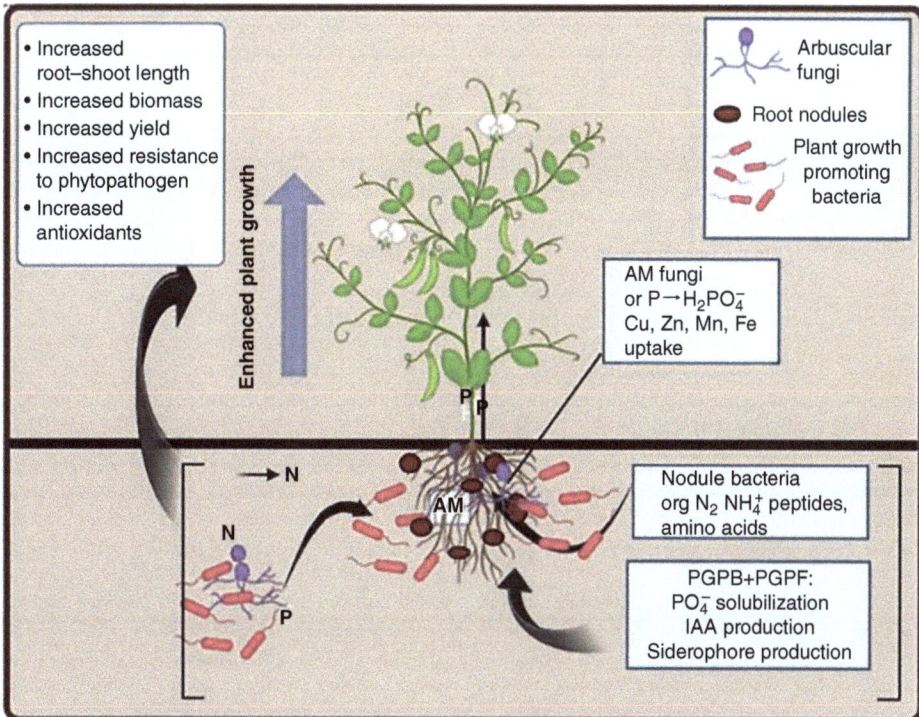

Fig. 1.3. Microbe-mediated nutrient uptake in leguminous plants.

(Aggarwal *et al.*, 2011). Under high soil P circumstances, a lower AMP per cent is realized due to variations in AMF growth. AMF symbiosis improved P nutrition in both upland and lowland rice. The amount of P that rice roots directly absorbed was much less than that of fungus hyphae. In both lowland and upland rice, AMF symbiosis improved P nutrition (Maiti *et al.*, 2017; Nopphakat *et al.*, 2021). Rice roots more readily absorbed P through fungus hyphae than by direct rice uptake. Polyphosphates (polyP; negatively charged liner phosphate polymers) are taken up by hyphae and assembled in rice cortical cells following hydrolysis of the polyP chain in AMF. Two transporter genes (*PT2* and *PT6*) involved in direct P absorption by roots had lower transcription levels in rice related to AMF (Konečný *et al.*, 2019). On the other hand, the AMF-specific P transporter gene (*PT11*) showed higher transcription levels. This explains why the AMF-mediated route significantly absorbs more P than roots do directly.

1.4 Ion Transport

Even though plant physiologists frequently use aquatic plants as model systems, little is known about their electrophysiological, biochemical or molecular ion transport pathways. Concurrently, ion transport in practically all essential biological functions, including nutrition uptake, growth, energy production and signalling, is mediated across membranes. There are two types of ion transport in plants.

1.4.1 Cation transport

In the freshwater-submerged plant *Egeria*, voltage, growth medium pH and phosphorylation/dephosphorylation all affected K^+ conductance. Additionally, it was found that unlike terrestrial plants, the control of K^+ uptake in response to external pH shifted to a more alkaline range (Natura and Dahse, 1998).

The seagrass *Posidonia oceanica*'s sheath protoplasts have a non-selective cation channel and K^+ inward and outward channels (Carpaneto *et al.*, 2004). The K^+ outward channel was impermeable to Na^+ and other monovalent cations and exhibited a high degree of selectivity for K^+.

It bore similarities to GORK (guard cell outward rectifying K) or SKOR (stellar K outward rectifier) channels, found in *A. thaliana* xylem parenchyma or guard cells. The voltage sensitivity of the *Posidonia* outward channel and the GORK/SKOR channels rely on the concentration of external K^+, but not the outward K^+ channel of the halophytic angiosperm *Zostera muelleri* (Garrill *et al.*, 1994). It has been suggested that in *Posidonia* this outer channel helps K^+ be released from sheath cells into xylem arteries.

1.4.2 Anion transport

Aquatic and other plants rely on active transport mechanisms for ion uptake, with early studies on higher plants indicating that high-growth, low-affinity anion transport systems require the import of NO_3 or H_2PO_4 alongside H^+ (Ajay *et al.*, 2015). The depolarization of plasma membrane caused by NO_3 and the simultaneous fluxes of H^+ and NO_3 have been the basis for this mechanism's proposal.

It has been proposed that the seagrass *Zostera marina* features high-affinity Na^+-dependent NO_3 and Pi transport systems, which function as H_2PO_4 and HPO_{42}. Using a Na^+ gradient that is directed from the apoplast to the cytosol in a highly salinized environment, this seagrass takes up NO_3 with a stoichiometry of one NO_3 for every two Na^+, one H_2PO_4 for every two Na^+, and one H_2PO_4 for every three Na^+ ions (García-Sánchez *et al.*, 2000).

1.5 Active and Passive Ion Transport

There are two methods by which ions are transported in plants: active and passive, each with unique properties. Ions travelling against concentration gradients define active ion transport (Stillwell, 2016; Tomkins *et al.*, 2021). Electrochemical potential gradients and attraction of cations and anions to negative and positive electropotentials, respectively, are the two factors that determine ion mobility (Alaoui *et al.*, 2022). The flow of ions down chemical gradients of potential energy or from higher to lower concentrations is categorized as passive ion transport. It should be noted that because of unequal charge distributions, electrochemical potentials are detected through membranes (Kisnieriene *et al.*, 2019). The degree of energy quality required is indicated by differences between membrane potentials and real potentials created by non-equilibrium allocations (Fageria *et al.*, 2006). Therefore, electrical charge might be calculated using the following modified Nernst equation, as Ting (1982) explained:

$$\Psi = \left(-RT/ZF/\ln\left(ai/a0\right)\right)$$

Where Ψ = electrochemical potential between root cells and external solutions in millivolts (mV); R = gas constant (8.3 J/mol K); T = absolute temperature (K); Z = net charge on ion (dimension less); F = Faraday constant (96,400 J/mol); ai = activity of ion inside a tissue and a0 = activity of ion outside a tissue. Remembering that RT/F = 26 mV will help you compute more quickly (Fageria *et al.*, 2006; Farhangi-Abriz and Ghassemi-Golezani, 2023).

1.6   Nutrient Homeostasis

Cellular metabolism, development and physiology depend on nutrient balance (Amtmann and Blatt, 2009). Organic macromolecule synthesis depends on nutrient elements, and different nutrient ions play structural or catalytic functions in essential proteins or function as the cofactors or molecules of signalling. The tremendous daily variations in surroundings that plants experience significantly impact their physiology and metabolism. For instance, daily patterns in transpiration rates modify the main nutrient transport channels via the xylem, and cyclical changes in nutrient needs power photosynthesis in chloroplasts. Thus, especially under situations of nutritional constraint, nutrients must be continually transported between tissues and organelles (Haydon *et al.*, 2015). Nutrient homeostasis in plants refers to the ability of plants to maintain a balanced and optimal internal concentration of essential nutrients despite variations in external nutrient availability. This process involves complex regulatory mechanisms at the cellular and whole-plant levels. Below is a detailed overview of the mechanisms of nutrient homeostasis in plants.

1.7   Ion Channels and Transporters

Ion channels and transporters are crucial for plant physiology, playing vital roles in nutrient uptake, signal transduction and stress responses. These proteins are embedded in cellular membranes, where they facilitate the selective movement of ions such as potassium (K^+), sodium (Na^+), calcium (Ca^{2+}) and chloride (Cl^-) across membranes (Pantoja, 2021). Ion channels typically allow rapid ion fluxes, responding to various stimuli like voltage changes, mechanical forces or chemical signals, while transporters often work through slower, energy-dependent mechanisms such as active transport or facilitated diffusion (Hedrich *et al.*, 2012). In roots, ion transporters are essential for nutrient absorption from the soil, while in leaves, they help regulate stomatal opening and closing, affecting gas exchange and water loss. Additionally, ion channels and transporters are integral to generating electrical signals that propagate within the plant, coordinating developmental processes and adaptive responses to environmental stressors such as drought or salinity (Joshi *et al.*, 2022).

1.7.1   Selective uptake

Selective uptake in plants refers to the ability of plants to absorb specific nutrients and elements from the soil, while excluding others, through a highly regulated process (Reid and Hayes, 2003). This selectivity is achieved via specialized transport proteins located in the root cell membranes, which facilitate the entry of essential ions such as nitrate, phosphate, potassium and magnesium. These transport proteins can differentiate between various ions based on their size, charge and chemical structure, ensuring that plants acquire necessary nutrients in optimal amounts (Mitra, 2017). Additionally, the rhizosphere, the soil region near plant roots, plays a crucial role in this process as it is influenced by root exudates that alter the chemical environment to favour the availability of specific nutrients (Pandey, 2015). This selective uptake mechanism is vital for plant growth, development and overall health as it helps maintain nutrient homeostasis and protects plants from toxic substances.

Understanding this process is key for improving agricultural practices and developing crops with enhanced nutrient efficiency and stress resistance (Huang *et al.*, 2020).

1.7.2 Nutrient-specific transport

Different transporters are responsible for the uptake of specific nutrients. For instance, nitrate transporters facilitate the uptake of nitrate ions, and iron transporters mediate the uptake of iron ions (Dubyak, 2004). Ion transport proteins generally fall into one of two categories: channels or transporters. Transporters are considered 'vectorial' enzymes from a functional standpoint (Stein and Litman, 2014). Their catalytic cycle consists of three steps: (i) selective binding of transported ion(s); (ii) conformational changes in transporter protein caused by ion(s); and (iii) coupling of these conformational changes to the actual physical movement of transported ion(s) across the membrane bilayer. On the other hand, channels are considered to be transport proteins that help ions move physically through processes involving minimal energy contact between the channel proteins and the ion(s) being carried (Ashrafuzzaman, 2021).

Extrinsic factors determine whether the channel protein is in a closed state (incapable of ion transport) or an open or 'gated' state (capable of ion transport), such as changes in the potential of a membrane or the binding of small regulatory molecules, such as external cells to neurotransmitters or intracellular second messengers (Drew *et al.*, 2021). Therefore, when different extrinsic factors govern those domains of channel protein that function as gates for regulating the availability of transported ions to a pore domain, the primary conformational changes in channels are generated. The pore domain is a conduit or route for ions travelling from one membrane end to another (Dubyak, 2004).

1.8 Signal Transduction Pathways

A cell's means of communication with another cell is called signal transduction. Maintaining physiological homeostasis depends on cell communication (Bi *et al.*, 2021). Cells create chemical ligands that serve as signals during signal transduction, which are transported to target cells. The cytoplasm and surfaces of these target cells contain receptors that react to various chemical cues. How a cell responds to a ligand depends on the receptor activated (Makvandi *et al.*, 2021). Plant physiology includes the ability to perform photosynthesis as a key component. A plant's ability to survive depends on the process of photosynthesis. There are four primary classes of signal-transducing receptors:

1. Receptors connected to enzymes: these are receptors that protrude through the target cell's plasma membrane and can operate as enzymes or activate or create enzymes.
2. G-protein-coupled receptors: these receptors are attached to G proteins within the cell.
3. Nuclear receptors: these change the target cell's gene expression and are found inside the nucleus.
4. Cationic ligand-gated channels: when their ligand is attached to these receptors, which function as ion channels, they open or close. The target cell's plasma membrane contains these receptors.

Plants can sense changes in nutrient concentrations in their environment. Receptor proteins on or within the cell membrane detect these changes and initiate signal transduction pathways. Signal transduction pathways involve a series of molecular events transmitting information about the cell's nutrient status. This information influences gene expression and cellular responses.

1.9 Nutrient Sensors

The primary elements that keep life going are amino acids, fats, carbohydrates and macronutrients. Cells must be able to detect changes in these nutrients and react accordingly to survive (Morris and Mohiuddin, 2020). Thus, nutrient-sensing pathways are established to control metabolic balance, cellular energy and various biological functions. As such, disruptions in various sensing routes are linked to many illnesses, particularly those related to metabolism. The primary component of these sensing pathways is molecular sensors, which can detect intracellular fluctuations in any nutrient either directly by attaching to it or indirectly by binding to its surrogate molecules. They also possess a certain degree of selectivity and affinity for each nutrient. Sensors that detect changes in nutrient levels set off signalling cascades that adjust cellular activities for energy and metabolic equilibrium, such as regulating nutrient absorption, *de novo* synthesis or catabolism (Sung *et al.*, 2023). Nutrient sensor proteins play a crucial role in monitoring internal nutrient concentrations. These sensors are sensitive to changes in nutrient levels and trigger signalling cascades accordingly. Changes in cytosolic calcium levels are often associated with nutrient signalling. Calcium ions act as secondary messengers in nutrient signal transduction pathways.

1.10 Vacuolar Storage and Remobilization

Many metabolic processes in vacuoles involve moving materials across the tonoplast. Depending on their origin, some solutes also need to pass through the plasma membrane. To enable their movement and accumulation against a concentration gradient, the cell is outfitted with an intricate network of transport mechanisms, cellular pathways and distinct intracellular habitats. These can handle various materials with different sizes, compositions and provenances. Solutes are transported from the cytoplasm to the vacuole by ion channels, solute/H^+ antiporters and ABC transporters bound to the tonoplast; however, solutes are transported from the apoplast to the plasma membrane by extra solute/H^+ symporters with fluid-phase endocytosis (Etxeberria *et al.*, 2012). Plants can store excess nutrients in vacuoles. This storage allows plants to withstand periods of nutrient scarcity by remobilizing stored nutrients when needed. During senescence or under nutrient-deficient conditions, plants can mobilize nutrients from older tissues to support the growth of younger, actively growing tissue.

1.11 Long-Distance Nutrient Transport

Like animals, plants use the vascular system to transmit signals across large distances to mediate communication between organs. Peptides released by xylem-mobile plants have garnered significant interest as long-range signalling molecules that plants use to communicate with each other and respond to changes in the nutritional status of their surroundings (Mattes, 2019). As part of the long-distance negative feedback loop surrounding nodule development, several leguminous CLE peptides generated by rhizobial inoculation function as 'satiety' signals. On the other hand, systemic 'hunger' signals produced by local nitrogen shortage in *Arabidopsis* CEP family peptides encourage compensatory nitrogen acquisition in other root sections. Additionally, xylem sap peptidomics suggest the presence of long-distance signalling peptides that are still unknown (Okamoto *et al.*, 2016).

1.11.1 Xylem and phloem

Nutrients are transported throughout the plant via the vascular tissues, such as the xylem for water and minerals and the phloem for organic nutrients, after being absorbed by the roots. For the nutrition of shoots and roots, the redistribution of vital elements between tissues during ontogeny, the preservation of charge balance in the leaves of nitrate-fed plants, the elimination of potentially toxic elements from leaf tissues and systemic signalling of plant nutritional status, long-distance transport of solutes in xylem and phloem is essential. The xylem and phloem vascular systems carry out long-distance movement of water and solutes, which includes elements and low-molecular-weight organic molecules (Kiri and Hausa, 2023). Long-distance transport from the roots to shoots primarily takes place in non-living xylem vessels. Coniferous trees rely on tracheids, which are non-living conducting cells with a length ranging from 2 to 6 mm. They lack a continuous system of xylem vessels (White and Ding, 2023). Tracheids can also impede long-distance transport in xylem vessels of annual plant species, as in the case of stem nodes or root–shoot junctions (Aloni and Griffith, 1991). Although these structures allow intense xylem–phloem solute transport, they also present an internal barrier to xylem volume flow.

Both the gradient in water potential and hydrostatic pressure, or root pressure, drive xylem movement. A water potential of zero is referred to as pure-free water. Water potential levels are, therefore, typically negative. The water potential gradient between the roots and the shoots is rather steep, especially during the day when the stomata are open. The following order causes values to become less negative: exterior solution of xylem sap, atmosphere, leaf cells and root cells. Hence, the flow of solutes in the xylem is unidirectional, going from roots to shoots. However, under specific circumstances in the shoots, there may also be a counterflow of water in the xylem, such as from low-transpiring fruits back to leaves (Lang and Thorpe, 1989).

1.12 Nutrient Partitioning

Nutrients are partitioned into different plant tissues according to physiological demands. Developing tissues may receive a higher allocation of nutrients compared with older, senescing tissues. One of the most significant nutrient conservation strategies perennial plants use is nutrient resorption, which is the transfer of nutrients from leaves to other plant tissues before abscission (Killingbeck, 1986). Resorption is thought to provide 31% and 40% of plants' yearly nitrogen and phosphorus needs worldwide, respectively. Resorption occurs before litter falls, and mineralization occurs after litter falls, significantly impacting the terrestrial nitrogen cycle. Plants can transfer photosynthates to roots in addition to reabsorbing nutrients from leaves. In alpine and Arctic environments, roots make for 70% or more of the total biomass of plants (Iversen *et al.*, 2015), which means that to save plant nutrients, effective nutrient allocation to the roots is also crucial (Poorter and Nagel, 2000; Freschet *et al.*, 2010). Consequently, to forecast how roots will contribute to plant nutrient conservation in future climate change scenarios, knowledge of how roots respond nutrient-wise to global change events, such as drought and nitrogen deposition, is essential (Porter *et al.*, 2013).

1.12.1 Nutrient interactions in plants

Understanding the dynamics of nutrient uptake, transport, absorption and their biological interactions is essential to ensuring greater agricultural plant productivity because the availability of nutrients has a significant impact on plant growth and development. Even though crop plants are frequently subjected to nutrient imbalances that adversely impact several metabolic processes, plants have developed defense mechanisms against dietary deficits. While many elements are naturally found in the soil, only 17 elements are currently recognized as crucial for appropriate crop plant growth and development. For the growth and development of crop plants, macronutrients – N, P, K, Ca, S and Mg – are required in larger quantities. The micronutrients – Fe, Zn, Cu, Mn, Mo, Cl and others – are required in smaller amounts (Kumar *et al.*, 2021). One of the main drivers of the Green Revolution of the 1960s was the production of enough food to feed the world's expanding population, and one of the main contributors to this success has been the usage of N and P fertilizers. According to reports, interactions between different nutrient elements may impact one another's absorption, transportation or assimilation. To better understand the sensing and signalling pathways triggered in response to the variable availability of nutrient elements, multi-level interactions between the nutrient elements must be investigated. The multi-level study contributes to our understanding of how plants reprogramme metabolic pathways in response to deficiencies, replenishments, excesses and/or adequacy of mineral nutrients by linking the transcriptome through enzymatic activities to metabolome.

This sheds light on how plants modify their metabolic pathways to ensure healthy growth and development without mineral nutrients (Amtmann and Armengaud, 2009). Being necessary macronutrients for plant growth, development and production, P and S interact with one another in cellular membranes during P-deficiency stress by replacing phospholipids with sulfolipids and galactolipids (Okazaki *et al.*, 2013). Even though these biological interactions between N, P and S are widely documented (Aulakh and Pasricha, 1977; Chotchutima *et al.*, 2016; Krouk and Kiba, 2020), there is still much to learn about the signalling pathways involved in nutritional availability or deficit reactions. Micronutrients like Zn and Fe also have homeostasis and exhibit biological interactions, proving that nutritional shortages and interactions are not limited to macroelements alone. Microelement deficiencies cause physiological issues affecting plant growth, development and productivity. While some understanding of these relationships has been gained at the physiological and molecular levels, further research is necessary to understand the complex nutritional cross-talks and to optimize crop output fully.

1.13 Environmental Factors Influencing Nutrient Homeostasis

Environmental factors play a crucial role in influencing nutrient homeostasis in organisms. Temperature, light, humidity and soil composition are key determinants of nutrient availability and uptake in plants. Temperature fluctuations can affect enzyme activity and metabolic rates, impacting nutrient absorption and utilization (Hanikenne *et al.*, 2021). Light intensity and quality influence photosynthesis, which in turn affects the production of organic compounds essential for nutrient assimilation. Soil pH and composition determine the solubility and availability of minerals, with certain pH levels favouring the uptake of specific nutrients. Water availability, dictated by humidity and rainfall patterns, is critical for nutrient transport within plants (Srivastava *et al.*, 2020). Additionally, biotic factors such as microbial activity in the soil can enhance or inhibit nutrient availability through processes like nitrogen fixation, phosphorus solubilization and organic matter decomposition (Feng *et al.*, 2020). Understanding these environmental influences is essential for optimizing nutrient management practices in agriculture and ensuring sustainable plant growth and productivity (Jia *et al.*, 2022).

1.13.1 pH and nutrient availability

Nutrient availability is impacted by soil pH. Plants can change their root exudates or develop symbiotic interactions in acidic or alkaline soil to improve nutrient uptake. Plant growth and development are restricted by soil alkalinity above pH 8.0 or acidity below pH 5.5 (Schubert *et al.*, 1990; Patil *et al.*, 2012). The apoplast and rhizosphere quickly become acidic or alkaline upon the uptake of NH_4^+ or NO_3^-, which results from transport and assimilation (Taylor and Bloom, 1998; Hinsinger *et al.*, 2003; Geilfus, 2017). It has been demonstrated that raising the external

pH to acidic levels in *Arabidopsis thaliana* can increase the expression of 20–41% of NH_4^+-responsive genes, indicating that apoplastic acidification is a part of NH_4^+-induced stress (Patterson *et al.*, 2010).

1.13.2 Temperature and water stress

Temperature extremes and water stress can impact nutrient uptake kinetics and assimilation processes. One of the primary environmental variables impeding plant growth and development and the dispersion of vegetation is extreme temperature, which is also a significant abiotic stress for plants (Jeon and Kim, 2013). Future agricultural production will be impacted by weather characterized by extreme temperature occurrences due to climate change. These days, abruptly low temperatures in crop production areas frequently result in a loss in yield (Lohani *et al.*, 2019). Thus, to maximize production, it is essential to comprehend how temperature can impact plants' movement of water and nutrients. Nitrogen is one of the most significant nutrients whose transit is affected by low temperatures. First, low temperatures change the soil's nitrogen form into NH_4^+ or NO_3^-.

Because NH_4^+ is less mobile at low temperatures, nitrifying bacteria become inactivated, increasing the NH_4^+/NO_3^- ratio (Laanbroek, 1990, Warren, 2009). Low temperatures have also been found to decrease NH_4^+ uptake by the high-affinity transport system (HATS) (Wang *et al.*, 1993). Additionally, a study conducted on *Brassica juncea* revealed that low temperatures reduce the expression of an NH_4^+ transporter (AMT2) as well as the primary NO_3^- transporters (NRT1;1, NRT1;2, NRT1;4, NRT1;5, NRT1;8 and NRT2;1) (Goel and Singh, 2015). A drop in temperature can also alter intracellular transport. For instance, abscisic acid can induce the overexpression of vacuolar transporter CAX1, which has been shown to increase the concentration of free Ca_2^+ in the cytosol (Catalá *et al.*, 2003).

1.13.3 Soil aeration

Living cells cannot thrive in many situations without oxygen. Due to their multicellular nature, roots need significant energy for their growth and metabolic functions. Oxygen is consequently required to create components rich in energy through the biological process of oxidation. Root absorption of necessary nutrients is reduced by inadequate soil aeration. Soil that has been wet or contains a lot of clay has so little air that the uptake of mineral plant nutrients is greatly hampered, and plant roots exposed to the environment lack oxygen. That nutrient acquisition is an energy-dependent process is amply demonstrated by how respiratory pollutants affect roots and change their ability to acquire nutrients.

1.14   Biofortification

The metal must pass through several membranes for micronutrients to be taken up from the environment, distributed to several tissues and organs, and subcellularly compartmentalized (Kathpalia and Bhatla, 2018). Complex retrograde signalling cascades connect the intracellular compartments to control cellular metal homeostasis. For plants to adapt to varying levels of micronutrient accessibility, intricate functional processes have been established. Both long- and short-distance signalling channels control these activities. In addition to their potential locations in the signalling cascade, several molecular components implicated in micronutrient uptake in the local and long-range regulatory pathways have been identified (Ruffel, 2018). Each transport step is tightly controlled at the transcriptional and post-translational phases. This short and extended incitation operates at the basic and functional levels to synchronize nutritional homeostasis at the individual and systemic levels. A few entities from many gene families are hypothesized to play important roles in the plant's ability to disperse micronutrients (Prathap *et al.*, 2022). To preserve nutritional homeostasis at individual and systemic levels, the integration of various transduction signals functions at both cellular and whole-plant levels. Increasing the micronutrient content of major crops or biofortifying them can greatly enhance global human nourishment (Afzal *et al.*, 2020). Biofortification aims to improve the nutrient content of edible plant parts. This involves breeding or genetic modification to enhance the accumulation of specific nutrients. Biofortification efforts target addressing micronutrient deficiencies, such as iron or zinc deficiencies, prevalent in certain regions.

The intricate web of mechanisms governing nutrient homeostasis in plants reflects the adaptability of plants to diverse environmental conditions (Ueda *et al.*, 2021). As our understanding of these processes deepens, researchers and agronomists can develop strategies to optimize nutrient management, enhance crop yields and contribute to sustainable agricultural practices (Salas-González *et al.*, 2021). Ongoing research continues to uncover new facets of nutrient homeostasis, enriching our knowledge of plant biology and offering potential solutions to global agricultural challenges.

1.15   Future Perspectives

As our understanding of plant nutrient requirements and homeostasis continues to deepen, several exciting prospects and avenues for future research emerge, promising advancements in agriculture, environmental sustainability and food security.

- **Precision nutrient management.** Future research will likely focus on refining precision nutrient management strategies. Technologies such as remote sensing, precision agriculture and advanced sensors offer the potential for real-time monitoring of plant nutrient status. Integrating these technologies with machine learning algorithms can optimize nutrient delivery, minimizing waste and environmental impact.
- **Molecular insights into homeostasis.** Advances in molecular biology and genomics will unravel further details of the molecular mechanisms governing nutrient homeostasis. Understanding how specific genes regulate nutrient uptake, transport and allocation will provide tools for developing crops with enhanced nutrient use efficiency and resilience to environmental stressors.
- **Biotechnological solutions.** Biotechnological approaches, including genetic engineering and synthetic biology, hold promise

for developing nutrient-efficient crops. Manipulating the expression of genes implicated in nutrient uptake and transport could lead to crops that thrive in nutrient-poor soils or under conditions of changing climate, contributing to global food security.

- **Microbiome and nutrient cycling.** Investigating the plant microbiome's role in nutrient cycling and acquisition is an emerging area of interest. Understanding how microbial communities influence nutrient availability and plant health opens possibilities for developing microbial-based solutions to enhance nutrient uptake and improve soil fertility.
- **Climate change adaptation.** With climate change affecting nutrient cycling and availability, future research will likely explore how plants adapt to altered environmental conditions. This includes investigating the impact of rising temperatures, changing precipitation patterns and increased carbon dioxide levels on nutrient homeostasis and plant metabolism.
- **Integrated nutrient management practices.** The future of nutrient management involves integrating various agricultural practices. This includes combining organic and conventional farming methods, cover cropping and agroforestry to create holistic approaches that enhance nutrient cycling, reduce environmental impact and promote sustainable agriculture.
- **Educational and outreach initiatives.** As research progresses, there is a need for educational initiatives to disseminate knowledge about sustainable nutrient management practices. Farmers, policy makers and agricultural stakeholders must be informed about the latest research findings to facilitate the adoption of sustainable practices on a global scale.

In summary, the future of plant nutrient management and homeostasis research holds immense promise for revolutionizing agriculture. By leveraging technological innovations, molecular insights and sustainable practices, researchers can contribute to a more resilient and efficient global food production system while addressing environmental challenges associated with nutrient management. The ongoing collaboration between scientists, farmers and policy makers will be crucial in translating research discoveries into practical and sustainable solutions for the benefit of future generations.

1.16 Conclusion

In conclusion, understanding plant nutrient requirements and the regulatory mechanisms governing nutrient homeostasis is essential for advancing sustainable agriculture and addressing global food security challenges. Continued research in this field is imperative to unravel the complexities of nutrient interactions, paving the way for innovative strategies to optimize nutrient management in crops and mitigate environmental impacts. Furthermore, some nutrients are acquired by plants through intrinsic connections, meaning that helpful microorganisms that can facilitate the uptake of one nutrient by plants may also indirectly aid in the uptake of another nutrient. Beneficial microorganisms are an important resource for creating environmentally acceptable instruments to supply crops with the necessary nutrients, whether chemical fertilization is used or not. Certain potential complications that may arise when using beneficial microbes are readily envisioned.

References

Afzal, S., Sirohi, P., Sharma, D. and Singh, N.K. (2020) Micronutrient movement and signalling in plants from a biofortification perspective. In: Aftab, T. and Hakeem, K.R. (eds) *Plant Micronutrients: Deficiency and Toxicity Management*. Springer, pp. 129–171.
Aggarwal, A., Kadian, N., Tanwar, A., Yadav, A. and Gupta, K.K. (2011) Role of arbuscular mycorrhizal fungi (AMF) in global sustainable development. *Journal of Applied and Natural Science* 3, 340–351.
Ahad, A. and Ferdous, A. (2019) *Dictionary of Ecology and Environmental Science*. Himachal Publication, Dhaka.

Ahmed, N., Zhang, B., Chachar, Z., Li, J., Xiao, G. *et al.* (2024) Micronutrients and their effects on horticultural crop quality, productivity and sustainability. *Scientia Horticulturae* 323, 112512.

Ajay, B.C., Singh, A.L., Kumar, N., Dagla, M.C., Bera, S.K. *et al.* (2015) *Role of Phosphorus Efficient Genotypes in Increasing Crop Production*. Daya Publishing House.

Al Eryani-Raqeeb, A., Mahmud, T.M.M., Syed Omar, S.R., Mohamed Zaki, A.R. and Al Eryani, A.R. (2009) Effects of calcium and chitosan treatments on controlling anthracnose and postharvest quality of papaya (*Carica papaya* L.). *International Journal of Agricultural Research* 4, 53–68.

Alaoui, I., El-ghadraoui, O., Serbouti, S., Ahmed, H. and Mansouri, I. (2022) The mechanisms of absorption and nutrients transport in plants: a review. *Tropical Journal of Natural Product Research* 6, 8–14.

Aloni, R. and Griffith, M. (1991) Functional xylem anatomy in root-shoot junctions of six cereal species. *Planta* 184, 123–129.

Amtmann, A. and Armengaud, P. (2009) Effects of N, P, K and S on metabolism: new knowledge gained from multi-level analysis. *Current Opinion in Plant Biology* 12, 275–283.

Amtmann, A. and Blatt, M.R. (2009) Regulation of macronutrient transport. *New Phytologist* 181, 35–52.

Ashrafuzzaman, M. (2021) *Biophysics and Nanotechnology of Ion Channels*. CRC Press, Boca Raton, Florida.

Aulakh, M.S. and Pasricha, N.S. (1977) Interaction effect of sulphur and phosphorus on growth and nutrient content of moong (*Phaseolus aureus* l.). *Plant and Soil* 47, 341–350.

Baker, J.M., Hawkins, N.D., Ward, J.L., Lovegrove, A., Napier, J.A. *et al.* (2006) A metabolomic study of substantial equivalence of field-grown genetically modified wheat. *Plant Biotechnology Journal* 4, 381–392.

Bhantana, P., Rana, M.S., Sun, X., Moussa, M.G., Saleem, M.H. *et al.* (2021) Arbuscular mycorrhizal fungi and its major role in plant growth, zinc nutrition, phosphorous regulation and phytoremediation. *Symbiosis* 84, 19–37.

Bi, D., Almpanis, A., Noel, A., Deng, Y. and Schober, R. (2021) A survey of molecular communication in cell biology: establishing a new hierarchy for interdisciplinary applications. *IEEE Communications Surveys and Tutorials* 23, 1494–1545.

Bisht, N. and Chauhan, P.S. (2020) Excessive and disproportionate use of chemicals cause soil contamination and nutritional stress. In: Larramendy, M.L. and Soloneski, S. (eds) *Soil Contamination – Threats and Sustainable Solutions*. IntechOpen, pp. 1–10.

Broadley, M.R., Hammond, J.P., White, P.J. and Salt, D.E. (2010) An efficient procedure for normalizing ionomics data for *Arabidopsis thaliana*. *New Phytologist* 186, 270–274.

Brown, P.H., Bellaloui, N., Wimmer, M.A., Bassil, E.S., Ruiz, J. *et al.* (2002) Boron in plant biology. *Plant Biology* 4, 205–223.

Carpaneto, A., Naso, A., Paganetto, A., Cornara, L., Pesce, E.R. *et al.* (2004) Properties of ion channels in the protoplasts of the mediterranean seagrass *Posidonia oceanica*. *Plant, Cell and Environment* 27, 279–292.

Catalá, R., Santos, E., Alonso, J.M., Ecker, J.R., Martinez-Zapater, J.M. *et al.* (2003) Mutations in the Ca2+/H+ transporter CAX1 increase CBF/DREB1 expression and the cold-acclimation response in Arabidopsis. *Plant Cell* 15, 2940–2951.

Chotchutima, S., Tudsri, S., Kangvansaichol, K. and Sripichitt, P. (2016) Effects of sulfur and phosphorus application on the growth, biomass yield and fuel properties of leucaena (*Leucaena leucocephala* (lam.) de wit.) as bioenergy crop on sandy infertile soil. *Resources* 50, 54–59.

Dhok, R. (2020) Effect of various elements on growth of plants. *Plant Nutrients* 223–227.

Drew, D., North, R.A., Nagarathinam, K. and Tanabe, M. (2021) Structures and general transport mechanisms by the major facilitator superfamily (MFS). *Chemical Reviews* 121, 5289–5335.

Drinkwater, L.E. and Snapp, S. (2007) Nutrients in agroecosystems: rethinking the management paradigm. *Advances in Agronomy* 92, 163–186.

Dubyak, G.R. (2004) Ion homeostasis, channels, and transporters: an update on cellular mechanisms. *Advances in Physiology Education* 28, 143–154.

Engel, R.E., Bruebaker, L. and Emborg, T.J. (2001) A chloride deficient leaf spot of durum wheat. *Soil Science Society of America Journal* 65, 1448–1454.

Etxeberria, E., Pozueta-Romero, J. and Gonzalez, P. (2012) In and out of the plant storage vacuole. *Plant Science* 190, 52–61.

Fageria, N.K., Baligar, V.C. and Clark, R.B. (2006) *Physiology of Crop Production*. CRC Press, Boca Raton, Florida.

Fageria, N.K., Baligar, V.C. and Li, Y.C. (2008) The role of nutrient efficient plants in improving crop yields in the twenty first century. *Journal of Plant Nutrition* 31, 1121–1157.

Farhangi-Abriz, S. and Ghassemi-Golezani, K. (2023) Improving electrochemical characteristics of plant roots by biochar is an efficient mechanism in increasing cations uptake by plants. *Chemosphere* 313, 137365.

Feng, H., Fan, X., Miller, A.J. and Xu, G. (2020) Plant nitrogen uptake and assimilation: regulation of cellular pH homeostasis. *Journal of Experimental Botany* 71, 4380–4392.

Freschet, G.T., Cornelissen, J.H.C., van Logtestijn, R.S.P. and Aerts, R. (2010) Substantial nutrient resorption from leaves, stems and roots in a subarctic flora: what is the link with other resource economics traits? *New Phytologist* 186, 879–889.

García-Sánchez, M.J., Jaime, M.P., Ramos, A., Sanders, D. and Fernández, J.A. (2000) Sodium-dependent nitrate transport at the plasma membrane of leaf cells of the marine higher plant *Zostera marina* L. *Plant Physiology* 122, 879–885.

Garrill, A., Tyerman, S.D. and Findlay, G.P. (1994) Ion channels in the plasma membrane of protoplasts from the halophytic angiosperm *Zostera muelleri*. *Journal of Membrane Biology* 142, 381–393.

Gayomba, S.R. and Muday, G.K. (2020) Flavonols regulate root hair development by modulating accumulation of reactive oxygen species in the root epidermis. *Development* 147, dev185819.

Geilfus, C.M. (2017) The pH of the apoplast: dynamic factor with functional impact under stress. *Molecular Plant* 10, 1371–1386.

George, T.S., Brown, L.K. and Bengough, A.G. (2021) Advances in understanding plant root hairs about nutrient acquisition and crop root function. In: Gregory, P. (ed.) *Understanding and Improving Crop Root Function*. Burleigh Dodds Science Publishing Ltd, Cambridge, UK, pp. 127–162.

Goel, P. and Singh, A.K. (2015) Abiotic stresses downregulate key genes involved in nitrogen uptake and assimilation in *Brassica juncea* L. *PloS One* 10, e0143645.

Hanikenne, M., Esteves, S.M., Fanara, S. and Rouached, H. (2021) Coordinated homeostasis of essential mineral nutrients: a focus on iron. *Journal of Experimental Botany* 72, 2136–2153.

Haydon, M.J., Román, Á. and Arshad, W. (2015) Nutrient homeostasis within the plant circadian network. *Frontiers in Plant Science* 6, 299.

Hedrich, R., Becker, D., Geiger, D., Marten, I. and Roelfsema, M.R.G. (2012) Role of ion channels in plants. In: Okada, Y (ed.) *Patch Clamp Techniques: From Beginning to Advanced Protocols*. Springer, pp. 295–322.

Hinsinger, P., Plassard, C., Tang, C. and Jaillard, B. (2003) Origins of root-mediated pH changes in the rhizosphere and their responses to environmental constraints: a review. *Plant and Soil* 248, 43–59.

Huang, X., Duan, S., Wu, Q., Yu, M. and Shabala, S. (2020) Reducing cadmium accumulation in plants: structure–function relations and tissue-specific operation of transporters in the spotlight. *Plants* 9, 223.

Iversen, C.M., Sloan, V.L., Sullivan, P.F., Euskirchen, E.S., McGuire, A.D. *et al.* (2015) The unseen iceberg: plant roots in arctic tundra. *New Phytologist* 205, 34–58.

Jain, A. and Zoncu, R. (2022) Organelle transporters and inter-organelle communication as drivers of metabolic regulation and cellular homeostasis. *Molecular Metabolism* 60, 101481.

Jeon, J. and Kim, J. (2013) Cold stress signaling networks in Arabidopsis. *Journal of Plant Biology* 56, 69–76.

Jia, Z., Giehl, R.F.H. and von Wirén, N. (2022) Nutrient-hormone relations: driving root plasticity in plants. *Molecular Plant* 15, 86–103.

Johnson, V.J. and Mirza, A. (2020) Role of macro and micronutrients in the growth and development of plants. *International Journal of Current Microbiology and Applied Sciences* 9(11), 576–587.

Joshi, S., Nath, J., Singh, A.K., Pareek, A. and Joshi, R. (2022) Ion transporters and their regulatory signal transduction mechanisms for salinity tolerance in plants. *Physiologia Plantarum* 174, e13702.

Karley, A.J. and White, P.J. (2009) Moving cationic minerals to edible tissues: potassium, magnesium, calcium. *Current Opinion in Plant Biology* 12, 291–298.

Karthika, K.S., Rashmi, I. and Parvathi, M.S. (2018) Biological functions, uptake and transport of essential nutrients in relation to plant growth. In: Hasanuzzaman, M., Fujita, M., Oku, H., Nahar, K. and Hawrylak-Nowak, B. (eds) *Plant Nutrients and Abiotic Stress Tolerance*. Springer, pp. 1–49.

Kathpalia, R. and Bhatla, S.C. (2018) Plant mineral nutrition. In: Bhatla, S.C. and Lal, M.A. (eds) *Plant Physiology, Development and Metabolism*. Springer, pp. 37–81.

Killingbeck, K.T. (1986) The terminological jungle revisited: making a case for use of the term resorption. *Oikos* 46, 263.

Kiri, I.Z. and Hausa, K. (2023) Mechanisms of nutrient uptake and assimilation processes in some plants: a review. *Dutse Journal of Pure and Applied Sciences* 9, 223–237.

Kisnieriene, V., Lapeikaite, I., Pupkis, V. and Beilby, M.J. (2019) Modeling the action potential in characeae *Nitellopsis obtusa*: effect of saline stress. *Frontiers in Plant Science* 10, 82.

Konečný, J., Hršelová, H., Bukovská, P., Hujslová, M. and Jansa, J. (2019) Correlative evidence for co-regulation of phosphorus and carbon exchanges with symbiotic fungus in the arbuscular mycorrhizal *Medicago truncatula. PloS One* 14, e0224938.

Krouk, G. and Kiba, T. (2020) Nitrogen and phosphorus interactions in plants: from agronomic to physiological and molecular insights. *Current Opinion in Plant Biology* 57, 1–16.

Kumar, S., Kumar, S. and Mohapatra, T. (2021) Interaction between macro- and micro-nutrients in plants. *Frontiers in Plant Science* 12, 665583.

Laanbroek, H.J. (1990) Bacterial cycling of minerals that affect plant growth in waterlogged soils: a review. *Aquatic Botany* 38, 109–125.

Lang, A. and Thorpe, M.R. (1989) Xylem, phloem and transpiration flows in a grape: application of a technique for measuring the volume of attached fruits to high resolution using Archimedes' principle. *Journal of Experimental Botany* 40, 1069–1078.

Lohani, N., Singh, M.B. and Bhalla, P.L. (2019) High temperature susceptibility of sexual reproduction in crop plants. *Journal of Experimental Botany* 71, 555–568.

Maiti, D., Toppo, N.N., Nitin, M. and Kumar, B. (2017) Arbuscular mycorrhizal technology based on ecosystem services rendered by native flora for improving phosphorus nutrition of upland rice: status and prospect. In: Varma, A., Prasad, R. and Tuteja, N. (eds) *Mycorrhiza – Eco-Physiology, Secondary Metabolites, Nanomaterials*. Springer, pp. 87–105.

Makvandi, P., Chen, M., Sartorius, R., Zarrabi, A., Ashrafizadeh, M. *et al.* (2021) Endocytosis of abiotic nanomaterials and nanobiovectors: inhibition of membrane trafficking. *Nano Today* 40, 101279.

Mattes, A. (2019) Advancement of natural products: Optimization of instrumentation and examples of their application to the isolation of new compounds. PhD thesis, University of Oklahoma, Norman, Oklahoma.

Mitra, D., Rad, K.V., Chaudhary, P., Ruparelia, J., Sagarika, M.S. *et al.* (2021) Involvement of strigolactone hormone in root development, influence and interaction with mycorrhizal fungi in plant: mini-review. *Current Research in Microbial Sciences* 2, 100026.

Mitra, G. (2017) Essential plant nutrients and recent concepts about their uptake. In: Naeem, M, Ansari, A.A. and Gill S.S. (eds) *Essential Plant Nutrients: Uptake, Use Efficiency, and Management*. Springer, pp. 3–36.

Morris, A.L. and Mohiuddin, S.S. (2020) Biochemistry, nutrients. In: *StatPearls [Internet]*. StatPearls Publishing, Treasure Island, Florida.

Nadeem, F., Hanif, M.A., Majeed, M.I. and Mushtaq, Z. (2018) Role of macronutrients and micronutrients in the growth and development of plants and prevention of deleterious plant diseases – a comprehensive review. *International Journal of Chemical and Biochemical Sciences* 14, 1–22.

Natura, G. and Dahse, I. (1998) Potassium conductance of Egeria leaf cell protoplasts: regulation by medium pH, phosphorylation and G-proteins. *Journal of Plant Physiology* 153, 363–370.

Nopphakat, K., Runsaeng, P. and Klinnawee, L. (2021) Acaulospora as the dominant arbuscular mycorrhizal fungi in organic lowland rice paddies improves phosphorus availability in soils. *Sustainability* 14, 31.

Okamoto, S., Tabata, R. and Matsubayashi, Y. (2016) Long-distance peptide signaling essential for nutrient homeostasis in plants. *Current Opinion in Plant Biology* 34, 35–40.

Okazaki, Y., Otsuki, H., Narisawa, T., Kobayashi, M., Sawai, S. *et al.* (2013) A new class of plant lipid is essential for protection against phosphorus depletion. *Nature Communications* 4, 1510.

Pandey, N. (2018) Role of plant nutrients in plant growth and physiology. In: Hasanuzzaman, M., Fujita, M., Oku, H., Nahar, K. and Hawrylak-Nowak, B. (eds) *Plant Nutrients and Abiotic Stress Tolerance*. Springer, pp. 51–93.

Pandey, R. (2015) Mineral nutrition of plants. In: Bahadur, B, Venkat Rajam, M., Sahijram, L. and Krishnamurthy, K. (eds) *Plant Biology and Biotechnology: Volume I: Plant Diversity, Organization, Function and Improvement*. Springer, pp. 499–538.

Pantoja, O. (2021) Recent advances in the physiology of ion channels in plants. *Annual Review of Plant Biology* 72, 463–495.

Patil, N.S., Apradh, V.T. and Karadge, B.A. (2012) Effects of alkali stress on seed germination and seedlings growth of Vigna aconitifolia (Jacq.) Marechal. *Pharmacognosy Journal* 4, 77–80.

Patterson, K., Cakmak, T., Cooper, A., Lager, I., Rasmusson, A.G. *et al.* (2010) Distinct signalling pathways and transcriptome response signatures differentiate ammonium- and nitrate-supplied plants. *Plant, Cell and Environment* 33, 1486–1501.

Pitsili, E. (2021) Functional analysis of a phloem cysteine protease in *Arabidopsis thaliana*. PhD thesis, University of Barcelona, Barcelona, Spain.

Poorter, H. and Nagel, O. (2000) The role of biomass allocation in the growth response of plants to different levels of light, CO2, nutrients and water: a quantitative review. *Functional Plant Biology* 27, 595–607.

Porter, E.M., Bowman, W.D., Clark, C.M., Compton, J.E., Pardo, L.H. and Soong, J.L. (2013) Interactive effects of anthropogenic nitrogen enrichment and climate change on terrestrial and aquatic biodiversity. *Biogeochemistry* 114, 93–120.

Prathap, V., Kumar, A., Maheshwari, C. and Tyagi, A. (2022) Phosphorus homeostasis: acquisition, sensing, and long-distance signaling in plants. *Molecular Biology Reports* 49, 8071–8086.

Puig, S. and Peñarrubia, L. (2009) Placing metal micronutrients in context: transport and distribution in plants. *Current Opinion in Plant Biology* 12, 299–306.

Rao, I.M. (2009) *Essential Plant Nutrients and their Functions*. International Center for Tropical Agriculture, Cali, Colombia.

Reid, R. and Hayes, J. (2003) Mechanisms and control of nutrient uptake in plants. *International Review of Cytology* 229, 73–114.

Ruffel, S. (2018) Nutrient-related long-distance signals: common players and possible cross-talk. *Plant and Cell Physiology* 59, 1723–1732.

Salas-González, I., Reyt, G., Flis, P., Custódio, V., Gopaulchan, D. *et al.* (2021) Coordination between microbiota and root endodermis supports plant mineral nutrient homeostasis. *Science* 371, eabd0695.

Saure, M.C. (2005) Calcium translocation to fleshy fruit: Its mechanism and endogenous control. *Scientia Horticulturae* 105, 65–89.

Schubert, S., Schubert, E. and Mengel, K. (1990) Effect of low pH of the root medium on proton release, growth, and nutrient uptake of field beans (*Vicia faba*). *Plant and Soil* 124, 239–244.

Srivastava, A.K., Shankar, A., Nalini Chandran, A.K., Sharma, M., Jung, K.-H. *et al.* (2020) Emerging concepts of potassium homeostasis in plants. *Journal of Experimental Botany* 71, 608–619.

Stéger, A. and Palmgren, M. (2022) Root hair growth from the pH point of view. *Frontiers in Plant Science* 13, 949672.

Stein, W.D. and Litman, T. (2014) *Channels, Carriers, and Pumps: An Introduction to Membrane Transport*. Elsevier.

Stillwell, W. (2016) *An Introduction to Biological Membranes: Composition, Structure and Function*. Elsevier.

Sung, Y., Yu, Y.C. and Han, J.M. (2023) Nutrient sensors and their cross-talk. *Experimental and Molecular Medicine* 55, 1076–1089.

Suryavanshi, S.S., Waikar, S.L. and Ajabe, M.A. (2020) Response of iron and zinc fortification on growth, yield and quality of spinach. *Journal of Pharmacognosy and Phytochemistry* 9, 2040–2043.

Tavakoli, M.T., Chenari, A.I., Rezaie, M., Tavakoli, A., Shahsavari, M. *et al.* (2014) The importance of micronutrients in agricultural production. *Advances in Environmental Biology* 8, 31–35.

Taylor, A.R. and Bloom, A.J. (1998) Ammonium, nitrate, and proton fluxes along the maize root. *Plant, Cell and Environment* 21, 1255–1263.

Tedersoo, L., Bahram, M. and Zobel, M. (2020) How mycorrhizal associations drive plant population and community biology. *Science* 367, eaba1223.

Ting, I.P. (1982) Plant mineral nutrition and ion uptake. In: *Plant Physiology*. Addison-Wesley, Reading, Massachusetts, pp. 331–363.

Tomkins, M., Hughes, A. and Morris, R.J. (2021) An update on passive transport in and out of plant cells. *Plant Physiology* 187, 1973–1984.

Tripathi, D.K., Singh, V.P., Chauhan, D.K., Prasad, S.M. and Dubey, N.K. (2014) Role of macronutrients in plant growth and acclimation: recent advances and future prospective. In: Ahmad, P., Wani, M., Azooz, M. and Phan Tran, L.S. (eds) *Improvement of Crops in the Era of Climatic Changes*. Springer, New York, pp. 197–216.

Tripathi, D.K., Singh, S., Singh, S., Mishra, S., Chauhan, D.K. *et al.* (2015) Micronutrients and their diverse role in agricultural crops: advances and future prospective. *Acta Physiologiae Plantarum* 37, 1–14.

Ueda, Y., Sakuraba, Y. and Yanagisawa, S. (2021) Environmental control of phosphorus acquisition: a piece of the molecular framework underlying nutritional homeostasis. *Plant and Cell Physiology* 62, 573–581.

Wahab, A., Muhammad, M., Munir, A., Abdi, G., Zaman, W. *et al.* (2023) Role of arbuscular mycorrhizal fungi in regulating growth, enhancing productivity, and potentially influencing ecosystems under abiotic and biotic stresses. *Plants* 12, 3102.

Waldrop, M.P. and Zak, D.R. (2006) Response of oxidative enzyme activities to nitrogen deposition affects soil concentrations of dissolved organic carbon. *Ecosystems* 9, 921–933.

Wang, M.Y., Siddiqi, M.Y., Ruth, T.J. and Glass, A.D.M. (1993) Ammonium uptake by rice roots (II. Kinetics of 13NH4+ influx across the plasmalemma). *Plant Physiology* 103, 1259–1267.

Warren, C.R. (2009) Why does temperature affect relative uptake rates of nitrate, ammonium and glycine: a test with Eucalyptus pauciflora. *Soil Biology and Biochemistry* 41, 778–784.

White, P.J. and Broadley, M.R. (2003) Calcium in plants. *Annals of Botany* 92, 487–511.

White, P.J. and Ding, G. (2023) Long-distance transport in the xylem and phloem. In: Rengel, Z., Cakmak, I. and White, P.J. (eds) *Marschner's Mineral Nutrition of Plants*. Academic Press, pp. 73–104.

2 The Role of Soil Health in Plant Defence Signalling

Vijayata Singh[1]*, Deepjyoti Singh[2], Anupriya Singh[3], Neha Sharma[4], Divya Chaudhary[5] and Naveen Chandra Joshi[5]

[1]Icahn School of Medicine at Mount Sinai, New York, USA; [2]North Carolina State University, Raleigh, NC, USA; [3]Delhi University, New Delhi, India; [4]Department of Microbiology, Faculty of Allied Health Sciences, Shree Guru Gobind Singh Tricentenary University, Budhera, Gurugram, Haryana, India; [5]Amity Institute of Microbial Technology, Amity University, Noida, Uttar Pradesh, India

Abstract

Soil health is essential for plant productivity and sustainable agriculture. Soil is a medium wherein nutrients and microorganisms interact to support plant growth and defence mechanisms. Poor soil conditions like inadequate moisture, poor structure and an imbalance of macro- and micronutrients can invite several soil-borne diseases and make plants susceptible to abiotic and biotic stress. The rhizosphere has a significant impact on plant defence expression and plant–pathogen interactions. Soil nutrients can reprogramme plant immunity through cross-talk between defence and nutrient signalling. This chapter discusses the facts, research and progress made over the years regarding soil health, including its physical, chemical and biological attributes, management practices and the effects of different plant defence signals.

Keywords: Rhizosphere, plant defence, root exudates, secondary metabolites, soil health

2.1 Introduction

The soil is a crucial component of our ecosystem. It serves multiple functions, such as recycling waste products and providing mechanical and nutritional support for plant growth. The most essential one is supplying a medium for plant growth. Organic matter and minerals are the major components of soils. The mineral components in soil comprise weathered rocks of different sizes, including clay, sand and silt. Meanwhile, the organic content of soil includes decomposing plant and microbial remains. The amount of pore space, mineral content and organic matter can vary greatly depending on the soil type (Moore and Bradley, 2018). Soil pH is a serious factor that plays a crucial role in the health of plants. For plants, the optimal pH level varies based on the soil type and its organic matter present. Soil pH has an important effect on the availability of plant nutrients. As soil pH decreases, nutrients such as copper (Cu), iron (Fe), manganese (Mn) and zinc (Zn) become more soluble. These nutrients are more readily available in acidic soil. At lower soil pH (5.5 or below) minerals like Mn, Zn or aluminum (Al) become

*Corresponding author: vijayata.cb@gmail.com

© CAB International 2025. *Soil Health and Nutrition Management*
(eds N.C. Joshi, T. Leustek and P.K. Singh)
DOI: 10.1079/9781800624597.0002

more soluble in soil water, which generally leads to plant toxicity (Weil and Brady, 2017; Moore and Bradley, 2018).

A plant needs 17 essential nutrients for their survival. Carbon (C), hydrogen (H) and oxygen (O) make up about 94% of a plant's total weight. The other 14 nutrients come from the soil and account for the rest of its weight. The three primary macronutrients are nitrogen (N), phosphorus (P) and potassium (K). The three secondary macronutrients are calcium (Ca), sulfur (S) and magnesium (Mg). The other eight elements are micronutrients because they are required in significantly smaller quantities than the macronutrients and include nickel, chlorine, iron, boron, zinc, molybdenum, copper and manganese (Moore and Bradley, 2018). The soil is a diverse ecosystem that houses various microorganisms including algae, bacteria, fungi, viruses, mites, nematodes and animals like insects, earthworms and moles. These soil-dwelling organisms interact with plants in different ways, such as competitive, exploitative, neutral, commensal, mutualistic and parasitic interactions (Bonkowski *et al.*, 2009; Müller *et al.*, 2016). Most plant science research has focused on minimizing the impact of pathogens, such as herbivory and infection, or alleviating abiotic stressors (Zhang *et al.*, 2013; Jacoby *et al.*, 2017). Soil organisms, both above and below the ground, profoundly affect plants. The microbiota present in plant roots can assist in enhancing the plant's ability to use micronutrients and macronutrients, particularly nitrogen and phosphorus. This results in a decrease in the quantity of nutrients lost due to leaching and immobilization, as well as a reduction in the production of greenhouse gases (Hamilton and Frank, 2001; Yoneyama *et al.*, 2007; Cardoso *et al.*, 2018). In addition to providing resistance against pathogens, root-associated microorganisms play an essential role in rhizo-remediation, phyto-stimulation and stress management (Guerrieri *et al.*, 2019).

Agricultural practices significantly impact the soil fauna and microbial community composition through pesticide and fertilizer application, tillage, monoculture and crop rotation, causing biodiversity loss and decreased biomass. Specifically, tillage changes the soil microhabitat and can disrupt the life cycle of earthworms, mites and other insects, leading to a decline in their population. It also substantially affects the diversity and distribution of bacteria and fungi. Additionally, tillage destroys the mycorrhizal hyphal networks, which causes a reduction in phosphorus uptake (Tsiafouli *et al.*, 2015; Le Guillou *et al.*, 2019).

> The impact of mineral fertilizers on plants is multifaceted. One of the benefits is an increase in plant productivity, which can lead to the enhanced organic material production that is released in the soil by means of root exudates and residues. This, in turn, promotes the growth of microbial communities using this carbon pool as their primary resource.
>
> (Geisseler and Scow, 2014)

Root exudates play a vital role in acquiring nutrients and interaction with soil microbiota, both chemically and biologically. They serve as a chemical signal between plants and other soil organisms. By releasing specific compounds, plants can manipulate the microorganisms in their immediate surroundings, recruiting beneficial microbes and fungi to increase nutrient uptake or protect themselves from pathogens and insects. This process helps plants create a healthier rhizosphere biome, leading to better growth and productivity (Berendsen *et al.*, 2012). The physical and chemical properties of soil and the availability of nutrients influences root exudate composition. Plants modify soil structure, pH and texture for enhancing penetration of roots into the soil and water and nutrient uptake (Read *et al.*, 2003, Naveed *et al.*, 2017). Surfactants such as phospholipids are released from plants to promote root growth and reduce the surface tension of soil. The sugars and organic acids found in the exudates of roots affect the texture of the soil by increasing soil aggregation and dispersion. Soil dispersion can help to release nutrients from the soil particles, while soil aggregation can create a more stable structure around the roots. There is a critical need to improve our understanding of these complex relationships to achieve sustainable and productive agriculture (Naveed *et al.*, 2017).

Plant defence phytohormones are a varied collection of a plant's secondary metabolites. They play a vital role in integrating the plant immune system's output and suppressing proliferation and growth of cells. Jasmonic acid (JA), ethylene (ET) and salicylic acid (SA) are the mediators of localized and systemic immune responses in plants

(Devoto and Turner, 2003; Han, 2016). Plants possess different defence mechanisms to safeguard themselves from various threats. For instance, systemic acquired resistance that is SA mediated provides non-specific immunity against disease-causing pathogens.

On the other hand, rhizobacteria that colonize plant roots can stimulate systemic resistance in leaves by releasing JA and ET. In the case of biotroph infection, SA and JA respond antagonistically to each other. Additionally, there are overlapping signalling pathways in defence-related phytohormones that regulate plant defence mechanisms through transcriptional and biosynthetic outputs (Vincent et al., 2022). The root microbiome is shaped in characteristic ways by plant defence phytohormones. Eliminating any of these defence phytohormone signalling system leads to abnormality in microbial communities in roots that might be linked to lower survival rate in wild-type soil. SA is the critical immune regulator present in leaves regulating root microbiome formation. Studies have shown that plants with altered signalling of SA have a root microbiome that differs from wild-type plants (Carvalhais et al., 2015; Chen et al., 2020). SA is a molecule that can be utilized in various ways by different bacterial strains. It can act as a signal for growth or a source of carbon. Its presence in the roots can affect the structure of the microbial community, and it happens because SA regulates the immune system outputs, which can, in turn, impact the bacterial communities in the root. Additionally, SA might affect the physiology of roots and microbe–microbe interactions, but this is still unclear. SA, a central regulator of plant immunity, directly influences the root microbiome composition. However, its role in the root remains largely unknown. Regulating the microbiome of root can improve the production of crop and its sustainability (Vincent et al., 2022).

2.2 Rhizosphere and Plant Signalling

The rhizosphere is a zone where plants and microbes interact with each other and exchange signals. As plants cannot move, they produce volatile compounds and root exudates in the rhizosphere to challenge microbes. In return, microbes release various phytohormones, quorum-sensing compounds and microbial volatile organic compounds and try to invade plant roots and stimulate defence signalling. These signal metabolites determine the soil structure and abundance of the rhizomicrobacteria. Besides microorganisms such as bacteria, fungi and oomycetes, other macroorganisms like oligochaeta, arthropods, mollusks and nematodes also exist in the rhizosphere, and they positively and negatively affect soil health. They help to create a suitable soil structure, aid in nutrient mineralization and balance the soil food web. Mineral nutrition is crucial in plant growth, development and defence systems. Nutrients activate enzymes that generate defence metabolites such as callose, phytoalexins, phenols, glucosinolates and lignin, which help plants defend against various threats (Huber and Thompson, 2007). Besides biochemical factors, nutrients impact plant shape, surface structure, hair, fibres and silicon play an essential role in the first line of defence (Broadley et al., 2012). It is challenging to make broad generalizations about how specific nutrients affect all plant–pathogen systems (Bateman, 1978). Several factors can determine the successful colonization of pests. These factors can be divided into two categories: biotic components (which include beneficial organisms and soil pathogens) and abiotic components (which include soil pH, temperature, nutrient availability and flood and drought conditions). All of these factors can influence the health of soil.

Further, we will explore the factors that impact soil health and the defence system and specifically focus on how plants communicate with microbes to influence the population of rhizosphere microbiota. This communication includes the roles of beneficial plant growth-promoting fungi (PGPF) and plant growth-promoting rhizobacteria (PGPR). We will also examine the processes of priming and eliciting immune responses that are induced systemic resistance (ISR) and systemic acquired resistance (SAR). The regulation of hormones such as SA, JA and ET is also considered. Finally, we will examine how signal exchanges occur through microbial phytohormones, volatile organic compounds and quorum sensing (QS) molecules.

2.2.1 Plant growth-promoting rhizobacteria

These are beneficial bacteria that inhabit or are located near plant roots and support the growth of plants. They are diverse in nature and function as biocontrol agents. PGPR help plants contain pathogens in two ways: directly and indirectly. PGPR synthesize phytohormones and facilitate the uptake of nutrients that plants can utilize. They can also provide local antagonism to soil-borne pathogens or induce ISR/SAR against pathogens throughout the plant (Glick, 1995). PGPR can indirectly reduce the pathogen population by producing substances such as antibiotics and lytic enzymes that are antagonistic towards them.

PGPR function as biocontrol agents by producing growth hormones like auxins or by fixing nitrogen in association with plant roots (Dobereiner, 1992; Patten and Glick, 2002). PGPF such as arbuscular mycorrhizal fungi (AMF), *Trichoderma* sp., *Penicillium digitatum*, *Aspergillus flavus*, *Gliocladium virens* and *Podospora bulbillosa*, can have direct or indirect effects on plant defence. They do this by regulating hormones, tolerating abiotic stressors, solubilizing nutrients, inhibiting phytopathogens and producing organic compounds and enzymes.

2.2.2 Antibiosis

Antibiotics are small organic compounds produced by microorganisms as a secondary metabolite. They help to suppress the growth of pathogens. Some bacterial species, such as *Pseudomonas* sp., *Bacillus* sp. and *Streptomyces* sp., are well-known for producing antibiotics. *Pseudomonas* secretes many antibiotics like phenazines, pyoluteorin, 2,4-diacetyl phloroglucinol, pyrrolnitrin and hydrogen cyanide, which have a wide range of antibacterial activity (Kenawy *et al.*, 2019). *Bacillus* sp. secrete antibiotics such as circulin, polymyxin and colistin that are effective against fungi as well as Gram-positive and -negative bacteria (Maksimov *et al.*, 2011). These antibiotics consist of different compounds that harm pathogens in the soil, thus protecting plants from them.

2.2.3 Lytic enzymes

Lytic enzymes produced by microorganisms can inhibit plant pathogens and are essential in the nutrient cycling process by breaking down organic matter in the ecosystem. Endophytes produce lytic enzymes that break down a variety of polymeric substances, such as cellulose, chitin, proteins and lipids (Tchameni *et al.*, 2020). A notable function of plant-associated endophytes is the secretion of enzymes that hydrolyse plant cell walls, including chitinase, β-1,3-glucanase, cellulase and protease (Dimkić *et al.*, 2022). Chitin is degraded by chitinase, which is a key element of fungal cell walls, thus weakening the structural integrity of the pathogen and aiding in plant defence. For example, chitinase produced by the endophytic bacterium *Streptomyces hygroscopicus* has been shown to inhibit the growth of various fungal species including *Hyaloperonospora parasitica*, *Aspergillus flavus*, *Fusarium oxysporum*, *A. niger*, *Ralstonia solani*, *Sclerotinia sclerotiorum* and *Botrytis cinerea*. Additionally, the chitinase-producing endophyte *Bacillus cereus* strain 65 has demonstrated protective effects on cotton seedlings against root disease caused by *R. solani* (Jha and Mohamed, 2023; Unuofin *et al.*, 2024).

2.2.4 Siderophores

A siderophore is an organic ligand specifically designed to bind and transport iron. These molecules facilitate iron acquisition through specialized uptake systems and their production is regulated by the availability of iron. While their primary role is iron transport, siderophores can also have other functions (Miethke and Marahiel, 2007). There are four main types of siderophores, classified by their iron-chelating groups: catecholate, phenolate, hydroxamate and carboxylate, although combinations of these types are also common (Hider and Kong, 2010). Usually, polyketide synthase (PKS) domains or non-ribosomal peptide synthetases (NRPS) working with NRPS modules are responsible for the synthesis of siderophores (Carroll and Moore, 2018). However, certain siderophores are produced by methods that are independent of PKS and NRPS. The synthesis of siderophores through NRPS is the most prevalent method. This process involves a complex assembly line of multiple enzymes, which incorporate hydroxy acids, amino acids and carboxylic acids, to form a peptide precursor. This precursor is then altered by NRPS or different enzymes to produce ultimate siderophore. An example of such alternation is the heterocycles generation, like the thiazoline ring, which results from the cysteine side chains cyclization (Crosa and Walsh, 2002). The release of siderophore is aided by efflux pumps and is an energy dependent mechanism.

Gram-positive and -negative bacteria have different ways to achieve siderophore influx. In case of Gram-positive bacteria, there is a lack of an outer membrane, which simplifies the translocation of iron-loaded siderophores (Kramer et al., 2020). Gram-negative bacteria, on the other hand, use a more intricate method that involves an outer membrane β-barrel receptor that specifically recognizes the iron-loaded siderophore. The receptor changes its conformation after binding, allowing the siderophore to be translocated into the periplasm (Schalk, 2008; Fritsch et al., 2022). A TonB complex drives this activity by supplying the required energy through the proton motive force. Iron-loaded siderophores are typically transported into the cytosol, where iron reduction occurs, via an ABC transporter located in the innermost membrane. Sometimes only the ferrous (Fe^{2+}) iron is transferred into the cytosol after iron reduction occurs in the periplasm (Singh et al., 2022). Siderophores are directly absorbed into Gram-positive bacteria, which lack receptors on the outer membrane, via ABC transporters that traverse the cell membrane. Some siderophores are recycled through specific mechanisms, while others undergo hydrolysis to release iron (Roskova et al., 2022).

2.3 Defence Priming

Certain compounds that are frequently linked to ISR (such as pathogenesis-related (PR) proteins) are generated in healthy tissues as a result of an initial infection (Ab Rahman et al., 2018). In the course of a later infection, additional substances are expressed, but only in plant regions where effective resistance is necessary. Biochemical

changes characteristic of plants expressing ISR becomes evident (Lanna-Filho et al., 2013). Numerous terms like conditioning, priming or sensitization are used to describe this process (Sticher et al., 1997). Chemical ISR inducers have the ability to provide priming effects (Chalupowicz et al., 2021). As a result, responses such as the lignification of cell walls or the synthesis of phytoalexins occur more rapidly and intensely during subsequent infections compared with the initial infection, enabling a stronger defence response (Saboki Ebrahim and Singh, 2011).

2.3.1 Induced systemic resistance

Certain beneficial rhizomicrobiomes, such as PGPR in ISR, can stimulate plant defence signalling against pests and phytopathogens. Like SAR, ISR can provide long-term and wide-ranging defence from these harmful agents (Bukhat et al., 2020). Despite beneficial microorganisms not having virulence genes, other microbe-associated molecular patterns (MAMPs) can still activate ISR. ISR mainly involves the ET and JA signalling pathways and is independent of SA (Fig. 2.1). However, research has suggested that SA may potentially contribute to ISR, resulting in the activation of PR proteins. Activating PR proteins activates enzymes such as chitinases and 1,3-glucanases, which contribute to lysis of pathogenic cells (dos Santos and Franco, 2023) and established plant cell walls and lead to cell death. Table 2.1 provides examples of ISR against plant pathogens caused by beneficial microbes, which help the soil provide a suppressive medium to stop the pathogen development through a variety of processes.

2.3.2 Systemic acquired resistance

Systemic acquired resistance is an immune response that protects plants against various pathogenic organisms, such as fungi, oomycetes, viruses and bacteria. It aims to develop a long-lasting immune response against these pathogens (Gozzo, 2003; Hossain, 2024). Upon pathogen infection, long-distance defence signals are triggered by pathogen-associated molecular pattern (PAMP) pattern-triggered immunity (PTI) and effector-triggered immunity (ETI) that induce defence

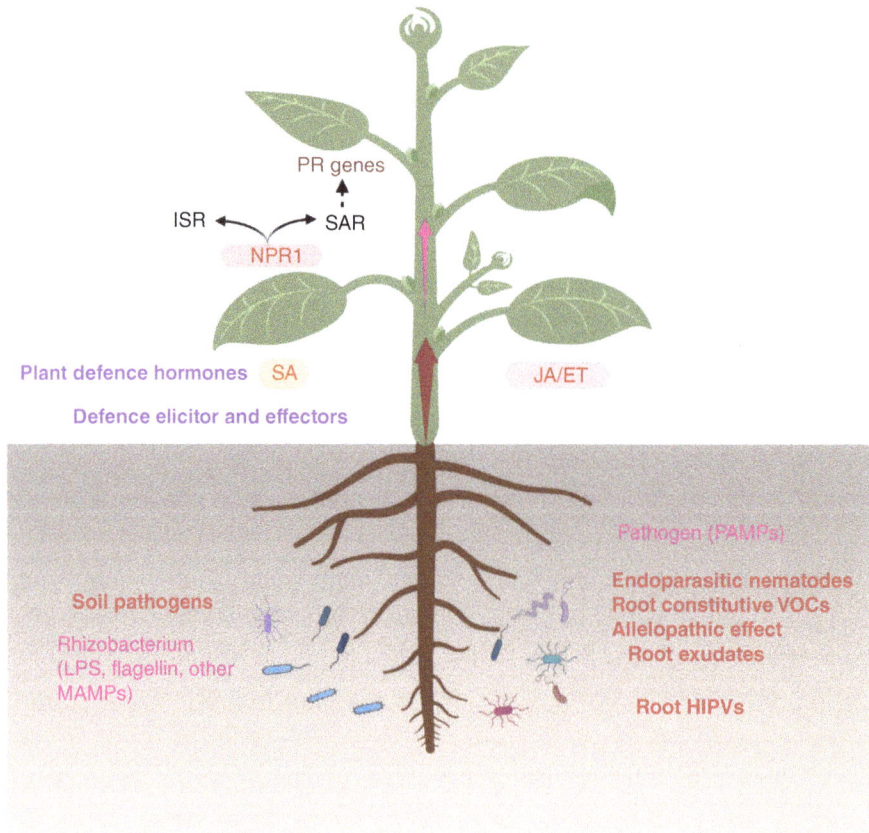

Fig. 2.1. Activation of defence signalling when pathogen infection is present in the soil. HIPVs, herbivore-induced plant volatiles; LPS, lipopolysaccharide; VOC, volatile organic compound; for other abbreviations, see text.

memory for future infections from broad-spectrum pathogens. This further activates PR genes. The process is dependent on the SA pathway (Ding *et al.*, 2022). Various rhizobacteria produce SA, which activates PR genes. SA binds with NPR1 (non-repressor of PR genes) and induces structural changes in its conformation, facilitating the binding of transcriptional factors for PR gene activation (Jamil *et al.*, 2022).

In JA signalling, COI1 (Coronatine insensitive 1) forms a ubiquitin complex that activates long-lasting SAR. In plants, ISR and SAR are two major defence mechanisms (Hönig *et al.*, 2023). ISR requires the signalling pathways of JA and ET, while SAR only requires the SA pathway. Also, some growth hormones like indole-3-acetic

acid (3-IAA), brassinosteroids (BR), abscisic acid (ABA) and gibberellins (GA) also contribute to defence signalling by interacting with SA, JA and ET pathways. Table 2.1 lists some pathogens inducing defence signalling (Poltronieri and Reca, 2019; Chieb and Gachomo, 2023).

2.4   Regulation of defence-related hormones

During pathogen infection, plants produce several defence signalling pathways to combat pathogens (Fig. 2.1). Some important signalling pathways are discussed here.

Table 2.1. Signalling pathways induced by some common plant pathogens.

Pathogen	Action mechanism	Host	Reference
Azospirillum sp. B510	The induction of signal transduction pathways triggered by ET	Tomato	Kusajima *et al.*, 2018
Enterobacter asburiae	Expression of defence-related genes and antioxidant enzymes increase N uptake	Sugarcane	Singh *et al.*, 2021
Paecilomyces variotii SJ1	Reactive oxygen species accumulation, increased SA and activated SA signalling pathway, N uptake	Rice	Wang *et al.*, 2022
Bacillus sp. 2P2	Higher activity of phenylalanine ammonia lyase, peroxidase, polyphenol oxidase, an ascorbate oxidase; upregulated expression of three pathogenesis-related genes, *PR1a*, *PR2a* and *PR3*	Wide range of hosts (tomato, tobacco, bell pepper, muskmelon, watermelon, sugar beet)	Sahu *et al.*, 2019
Enterobacter asburiae	Expression of defence-related genes and antioxidant enzymes	Lettuce	Kashyap *et al.*, 2023
Fusarium Fo47	Production of SA, JA and ET	Cucumber	Benhamou *et al.*, 2002

ET, ethylene; JA, jasmonic acid; N, nitrogen; SA, salicylic acid.

2.4.1 Salicylic acid-mediated signalling

Salicylic acid is a standard plant compound signalling pathway in flavonoids, specifically catechin and anthocyanidins. It inhibits the growth of *Melampsora larici-populina*, a foliar rust fungus that infects poplar trees. Methyl jasmonate (MeJA) and SA also activate defence-related genes and proteins that protect cassava plants from *Xanthomonas axonopodis* pv. *manihotis*, a disease-causing pathogen (Peng *et al.*, 2021). In order to protect tomato plants from the cucumber mosaic virus, *Trichoderma harzianum* must cause systemic resistance via the JA, ET and SA pathways. At the same time, *T. viride* provides resistance against *Phytopthora* infection in potato through SA signalling (Rai *et al.*, 2020a).

2.4.2 Jasmonic acid-mediated signalling

The JA pathway is generally activated by conjugating JA to L-isoleucine to produce JA-Ile. As a coreceptor of JA-Ile, COI1 binds to JA-Ile and targets proteins with the jasmonate ZIM domain, which causes ubiquitin to break down the targeted protein (Chini *et al.*, 2007; Yan *et al.*, 2009; Sheard *et al.*, 2010). This action allows transcription factor MYC2 to be released to activate

or repress genes that respond to JA and ET (Chini *et al.*, 2007; Kazan and Manners, 2013). Metabolite profiling has discovered that the leaves of plants have lower sugar content, total soluble protein and hydrolyzable amino acids but higher amount of JA and 12-oxo-phytodienoic acid (OPDA). By releasing bioactive molecules into the rhizosphere, plants can modify the soil microbiota. These root exudates typically include primary and secondary metabolites. For example, maize roots secrete benzoxazinoids (BXs) that encourage the colonization of the rhizosphere by the bacterium *Pseudomonas putida*, that promotes growth of the plant and reduces the pathogenicity of *Agrobacterium tumefaciens* (Hu *et al.*, 2018; Thoenen *et al.*, 2023).

2.4.3 Ethylene-mediated signalling

Ethylene is a gaseous phytohormone exhibiting a crucial function in defence mechanisms in multiple ways. In stressful conditions, the biosynthesis of ET increases through the activation of aminocyclopropane-1-carboxylic acid (ACC) synthase (ACS) and ACC oxidase (ACO) enzymes (Khan *et al.*, 2024). ET signalling is regulated through the APETALA2/Ethylene responsive factor (AP2/ERF) transcription factors, which control the expression of PR genes. ET biosynthesis is further regulated by nitric oxide (NO), hydrogen sulfide (H_2S), reactive oxygen species (ROS) and auxin biosynthesis. ET helps in the production of ROS under abiotic stresses such as heavy metal toxicity, salinity, drought and low temperature (Iqbal *et al.*, 2017), while NO negatively affects ET biosynthesis (Zhu *et al.*, 2006). In rice, waterlogging increases ET levels, which antagonizes ABA. ET regulates stomatal closure by producing flavanols in guard cells during drought (Watkins *et al.*, 2017). Depending on the timing of the pathogen attack, ET signalling can have either synergistic or antagonistic effects on biotic stress (Bleecker and Schaller, 1996).

2.4.4   Abscisic acid-mediated signalling

Abscisic acid is a hormone involved in plant responses to stress, especially osmotic stress. It works in coordination with other defence hormones to help plants resist different types of biotic and abiotic stresses (Bharath *et al.*, 2021). ABA works together with SA to regulate osmotic stress caused by drought or flooding, but it has an antagonistic relationship with ET in this process. ABA also works with JA to help plants cope with stress. The amount of ABA in plants can be impacted by a number of soil parameters, like moisture, pH and the presence of pollutants or fertilizers. For example, ABA levels are higher in acidic soils with low moisture and plants growing in saline conditions. ABA helps plants close their stomata and enhances their immune system against bacterial infections (Adie *et al.*, 2007).

2.4.5   Signal exchange through quorum sensing

All living things depend on basic processes called substance and energy metabolism, which support key functions like growth, reproduction, genetic variety and species development (Sattley and Madigan *et al.*, 2021). In a microbial community, these metabolic processes are governed by enzyme activity, gene expression and interactions with cohabitating microorganisms, which impact metabolism through cell to cell signalling and the exchange of substances (Mihailović *et al.*, 2011). Further, QS modulates these pathways where microbes regulate gene expression based on cell density, thereby influencing their metabolism (Fuqua *et al.*, 1994).

QS synthase synthesizes QS molecules (QSMs) and secretes them in the environment. As cell density increases, the concentration of QSMs rises. Once a threshold concentration is reached, QSMs traverse across the plasma membrane, bind to the specific receptors and trigger gene expression that regulates various metabolic processes (Fuqua *et al.*, 1994; Miller and Bassler, 2001). QS is crucial for the production of biofilm, synthesis of antimicrobial substances and secretion of exoenzymes (Vattem *et al.*, 2007; Chong *et al.*, 2012; Ruhal and Kataria, 2021). As a result, QS influences microbial interactions and directly influences the production of enzymes involved in metabolic processes.

QS is a highly prevalent mechanism in microbial populations (Waters and Bassler, 2005; Ryan *et al.*, 2008). Through QS, microbes produce, release and detect QSMs in order to perform inter- and intraspecific communication to make environmental changes (Bassler, 1999). These QSMs, also known as autoinducers, are autoinducer-2 (AI-2), oligopeptides and N-acyl homoserine lactones (AHLs) (Miller and Bassler, 2001). Typically, in Gram-negative bacteria, AHLs are the primary QSMs, while oligopeptides in Gram-positive bacteria facilitate intraspecific communication. The signal for interspecific communication in both Gram-positive and -negative bacteria, however, is provided by AI-2 (Bassler, 1999). Among these systems, AHL-mediated QS (AHL-QS) is notably distributed and plays a crucial role in regulating metabolic processes within bacterial communities (Liu *et al.*, 2022). Understanding how bacteria use AHL-QS for regulation of gene expression and metabolic activities is pivotal for exploring various cycles in ecosystems. Moreover, insights into the interaction of AHL-QS and metabolic pathways have significant implications for various applications, including prevention of diseases, fermentation and rehabilitation of the environment.

2.5   Role of Earthworms and Nematodes in Soil Health and Plant Defence

In addition to microorganisms like protozoa, fungi and bacteria, other organisms such as nematodes, arthropods and earthworms are important in establishing the relationship between soil and plants. It is crucial to ensure a well-balanced nutritional requirement in the soil for optimal plant growth. Most of these nutrients consist of organic compounds that plants cannot directly absorb (Xiao *et al.*, 2017). Composting and mineralizing complex organic matter are crucial for making nutrients available to plants. As soil engineers, earthworms can aid in nutrient recycling by breaking down complex organic matter and the release of essential macronutrients like nitrogen, phosphorus and potassium in the soil (Xiao *et al.*, 2017). Earthworms are beneficial for soil health in multiple ways. They enhance soil structure by burrowing and creating channels that improve aeration. These channels increase the flow of oxygen and provide a habitat for helpful microorganisms like bacteria and fungi. These microbes assist in the breakdown of organic materials, providing nutrients that are necessary for the development of plants (Bertrand *et al.*, 2015; van Groenigen *et al.*, 2015; Bottino *et al.*, 2016). Earthworms play a significant role in maintaining soil health but also directly or indirectly impact plant defence mechanisms. Earthworms can create an environment that is less favourable to harmful pathogens, which helps to suppress soil-borne diseases. Furthermore, some reports suggest earthworms are able to enhance plant tolerance to herbivory by increasing the development of secondary metabolites such as tannins, terpenes and alkaloids. These secondary metabolites harm insects and herbivores (Wurst, 2013; Trouvé *et al.*, 2014; Xiao *et al.*, 2017).

Nematodes, like earthworms, are crucial for maintaining a healthy soil ecosystem. They can be either helpful or harmful to plants. Free-living nematodes break down complex organic matter and release nutrients into the soil, thus contributing to plant health. Additionally, they feed on microbes, including bacteria, fungi and other nematodes, which also aid in plant growth (Grewal and Georgis, 1999). Nematodes living freely in soil can carry helpful and harmful microorganisms to plants. However, parasitic nematodes harm plants' roots, creating conditions that can lead to the colonization of both pathogenic and beneficial microorganisms. Like free-living nematodes, parasitic nematodes can also act as carriers of harmful and helpful microorganisms (Li *et al.*, 2024). Nematodes can induce systemic resistance in plants by transmitting electrical signals and ROS to systemic tissues. These signals activate MPK1/2, which in turn induces JA signalling to activate defence in roots against nematodes (Wang *et al.*, 2019; Li *et al.*, 2024).

2.6 Abiotic Factors Influencing Soil Health and Defence Systems

Approximately 70% of plant damage is the result of various abiotic factors like temperature fluctuations, salt stress, nutrient deficiency, drought, metal toxicity, waterlogging and salinity (Kumar, 2020). The rhizomicrobiome counters abiotic stress by inducing systemic tolerance through plant–microbe signalling, leading to plant physiological and biochemical changes (Swapnil *et al.*, 2023). When subjected to stress, plants activate defence-related mechanisms including biofilm formation, antibiotic production, phytohormone production, siderophores, quorum quenching compounds and hydrolytic enzymes. The most common rhizobacteria that help mitigate abiotic stress are *Caulobacter, Serratia, Rhizobia, Erwinia, Burkholderia, Flavobacterium, Methylobacterium, Trichoderma, Micrococcus, Chromobacterium* and *Pseudomonas* (Chakraborty *et al.*, 2015; Rai *et al.*, 2020b).

Plants receive abiotic signal stresses through receptor-like kinases in their cell walls. The signal transmits into a secondary messenger that activates JA, ET, SA and ABA signalling (Osakabe *et al.*, 2013; Rehman *et al.*, 2021). Due to stress, plant roots also secrete several peptides such as clavate (CLV), inflorescence deficient in abscission (IDA), phytosulfokine-α (PSK-α), CAP-derived peptide-1 (CAPE-1), rapid alkalinization factor (RALF) and IDA-like peptides, which further activate the defence hormones (Nam *et al.*, 2001). In many cases, PGPR also produces phytohormones, mainly auxin, gibberellins, cytokinin, ABA, JT, ET and SA, for regulation of the abiotic stress response in plants (Tsukanova *et al.*, 2017). Furthermore, it leads to ROS and activates several enzymatic pathways, such as protein kinases and phosphatases including mitogen-activated protein kinases (MAPKs) and calcium-dependent protein kinases (CDPKs) (Schulz *et al.*, 2013; Mohanta *et al.*, 2018). PGPR-mediated stress resistance modulates transcription factors to combat the stress response. For example, *Bacillus amyloliquefaciens* combats high salt stress by upregulating MAPK and CDPK in *Oryza sativa* pathways. Table 2.2 describes some rhizobacteria that protect plants from abiotic stress and induce defence signalling (Chauhan *et al.*, 2019).

2.7   Nutrient Acquisition and Soil Health

Pathogens infect host plants to acquire nutrients. Nutrient availability depends on conditions such as pH of soil, rhizomicrobiome, herbicide programme and ratios with other mineral nutrients (as shown in Fig. 2.2). These pathogens first colonize the rhizosphere and phyllosphere and later they invade and gain access to the vascular elements (Fatima and Senthil-Kumar, 2015). Majorly, they access the apoplast, xylem, phloem and cell organelles (niches) for nutrient uptake and to avoid alterations in the surrounding (Vorholt, 2012; Griffin and Carson, 2015). Biotrophs employ a type III secretion system (TTSS) to inject effector chemicals that modify plant metabolism, releasing nutrients (Göhre and Robatzek, 2008; Cunnac *et al.*, 2009). The appropriate percentages of macro- and micronutrients used by plants and their impact on defence are listed in Table 2.3.

2.7.1   Macronutrients and plant defence

To grow, plants require macronutrients. The basic macronutrients are N, P and K; the secondary

nutrients are Ca, Mg and S. Some common diseases due to macronutrient imbalance are listed in Table 2.4.

Nitrogen

Nitrogen is a crucial macronutrient for plant growth, development and defence. It plays an important role in creating chlorophyll, which is essential for photosynthesis. Nitrogen is also a building block for proteins necessary for a variety of cellular functions. Plants obtain nitrogen from either the soil in nitrates or through symbiotic fixation by microbes in the atmosphere (Tripathi *et al.*, 2022). Nitrogen may impact plant defence in both favourable and unfavourable ways.

Interestingly, adding nitrogen can compromise certain plants' resistance to various diseases. For instance, rice, grapevines, *Medicago*, potato and apple trees can all have their resistance compromised by the addition of N. Specifically, rice becomes more susceptible to sheath rot, grapevines to powdery mildew, *Medicago truncatula* to *Aphanomyces euteiches*, potatoes to *Phytophthora infestans* and apple trees to *Venturia inaequalis* (Keller *et al.*, 2003; Leser and Treutter, 2005; Mittelstraß *et al.*, 2006; Thalineau *et al.*, 2018). Adding nitrogen to crops has been found to make wheat more resistant to leaf spot and tomatoes resistant to *Botrytis cinerea*, according to a study by Lecompte *et al.* (2010). Nitrogen can affect plants in various ways to protect them against pathogens, such as by

Table 2.2. Rhizo-microorganism protection against abiotic stresses using different signalling pathways in different plant species. Adopted from Jamil *et al.*, 2022.

Abiotic stress	Crop plant	Microorganism	Signalling pathways	Reference
Drought	*Oryza sativa*	*Pseudomonas fluorescens*	ABA	Chieb and
	Triticum aestivum, *Zea mays*	*Bacillus* sp. and *Enterobacter* sp.	IAA, SA	Gachomo, 2023
Salinity	*Triticum aestivum*	*Dietzia natronolimnaea*	ABA	Boamah *et al.*, 2022
	Cucumis sativus	*Trichoderma asperellum*	IAA, GA, ABA	
High temperature	*Lycopersicum esculentum*	*Bacillus cereus*	ET	Jalal *et al.*, 2023)
	Triticum aestivum	*Funneliformismosseae* and *Paraburkholderia graminis*	ROS	
Low temperature	*Oryza sativa*	*Bacillus amyloliquefaciens*	ABA, SA, JA, ET	Tomar *et al.*, 2021
	Triticum aestivum	*Bacillus velezensis*	ROS, ABA	

ABA, abscisic acid; ET, ethylene; GA, gibberellin; IAA, indole acetic acid; JA, jasmonic acid; ROS, reactive oxygen species; SA, salicylic acid.

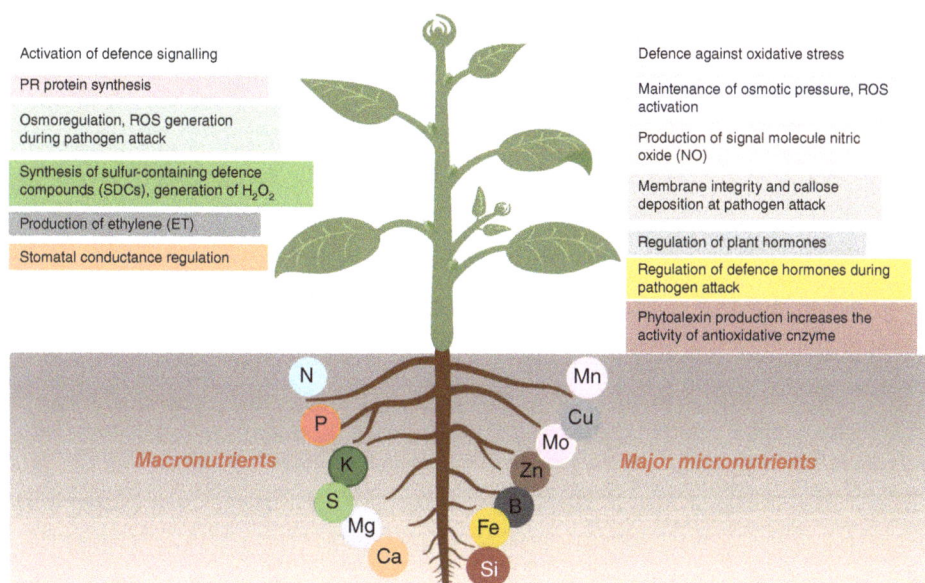

Fig. 2.2. Role of different macro- and micronutrients in defence-related mechanisms. For abbreviations, see text.

altering the pathogen's infection strategy, producing ROS, regulating the expression of defence-related genes or activating signalling pathways through hormones and amino acids. Nitrogen can also influence SAR by affecting NO production. Plants absorb nitrogen as nitrate and ammonium ions, and the right concentration and form of nitrogen are crucial for effective plant defence (González-Hernández et al., 2019). The balance between these two ions is essential for plant health. Higher levels of NO_3^- can increase NO, SA and PR gene expression. In comparison, higher levels of NH_4^+ can reduce SA and PR expression and plant resistance.

Phosphorus

Phosphorus serves as a building block in metabolites, and signalling molecules (Lambers and Plaxton, 2015). Phosphorus is essential for nucleic and physiological compounds and ADP phosphorylation. Plants detect extracellular ATP as damage-associated molecular patterns (DAMP) during pathogen infection (Tanaka et al., 2014). It is considered a defence signalling molecule. Phosphorus starvation conditions alleviate symbiosis with AMF-like fungi in Arabidopsis (Hiruma

et al., 2016), which provides soluble phosphorus to plants. In rice, phosphorus deficiency alleviates SA biosynthesis through phenylalanine ammonia-lyase (PAL) gene activation. Further, it activates IAA-NO pathways to enhance immunity (Wu et al., 2022).

Potassium

Potassium is an essential nutrient for plants. It helps plants absorb other vital minerals such as nitrogen, phosphorus and magnesium. Potassium also functions as a cofactor for various enzymes that catalyse processes like photosynthesis, protein synthesis and energy production. Overall, potassium is crucial for the health and growth of plants, and it also regulates osmotic regulation and stomatal movement (Broadley et al., 2012). Potassium is essential for plants as it helps them deal with environmental stresses such as salinity, drought, extreme temperatures, toxic metals, light and waterlogging. Plants that lack potassium are more vulnerable to diseases, and having an adequate supply of potassium can significantly reduce the frequency of fungal, bacterial, viral and nematodes diseases by 70%, 69%, 415% and 33%, respectively (Amtmann

Table 2.3. Relative amounts (out of 100) of the essential nutrients required by most plants.

Nutrient	Amount (%)	Activity
Macronutrient		
Carbon (C)	45	Modulate basal, effector-triggered and systemic immunity
Oxygen (O)	45	Orchestrates antioxidants activity
Hydrogen (H)	6	Regulating glutathione metabolism, inducing expression of PR (pathogenesis-related) and other defence-related genes, modulating enzyme activity through post-translational modifications and interacting with other phytohormones
Nitrogen (N)	1.5	Generation of nitric oxide (NO)
Phosphorus (P)	0.2	Regulating enzymatic reactions and protein phosphorylation through ATP regulation
Potassium (K)	1.0	Upregulation of jasmonic acid biosynthesis gene *LOX2*
Micronutrient		
Calcium (Ca)	0.5	Works as secondary messenger in defence signalling, regulation of calcium-regulated protein like calmodulin, Ca channels
Magnesium (Mg)	0.2	Activates antioxidative defense enzymes
Sulfur (S)	0.1	Regulation of glutathione, regulation of ROS (reactive oxygen species)
Iron (Fe)	0.01	Accumulation of reactive ferric iron which regulates H_2O_2 production
Chlorine (Cl)	0.01	Cl$^-$ coupled with K$^+$/Na$^+$ regulates ion transportation through plasma membrane which regulates homeostasis
Manganese (Mn)	0.005	Works as a cofactor of various enzymes, including superoxide dismutase (MnSOD), catalase (MnCAT) and decarboxylases
Boron (B)	0.001	Regulation of phytohormones like abscisic acid and salicylic acid, and ROS induction
Zn (Zn)	0.001	Priming of several defence signalling
Copper (Cu)	0.0006	Activates ethylene signalling suppressing abscisic acid biosynthesis
Molybdenum (Mo)	0.00001	Regulation of many enzymes, such as CAT, POD and SOD (superoxide dismutase)
Cobalt (Co)	–	Gibberellic acid-induced programmed cell death
Nickel (Ni)	–	Glutathione (GSH) homeostasis, regulation of enzyme urease

et al., 2008). Excessive use of potassium and nitrogen in strawberry cultivation can make the plants more vulnerable to anthracnose disease (Nam *et al.*, 2006). Potassium plays a crucial role in maintaining membrane potential, voltage-gated ion channels, intracellular calcium signalling and activation of several transcription factors. It also facilitates the transport of hormones and metabolites for systemic signalling. It moves quickly in the cell apoplast, making it necessary to maintain homeostasis.

Calcium

Calcium (Ca^{2+}) is an essential component of plant cell walls and it functions as a crucial second messenger in response to defence mechanisms. The concentration of free Ca^{2+} in the cytosol of plant cells changes dynamically and plays both positive and negative roles in maintaining calcium signatures. Calcium is vital for PTI and ETI. It provides physical strength to the cell wall during pathogen attacks, producing ROS and post-translational modifications via activation of MAPK. Calcium ion channels and Ca^{2+}-binding proteins (including calcineurin B-like proteins, calmodulin, calmodulin-like proteins and calcium-dependent protein kinases) sense changes in calcium concentration and decode the calcium signals into physiological and cellular responses so that plants are able to survive the harsh challenges posed by pathogens (Huber and Graham, 1999).

Magnesium

Magnesium is an essential factor for the growth of plants and microbes as it affects disease directly or indirectly. It acts as a cofactor for enzymes, an essential component of chlorophyll

Table 2.4. Pathogen infection due to imbalance of nutrients.

Nutrients	Crop	Disease	Pathogen	Cause	Reference
Macronutrients					
Nitrogen (N)	Potato	Early blight	*Alternaria solani*	High N reduces disease severity	MacKenzie, 1981
Potassium (K)	Rice	Rice blast	*Magnaportheoryzae*	High P level	
Phosphorus (P)	Rice	Blast disease	*Magnaporthe orygae*	High P content increase disease	Campos-Soriano et al., 2020
Magnesium (Mg)	Corn	Corn stunt disease	*Spiroplasmakunkelii*	Low Mg supply	MacKenzie, 1981
	Tomato	Tomato bacterial spot	*Xanthomonas campestris pv. vesicatoria*	High Mg supply	Woltz and Jones, 1979
Calcium (Ca)	Crucifers	Club root disease	*Plasmodiophorabrassicae*	Low Ca leads to imbalance in pH	McGrann et al., 2016
Sulfur (S)	Grapes	Powdery mildew	*Uncinulanecator*	S application reduces the disease symptoms	Magarey, 1992
Micronutrients					
Boron (B)	Crucifers	Club root disease	*Plasmodiophorabrassicae*	B supplement reduces the disease symptom	Deora et al., 2011
Zinc (Zn)	Wheat	Fusarium head blight	*Fusarium graminearum*	Zn application reduces the disease severity	Alsamir et al., 2020
Copper (Cu)	Potato	Potato late blight	*Phytophthora infestons*	Low Cu	Bangemann et al., 2014
Silicon (Si)	Wheat	Root rot	*Pyriculariaorygae*	Si application reduces the disease severity	Debona et al., 2014
Iron (Fe)	Apple	Apple canker	*Sphaeropsismalorum*	Fe application increases disease resistance	Tripathi et al., 2022
Manganese (Mn)	Potato	Common scab	*Streptomyces scabies*	Mn application increases resistance to disease	Kopecky et al., 2021

and middle lamella. The magnesium availability is dependent on environmental conditions such as rhizomicrobiome activity, soil pH and the nature of herbicides used for weed control. The ratio of other mineral nutrients, such as calcium, potassium and manganese, also affects magnesium availability. Plants can absorb magnesium in the various forms like sulfate (SO_4^-), chlorine (Cl), carbonate (CO), nitrate (NO_3^-) and monovalent oxygen (O). The anion component of these salts affects magnesium solubility in the soil or the physiological function of magnesium in the plant. For example, CO_3 changes soil pH, while Cl, NO_3^-, S, and SO_4^- affect the physiological function of magnesium in the plant; an imbalance of magnesium results in the formation of ROS and further activating PR genes. Plants grown in acidic soils with low pH are more susceptible to diseases such as bacterial soft rots, cabbage clubroot and *Fusarium* wilts as they absorb less magnesium, calcium, and molybdenum (Huber and Graham, 1999). Applying dolomitic lime ($CaCO_3$ + $MgCO_3$) may increase soil pH and lead to more magnesium availability to reduce these diseases (Huber and Jones, 2013). Further increasing magnesium levels significantly increased susceptibility to bacterial spots caused by *Xanthomonas campestris* pv. *vesicatoria* in tomato and pepper.

Sulfur

Sulfur is an essential element in the resistance of plants against diseases. The defence mechanisms against pathogens involve sulfur-containing defence compounds (SDCs), which interact with pathogen recognition and activate cell signalling pathways. SDCs lead to various cell defence processes that are regulated by ET, JA, SA and ROS (Künstler *et al.*, 2020). Certain plant compounds, known as sulfur-containing amino acids (SAA), including methionine, glutathione, cysteine, thionins, hydrogen sulfide, phytoalexins, defenisns and glucosinolates, play a crucial role in plant defence mechanisms (Künstler *et al.*, 2020). In plants, SA signalling is regulated positively by glutathione (GSH) and NO. This further upregulates isochorismate synthase expression, a critical enzyme in SA biosynthesis, leading to an accumulation of SA (Vega *et al.*, 2022; Wawrzyńska *et al.*, 2024). The reduction of cysteine residues and the release of NPR1 monomer result from

this accumulation. The activated NPR1 then translocates from the cytoplasm to the nucleus, facilitated by S-nitrosoglutathione, and enhances PR gene activation (Borrowman *et al.*, 2023).

On the other hand, JA signalling depends on the oxidative status of GSH/GSSG (Rai *et al.*, 2023). Infection of plants with *Alternaria brassicicola* or *Botrytis cinerea* increased the expression of plant defensin gene (*PDF 1.2*). ET biosynthesis requires sulfur via S-adenosyl-1-methionine (SAM), an activated form of Met. In addition, GSH influences ET biosynthesis through 1-aminocyclopropane-1-carboxylate synthase (ACS), 1-aminocyclopropane-1-carboxylate oxidase (ACO) and SAM synthase (SAM1) (Li *et al.*, 2021; Lee *et al.*, 2023). Elevated GSH contents increase resistance to *B. cinerea* and *Pseudomonas syringae* pv. *tabaci* which indicates that both ET and SA levels are synergistically elevated by GSH (Künstler *et al.*, 2020).

2.7.2 Micronutrients and plant defence

Micronutrients are important in plant metabolism, transport and different pathogen responses. Some common diseases due to micronutrient imbalance are listed in Table 2.4.

Boron

Boron is a micronutrient essential for plants, but its role in plant physiology and metabolism is not yet fully understood. Boron helps plants reduce their susceptibility to diseases by performing various functions, such as strengthening the structure of cell walls by producing carbohydrate–borate complexes, controlling cell membrane permeability and aiding in lignin and phenolics metabolism (Broadley *et al.*, 2012). In certain conditions, when there is exposure of plants to high boron levels, two hormones – SA and ABA – are produced that help regulate the plant's response to stress. Boron-deficient conditions can lead to debilitated cell walls that allow for better water retention, making it easier for pathogens to infect the plant. Some soil-borne pathogens, like *Pantoea*, *Ralstonia* and *Pectobacterium*, can play a beneficial role in maintaining boron homeostasis and promoting adequate plant uptake (Moustafa-Farag *et al.*, 2020; Choudhary *et al.*, 2021).

Zinc

Zinc acts as a vital cofactor for numerous enzymes (Hambidge *et al.*, 2000). It plays a role in the binding of transcription factors to DNA and protein–protein interactions. Both excess and deficiency of zinc make plants more susceptible to pests (Sinclair and Krämer, 2012) due to the regulation of zinc by superoxide dismutase (SOD) and the zinc finger domain. In conditions of zinc deficiency, ROS are generated, which activates defence signalling. The zinc finger domain is found in the resistance protein NBS-LRRs of plants, which are directly involved in effector-triggered immunity (Dai *et al.*, 2010). About 37% of proteins involved in defence contain the zinc finger domain. The *Pi54* gene provides resistance to *Magnaporthe oryzae*. *Pi54*-containing transgenic rice upregulates the expression of transcription factors linked to defence, including NAC6, MADS-box and WRKY (Gupta *et al.*, 2012).

Copper

Copper serves as a crucial cofactor in the function of many metalloproteins and also regulates various biochemical and physiological processes. However, an excess of copper leads to oxidative outburst and increase in ROS production. Plants control their physiology by only absorbing a limited amount of copper. Copper is an essential component of many enzymes and proteins, including cytochrome c oxidase (COX), polyphenol oxidase (PPO), plastocyanin, multicopper oxidase (laccase) and Cu/Zn superoxide dismutase (Yruela, 2005). Copper primarily influences ET signalling in response to pathogen attacks. In potatoes, Cu^{2+} triggers the defence response to potato late blight by activating ET biosynthesis and inhibiting the biosynthesis of ABA (Liu *et al.*, 2020).

Manganese

Manganese is a crucial micronutrient essential for plant metabolic processes. It exists in many oxidation states (0, II, III, IV, VI and VII), but mostly appears as II, III and IV in biological systems. The most soluble form, MnO_2, plays an important role in central exchange within the system. However, the uptake and transport of manganese depend on factors such as soil pH, redox conditions and the population of rhizobacteria in the rhizosphere. Manganese has a dual role in plant health, as both deficiency and excess can be harmful, leading to susceptibility and toxicity, respectively. Manganese serves as a cofactor for SOD, helping to reduce oxidative stress resulting from ROS and free radicals. However, excessive levels of these enzymes can also be detrimental to plants (Ducic and Polle, 2005).

Iron

Iron is a crucial micronutrient that functions as a catalyst with hydrogen peroxide in the Fenton reaction. This reaction generates ROS and hydroxide ions. These potential oxidizers first harm DNA, RNA, proteins and lipids, which contributes to apoptosis. Controlled iron absorption is essential for plant survival. Iron deficiency can be fatal to plants, while excess iron can cause toxicity during interactions with pathogens. In iron-deficient conditions, plants engage with various other microorganisms (PGPRs) that produce siderophores. These siderophores help combat the deficiency and aid in the plant's survival (Scavino and Pedraza, 2013). Siderophores are small molecules created by PGPRs. They play a critical role in improving iron uptake in plants, competing with pathogens and displaying antibiotic activity against plant pathogens. In comparison to iron starvation, plants infected with *Dickeya dadantii* and *B. cinerea* exhibit fewer disease symptoms (Kieu *et al.*, 2012).

Silicon

Silicon contributes significantly to plant defence in a variety of ways.. It reinforces plant structures, induces SAR, produces antimicrobial compounds and activates defence-related genes. Silicon accumulates on wax, epidermal cells and cuticles, creating a physical barrier that prevents pathogen penetration (Ma *et al.*, 2006). In addition to providing a physical barrier, silicon also enhances biochemical resistance in plants by increasing the defence-related enzyme activity including chitinase, peroxidase, PAL, ascorbate peroxidase, glutathione reductase, catalase and lipooxygenase. Furthermore, silicon induces the synthesis of antimicrobial compounds like PR proteins, phytoalexins and flavonoids and regulates signalling molecules like SA, ET and JA (Fauteux *et al.*, 2006; Van Bockhaven *et al.*, 2013). Experiments with *Arabidopsis* infected with powdery mildew (*Erysiphe cichoracearum*) have shown elevated levels of SA, JA and ET signalling. Additionally, studies on tomatoes infected with *Ralstonia solanacearum* have demonstrated that silicon activates the JA and ET pathways (Zhang *et al.*, 2004; Chen *et al.*, 2010). *Arabidopsis* infected with *E. cichoracearum* shows elevated SA, JA and ET signalling levels.

2.8 Conclusion and Future Perspectives

Soil health is vital for plant growth since it controls water, promotes plant and animal life, filters and buffers possible contaminants, cycles nutrients, provides physical stability and supports defence mechanisms. It is essential to manage soil health by adopting techniques like organic amendment, cover crops, applying green manures, crop rotation, nutrient management, enhancing the growth of PGPR, beneficial organisms (bioagents like antagonist fungi and bacteria), vesicular arbuscular mycorrhiza and disease-suppressive soil. These practices help plants combat pathogens and protect themselves against soil erosion, improving water infiltration and balancing nutrient cycling to help rebuild the ecosystem and create a sustainable approach to agriculture. Managing soil health by promoting more environmentally friendly agronomic techniques helps reduce the greenhouse effect and combat global warming in the long run.

Acknowledgements

We thank Dr Deepak K. Singh (Albert Einstein College of Medicine, New York) and Dr Shantipal Gangwar (Columbia University, New York) for their critical reading and suggestions for this chapter.

References

Ab-Rahman, S.F.S., Singh, E., Pieterse, C.M. and Schenk, P.M. (2018) Emerging microbial biocontrol strategies for plant pathogens. *Plant Science* 267, 102–111.

Adie, B.A.T., Pérez-Pérez, J., Pérez-Pérez, M.M., Godoy, M., Sánchez-Serrano, J.-J. *et al.* (2007) ABA is an essential signal for plant resistance to pathogens affecting JA biosynthesis and the activation of defenses in *Arabidopsis*. *Plant Cell* 19, 1665–1681.

Alsamir, M., Abass, M., Trethowan, R. and Al-Samir, E. (2020) The application of zinc fertilizer reduces fusarium infection and development in wheat. *Australian Journal of Crop Science* 14, 1088–1094.

Amtmann, A., Troufflard, S. and Armengaud, P. (2008) The effect of potassium nutrition on pest and disease resistance in plants. *Physiologia Plantarum* 133, 682–691.

Bangemann, L.-W., Westphal, A., Zwerger, P., Sieling, K. and Kage, H. (2014) Copper reducing strategies for late blight (*Phytophthora infestans*) control in organic potato (*Solanum tuberosum*) production. *Journal of Plant Diseases and Protection* 121, 105–116.

Bassler, B.L. (1999) How bacteria talk to each other: regulation of gene expression by quorum sensing. *Current Opinion in Microbiology* 2, 582–587.

Bateman, D.F. (1978) The dynamic nature of disease. In: Horsfall, J.G. and Cowling, E.B. (eds) *How Plants Suffer from Disease*. Academic Press.

Benhamou, N., Garand, C. and Goulet, A. (2002) Ability of nonpathogenic *Fusarium oxysporum* strain fo47 to induce resistance against *Pythium ultimum* infection in cucumber. *Applied and Environmental Microbiology* 68, 4044–4060.

Berendsen, R.L., Pieterse, C.M.J. and Bakker, P.A.H.M. (2012) The rhizosphere microbiome and plant health. *Trends in Plant Science* 17, 478–486.

Bertrand, M., Barot, S., Blouin, M., Whalen, J., de Oliveira, T. *et al.* (2015) Earthworm services for cropping systems. a review. *Agronomy for Sustainable Development* 35, 553–567.

Bharath, P., Gahir, S. and Raghavendra, A.S. (2021) Abscisic acid-induced stomatal closure: an important component of plant defense against abiotic and biotic stress. *Frontiers in Plant Science* 12, 615114.

Bleecker, A.B. and Schaller, G.E. (1996) The mechanism of ethylene perception. *Plant Physiology* 111, 653.

Boamah, S., Zhang, S., Xu, B., Li, T., Calderón-Urrea, A. et al. (2022) Trichoderma longibrachiatum TG1 increases endogenous salicylic acid content and antioxidants activity in wheat seedlings under salinity stress. PeerJ 10, e12923.

Bonkowski, M., Villenave, C. and Griffiths, B. (2009) Rhizosphere fauna: the functional and structural diversity of intimate interactions of soil fauna with plant roots. Plant and Soil 321, 213–233.

Borrowman, S., Kapuganti, J.G. and Loake, G.J. (2023) Expanding roles for S-nitrosylation in the regulation of plant immunity. Free Radical Biology and Medicine 194, 357–368.

Bottino, F., Cunha-Santino, M.B. and Bianchini, I. (2016) Decomposition of particulate organic carbon from aquatic macrophytes under different nutrient conditions. Aquatic Geochemistry 22, 17–33.

Broadley, M., Brown, P., Cakmak, I., Rengel, Z. and Zhao, F. (2012) Function of nutrients: micronutrients. In: Marschner, P. (ed.) Marschner's Mineral Nutrition of Higher Plants, 3rd edn. Academic Press, San Diego, California, pp. 191–248.

Bukhat, S., Imran, A., Javaid, S., Shahid, M., Majeed, A. et al. (2020) Communication of plants with microbial world: exploring the regulatory networks for PGPR mediated defense signaling. Microbiological Research 238, 126486.

Campos-Soriano, L., Bundó, M., Bach-Pages, M., Chiang, S.F., Chiou, T.J. et al. (2020) Phosphate excess increases susceptibility to pathogen infection in rice. Molecular Plant Pathology 21, 555–570.

Cardoso, P., Alves, A., Silveira, P., Sá, C., Fidalgo, C. et al. (2018) Bacteria from nodules of wild legume species: phylogenetic diversity, plant growth promotion abilities and osmotolerance. Science of the Total Environment 645, 1094–1102.

Carroll, C.S. and Moore, M.M. (2018) Ironing out siderophore biosynthesis: a review of non-ribosomal peptide synthetase (NRPS)-independent siderophore synthetases. Critical Reviews in Biochemistry and Molecular Biology 53, 356–381.

Carvalhais, L.C., Dennis, P.G., Badri, D.V., Kidd, B.N., Vivanco, J.M. et al. (2015) Linking jasmonic acid signaling, root exudates, and rhizosphere microbiomes. Molecular Plant–Microbe Interactions 28, 1049–1058.

Chakraborty, U., Chakraborty, B., Dey, P. and Chakraborty, A.P. (2015) Role of microorganisms in alleviation of abiotic stresses for sustainable agriculture. In: Chakraborty, U. and Chakraborty, B. (eds) Abiotic Stresses in Crop Plants. CAB International, Wallingford, UK, pp. 232–253.

Chalupowicz, L., Manulis-Sasson, S., Barash, I., Elad, Y., Rav-David, D. et al. (2021) Effect of plant systemic resistance elicited by biological and chemical inducers on the colonization of the lettuce and basil leaf apoplast by Salmonella enterica. Applied and Environmental Microbiology 87, e0115121.

Chauhan, P.S., Lata, C., Tiwari, S., Chauhan, A.S., Mishra, S.K. et al. (2019) Transcriptional alterations reveal Bacillus amyloliquefaciens–rice cooperation under salt stress. Scientific Reports 9, 11912.

Chen, X., Marszałkowska, M. and Reinhold-Hurek, B. (2020) Jasmonic acid, not salicyclic acid restricts endophytic root colonization of rice. Frontiers in Plant Science 10, 1758.

Chen, Y., Zhang, W.-Z., Liu, X., Ma, Z.-H., Li, B. et al. (2010) A real-time PCR assay for the quantitative detection of Ralstonia solanacearum in the horticultural soil and plant tissues. Journal of Microbiology and Biotechnology 20, 193–201.

Chieb, M. and Gachomo, E.W. (2023) The role of plant growth promoting rhizobacteria in plant drought stress responses. BMC Plant Biology 23, 407.

Chini, A., Fonseca, S., Fernandez, G., Adie, B., Chico, J. et al. (2007) The JAZ family of repressors is the missing link in jasmonate signalling. Nature 448, 666–671.

Chong, G., Kimyon, O., Rice, S.A., Kjelleberg, S. and Manefield, M. (2012) The presence and role of bacterial quorum sensing in activated sludge. Microbial Biotechnology 5, 621–633.

Choudhary, S., Zehra, A., Mukarram, M., Wani, K.I., Naeem, M. et al. (2021) Salicylic acid-mediated alleviation of soil boron toxicity in Mentha arvensis and Cymbopogon flexuosus: growth, antioxidant responses, essential oil contents and components. Chemosphere 276, 130153.

Crosa, J.H. and Walsh, C.T. (2002) Genetics and assembly line enzymology of siderophore biosynthesis in bacteria. Microbiology and Molecular Biology Reviews 66, 223–249.

Cunnac, S., Lindeberg, M. and Collmer, A. (2009) Pseudomonas syringae type III secretion system effectors: repertoires in search of functions. Current Opinion in Microbiology 12, 53–60.

Dai, L., Wu, J., Li, X., Wang, X., Liu, X. et al. (2010) Genomic structure and evolution of the Pi2/9 locus in wild rice species. Theoretical and Applied Genetics 121, 295–309.

Debona, D., Rodrigues, F., Rios, J., Nascimento, K. and Silva, L. (2014) The effect of silicon on antioxidant metabolism of wheat leaves infected by Pyricularia oryzae. Plant Pathology 63, 581–589.

Deora, A., Gossen, B.D., Walley, F. and McDonald, M.R. (2011) Boron reduces development of clubroot in canola. *Canadian Journal of Plant Pathology* 33, 475–484.

Devoto, A. and Turner, J.G. (2003) Regulation of jasmonate-mediated plant responses in Arabidopsis. *Annals of Botany* 92, 329–337.

Dimkić, I., Janakiev, T., Petrović, M., Degrassi, G. and Fira, D. (2022) Plant-associated *bacillus* and pseudomonas antimicrobial activities in plant disease suppression via biological control mechanisms – a review. *Physiological and Molecular Plant Pathology* 117, 101754.

Ding, L.N., Li, Y.T., Wu, Y.Z., Li, T., Geng, R. *et al.* (2022) Plant disease resistance-related signaling pathways: recent progress and future prospects. *International Journal of Molecular Sciences* 23, 16200.

Dobereiner, J. (1992) History and new perspectives of diazotrophs in association with non-leguminous plants. *Symbiosis* 13, 1–3.

Dos Santos, C. and Franco, O.L. (2023) Pathogenesis-related proteins (PRS) with enzyme activity activating plant defense responses. *Plants* 12, 2226.

Ducic, T. and Polle, A. (2005) Transport and detoxification of manganese and copper in plants. *Brazilian Journal of Plant Physiology* 17, 103–112.

Fatima, U. and Senthil-Kumar, M. (2015) Plant and pathogen nutrient acquisition strategies. *Frontiers in Plant Science* 6, 750.

Fauteux, F., Chain, F., Belzile, F., Menzies, J.G. and Bélanger, R.R. (2006) The protective role of silicon in the *Arabidopsis*–powdery mildew pathosystem. *Proceedings of the National Academy of Sciences USA* 103, 17554–17559.

Fritsch, S., Gasser, V., Peukert, C., Pinkert, L., Kuhn, L. *et al.* (2022) Uptake mechanisms and regulatory responses to MECAM-and DOTAM-based artificial siderophores and their antibiotic conjugates in *Pseudomonas aeruginosa*. *ACS Infectious Diseases* 8, 1134–1146.

Fuqua, W.C., Winans, S.C. and Greenberg, E.P. (1994) Quorum sensing in bacteria: the LuxR-LuxI family of cell density-responsive transcriptional regulators. *Journal of Bacteriology* 176, 269–275.

Geisseler, D. and Scow, K.M. (2014) Long-term effects of mineral fertilizers on soil microorganisms – a review. *Soil Biology and Biochemistry* 75, 54–63.

Glick, B.R. (1995) The enhancement of plant growth by free-living bacteria. *Canadian Journal of Microbiology* 41, 109–117.

Göhre, V. and Robatzek, S. (2008) Breaking the barriers: microbial effector molecules subvert plant immunity. *Annual Review of Phytopathology* 46, 189–215.

González-Hernández, A.I., Fernández-Crespo, E., Scalschi, L., Hajirezaei, M.R., Wirén, N. *et al.* (2019) Ammonium mediated changes in carbon and nitrogen metabolisms induce resistance against *Pseudomonas syringae* in tomato plants. *Journal of Plant Physiology* 239, 28–37.

Gozzo, F. (2003). Systemic acquired resistance in crop protection: from nature to a chemical approach. *Journal of Agricultural and Food Chemistry* 51, 4487–4503.

Grewal, P. and Georgis, R. (1999) *Entomopathogenic nematodes*. In: Hall, F.R. and Menn, J.J. (eds) *Methods in Biotechnology, Vol. 5. Biopesticides: Use and Delivery*. Humana Press, Totowa, New Jersey, pp. 271–299.

Griffin, E.A. and Carson, W.P. (2015) The ecology and natural history of foliar bacteria with a focus on tropical forests and agroecosystems. *Botanical Review* 81, 105–149.

Guerrieri, A., Dong, L. and Bouwmeester, H.J. (2019) Role and exploitation of underground chemical signaling in plants. *Pest Management Science* 75, 2455–2463.

Gupta, S.K., Rai, A.K., Kanwar, S.S., Chand, D., Singh, N.K. *et al.* (2012) The single functional blast resistance gene pi54 activates a complex defence mechanism in rice. *Journal of Experimental Botany* 63, 757–772.

Hambidge, M., Cousins, R.J. and Costello, R.B. (2000) Zinc and health: current status and future directions. In: *Proceedings of a Workshop held November 4–5*, American Society for Nutritional Sciences.

Hamilton, E.W. and Frank, D.A. (2001) Can plants stimulate soil microbes and their own nutrient supply? Evidence from a grazing tolerant grass. *Ecology* 82, 2397–2402.

Han, G.-Z. (2016) Evolution of jasmonate biosynthesis and signaling mechanisms. *Journal of Experimental Botany* 68, 1323–1331.

Hider, R.C. and Kong, X. (2010) Chemistry and biology of siderophores. *Natural Product Reports* 27, 637–657.

Hiruma, K., Gerlach, N., Sacristán, S., Nakano, R.T., Hacquard, S. *et al.* (2016) Root endophyte Colletotrichum tofieldiae confers plant fitness benefits that are phosphate status dependent. *Cell* 165, 464–474.

Hönig, M., Roeber, V.M., Schmülling, T. and Cortleven, A. (2023) Chemical priming of plant defense responses to pathogen attacks. *Frontiers in Plant Science* 14, 1146577.

Hossain, M.M. (2024). Upscaling plant defense system through the application of plant growth-promoting fungi (PGPF). In: Kumar, V. and Iram, S. (eds) *Microbial Technology for Agro-Ecosystems*. Academic Press, pp. 61–95.

Hu, L., Robert, C.A.M., Cadot, S., Zhang, X., Ye, M. *et al.* (2018) Root exudate metabolites drive plant-soil feedbacks on growth and defense by shaping the rhizosphere microbiota. *Nature Communications* 9, 2738.

Huber, D. and Graham, R. (1999) The role of nutrition in crop resistance and tolerance to diseases. In: Rengel, Z. (ed.) *Mineral Nutrition of Crops: Fundamental Mechanisms and Implications*. Food Product Press, New York, pp. 169–204.

Huber, D.M. and Jones, J.B. (2013) The role of magnesium in plant disease. *Plant and Soil* 368, 73–85.

Huber, D.M. and Thompson, I.A. (2007) Nitrogen and plant disease. In: Datnoff, L.E., Elmer, W.H. and Huber, D.M. (eds). Mineral *Nutrition and Plant Disease*, Vol. 55. The American Phytopathological Society, St Paul, Minnesota, pp. 31–44.

Iqbal, N., Khan, N.A., Ferrante, A., Trivellini, A., Francini, A. *et al.* (2017) Ethylene role in plant growth, development and senescence: Interaction with other phytohormones. *Frontiers in Plant Science* 8, 475.

Jacoby, R., Peukert, M., Succurro, A., Koprivova, A. and Kopriva, S. (2017) The role of soil microorganisms in plant mineral nutrition – current knowledge and future directions. *Frontiers in Plant Science* 8, 1617.

Jalal, A., Oliveira, C.E. da S., Rosa, P.A.L., Galindo, F.S., Teixeira Filho, M.C.M. *et al.* (2023) Beneficial microorganisms improve agricultural sustainability under climatic extremes. *Life* 13, 1102.

Jamil, F., Mukhtar, H., Fouillaud, M. and Dufossé, L. (2022) Rhizosphere signaling: insights into plant–rhizomicrobiome interactions for sustainable agronomy. *Microorganisms* 10, 899.

Jha, Y. and Mohamed, H.I. (2023) Enhancement of disease resistance, growth potential, and biochemical markers in maize plants by inoculation with plant growth-promoting bacteria under biotic stress. *Journal of Plant Pathology* 105, 729–748.

Kashyap, A.S., Manzar, N., Meshram, S. and Sharma, P.K. (2023) Screening microbial inoculants and their interventions for cross-kingdom management of wilt disease of solanaceous crops – a step toward sustainable agriculture. *Frontiers in Microbiology* 14, 1174532.

Kazan, K. and Manners, J.M. (2013) MYC2: The master in action. *Molecular Plant* 6, 686–703.

Keller, M., Rogiers, S.Y. and Schultz, H.R. (2003) Nitrogen and ultraviolet radiation modify grapevines' susceptibility to powdery mildew. *Vitis-Geilweilerhof* 42, 87–94.

Kenawy, A., Dailin, D.J., Abo-Zaid, G.A., Malek, R.A., Ambehabati, K.K. *et al.* (2019) Biosynthesis of antibiotics by PGPR and their roles in biocontrol of plant diseases. In: Sayyed, R.Z. (ed.) *Plant Growth Promoting Rhizobacteria for Sustainable Stress Management, Vol. 2. Rhizobacteria in Biotic Stress Management*. Springer, Singapore, pp. 1–35.

Khan, S., Alvi, A.F., Saify, S., Iqbal, N. and Khan, N.A. (2024) The ethylene biosynthetic enzymes, 1-aminocyclopropane-1-carboxylate (ACC) synthase (ACS) and ACC oxidase (ACO): the less explored players in abiotic stress tolerance. *Biomolecules* 14, 90.

Kieu, N.P., Aznar, A., Segond, D., Rigault, M., Simond-Côte, E. *et al.* (2012) Iron deficiency affects plant defence responses and confers resistance to *Dickeya dadantii* and *Botrytis cinerea*. *Molecular Plant Pathology* 13, 816–827.

Kopecky, J., Rapoport, D., Sarikhani, E., Stovicek, A., Patrmanova, T. *et al.* (2021) Micronutrients and soil microorganisms in the suppression of potato common scab. *Agronomy* 11, 383.

Kramer, J., Özkaya, Ö. and Kümmerli, R. (2020) Bacterial siderophores in community and host interactions. *Nature Reviews. Microbiology* 18, 152–163.

Kumar, S. (2020) Abiotic stresses and their effects on plant growth, yield and nutritional quality of agricultural produce. *International Journal of Food Science and Agriculture* 4, 367–378.

Künstler, A., Gullner, G., Ádám, A.L., Kolozsváriné Nagy, J. and Király, L. (2020) The versatile roles of sulfur-containing biomolecules in plant defense – a road to disease resistance. *Plants* 9, 1705.

Kusajima, M., Shima, S., Fujita, M., Minamisawa, K., Che, F.-S. *et al.* (2018) Involvement of ethylene signaling in Azospirillum sp. B510-induced disease resistance in rice. *Bioscience, Biotechnology, and Biochemistry* 82, 1522–1526.

Lambers, H. and Plaxton, W.C. (2015) Phosphorus: back to the roots. *Annual Plant Reviews* 48, 1–22.

Lanna-Filho, R., Souza, R.M., Magalhães, M.M., Villela, L., Zanotto, E. *et al.* (2013) Induced defense responses in tomato against bacterial spot by proteins synthesized by endophytic bacteria. *Tropical Plant Pathology* 38, 295–302.

Le Guillou, C., Chemidlin Prévost-Bouré, N., Karimi, B., Akkal-Corfini, N., Dequiedt, S. *et al.* (2019) Tillage intensity and pasture in rotation effectively shape soil microbial communities at a landscape scale. *MicrobiologyOpen* 8, e00676.

Lecompte, F., Abro, M.A. and Nicot, P.C. (2010) Contrasted responses of *Botrytis cinerea* isolates developing on tomato plants grown under different nitrogen nutrition regimes. *Plant Pathology* 59, 891–899.

Lee, Y.H., Ren, D., Jeon, B. and Liu, H.W. (2023) *S*-adenosylmethionine: more than just a methyl donor. *Natural Product Reports* 40, 1521–1549.

Leser, C. and Treutter, D. (2005) Effects of nitrogen supply on growth, contents of phenolic compounds and pathogen (scab) resistance of apple trees. *Physiologia Plantarum* 123, 49–56.

Li, J., Sun, C., Cai, W., Li, J., Rosen, B.P. *et al.* (2021) Insights into S-adenosyl-l-methionine (SAM)-dependent methyltransferase related diseases and genetic polymorphisms. *Mutation Research. Reviews in Mutation Research* 788, 108396.

Li, T., Li, X., Zheng, L. and Li, H. (2024) Stable body sizes in soil nematodes across altitudes: The role of intrageneric variation in community assembly. *Ecology and Evolution* 14, e70025.

Liu, H.F., Xue, X.J., Yu, Y., Xu, M.M., Lu, C.C. *et al.* (2020) Copper ions suppress abscisic acid biosynthesis to enhance defence against phytophthora infestans in potato. *Molecular Plant Pathology* 21, 636–651.

Liu, L., Zeng, X., Zheng, J., Zou, Y., Qiu, S. and Dai, Y. (2022) AHL-mediated quorum sensing to regulate bacterial substance and energy metabolism: a review. *Microbiological Research* 262, 127102.

Ma, J.F., Tamai, K., Yamaji, N., Mitani, N., Konishi, S. *et al.* (2006) A silicon transporter in rice. *Nature* 440, 688–691.

MacKenzie, D.R. (1981) Association of potato early blight, nitrogen fertilizer rate, and potato yield. *Plant Disease* 65, 575.

Magarey, P.A. (1992) Softer strategies for diseases and pests of grapes. *Australian Grapegrower and Winemaker* 345, 37–39.

Maksimov, I.V., Abizgil'dina, R.R. and Pusenkova, L.I. (2011) Plant growth promoting rhizobacteria as alternative to chemical crop protectors from pathogens (review). *Applied Biochemistry and Microbiology* 47, 333–345.

McGrann, G.R.D., Gladders, P., Smith, J.A. and Burnett, F. (2016) Control of clubroot (*Plasmodiophora brassicae*) in oilseed rape using varietal resistance and soil amendments. *Field Crops Research* 186, 146–156.

Miethke, M. and Marahiel, M.A. (2007) Siderophore-based iron acquisition and pathogen control. *Microbiology and Molecular Biology Reviews* 71, 413–451.

Mihailović, D.T., Budinčević, M., Balaž, I. and Mihailović, A. (2011) Stability of intercellular exchange of biochemical substances affected by variability of environmental parameters. *Modern Physics Letters B* 25, 2407–2417.

Miller, M.B. and Bassler, B.L. (2001) Quorum sensing in bacteria. *Annual Review of Microbiology* 55, 165–199.

Mittelstraß, K., Treutter, D., Plessl, M., Elstner, E.F., Heller, W. *et al.* (2006) Modification of primary and secondary metabolism of potato plants by nitrogen application differentially affects resistance to *Phytophthora infestans* and *Alternaria solani*. *Plant Biology* 8, 653–661.

Mohanta, T.K., Bashir, T., Hashem, A., Abd Allah, E.F., Khan, A.L. *et al.* (2018) Early events in plant abiotic stress signaling: Interplay between calcium, reactive oxygen species and phytohormones. *Journal of Plant Growth Regulation* 37, 1033–1049.

Moore, K. and Bradley, L.K. (eds) (2018) *North Carolina Extension Gardener Handbook*. College of Agriculture and Life Sciences, NC State University, North Carolina.

Moustafa-Farag, M., Mohamed, H.I., Mahmoud, A., Elkelish, A., Misra, A.N. *et al.* (2020) Salicylic acid stimulates antioxidant defense and osmolyte metabolism to alleviate oxidative stress in watermelons under excess boron. *Plants* 9, 724.

Müller, D.B., Vogel, C., Bai, Y. and Vorholt, J.A. (2016) The plant microbiota: systems-level insights and perspectives. *Annual Review of Genetics* 50, 211–234.

Nam, K.W., Jee, H.J., Lee, H.S. and Yeh, W.H. (2001) Effective chemicals on control of sweet pepper blight caused by *Phytophthora capsici* in hydroponic culture. *Plant Pathology Journal* 17, 369–3.

Nam, M.H., Jeong, S.K., Lee, Y.S., Choi, J.M. and Kim, H.G. (2006) Effects of nitrogen, phosphorus, potassium and calcium nutrition on strawberry anthracnose. *Plant Pathology* 55, 246–249.

Naveed, M., Brown, L.K., Raffan, A.C., George, T.S., Bengough, A.G. *et al.* (2017) Plant exudates may stabilize or weaken soil depending on species, origin and time. *European Journal of Soil Science* 68, 806–816.

Osakabe, Y., Yamaguchi-Shinozaki, K., Shinozaki, K. and Tran, L.S.P. (2013) Sensing the environment: key roles of membrane-localized kinases in plant perception and response to abiotic stress. *Journal of Experimental Botany* 64, 445–458.

Patten, C.L. and Glick, B.R. (2002) Role of *Pseudomonas putida* indoleacetic acid in development of the host plant root system. *Applied and Environmental Microbiology* 68, 3795–3801.

Peng, Y., Yang, J., Li, X. and Zhang, Y. (2021) Salicylic acid: biosynthesis and signaling. *Annual Review of Plant Biology* 72, 761–791.

Poltronieri, P. and Reca, I.B. (2019) Microbial products and secondary metabolites in plant health. In: Poltronieri, P. and Hong, Y. (eds) *Applied Plant Biotechnology for Improving Resistance to Biotic Stress*. Academic Press, pp. 189–202.

Rai, G.K., Kumar, P., Choudhary, S.M., Singh, H., Adab, K. *et al.* (2023) Antioxidant potential of glutathione and crosstalk with phytohormones in enhancing abiotic stress tolerance in crop plants. *Plants* 12, 1133.

Rai, K.K., Pandey, N. and Rai, S.P. (2020a) Salicylic acid and nitric oxide signaling in plant heat stress. *Physiologia Plantarum* 168, 241–255.

Rai, P.K., Singh, M., Anand, K., Saurabh, S., Kaur, T. *et al.* (2020b) Role and potential applications of plant growth-promoting rhizobacteria for sustainable agriculture. In: Rastegari, A.A., Yadav, A.N. and Yadav, N. (eds) *New and Future Developments in Microbial Biotechnology and Bioengineering*. Elsevier, Amsterdam, pp. 49–60.

Read, D.B., Bengough, A.G., Gregory, P.J., Crawford, J.W., Robinson, D. *et al.* (2003) Plant roots release phospholipid surfactants that modify the physical and chemical properties of soil. *New Phytologist* 157, 315–326.

Rehman, A., Azhar, M.T., Hinze, L., Qayyum, A., Li, H. *et al.* (2021) Insight into abscisic acid perception and signaling to increase plant tolerance to abiotic stress. *Journal of Plant Interactions* 16, 222–237.

Roskova, Z., Skarohlid, R. and McGachy, L. (2022) Siderophores: an alternative bioremediation strategy? *Science of the Total Environment* 819, 153144.

Ruhal, R. and Kataria, R. (2021) Biofilm patterns in gram-positive and gram-negative bacteria. *Microbiological Research* 251, 126829.

Ryan, R.P., Fouhy, Y., Garcia, B.F., Watt, S.A., Niehaus, K. *et al.* (2008) Interspecies signalling via the *Stenotrophomonas maltophilia* diffusible signal factor influences biofilm formation and polymyxin tolerance in pseudomonas aeruginosa. *Molecular Microbiology* 68, 75–86.

Saboki Ebrahim, K.U. and Singh, B. (2011) Pathogenesis related (PR) proteins in plant defense mechanism. *Science against Microbial Pathogens* 2, 1043–1054.

Sahu, P.K., Singh, S., Gupta, A., Singh, U.B., Brahmaprakash, G.P. *et al.* (2019) Antagonistic potential of bacterial endophytes and induction of systemic resistance against collar rot pathogen *Sclerotium rolfsii* in tomato. *Biological Control* 137, 104014.

Sattley, W.M. and Madigan, M.T. (2021) Bacteriology. *eLS* 1, 821–829.

Scavino, A.F. and Pedraza, R.O. (2013) The role of siderophores in plant growth-promoting bacteria. In: Maheshwari, D., Saraf, M. and Aeron, A. (eds) *Bacteria in Agrobiology: Crop Productivity*. Springer, pp. 265–285.

Schalk, I.J. (2008) Metal trafficking via siderophores in gram-negative bacteria: specificities and characteristics of the pyoverdine pathway. *Journal of Inorganic Biochemistry* 102, 1159–1169.

Schulz, P., Herde, M. and Romeis, T. (2013) Calcium-dependent protein kinases: hubs in plant stress signaling and development. *Plant Physiology* 163, 523–530.

Sheard, L.B., Tan, X., Mao, H., Withers, J., Ben-Nissan, G. *et al.* (2010) Jasmonate perception by inositol-phosphate-potentiated COI1–JAZ co-receptor. *Nature* 468, 400–405.

Sinclair, S.A. and Krämer, U. (2012) The zinc homeostasis network of land plants. *Biochimica et Biophysica Acta* 1823, 1553–1567.

Singh, P., Singh, R.K., Li, H.-B., Guo, D.-J., Sharma, A. *et al.* (2020) Diazotrophic bacteria *Pantoea dispersa* and *Enterobacter asburiae* promote sugarcane growth by inducing nitrogen uptake and defense-related gene expression. *Frontiers in Microbiology* 11, 600417.

Singh, P., Chauhan, P.K., Upadhyay, S.K., Singh, R.K., Dwivedi, P. *et al.* (2022) Mechanistic insights and potential use of siderophores producing microbes in rhizosphere for mitigation of stress in plants grown in degraded land. *Frontiers in Microbiology* 13, 898979.

Sticher, L., Mauch-Mani, B. and Métraux, J.P. (1997) Systemic acquired resistance. *Annual Review of Phytopathology* 35, 235–270.

Swapnil, P., Meena, M., Marwal, A., Vijayalakshmi, S. and Zehra, A. (eds) (2023) *Plant-Microbe Interaction – Recent Advances in Molecular and Biochemical Approaches*, Vol. 1. *Overview of Biochemical and Physiological Alteration during Plant-Microbe Interaction*. Elsevier.

Tanaka, K., Choi, J., Cao, Y. and Stacey, G. (2014) Extracellular ATP acts as a damage-associated molecular pattern (DAMP) signal in plants. *Frontiers in Plant Science* 5, 446.

Tchameni, S.N., Cotârleţ, M., Ghinea, I.O., Bedine, M.A.B., Sameza, M.L. *et al.* (2020) Involvement of lytic enzymes and secondary metabolites produced by *Trichoderma* spp. in the biological control of *Pythium myriotylum*. *International Microbiology* 23, 179–188.

Thalineau, E., Fournier, C., Gravot, A., Wendehenne, D., Jeandroz, S. *et al.* (2018) Nitrogen modulation of *Medicago truncatula* resistance to *Aphanomyces euteiches* depends on plant genotype. *Molecular Plant Pathology* 19, 664–676.

Thoenen, L., Giroud, C., Kreuzer, M., Waelchli, J., Gfeller, V. *et al.* (2023) Bacterial tolerance to host-exuded specialized metabolites structures the maize root microbiome. *Proceedings of the National Academy of Sciences USA* 120, e2310134120.

Tomar, S., Babu, M.S., Gaikwad, D.J. and Maitra, S. (2021) A review on molecular mechanisms of wheat (*Triticum aestivum* L.) and rice (*Oryza sativa* L.) against abiotic stresses with special reference to drought and heat. *International Journal of Agriculture Environment and Biotechnology* 14, 215–222.

Tripathi, R., Tewari, R., Singh, K.P., Keswani, C., Minkina, T. *et al.* (2022) Plant mineral nutrition and disease resistance: a significant linkage for sustainable crop protection. *Frontiers in Plant Science* 13, 883970.

Trouvé, R., Drapela, T., Frank, T., Hadacek, F. and Zaller, J.G. (2014) Herbivory of an invasive slug in a model grassland community can be affected by earthworms and mycorrhizal fungi. *Biology and Fertility of Soils* 50, 13–23.

Tsiafouli, M.A., Thébault, E., Sgardelis, S.P., de Ruiter, P.C., van der Putten, W.H. *et al.* (2015) Intensive agriculture reduces soil biodiversity across Europe. *Global Change Biology* 21, 973–985.

Tsukanova, K.A., Chebotar, V.K., Meyer, J.J.M. and Bibikova, T.N. (2017) Effect of plant growth-promoting rhizobacteria on plant hormone homeostasis. *South African Journal of Botany* 113, 91–102.

Unuofin, J.O., Odeniyi, O.A., Majengbasan, O.S., Igwaran, A., Moloantoa, K.M. *et al.* (2024) Chitinases: expanding the boundaries of knowledge beyond routinized chitin degradation. *Environmental Science and Pollution Research International* 31, 38045–38060.

Van Bockhaven, J., De Vleesschauwer, D. and Höfte, M. (2013) Towards establishing broad-spectrum disease resistance in plants: silicon leads the way. *Journal of Experimental Botany* 64, 1281–1293.

Van Groenigen, K.J., Xia, J., Osenberg, C.W., Luo, Y. and Hungate, B.A. (2015) Application of a two-pool model to soil carbon dynamics under elevated CO_2. *Global Change Biology* 21, 4293–4297.

Vattem, D.A., Mihalik, K., Crixell, S.H. and McLean, R.J.C. (2007) Dietary phytochemicals as quorum sensing inhibitors. *Fitoterapia* 78, 302–310.

Vega, A., Delgado, N. and Handford, M. (2022) Increasing heavy metal tolerance by the exogenous application of organic acids. *International Journal of Molecular Sciences* 23, 5438.

Vincent, S.A., Ebertz, A., Spanu, P.D. and Devlin, P.F. (2022) Salicylic acid-mediated disturbance increases bacterial diversity in the phyllosphere but is overcome by a dominant core community. *Frontiers in Microbiology* 13, 809940.

Vorholt, J.A. (2012) Microbial life in the phyllosphere. *Nature Reviews Microbiology* 10, 828–840.

Wang, G., Hu, C., Zhou, J., Liu, Y., Cai, J. *et al.* (2019) Systemic root-shoot signaling drives jasmonate-based root defense against nematodes. *Current Biology* 29, 3430–3438.

Wang, Y., Liu, H., Fu, G., Li, Y., Ji, X. *et al.* (2022) Paecilomyces variotii extract increases lifespan and protects against oxidative stress in Caenorhabditis elegans through SKN-1, but not DAF-16. *Arabian Journal of Chemistry* 15, 104073.

Waters, C.M. and Bassler, B.L. (2005) Quorum sensing: cell-to-cell communication in bacteria. *Annual Review of Cell and Developmental Biology* 21, 319–346.

Watkins, J.M., Chapman, J.M. and Muday, G.K. (2017) Abscisic acid-induced reactive oxygen species are modulated by flavonols to control stomata aperture. *Plant Physiology* 175, 1807–1825.

Wawrzyńska, A. and Sirko, A. (2024) Sulfate availability and hormonal signaling in the coordination of plant growth and development. *International Journal of Molecular Sciences* 25, 3978.

Weil, R. and Brady, N. (2017) *The Nature and Properties of Soils*, 15th edn. Pearson Education.

Woltz, S. and Jones, J.P. (1979) Effects of magnesium on bacterial spot of pepper and tomato and on the in vitro inhibition of *Xanthomonas vesicatoria* by Streptomycin. *Plant Disease Reporter* 63, 182–184.

Wu, Q., Jing, H.-K., Feng, Z.-H., Huang, J., Shen, R.-F. *et al.* (2022) Salicylic acid acts upstream of auxin and nitric oxide (NO) in cell wall phosphorus remobilization in phosphorus deficient rice. *Rice* 15, 42.

Wurst, S. (2013) Plant-mediated links between detritivores and aboveground herbivores. *Frontiers in Plant Science* 4, 380.

Xiao, L., Sun, Q., Yuan, H. and Lian, B. (2017) A practical soil management to improve soil quality by applying mineral organic fertilizer. *Acta Geochimica* 36, 198–204.

Yan, J., Zhang, C., Gu, M., Bai, Z., Zhang, W. *et al.* (2009) The *Arabidopsis* CORONATINE INSENSITIVE1 protein is a jasmonate receptor. *Plant Cell* 21, 2220–2236.

Yoneyama, K., Xie, X., Kusumoto, D., Sekimoto, H., Sugimoto, Y. *et al.* (2007) Nitrogen deficiency as well as phosphorus deficiency in sorghum promotes the production and exudation of 5-deoxystrigol, the host recognition signal for arbuscular mycorrhizal fungi and root parasites. *Planta* 227, 125–132.
Yruela, I. (2005) Copper in plants. *Brazilian Journal of Plant Physiology* 17, 145–156.
Zhang, H., Zhang, D., Chen, J., Yang, Y., Huang, Z. *et al.* (2004) Tomato stress-responsive factor TSRF1 interacts with ethylene responsive element GCC box and regulates pathogen resistance to *Ralstonia solanacearum*. *Plant Molecular Biology* 55, 825–834.
Zhang, Y., Lubberstedt, T. and Xu, M. (2013) The genetic and molecular basis of plant resistance to pathogens. *Journal of Genetics and Genomics* 40, 23–35.
Zhu, S., Liu, M. and Zhou, J. (2006) Inhibition by nitric oxide of ethylene biosynthesis and lipoxygenase activity in peach fruit during storage. *Postharvest Biology and Technology* 42, 41–48.

3 Impact of Pesticides on Soil Health of Agroecosystems and Plant Nutrition: Challenges and Sustainable Management

Jiya Navshree[1], Nikita Wadhwa[2], Manika Bhatia[3], Prashansa Sharma[4], Linthoi Khomdram[2], Arti Mishra[5,6] and Shalini Kaushik Love[7]*

[1]*South Asian University, Rajpur Road, Maidan Garhi, New Delhi, India;* [2]*Hansraj College, University of Delhi, North Campus, New Delhi, India;* [3]*TERI School of Advanced Studies, New Delhi, India;* [4]*Department of Plant Molecular Biology, University of Delhi, South Campus, New Delhi, India;* [5]*Department of Botany, Hansraj College, University of Delhi, India;* [6]*Umeå Plant Science Center, Department of Plant Physiology, Umea University, Umeä, Sweden;* [7]*DESM (Department of Education Science and Mathematics), Regional Institute of Education, Bhubaneswar, Odisha, India*

Abstract

Pesticides play an important role in increasing crop productivity and ensuring food security. However, the extensive and irrational use of pesticides has affected soil health globally. The physicochemical and biological properties of soil determine soil health. Soil microbiota and soil fauna play an important role in the ecological functioning of soils and nutrient cycling in agroecosystems. Pesticides have been shown to negatively affect soil microbiota, structure, composition and fertility. As a consequence, the availability and absorption of nutrients for plants are affected. In this chapter, we address comprehensively the impact of pesticides on soil health and plant nutrition. Also discussed are the probable approaches and challenges faced while addressing the issue of excessive pesticide use in agriculture, including policy and regulatory approaches, biotechnology and technology-based solutions, practising sustainable agriculture based upon principles of agroecology, and integrated pest management (IPM).

Keywords: Pesticides, soil health, microbiota, plant nutrition, integrated pest management

3.1 Introduction

Pesticides are the chemicals used to deal with pests in agroecosystems. Based on the target pest, they can be classified into fungicides, nematicides, molluscicides, rodenticides, herbicides/weedicides (which may act as plant growth regulators) and other chemical substances. Pesticides are essential for crop protection and preventing vector-borne diseases, ultimately improving crop productivity (Pathak *et al.*, 2022). Despite being vital for pest control and increasing crop productivity, applying pesticides substantially impacts the soil ecosystem and the plant's nutritional dynamics. Thus, a delicate balance is required to maintain soil health, providing plants with the

---

*Corresponding author: shalinikaushiklove@gmail.com

best nutrition and managing pests efficiently (Alengebawy *et al.*, 2021).

Extensive and prolonged use of pesticides has been found to affect the soil environment significantly. Once they enter the soil, pesticides undergo various physical and biological decomposition processes, considerably altering the soil's physicochemical characteristics. Scientific evidence points to the use of pesticides and synthetic inorganic fertilizers altering soil pH, ultimately disturbing the soil microbial diversity (Rasool *et al.*, 2014). Apart from that, pesticides can affect the complex balance and functioning of soil microbial communities involved in various roles, such as essential elements in the cycling of nutrients and preservation of soil fertility. These microbial populations can be disrupted, which can change the properties of the soil and affect the availability of nutrients. Beneficial soil microorganisms and biota are negatively impacted by the indiscriminate use of different agrochemicals (Meena *et al.*, 2020; Yadav *et al.*, 2023). Pesticides have, nevertheless, also shown beneficial effects, chiefly linked to increased plant biomass and rhizodeposition. These are reported when pesticides are used at a greater dilution and may not be accurate for agroecosystems where these are used indiscriminately. Moreover, the adverse impacts of these herbicides on the soil ecosystem far outweigh the minor growth increase, if any (Khmelevtsova *et al.*, 2023).

Implementing sustainable management in agroecosystems is fraught with difficulties (Fenibo *et al.*, 2021), including lack of proper knowledge, financial disparity, government policies and measures, insufficient adoption of biopesticides and organic fertilizers and technological limitations (Lykogianni *et al.*, 2021). To minimize the detrimental effects on soil health and maintain optimal plant nutrition while guaranteeing agricultural productivity, it is essential to comprehend the difficulties involved with applying pesticides and identify sustainable management strategies (Tudi *et al.*, 2021). This chapter summarizes the available literature and draws attention to the difficulties associated with the use of pesticides. It also provides insights into sustainable management practices essential for preserving soil health, maximizing plant nutrition and sustaining agricultural productivity. The complex interplay among plant nutrition,

soil health in agroecosystems and pesticides represents a crucial aspect of contemporary agriculture.

3.2 Worldwide Dependence on Pesticides

According to United Nations estimates, it is projected that by 2050 there will be 9.8 billion people on the planet. Therefore, the demand for food will subsequently increase (UN, 2017). There is a worldwide burden of crop pests and pathogens in agroecosystems for crop production (Savary *et al.*, 2019). The use of various pesticides to control pests becomes crucial, considering that 45% of the crops get destroyed yearly due to pests (Sharma *et al.*, 2019). Without using pesticides, the agricultural industry would be expected to lose 78% of its fruit, 54% of its vegetables and 32% of its grain production (Parween *et al.*, 2016). Most countries are thus compelled to use different types of pesticides to increase productivity in agriculture and meet their present-day food needs (Schutter and Vanloqueren, 2011). As per the report published by the Food and Agriculture Organization (FAO), the global use of pesticides in agriculture has almost doubled from the 1990s to the present decade. There has also been a sharp rise in the trade of agrochemicals (imports and exports) since then (FAOSTAT, 2022). Pesticide consumption in various continents as a percentage of their global use stands at 32.6% for Asia, 18.3% for Europe, 1.8% for Oceania, 4.4% for Africa and 42.9% for the Americas (FAOSTAT, 2022).

3.3 Negative Impacts of Pesticides

Even though pesticides help ensure food security, their extensive use can pose serious threats (Sharma *et al.*, 2019). For instance, frequent and excessive use of pesticides in the rice fields to cope with the pest load has accelerated soil, water and air pollution (Fahad *et al.*, 2021).

The imprudent and persistent application of pesticides harms the environment and human health (Sharma *et al.*, 2019). A large-scale study in eight European countries revealed that the application of pesticides, especially fungicides and insecticides, most severely impacted the diversity of wild plant species and birds compared with other parameters of agricultural intensification (Geiger *et al.*, 2010). Exposure to pesticides such as organochlorine and organophosphates has been associated with the cause of many diseases, including Parkinson's and Alzheimer's diseases in human beings (Blair *et al.*, 2015). The application of pesticides irrationally also confers pesticide resistance, causing changes in plants' physiological and morphological responses. For example, a study reported boll weevil larvae to be almost three times higher after excessive application of pesticides (Altieri and Nicholls, 2003). Pesticides are also known to harm non-target beneficial organisms like predators and pollinators, thus disbalancing the ecosystem (Serrão *et al.*, 2022). They further aggravate the situation by accumulating residues in the environment. The areas where the concentrations of pesticide residues increase their 'no effect' limit poses a severe risk of environmental pollution. According to one estimate, 31% of agricultural land is hazardous to chemical pollution and 64% of agricultural land worldwide is at risk (Tang *et al.*, 2021).

3.4 Impact of Pesticides on Soil Physicochemical Characteristics

Pesticides have complex and extensive impacts on the physicochemical characteristics of soil. They are essential instruments in modern agriculture for insect control and crop protection (Damalas and Eleftherohorinos, 2011; Tudi *et al.*, 2021). They go through a variety of biodegradation processes once they penetrate the soil, including chemical decomposition driven by variables like pH, humidity and temperature, as well as biological degradation aided by enzymes made by microorganisms (Nannipieri *et al.*, 2003; Rasool *et al.*, 2022). Furthermore, via procedures like adsorption, leaching, volatilization, spray drift and runoff, these pesticides can disperse from their intended target areas to other environmental elements or non-target plants (Pathak *et al.*, 2022). These compounds behave differently in the environment due to their distinctive properties. Organochlorine substances, such as DDT, have modest acute toxicity but a remarkable

potential for bioaccumulation and persistence, resulting in long-lasting environmental effects even after their sale has been outlawed in many nations (Goswami *et al.*, 2013).

Pesticides can change the pH of the soil, which is a crucial component in nutrient availability and microbial activity. Pesticides used and disposed of often end up in the environment. After entering the environment, they undergo various transformation processes, which may lead to new chemical compounds. This transformation might change pH, such as producing acidic or alkaline breakdown products (Tudi *et al.*, 2021).

For instance, 2,4-D herbicides can cause soil to become more acidic, but some organophosphate insecticides can cause pH levels to rise. These pH changes may result in imbalances in the soil ecosystem by substantially affecting nutrient solubility and microbial diversity (Tudi *et al.*, 2021). A study conducted by Marioara Nicoleta *et al.* (2015) revealed the damaging effects of the insecticides cypermethrin and thiamethoxam on soil, based on the soil's physical characteristics, in addition to biochemical and microbiological activity. Thiomethoxam caused a 6.5% decrease in phosphatase activity compared with the control group. Additionally, there was a 58.1% decrease in the population of nitrifying bacteria. On the other hand, cypermethrin caused a significant decline in dehydrogenase enzyme activity, specifically by 32.8%, along with a 74% decrease in the quantity of nitrifying bacteria. Additionally, the humidity and pH values showed direct proportional correlations with urease and dehydrogenase activities and the population of nitrifying bacteria with correlation coefficients above +0.9 and +0.8, respectively. It is also known that cypermethrin and other pyrethroid insecticides cannot be degraded naturally. Hence, an intermediate metabolite called 3-phenoxy benzoic acid is produced during their breakdown process, leading to secondary contamination issues in agricultural goods and related problems (Huang *et al.*, 2018). They also affect soil respiration and biomass (Goswami *et al.*, 2013).

A key component of soil health, soil organic matter (SOM) is essential for soil structure, water retention and nutrient cycling (Fig. 3.1). Pesticides' interactions with soil microbial communities can impact SOM content, soil carbon cycle

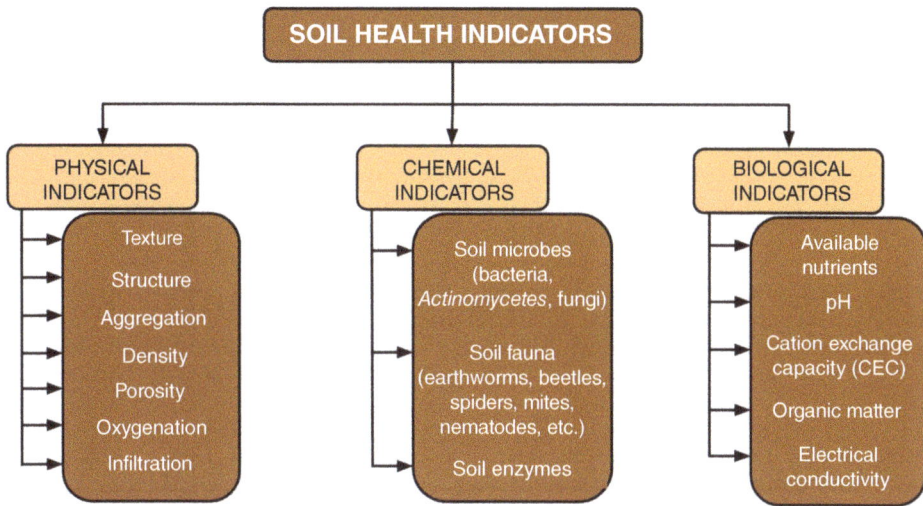

Fig. 3.1. Major soil health indicators.

and structure stabilization (Sim *et al.*, 2022). Pesticides such as neonicotinoid insecticides can have an indirect impact on SOM by causing harm to organisms that live in the soil, such as earthworms, who are involved in the creation of SOM through their burrowing activities (Mishra *et al.*, 2022). Likewise, glyphosate has been shown to impede microbial activity, potentially slowing the pace of SOM decomposition. At the same time, glyphosate-based herbicides (GBHs) and glyphosate alone (GLYs) have an impact on earthworms (Schmidt *et al.*, 2023).

3.5 Impact of Pesticides on the Soil Microbiome

Bacteria, *Actinomycetes* and fungi constitute a major part of the soil microbiome, with bacteria constituting almost 15% of the total living biomass (Banerjee and van der Heijden, 2023). Several studies show the harmful effects of pesticides on these organisms (Yousaf *et al.*, 2013; Arora and Sahni, 2016; Al-Ani *et al.*, 2019; Borowik *et al.*, 2023). These effects may vary according to the type, concentration, degradability and toxicity of the pesticide applied. For instance, chlorpyrifos, at high concentrations, inhibited the growth of aerobic dinitrogen-fixing

bacteria (Supreeth *et al.*, 2016). Other factors like soil texture, organic matter content, tillage system and vegetation also play an important role in determining these effects.

Pesticides significantly reduce soil microbial diversity. Applying urea herbicides like diuron and linuron causes a reduction in bacterial diversity by altering their functional abilities (Lo, 2010). Sulfonylurea herbicides, like metsulfuron methyl, chlorsulfuron and thifensulfuron methyl, have been found to inhibit the synthesis of branched-chain amino acids valine, leucine and isoleucine in bacteria by targeting the enzyme acetolactate synthase (ALS; the primary enzyme involved in the synthesis of these amino acids in plants) (Johnsen *et al.*, 2001; Lo, 2010). In a study on the effects of chlorpyrifos on soil microbes in the sub-humid tropical rice–rice system, it was observed that chlorpyrifos significantly reduced the number of Gram-positive, oligotrophic and copiotrophic bacteria (Kumar *et al.*, 2017). Mancozeb, a fungicide, caused a drastic reduction in the total population of fungi, *Actinomycetes* and *Pseudomonas* bacteria. Tebuconazole, another fungicide, was found to inhibit ergosterol biosynthesis in fungi. Nematicides might not have had a direct effect, but they caused temporary fluctuations in the soil microbial diversity. The protein synthesis in fungi, bacteria and plants, which takes place via the shikimic acid pathway, was inhibited by glyphosate, a commonly used herbicide (Lo, 2010). Triazole fungicides, like difenoconazole, hexaconazole and paclobutrazol, were also found to reduce the overall soil microbial diversity (Roman *et al.*, 2021). The adverse effects of pesticides on soil bacteria have been well known for a long time. The nitrogen-metabolizing bacteria are the most affected among the several different populations in soil. However, the population of phosphorus-solubilizing and sulfur-oxidizing bacteria and other nutrient-mineralizing ones are unaffected (Kumar *et al.*, 2017; Storck *et al.*, 2018).

3.5.1   Impact of pesticides on bacteria involved in the nitrogen cycle

The nitrogen cycle greatly depends on the variety of bacteria involved in various stages, that is, nitrogen fixation, nitrification and denitrification. Pesticides are known to affect these processes by affecting the bacteria engaged at different stages. Nitrogen fixation can be affected by reducing the growth and number of free-living bacteria such as *Azospirillum* and symbionts of the legume such as *Rhizobium*. The growth population of *Azospirillum* and other aerobic nitrogen fixers in non-flooded sandy loam alluvial soil was found to be reduced by the application of the herbicide butachlor (Lo, 2010). The herbicide alachlor affected the growth of diazotrophs such as *Azospirillum bradense* (Pozo *et al.*, 1994). Insecticides like chlorpyrifos and methylpyrimifos also reduced the growth of heterotrophic nitrogen fixers such as *Azospirillum* sp. (Martinez-Toledo *et al.*, 1992; Kumar *et al.*, 2017). Herbicide 2,4-D affects the growth of blue-green algae (BGA) (Meena *et al.*, 2020). Malathion was observed to reduce the number of diazotrophs, like *Azotobacter chroococcum* and *Rhizobium trifolii*, in agricultural soils (Borowik *et al.*, 2023).

Insecticides like DDT, methyl parathion and pentachlorophenol interfere with the legume *Rhizobium* chemical signalling, thus reducing N_2 fixation (Arora and Sahni, 2016). A study depicting *Sinorhizobium meliloti* in symbiosis with alfalfa in farm soils showed that methyl parathion and pentachlorophenol inhibited NodD signalling by 90% and that DDT cut signalling by 45% (Reuhs *et al.*, 1998). These pesticides also disrupted NodD signalling in *Rhizobium* sp. strain NGR234 (Potera, 2007).

The herbicides 2,4-D and pendimethalin have been found to affect the N_2-fixing capacity of *Rhizobium* sp. in crop plants (Meena *et al.*, 2020). The N_2-fixing ability of soil microflora has been dramatically affected by chlorpyrifos, alone or in combination with lindane (Martinez-Toledo *et al.*, 1992). Mancozeb was found to have a long-term inhibitory effect on aerobic N_2 fixers in natural soil (Johnsen *et al.*, 2001).

Insecticides such as fenamiphos, methylpyrimifos and chlorpyrifos were detrimental to nitrifying bacteria growth (Lo, 2010; Kumar *et al.*, 2017). Insecticides permethrin and cypermethrin also reduced the number of ammonifying and nitrifying bacteria in soil (Borowik *et al.*, 2023). Fungicides such as mancozeb, etridiazol and dimethomorph inhibited the growth of ammonium-oxidizing nitrifiers in the long term (Johnsen *et al.*, 2001; Yang *et al.*, 2011).

Generally, denitrifying bacteria (DNB) are more tolerant to pesticides than N_2-fixing and nitrifying bacteria (Kumar *et al.*, 2017). However, reports indicate the adverse impact of pesticides on DNB. In the wetland systems, fungicide carboxin was found to affect the DNB significantly (Yang *et al.*, 2011). In the dry season, chlorpyrifos reduced the number of DNB in the sub-humid tropical rice–rice system (Kumar *et al.*, 2017).

Interestingly, a few pesticides have been found to affect soil bacteria positively. Many bacteria use pesticides as a source of energy and grow in number. *Flavobacterium* sp., *Pseudomonas* sp. and *Arthrobacter* sp. utilize malathion, parathion, chlorpyrifos and many other pesticides as a carbon source for their growth and proliferation (Diğrak and Kazanici, 2001). *Streptomyces* sp. HP-11 also utilizes chlorpyrifos as a carbon source (Supreeth *et al.*, 2016). Pyrethroid insecticides permethrin and cypermethrin were found to stimulate the growth of organotrophic bacteria and non-symbiotic N_2-fixing bacteria, as they act as carbon sources for many autochthonous microbes (Borowik *et al.*, 2023). Carbofuran stimulated the growth of *Azospirillum* and other anaerobic nitrogen fixers in flooded and non-flooded soil (Lo, 2010). The herbicide alachlor also enhanced the growth of denitrifying and N_2-fixing bacteria (Pozo *et al.*, 1994).

3.5.2 Impact of pesticides on *Actinomycetes*

Actinomycetes are Gram-positive bacteria with bacterial and fungal characteristics (Supreeth et al., 2016). Pesticides have stimulatory as well as inhibitory effects on them. In one study, spraying chlorpyrifos stimulated the *Actinomycetes* populations in soil (Supreeth et al., 2016). Glyphosate also favoured their growth (Arora and Sahni, 2016). A few pesticides like mancozeb, metham sodium, 2,4-D and paclobutrazol were found to reduce the number of *Actinomycetes* in soil (Karpouzas et al., 2005; Lo, 2010; Roman et al., 2021; Rillig et al., 2023). In dry seasons, chlorpyrifos inhibited *Actinomycetes* growth in the sub-humid tropical rice–rice systems (Kumar et al., 2017).

3.5.3 Impact of pesticides on fungi

The application of pesticides mostly had inhibitory effects on fungal populations, including a reduction in their growth, number and diversity or hampering the nutrient-supplying abilities of beneficial genera. In a 30-day study, the application of a fungicide, captan, showed a substantial decrease in soil fungi. In another study, herbicide 2,4-D was also found to deplete the fungal population by half. Furthermore, chloropicrin and methyl bromide were found to completely inhibit the growth of *Glomus mosseae* and *G. fasciculatum* on citrus (Rillig et al., 2023). Permethrin and cypermethrin notably affected the number and structure of the fungal genus *Ascomycota*, respectively (Borowik et al., 2023). At higher concentrations, sulphonylureas depleted fungal growth (Johnsen et al., 2001). In dry seasons, fungi were found to be distressed by applying chlorpyrifos (Kumar et al., 2017). However, some exceptions exist; for instance, the growth of *Aspergillus fumigatus* was stimulated by phorate (Gonzalez-Lopez et al., 1992). In cotton, the mycorrhizal infection was boosted by 1,3-dichloropropane, a nematicide (Rillig et al., 2023). Arbuscular mycorrhizal colonization is sometimes favoured by fungicides applied on seeds in combination with fludioxonil (Meena et al., 2020).

3.5.4 Impact of pesticides on mycorrhizae

Mycorrhizae play a remarkable role in the ecosystem by providing almost 90% of the land plants with a diverse array of benefits (Banerjee and van der Heijden, 2023) such as increasing the surface area, thereby facilitating the adsorption of mineral nutrients and providing protection from stress and pathogens (Tedersoo et al., 2020). Sadly, they are one of the major victims of pesticide application, especially when applied to the seeds or the soil directly (Arora and Sahni, 2016). Fungicides have a huge impact on non-target fungal associations. They show toxicity towards mycorrhizal hyphal growth, ultimately hindering root colonization. Emisan and carbendazim had a damaging effect on arbuscular mycorrhizal fungi (AMF) in groundnut. On benzoyl application, mycorrhizal associations were disrupted for a long time (Meena et al., 2020). Benomyl, phenylpyrazole and chlorpyrifos (only at higher concentrations) also had ill effects on mycorrhiza (Arora and Sahni, 2016).

3.6 Impact of Pesticides on Plant Nutrition

Of most of the pesticides applied to crops, merely 0.1% is absorbed by pests, thus indicating a substantial impact on a wide range of non-target organisms, including the plants themselves; thereby, its chemical constituents interfere with many of the plant's biochemical pathways (Homayoonzadeh et al., 2020). Pesticides enter plants through their root system by either active or passive absorption. Once absorbed, they can either be metabolized by the plants or they enter the food chain, causing biomagnification. They impact plants through various mechanisms, including the generation of excessive reactive oxygen species (ROS), oxidative stress, DNA damage, reduced photosynthesis, necrosis, chlorosis, and leaf deformities, eventually leading to plant death (Sharma et al., 2017). Pesticides also control or eliminate plants by inhibiting biological processes like photosynthesis, cell division and enzyme function, which affect leaf formation and root growth. They interfere with pigment production, and proteins alter DNA, disrupt cell membranes and stimulate uncontrolled growth (Parween *et al.*, 2016). Pesticides can affect plants' early germination and growth, leading to changes in biochemical, physiological, enzymatic and non-enzymatic antioxidation pathways. These alterations ultimately impact plant yield and result in the accumulation of pesticide residues within plants, vegetables, fruits and unintended organisms (Homayoonzadeh *et al.*, 2020; Sharma *et al.*, 2017).

Weedicides, fungicides and insecticides of different chemical classes affect plants in several ways, as listed here.

3.6.1 Growth and development

Pesticides adversely impact the growth and development of different crop plants. An insecticide called dimethoate was observed to reduce root and shoot growth significantly. The reduction in root growth was more pronounced than that of the shoot. This could be due to the direct contact of dimethoate with the root (Parween *et al.*, 2016). Groundnut (*Arachis hypogaea* L.) growth was negatively impacted by pesticides (alachlor, butachlor and oxyfluorfen) by reducing the height, fresh weight, seed characteristics, biomass, yield and oil content at various concentrations (Hatamleh *et al.*, 2022). Pesticide application also affects the plant growth components; for instance, topsin M, benlate, demacron, chlorsulfuron and cypermethrin, when applied at high concentration, cause a decrease in leaf area ratio, leaf growth rate (LGR), leaf weight ratios (LWR), crop growth rate (CGR) and relative growth rate (RGR) of soybean (Siddiqui and Ahmed, 2006).

Pesticides also inhibit the key growth-determining physiological processes, including photosynthesis and nitrogen metabolism. Fusillade, a herbicide, accumulates in actively growing plant areas, interfering with energy production and cell metabolism, consequently reducing plant growth by disrupting metabolic processes like photosynthesis and nitrogen metabolism (Parween *et al.*, 2016). Furthermore, herbicides like chlorotoluron and the fungicide mancozeb have been found to interfere with higher plant photosynthetic electron transport and chlorophyll

content. Mancozeb, when applied on *Solanum lycopersicum*, leads to a reduction of carotenoids by 91% and chlorophyll a and b contents by 91% and 80%, respectively (Hatamleh *et al.*, 2022). Metribuzin obstructs photosynthesis by blocking electron transfer in photosystem II, which results in pigment deficiency in plant foliage (Parween *et al.*, 2016). Acetachlor and sulfonyl bensulfuron inhibit cell division, decreasing chlorophyll content and impeding plant growth. Additionally, an excessive application of cadmium inhibits nitrate reductase action, restricting nitrate absorption necessary for nitrogen assimilation within the plant body (Huang and Xiong, 2009).

The accumulation of sugar is crucial for driving plant germination, growth and development (Lastdrager *et al.*, 2014). Pesticides, including imidacloprid, dimethoate and fusillade, disrupt this process in different crop species, including cereals and pulses (Parween *et al.*, 2016). Extensive imidacloprid seed treatment has been reported to adversely impact crop germination and early growth, including leek, white cabbage and sweet corn. In addition, plants exposed to the chemical insecticide kitazin hamper the germination of *Pisum sativum* in both *in vitro* and *in vivo* conditions (Hatamleh *et al.*, 2022).

3.6.2 Nutrient uptake

Pesticides interfere with the uptake of micronutrients and macronutrients by reducing water potential in soil, ultimately causing many diseases in plants. Imazethapyr, cadmium, butachlor and benzosulfurol methyl impair nitrogen assimilation and metabolism by inhibiting the functions of the main enzymes involved, including nitrogen translocase and nitrogen reductase (Huang and Xiong, 2009). Pesticides like sodium arsenite interfere with the accumulation, absorption and assimilation of macronutrients, mainly nitrogen, phosphorus, potassium, calcium and magnesium in *Phaseolus vulgaris* and *Glycine max*. Arsenic from sodium arsenate competes with nitrogen and interferes with its uptake. The application of sodium arsenic is also known to cause root discoloration, plasmolysis and leaf wilting, besides uncoupling oxidative phosphorylation, impeding cell division, decreasing transpiration and reducing yield. All these factors can result in the death of the plant (Carbonell-Barrachina *et al.*, 1997).

Herbicide residues such as chlorsulfuron, diclofop, haloxyfop and fluazifop have particular impacts on the root cells' plasma membrane and micronutrient transport system, which in turn affects the uptake of cations like zinc, copper and magnesium (Siddiqui and Ahmed, 2006).

3.6.3 Oxidative stress

Pesticide load leads to the induction of oxidative stress in the crop and other components of the agroecosystem and reduces overall productivity. An unwanted increase in metals from pesticides like arsenite from sodium arsenite in soybean, and likewise pesticides diazinon, imidacloprid and mancozeb on *Solanum lycopersicum* cause oxidative stress (Hatamleh *et al.*, 2022). Oxidative stress in plants stimulates the overproduction of ROS. In response to this stress, the plant's defence mechanism kicks in. Antioxidant enzymes that break down ROS, like glutathione peroxidase (GPX), catalase (CAT) and superoxide dismutase (SOD), as well as other non-enzymatic antioxidants like carotenoids, phenolic compounds, ascorbate and different nitrogenous metabolites like amino acids, particularly proline from plants, work to neutralize the ROS (Sharma *et al.*, 2017). ROS-induced oxidative stress impairs the majority of proteins and nucleic acids and the photosynthetic electron transport, and causes lipid peroxidation, membrane breakdown and enzyme deactivation, all of which lead to cell death and adversely affect plant development (Homayoonzadeh *et al.*, 2020).

3.7 Impact of Pesticides on Soil Enzymes

Soil enzymes are an inevitable part of soil health. The diversity and concentration of enzymes are indicators of soil quality, fertility, equilibrium between different biogeochemical cycles and soil pollution. They may occur in free form, within microbial cells or as immobilized extracellular enzymes.

Pesticides negatively affect soil enzymes like dehydrogenase, hydrolase, nitrogenase, phosphatase, etc. (Table 3.1). The impacts of pesticides on some crucial soil enzymes are discussed here (Arora and Sahni, 2016).

3.7.1 Dehydrogenase

Dehydrogenase occurs in all microbial cells and contributes to their respiratory processes. Therefore, dehydrogenase activity is one of the indicators of soil microbial activity (Arora and Sahni, 2016). Pesticides may have either inhibitory or neutral impacts on dehydrogenase activity. Applying only endosulfan and mancozeb stimulates dehydrogenase activity when applied at a very high concentration (100–200 times the standard values). The dehydrogenase activity is inhibited after 90 days of using quinaphol (Arora and Sahni, 2016).

3.7.2 Cellulase

Cellulase is an enzyme present in soil microbes and is responsible for transforming organic matter with glucose as one of the final products in soil. In general, the fungicides and herbicides seem to have no effect or are inhibitory to the enzyme cellulase; however, the insecticides belonging to the family of organophosphates (monocrotophos, quinalphos, profenofos and celcron) have a stimulating effect on its activity. These molecules destroy soil insects and make substrates available to stimulate cellulase activity, which herbicides inhibit (Riah *et al.*, 2014).

3.7.3 Phosphatase

Among the five types of phosphatases known, phosphomonoesterase (consisting of acid and alkaline phosphatases) is the most abundant in soil (Riah *et al.*, 2014). These enzymes are indicators of soil fertility and are of paramount importance in the phosphorus cycle in soil ecosystems (Dick *et al.* 2000; Schneider *et al.* 2001). They are present as exoenzymes secreted by microorganisms in the soil. They catalyse the hydrolysis of ester and anhydrides of phosphoric acid and convert the organic phosphate

Table 3.1. Implications of pesticides on the catalytic function of enzymes.

Pesticide	Category	Enzyme	Enzyme class	Effect of pesticide on enzyme activity[a]	Reference
Cloransulam methyl, diclosulam	Herbicide	β-glucosidase	Hydrolases	P	Zhang *et al.*, 2021b
		Urease	Hydrolases	P	
Azoxystrobin	Fungicide	Catalase	Oxidoreductase	P	Wang *et al.*, 2020
		Urease	Hydrolases	I	
		Invertase	Hydrolases	I	
		Phosphatase	Hydrolases	I	
QuadrisR	Fungicide	Dehydrogenase	Oxidoreductase	I	Boteva *et al.*, 2022
S-Metolachlor	Herbicide	Dehydrogenase	Oxidoreductase	I	Filimon *et al.*, 2021
		Protease	Hydrolases	P	
		Phosphatase	Hydrolases	I	
		Urease	Hydrolases	I; S-metolachlor cannot bind to the catalytic cavity of urease, but can bind to other sites and cause inhibition	
Aclonifen	Herbicide	Dehydrogenase	Oxidoreductase	I	Caraba *et al.*, 2023
Trifloxystrobin	Fungicide	Urease	Hydrolases	Initially promoted but inhibited eventually	Xiao *et al.*, 2023
		Dehydrogenase	Oxidoreductase	I (after 7 days of exposure); P (after 21 days)	
		β-glucosidase	Hydrolases	P	
Thiophanate methyl (Tobist)	Fungicide	Dehydrogenase	Oxidoreductases	I (1st week), P (2nd week) and returned to normal by 28th day.	Aly *et al.*, 2023
		Peroxidase	Oxidoreductases	P	
		Polyphenol oxidase (PPO)	Cresolases	I	
Carbendazim	Fungicide	Dehydrogenase	Oxidoreductases	P	Chauhan *et al.*, 2023
		Acid phosphatase	Hydrolases	P	
		Alkaline phosphatase	Hydrolases	P	
		β-glucosidase	Hydrolases	P	
		Urease	Hydrolases	P	
		Protease	Hydrolases	P	
Chlorothalonil	Fungicide	Acid phosphatase	Hydrolases	I	Baćmaga *et al.*, 2018
Novaluron, indoxacarb, thiophanate methyl, propineb, novaluron + thiophanate methyl	Insecticide	Urease	Hydrolases	P	Meghana *et al.*, 2023

Pesticide	Type	Enzyme	Enzyme class	Effect	Reference
Organophosphate insecticide (QOI)	Insecticide	Acetylcholinesterase	Cholinesterase	I (highest during post-clitellar stage; lowest during pre-clitellar stage)	Sujeeth et al., 2023
		Carboxylesterase	Carboxylesterase	I	
		Superoxide dismutase	Oxidoreductases	I	
		Glutathione S transferase		P	
Fluopimomide	Fungicide	Catalase	Oxidoreductases	P (post-clitellar > clitellar > pre-clitellar)	Jiang et al., 2022
		Lipid peroxidase	Lipoxygenases	P (pre-clitellar > clitellar > post-clitellar)	
		Dehydrogenase	Oxidoreductases	P	
		Phosphatase	Hydrolases	P	
		Urease	Hydrolases	P	
		Invertase	Hydrolases	P (day 20), I (day 40) and returned to control level (day 60)	
Acephate, acetamiprid, imidacloprid	Insecticide	Dehydrogenase	Oxidoreductases	I	Singh and Singh, 2023
		Urease	Hydrolases	I	
Terbuthylazine (HumiAgra)	Herbicide	Catalase	Oxidoreductases	I	Baćmaga et al., 2022
		Acid phosphatase	Hydrolases	I	
		Urease	Hydrolases	I (day 37), P (day 111)	
		Dehydrogenase	Oxidoreductases	P (day 37), I (days 74 and 111)	
		Arylsulfatase	Hydrolases	I (day 37), P (day 111)	
		β-glucosidase	Hydrolases	P (day 37), I (days 74 and 111)	
Permethrin, cypermethrin	Insecticide	Dehydrogenase	Oxidoreductases	I	Borowik et al., 2023
		Phosphatase	Hydrolases	I	
		Catalase	Oxidoreductases	I	
		Urease	Hydrolases	I	
		Arylsulfatase	Hydrolases	I	
		β-glucosidase	Hydrolases	I	
Triazophos, spinosad, cypermethrin	Insecticide	Catalase	Oxidoreductases	P	Sarkar et al., 2022
Tribenuron methyl (TBM), tebuconazole (TEB)	Herbicide	Superoxide dismutase	Oxidoreductases	P (at 0.05 lethal concentration (LC)P), I (at 0.01 LC)	Chen et al., 2018
				TEB enhanced SOD activity at moderate and high concentrations	
				Combined exposure also escalated the activity as the pesticide mixture facilitated interaction between TBM and TEB	

Continued

Table 3.1. Continued.

Pesticide	Category	Enzyme	Enzyme class	Effect of pesticide on enzyme activity[a]	Reference
		Catalase	Oxidoreductases	P (TEB + TBM treatment); impact of TEB was prolonged. Combined exposure also enhanced the activity	
Atrazine, pendimethalin	Herbicide	Cellulase	Hydrolases	I (TBM and combined treatment of TBM + TEB)	Kumari *et al.*, 2020
Tembotrione, topramezone	Herbicide	L-asparaginase	Hydrolases	P	Kumari *et al.*, 2020
	Herbicide	L-asparaginase	Hydrolases	P (up to 45 days days after treatment (DAT) with tembotrione)); P (up to 15 DAT with topramezone and reduced thereafter	
Chlorpyrifos, cypermethrin	Insecticide	Dehydrogenase	Oxidoreductases	I (cypermethrin at low concentration); P (chlorpyrifos at low concentration)	Nistala and Kumar, 2023
Mesotrione	Herbicide	β-glucosidase	Hydrolases	—	Du *et al.*, 2018
		Acid phosphatase	Hydrolases	P	
2,4-D + glyphosate	Herbicide	β-glucosidase	Hydrolases	P	Tyler, 2022
		Cellobiohydrolase	Hydrolases	P	
		Phosphatase	Hydrolases	P	
Bensulfuronm ethyl, pretilachlor 60, azimsulfuron, pyrazosulfuron ethyl, oxyfluorfen	Herbicide	Dehydrogenase	Oxidoreductases	P (after initial decline)	Arya and Ameena, 2016
		Urease	Hydrolases	No effect	
Mesotrione	Herbicide	Urease	Hydrolases	—	Sun *et al.*, 2020
		Catalase	Oxidoreductase	P (at 20 mg/kg); I (at 100 mg/kg)	
		Invertase	Hydrolases	P (at 20 mg/kg) and decreased thereafter	
Imidacloprid	Insecticide	Dehydrogenase	Oxidoreductase	—	Mahapatra *et al.*, 2017
		Alkaline phosphatase	Hydrolases	—	
		β-glycosidase	Hydrolases	—	
		Fluorescein diacetate hydrolase (FDA)	Hydrolases	P	
		Acid phosphatase	Hydrolases	I (attenuated)	
		Urease	Hydrolases	P	
Oxyfluorfen, chlorpyrifos	Insecticide	Dehydrogenase	Oxidoreductase	—	Franco-Andreu *et al.*, 2016
		Urease	Hydrolases	I (initial 90 days)	
		β-glycosidase	Hydrolases	I (oxyfluorfen)	
		Phosphatase	Hydrolases		

Pesticide	Type	Enzyme	Enzyme class	Effect	Reference
Pretilachlor	Herbicide	FDA	Hydrolases	—	Sahoo et al., 2017
		β-glucosidase	Hydrolases	Non-significant difference in enzyme activity	
		Urease	Hydrolases	—	
		Phosphatase	Hydrolases	No significant difference was observed	
		Dehydrogenases	Oxidoreductase	—	
Falcon 460 EC	Fungicides	Dehydrogenases	Oxidoreductase	—	Baćmaga et al., 2016
Thiamethoxam	Insecticide	Dehydrogenases	Oxidoreductase	I (with increasing concentration of insecticide)	Jyot et al., 2015
		Phosphatase	Hydrolases	—	
		Urease	Hydrolases	I (with increasing concentration of insecticide)	
Alachlor	Herbicide	Dehydrogenase	Oxidoreductase	No effect (for the initial 16 days), then increased by 26.1% and was inhibited at 250 mg/kg or more	Tejada et al., 2017
		Urease	Hydrolases	No effect	
		Alkaline phosphatase	Hydrolases	Similar to dehydrogenase	
		β-glucosidase	Hydrolases	P	
Prothioconazole	Fungicide	Dehydrogenase	Oxidoreductase	—	Zhai et al., 2022
		Catalase	Oxidoreductase	—	
		Urease	Hydrolases	—	

[a] I, inhibited the enzyme activity; P, promoted the enzyme activity.
[b] LC50 (lethal concentration 50) is the quantity of pesticide sufficient to eliminate 50% of the test organisms in a single treatment.

compounds to inorganic ones (Riah *et al.*, 2014; Arora and Sahni, 2016).

Fungicides either inhibit or do not adversely affect phosphatase enzymes (Yan *et al.* 2011). The alkaline phosphatase activity was found to be inhibited by the application of mefenoxam and metalaxyl in the soil, whereas the reverse was true for acid phosphatase activity (Monkiedje and Spiteller, 2002).

Insecticides have an inhibitory effect on the physiological activities of enzymes even when applied as per instruction and proper doses. However, cadusafos (an organophosphate insecticide), when used even ten times the recommended dose, was found not to affect phosphatase activities.

Herbicides may have either stimulatory (imazethapyr and butachlor) or inhibitory impacts on enzyme activity at high doses. In an experiment by Sannino and Gianfreda (2001), the herbicide glyphosate was used, and the impact was observed on 30 different soil samples. It was found that due to the presence of the phosphoric group, glyphosate decreased the activity of phosphate in the range of 5–90%. Overall, pesticides mainly harm phosphatases (Riah *et al.*, 2014).

3.7.4 Urease

Urease catalysis is the hydrolysis of urea into carbon dioxide and ammonia. It originates from plants and is also present in soil microbes and fungi as intracellular or extracellular enzymes (Arora and Sahni, 2016). Applying herbicides and fungicides may have inhibitory, stimulatory or no effect on urease activity. In an experiment performed by Sannino and Gianfreda (2001), it was found that out of 30 soil samples, the activities of urease decreased by 10–80% in six soil samples after the application of glyphosate. However, fungicides like carbendazim and validamycin enhanced urease activity by 70% and 21–30%, respectively (Riah *et al.*, 2014). The enhanced urease activity could disturb the nitrogen balance in the soil and its availability to the plants (Chein *et al.*, 2009).

3.7.5 Nitrogenase

Nitrogenase enzyme is present in diazotrophs, including the free-living *Azotobacter* sp. and legume-associated *Rhizobium* sp. Because of this enzyme, the unavailable atmospheric nitrogen is converted into plant-available ammonia by a process known as biological nitrogen fixation. Because of its life-sustaining role, it is one of the most well-characterized enzymes. Pesticide application affects the efficiency and activity of nitrogenase in nitrogen-fixing bacteria, purple non-sulfur bacteria, methylotrophic bacteria and cyanobacteria (Riah *et al.*, 2014).

The experiments by Niewiadomska and Klama (2005) found that applying carbendazim, imazethapyr and thiram reduced nitrogenase activity. Kanungo *et al.* (1995) found that repeated application of carbofuran significantly stimulated rhizosphere-associated nitrogenase activity (Hussain *et al.*, 2009).

3.7.6 Hydrolase

Hydrolase is important in soil nitrogen, phosphorus, sulfur and carbon cycles. For example, fluorescein diacetate hydrolase is a substrate hydrolysed by many enzymes like protease, lipase and esterase. The availability of pesticides belonging to the imidazoline (imazethapyr) and organochlorine (endosulfan) groups activated fluorescein diacetate hydrolase. The treatment of organophosphate (ethion and chlorpyrifos) at varying concentrations had an identical impact on hydrolase's enzymatic activity (Hussain *et al.*, 2009).

3.8 Impact of Pesticides on Soil Fauna

Soil fauna, including the soil invertebrates and other soil microorganisms, sustain the soil ecosystem, decompose organic matter and recycle nutrients. Based on their size, they can be classified into microfauna, mesofauna and macrofauna. The microfauna is less than 0.2 mm and includes protists and nematodes. In contrast, mesofauna and macrofauna are more than 0.2 and 2 mm in size, respectively, and include organisms like mites, earthworms, beetles, spiders, snails, nematodes, collembola (springtails), enchytraeids (Bünemann *et al.*, 2018; Bart *et al.*, 2019b; Hedĕnec *et al.*, 2022). The diversity and quantity of various soil organisms is an important indicator of soil health. These organisms support the soil and perform various ecosystem services, enabling it to become self-sufficient (Gunstone et al., 2021). Litter consumption by soil fauna such as bacterial, fungal and litter feeders is known across different world biomes such as tropical, temperate, grasslands, etc. (Heděnec et al., 2022).

The agroecosystems increase soil fertility and soil aeration by making soil aggregates. For example, the earthworms have been called soil ecosystem engineers (Coleman, 2015; Hashimi et al., 2020). They improve the physical properties of soil and stimulate its microbial and enzymatic activity (Bart et al., 2019b; Gunstone et al., 2021).

Earthworms also increase the tolerance capacity of microbes to pesticides and provide favourable microhabitats for the microbiota (Bart et al., 2019b). Therefore, these worms stimulate soil health and plant yield, proving to be a boon to farmers. Overall, soil fauna contributes to plant growth and primary productivity and forms a part of the biogeochemical cycles. Being a crucial part of the food web, these organisms are important for the ecosystem's vitality (Bart et al., 2019b).

Pesticides can accumulate from the soils in higher level food chain organisms. They can negatively influence the non-target organisms' number, growth, lifespan and metabolism, or even their genetic makeup (Edwards, 2002). Modern agricultural practices tend to destroy their habitat or make fewer food sources accessible to them (Bart et al., 2019b).

As a result, it has been found that intensive agriculture reduces soil biodiversity by almost 60% (Gunstone et al., 2021). Pesticides significantly impede the survival and proliferation of soil communities. For example, pendimethalin, a commonly used herbicide, reduces the number of soil nematodes and other invertebrates (Bünemann et al., 2018). Fludioxonil, acetochlor and terbuthylazine also reduce the number of nematode species in the soil (Gunstone et al., 2021). Dimethoate leads to a temporary reduction in microarthropods like Collembola. While malathion impacts earthworm reproduction in the short term, butachlor and copper-based fungicides are highly toxic to earthworms (Bünemann et al., 2018). Many pesticides, on application to

the soil to kill pests, also affect non-target organisms – natural predators and pollinators like bees, spiders, beetles, butterflies and moths – leading to an overall loss in biological diversity (Isenring, 2010; Yadav and Devi, 2017). An investigation revealed that carbamates (aldicarb, benomyl, carbofuran, methiocarb), organophosphates (chlorpyrifos, diazinon, dimethoate, fenitrothion), pyrethroids (cyfluthrin, cyhalothrin) and neonicotinoids (imidacloprid, thiamethoxam, clothianidin) are detrimental to insects (Isenring, 2010). Neonicotinoids and cypermethrin have been found to be harmful to bees (Hashimi et al., 2020). Apart from insects, the adverse effects of pesticides are also visible in higher organisms like birds, mammals, fish, amphibians and reptiles (Hashimi et al., 2020). Highly toxic pesticides can cause wildlife poisoning, resulting in a major decline in soil population (Isenring, 2010).

3.9 Challenges and Sustainable Management

Pesticides have increased food production globally, but their impact on the environment is disheartening. As a result, people are now becoming more concerned about shifting to sustainable management of agroecosystems. However, this is a difficult task as implementing it has many challenges.

3.9.1 Lack of correct knowledge

Most farm workers, especially in the developing world, are unaware of how to handle their farmwork safely. A study was conducted in Sierra Leone, where 500 farmers working in rice fields were interviewed. More than 70% of them were not aware of the proper dosages for the application of pesticides, and they had never been trained in their use. Furthermore, they were unaware of various pesticides and could not distinguish between the types, especially liquid ones. Most of them relied on illiterate vendors, using pictures of the target pests on the containers, to explain it to them (Sankoh et al., 2016).

Farmers often apply pesticides in an unsafe manner due to a lack of knowledge of hazards and their associated adverse effects on health and the environment. Risky handling practices have a high potential for human exposure, injuries and illness. Farmers unknowingly buy illegal forms of registered products. The products are unregistered, so their labels have not been checked for clear directions and safety warnings. Moreover, disintegrated and unorganized pesticide legislation has led to inefficient pesticide management systems, which, in turn, allow easy access to highly hazardous pesticides to users and the general public who have little or no knowledge of pesticide hazards (Ngowi *et al.*, 2016) (Table 3.2).

3.9.2 Financial disparity

The shift from conventional agriculture to sustainable methods is more or less difficult for countries based on their financial status. More than half of the population of developing nations is directly or indirectly involved in agriculture and they are far from reaching the agricultural standards of developed countries. The population of developed countries involved in agriculture is less than 2% and performs much better. For instance, Australia, the USA and most European countries have employed various improved tools, methods, integrated systems and techniques, which have improved the quality and yield quantity of agriculture compared with that of the rest of the developing and underdeveloped countries. The use of sophisticated technologies like nanotechnology applications, which track products and nutrient levels to increase productivity without contaminating soils and waters and to protect against several insect pests and microbial diseases, is limited to only some developed European countries (Prasad *et al.*, 2017). These differences demonstrate that advanced technology and sophisticated techniques make the agricultural field ecologically balanced and more productive, which further depends on the economic status of a country.

3.9.3 Government policy and measures

Government policies play a crucial role in agriculture. Although state policies support various crops to be grown with subsidies, there are instances when these crops are not ecologically favourable. Due to the limited range of crops the state supports, farmers are compelled to grow them, even if they are unfavourable. For example, farmers from water-scarce regions of Punjab insisted on growing water-intensive crops as they were supported by the state, instead of more ecologically supported crops such as pulses and millets. This has been the reason behind the water distress in Punjab (Prasad *et al.*, 2017).

The European Union (EU) has defended research and implementation of integrated pest management (IPM) through National Action Plans (European Commission, 2020) based on the idea that pesticide usage can be significantly reduced by developing IPM at a large scale. However, there are challenges in implementing IPM. For the use of IPM in soybean cultivation, the worry of significant yield loss in the absence of insecticides, the refusal to fully adopt eco-friendly approaches and the substantial amount of work required for insect monitoring are some of the challenges hindering the implementation of sustainable management (Jacquet *et al.*, 2022).

3.9.4 Use of biopesticides

Biopesticides are the sole sustainable management system inhibiting pests through natural biochemical processes. However, the adoption of biopesticides is limited by insufficient supply to meet farmers' demand, high costs (especially when produced commercially) and slow effects on crop yields. Additionally, there are challenges in meeting global demand, discrepancies in standard preparation and guidelines, determining the appropriate amount of active ingredients, and susceptibility to various environmental factors (Fenibo *et al.*, 2021).

3.9.5 Use of organic farming

Organic farming efficiently reduces synthetic pesticide use while maintaining soil fertility and nutrient cycles. Organic agriculture has spread in the EU, with a 74% rise in organic farmland over a decade from 2008 to 2018. But, it still accounts for only 8% of all farmland. Although

Table 3.2. Effect of pesticides on agroecosystems.

Affected organisms (Family)	Common name of pesticide	Category	Class	Adverse effects on organisms	Reference
Plants					
Lonicerae japonicae flos (Caprifoliaceae)	Imidacloprid (IMI)	Insecticide	Neonicotinoids	Components of *Lonicerae japonicae flos* decreased across the flowering development at the third green stage (TGS), the second white stage (SWS) and the complete white stage (CWS)	Pan *et al.*, 2021
	Compound flonicamid and acetamiprid (CFA)	Insecticide	Neonicotinoids	Components of *Lonicerae japonicae flos* decreased across the flowering development at the TGS, SWS and CWS	
Oryza sativa (Poaceae)	Chlorpyrifos	Insecticide	Organophosphorothionate	No negative impact	Saengsanga and Phakratok, 2023
Solanum lycopersicum (Solanaceae)	Fluopimomide	Fungicide	Cramer	No negative impact	Jiang *et al.*, 2022
Prunus serotina (Rosaceae)	Chikara 25 WG, Chwastox turbo 340 sl, Logo 310 WG + Mero 842 EC, Mustang Forte 195 SE, Roundup Flex 480 FL	Insecticide	Sulfonylurea chlorophenoxya, benzamides, sulfonylurea, triazolopyrimidine sulfonanilide, pyridine carboxylic acid, organic acid	Reduced the size of pollen grains	Wrońska–Pilarek *et al.*, 2023
Phaseolus vulgaris (Fabaceae)	Pendimethalin	Herbicide	Dinitroaniline	44% inhibition of nitrogen fixation rate	Paniagua-López *et al.*, 2023
Medicago sativa (Fabaceae)	Clethodim	Herbicide	Cyclohexanedione, pendimethalin	Cyclohexanedione caused a 30% reduction in nodulation, while pendimethalin inhibited it	Paniagua-López *et al.*, 2023

Continued

Table 3.2. Continued.

Affected organisms (Family)	Common name of pesticide	Category	Class	Adverse effects on organisms	Reference
Solanum lycopersicum (Solanaceae)	Diazinon (DIZN), IMI and mancozeb (MNZB)	Insecticide; insecticide and fungicide	Organophosphate, neonicotinoids and ethylene bisdithiocarbamate	200 µg/ml mancozeb alleviated the time required for sprouting of seeds, amount of chlorophyll a and b, carotenoid and lycopene, flowering and fruit development and root biomass. 200 µg/ml diazinon caused a dose-dependent reduction in root cells which can potentially germinate, and an overall soluble quantity of sugar and protein. Increased reactive oxygen species (ROS) and thiobarbituric acid amounts caused membrane decay and cell death	Hatamleh *et al.*, 2022
Brassica rapa (Brassicaceae)	Dinotefurant	Insecticide	Neonicotinoids	It caused oxidative stress and upregulated proline, betaine, spermidine, phenylalanine and some phenolic acids and intermediates in the tricarboxylic acid (TCA) cycle. It inhibited the release of glucosinolates and induced the accumulation of lactic acid and 3-phenyllactic acid	Li *et al.*, 2021
Solanum tuberosum (Solanaceae)	Dithianon, difenoconazole, fenazaquin, pyridaben and tolfenpyrad	Insecticide	Naphthodithiin, dioxolanes, quinazoline, pyridazinone and pyrazole	Inhibited mitochondrial respiration rate and rate of oxygen consumption during succinate-supported respiration. Dithianon had no adverse effect	Gureev *et al.*, 2022
Lens culinaris (Fabaceae)	Imazethapyr	Herbicide	Imidazole compound	Hindered plant development and biogenesis of amino acids with branched side chains and hampered source–sink relationship	Shivani Grewal *et al.*, 2023

Species (Family)	Pesticide	Type	Chemical class	Effect	Reference
Brassica napus (Brassicaceae)	Clopyralid, benazolin and clethodim	Insecticide	Organophosphate	Inhibited growth, repressed photosynthetic efficiency, amount of chlorophyll and fluorescence value, reduced catalytic activity of enzymes, increased rate of peroxide and hydroperoxide formation from lipids, enhanced ROS production leading to degradation of chloroplastic and mitochondrial structure in mesophyll tissue of leaves	(Hong et al., 2023)
Worms					
Lumbricus terrestris (Lumbricidae)	Glyphosate-based herbicide	Herbicide	Glycine derivative	Increased earthworm activity	Brandmaier et al., 2023
Eisenia fetida (Lumbricidae)	Diuron	Herbicide	Benzenoids	Induced slight oxidative stress and DNA damage	Wang et al., 2023
Eisenia fetida (Lumbricidae)	Clothianidin	Insecticide	Neonicotinoids	Decreased body weight, cocoon number, cocoon weight and hatchling production	Chowdhary et al., 2023
Eudrilus eugeniae (Eudrilidae)	Quinalphos and organophosphate insecticide (QOI)]	Insecticide	Organothiophosphate	No negative effect. Morphological changes include curing body parts, compression of the clitellar region, disintegration and inflammation of body parts and blood loss	Sujeeth et al., 2023
Bursaphelenchus xylophilus (Para-sitaphelenchidae)	Abamectin	Insecticide	Macrocyclic lactone	Paralysis on treatment with abamectin at 0.06 µg/ml or more. Increased mortality and inhibited reproduction	Lee et al., 2023
Aporrectodea caliginosa (Lumbricidae)	Atrazine and metribuzin	Herbicide	Triazine	Toxic to earthworms	Fouad et al., 2023
Lampito mauritii (Megascolecidae)	Monocrotophos	Insecticide	Organophosphate	It showed synthesis of vacuoles within cells, nuclear disintegration, blocked blood sinuses, reduced proliferation and degradation of monocrotophos	Khan et al., 2021

Continued

Table 3.2. Continued.

Affected organisms (Family)	Common name of pesticide	Category	Class	Adverse effects on organisms	Reference
Eisenia andrei (Lumbricidae)	Imazalil	Fungicide	Imidazoles	It does not cause immediate earthworm death, but death occurs due to contaminated soil. Immunocompetent cellular viability and density were reduced, leading to oxidative stress-induced irreversible cellular damage	Pereira *et al.*, 2020
Eisenia andrei (Lumbricidae)	IMI, thiacloprid, and clothianidin	Insecticide	Neonicotinoids	Affected earthworm growth	Van Loon *et al.*, 2022
Aporrectodea caliginosa (Lumbricidae)	Cuprafor micro® and Swing Gold®	Fungicide	Copper oxychloride and phenylpyrroles	Decreased cocoon production and hatching success	Bart *et al.*, 2019a
Caenorhabditis elegans (Rhabditidae)	Chlordecone	insecticide	Organochlorine	Induces progressive loss of dopamine neurons, linked to locomotor deficits, mild alterations in food perception, and cholinergic and serotoninergic neuronal cells. The treatment promotes phosphorylation of the aggregation-prone protein tau, which indicates a form of Parkinson's disease	Parrales-Macias *et al.*, 2023
Lumbricus terresteris (Lumbricidae)	Valette and Oscar	Fungicide and herbicide	Benzimidazole and sulfonylurea	Reduced weight and no effect on epidermal cells and muscle fibres	Berrouk *et al.*, 2023
Caenorhabditis elegans (Rhabditidae)	Deltamethrin	Insecticide	Pyrethroid	Delayed development and reduced UPRER (endoplasmic reticulum unfolded protein response) activation IRE-1/XBP-1 pathway	Chen *et al.*, 2023
Caenorhabditis elegans (Rhabditidae)	Fluopimomide	Fungicide	Benzimidazole	Decreased pharyngeal pumping and increased level of ROS, which downregulated mitochondrial electron transport chain-associated genes *mev-1* and *isp-1*. Reduced activity of succinate dehydrogenase and superoxide dismutase reduced oxygen consumption	Liu *et al.*, 2023

Other organisms

Organism	Type	Pesticide	Class	Effect	Reference
Lexandrium catenella (Ostreopsidaceae)	Herbicide	Terbutryn	Phenylethanolamines	Reduced the growth rate by inhibiting photosynthesis, which may affect the photosystem II repair by repressing the D1 protein gene. No effect on the maximum relative electron transport rate of photosynthesis and saxitoxin production	Xing et al., 2023
Mamestra brassicae (Noctuidae)	Insecticide	Chlorantraniliprole and indoxacarb	Pyrazolylpyridine and organochlorine	Prolonged larval and pupal stage duration and decreased female pupal weight while only indoxacarb affected male pupa	Moustafa et al., 2023
Monopterus albus (Synbranchidae)	Herbicide	Metamifop	Aryloxyphenoxypropionate	Increased vitellogenin (VTG) levels in the liver and plasma; did not affect plasma sex hormone levels. Elevated CYP19A1b manifestation was observed along with suppression of action of CYP17 and YP19A1a, CYP17, FSHR, LHCGR, hsd11b2 and 3β-HSD	Zhang et al., 2023
Clarias batrachus (Clariidae)	Herbicide	Pretilachlor	Chloroacetanilide	Increased plasma 17β-estradiol, plasma vitellogenin amount and gonadal aromatase activity in male fish. In female fish, it decreased plasma concentration of testosterone while plasma 17β-estradiol concentrations, plasma vitellogenin concentration and gonadal aromatase activity remained unaffected	Soni and Verma, 2020
Trogoderma granarium (Dermestidae)	Insecticide	Phosphine	Organophosphate	Induced slow larval development and faster hatching on days 1 and 3, and the hatching rate reduced on days 4 and 6	Lampiri and Athanassiou, 2021

Continued

Table 3.2. Continued.

Affected organisms (Family)	Common name of pesticide	Category	Class	Adverse effects on organisms	Reference
Bacterial community	Trifloxystrobin	Fungicide	β-methoxy acryl ester	Inhibited both nitrification and denitrification and diminished carbon sequestration ability	Xiao et al., 2023
Aulosira fertilissima (Fortieaceae)	Butachlor	Herbicide	Acetanilide	Organic carbon content and dehydrogenase activity reduced	Arora et al., 2019
Anabaena sp. and Nostoc muscorum (Nostocaceae)	Pretilachlor	Herbicide	Chloroacetanilide	Growth was limited by light. Reduction in the amount of photosynthetic pigment	Kumar et al., 2018
Mesorhizobium sp. (Rhizobiaceae)	Metribuzin and glyphosate	Herbicide	Triazines and glycine derivative	Inhibitory to siderophore zone. Glyphosate greatly hampered the biosynthesis of salicylic acid (SA) and 2,3-dihydroxy benzoic acid (DHBA)	Ahemad and Khan, 2012
	IMI and thiamethoxam	Insecticides	Neonicotinoids	Inhibitory to siderophore zone. Thiamethoxam exhibited the most harmful effects on biosynthesis of SA and DHBA	
	Hexaconazole, metalaxyl and kitazin	Fungicides	Triazole, acyl alanine and thiophosphate	Hexaconazole reduced the siderophore zone. Hexaconazole and thiamethoxam displayed damaging effects on the formation of SA and DHBA based on treatment dose	
Apis mellifera (Apidae)	Carbendazim	Fungicides	Benzimidazoles	Reduced pollen consumption causes nutritional imbalance and suppresses their immunity, making them vulnerable to pathogens	Wang et al., 2022
Daphnia magna (Daphniidae)	Profenofos	Insecticide	Organophosphate	Decreased life expectancy and negatively affected development and reproduction. Regulated the expression of growth-related cell cycle protein and increased transcription of developmental genes	Li et al., 2023a

Organism	Compound	Class	Type	Effect	Reference
Penicillium digitatum (Trichocomaceae)	Propiconazole	Fungicides	Conazole	Reduced conidia virulence, impaired conidia membrane integrity, stimulated ROS generation and increased expression of the CYP51A gene were observed	Zhang et al., 2021a
Rhizophagus irregularis (Glomeraceae)	Albendazole	Antihelmintics	Benzimidazole	Impaired development of arbuscules and their functionality and the symbiotic organelle of arbuscular mycorrhizal fungi (AMF). Reduced the transcription of genes that catalyse the formation of arbuscules, phosphorus and nitrogen assimilation, obstructing the colonization capacity and function	Gkimprixi et al., 2023
Apis mellifera (Apidae)	IMI	Insecticide	Neonicotinoids	Delayed growth and development by disturbing the regulation of moult development. The damaged gut constrains the metabolic processes and metabolism, stimulating oxidative stress and programmed cell death	Li et al., 2023b

certified organic products fetch higher prices and require lower input use, they still yield less than conventional methods. This shortcoming is, however, equalized at the farm scale by the premiums paid for environmental benefits in agriculture in some countries. Nevertheless, some technical problems, particularly the ones concerned with weed management, remain unsolved (Jacquet *et al.*, 2022).

3.9.6 Use of advanced technology

The use of advanced technology can prove to be extremely useful in the field of agriculture. Internet services can teach farmers about the careful and correct use of many pesticides and fertilizers, including even those in the remotest areas. Modern equipment like drones can greatly help to spray pesticides in a controlled manner. The main challenges faced in this case include a disrupted market, poor connectivity and coverage, investment and lack of innovative technology and skilled manpower.

3.9.7 Incentives

Farmer incentives are one of the core issues in an attempt to move towards sustainable agriculture. Farmers grow crops to cater to their family needs and to profit from the competitive market economy so they usually prioritize profits over sustainability in their agricultural approaches. If policy makers influenced the government to incentivize farmers to adopt agroecological environment-friendly approaches, the issue of pesticide abuse could be addressed. Some countries, like Australia, Canada, Japan, Norway, Switzerland, the USA and the EU, have initiated 'green payments', which are payments to farmers who adopt sustainable or environmentally benign farming practices (Edlinger *et al.*, 2022). Such incentives should exist across the globe.

3.10 Future Perspectives

Given the significant risks associated with using chemical pesticides in agriculture, it is essential to find alternative suitable agroecological approaches to boost crop productivity at each local level of agroecosystems. This can be done through community involvement. This would help achieve food safety and security in a more eco-friendly and sustainable way. Practising organic farming and IPM and promoting biopesticides over synthetic agrochemicals can help us achieve this (Carvalho, 2017). IPM aims at farm sustainability and improving human health and farm profits (Baker *et al.*, 2020). Recent technological advancements such as precision agriculture and biotechnological approaches, including the CRISPR/Cas system, are a way forward (Ahmad *et al.*, 2020). We may face certain challenges in adopting IPM and other newer innovative measures in the agricultural system, but our own will would surely help us implement and achieve the goal of pesticide-free agriculture.

References

Ahemad, M. and Khan, M.S. (2012) Effects of pesticides on plant growth promoting traits of Mesorhizobium strain MRC4. *Journal of the Saudi Society of Agricultural Sciences* 11, 63–71.

Ahmad, S., Wei, X., Sheng, Z., Hu, P. and Tang, S. (2020) CRISPR/cas9 for development of disease resistance in plants: recent progress, limitations and future prospects. *Briefings in Functional Genomics* 19, 26–39.

Al-Ani, M.A.M., Hmoshi, R.M., Kanaan, I.A. and Thanoon, A.A. (2019) Effect of pesticides on soil microorganisms. *Journal of Physics* 1294, 072007.

Alengebawy, A., Abdelkhalek, S.T., Qureshi, S.R. and Wang, M.-Q. (2021) Heavy metals and pesticides toxicity in agricultural soil and plants: ecological risks and human health implications. *Toxics* 9, 42.

Altieri, M.A. and Nicholls, C.I. (2003) Soil fertility management and insect pests: harmonizing soil and plant health in agroecosystems. *Soil and Tillage Research* 72, 203–211.

Aly, M., El-Aswed, A., Alsahaty, S. and Basyony, A. (2023) Side-effect of soil fumigation and soil drench by some methyl bromide alternatives on three soil enzyme activities. *Alexandria Science Exchange Journal* 44, 1–13.

Arora, S. and Sahni, D. (2016) Pesticides effect on soil microbial ecology and enzyme activity – an overview. *Journal of Applied and Natural Science* 8, 1126–1132.

Arora, S., Arora, S., Sahni, D., Sehgal, M., Srivastava, D.S. *et al.* (2019) Pesticides use and its effect on soil bacteria and fungal populations, microbial biomass carbon and enzymatic activity. *Current Science* 116, 643.

Arya, S.R. and Ameena, M. (2016) Herbicides effect on soil enzyme dynamics in direct-seeded rice. *Indian Journal of Weed Science* 48, 316.

Baćmaga, M., Wyszkowska, J. and Kucharski, J. (2016) The effect of the falcon 460 EC fungicide on soil microbial communities, enzyme activities and plant growth. *Ecotoxicology* 25, 1575–1587.

Baćmaga, M., Wyszkowska, J. and Kucharski, J. (2018) The influence of chlorothalonil on the activity of soil microorganisms and enzymes. *Ecotoxicology* 27, 1188–1202.

Baćmaga, M., Wyszkowska, J., Kucharski, J., Borowik, A. and Kaczyński, P. (2022) Possibilities of restoring homeostasis of soil exposed to terbuthylazine by its supplementation with HumiAgra preparation. *Applied Soil Ecology* 178, 104582.

Baker, B.P., Green, T.A. and Loker, A.J. (2020) Biological control and integrated pest management in organic and conventional systems. *Biological Control* 140, 104095.

Banerjee, S. and van der Heijden, M.G.A. (2023) Soil microbiomes and one health. *Nature Reviews Microbiology* 21, 6–20.

Bart, S., Barraud, A., Amossé, J., Péry, A.R.R., Mougin, C. *et al.* (2019a) Effects of two common fungicides on the reproduction of Aporrectodea caliginosa in natural soil. *Ecotoxicology and Environmental Safety* 181, 518–524.

Bart, S., Pelosi, C., Barraud, A., Péry, A.R.R., Cheviron, N. *et al.* (2019b) Earthworms mitigate pesticide effects on soil microbial activities. *Frontiers in Microbiology* 10, 1535.

Berrouk, H., Necib, A., Meraiahia, A., Boumaza, N.-E. and Hmaidia, K. (2023) Individual and combined effects of a fungicide (Valette) and an herbicide (Oscar) on the growth and mortality of earthworm Lumbricus terresteris (Linnaeus, 1758). *Animal Research International* 20, 4694–4704.

Blair, A., Ritz, B., Wesseling, C. and Beane Freeman, L. (2015) Pesticides and human health. *Occupational and Environmental Medicine* 72, 81–82.

Borowik, A., Wyszkowska, J., Zaborowska, M. and Kucharski, J. (2023) The impact of permethrin and cypermethrin on plants, soil enzyme activity, and microbial communities. *International Journal of Molecular Sciences* 24, 2892.

Boteva, S.B., Kenarova, A.E., Petkova, M.R., Georgieva, S.S., Chanev, C.D. *et al.* (2022) Soil enzyme activities after application of fungicide QuadrisR at increasing concentration rates. *Plant, Soil and Environment* 68, 382–392.

Brandmaier, V., Altmanninger, A., Leisch, F., Gruber, E., Takács, E. *et al.* (2023) Glyphosate-based herbicide formulations with greater impact on earthworms and water infiltration than pure glyphosate. *Soil Systems* 7, 66.

Bünemann, E.K., Bongiorno, G., Bai, Z., Creamer, R.E., De Deyn, G. *et al.* (2018) Soil quality – a critical review. *Soil Biology and Biochemistry* 120, 105–125.

Caraba, M.N., Roman, D.L., Caraba, I.V. and Isvoran, A. (2023) Assessment of the effects of the herbicide aclonifen and its soil metabolites on soil and aquatic environments. *Agriculture* 13, 1226.

Carbonell-Barrachina, A.A., Burló-Carbonell, F. and Mataix-Beneyto, J. (1997) Effect of sodium arsenite and sodium chloride on bean plant nutrition (macronutrients). *Journal of Plant Nutrition* 20, 1617–1633.

Carvalho, F.P. (2017) Pesticides, environment, and food safety. *Food and Energy Security* 6, 48–60.

Chauhan, S., Yadav, U., Bano, N., Kumar, S., Fatima, T. *et al.* (2023) Carbendazim modulates the metabolically active bacterial populations in soil and rhizosphere. *Current Microbiology* 80, 280.

Chein, S.H., Prochnow, L.I. and Cantarella, H. (2009) Recent developments of fertilizer production and use to improve nutrient efficiency and minimize environmental impacts. In: Sparks, D.L. (ed.) *Advances in Agronomy*, Vol. 102. Elsevier, pp. 267–322.

Chen, C., Deng, Y., Liu, L., Zou, Z., Jin, C. *et al.* (2023) High-dose deltamethrin induces developmental toxicity in Caenorhabditis elegans via IRE-1. *Molecules* 28, 6303.

Chen, J., Saleem, M., Wang, C., Liang, W. and Zhang, Q. (2018) Individual and combined effects of herbicide tribenuron-methyl and fungicide tebuconazole on soil earthworm Eisenia fetida. *Scientific Reports* 8, 2967.

Chowdhary, A.B., Singh, J., Quadar, J., Singh, S., Dutta, R. *et al.* (2023) Earthworm's show tolerance and avoidance response to pesticide clothianidin: effect on antioxidant enzymes. *International Journal of Environmental Science and Technology* 20, 4245–4254.

Coleman, D. (2015) Soil fauna: occurrence, biodiversity, and roles in ecosystem function. In: Paul, E.A. (ed.) *Soil Microbiology, Ecology, and Biochemistry.* Academic Press, pp. 111–149.

Damalas, C.A. and Eleftherohorinos, I.G. (2011) Pesticide exposure, safety issues, and risk assessment indicators. *International Journal of Environmental Research and Public Health* 8, 1402–1419.

Dick, W.A., Cheng, L. and Wang, P. (2000) Soil acid and alkaline phosphatase activity as pH adjustment indicators. *Soil Biology and Biochemistry* 32, 1915–1919.

Diğrak, M. and Kazanici, F. (2001) Effect of some organophosphorus insecticides on soil microorganisms. *Turkish Journal of Biology* 25, 6.

Du, Z., Zhu, Y., Zhu, L., Zhang, J., Li, B. *et al.* (2018) Effects of the herbicide mesotrione on soil enzyme activity and microbial communities. *Ecotoxicology and Environmental Safety* 164, 571–578.

Edlinger, A., Garland, G., Hartman, K., Banerjee, S., Degrune, F. *et al.* (2022) Agricultural management and pesticide use reduce the functioning of beneficial plant symbionts. *Nature Ecology and Evolution* 6, 1145–1154.

Edwards, C.A. (2002) Assessing the effects of environmental pollutants on soil organisms, communities, processes and ecosystems. *European Journal of Soil Biology* 38, 225–231.

European Commission (2020) *Report on the Experience Gained by Member States on the Implementation of National Targets Established in their National Action Plans and on Progress in the Implementation of Directive 2009/128/EC on the Sustainable Use of Pesticides.* European Commission, Brussels, Belgium.

Fahad, S., Saud, S., Akhter, A., Bajwa, A.A., Hassan, S. *et al.* (2021) Bio-based integrated pest management in rice: an agro-ecosystems friendly approach for agricultural sustainability. *Journal of the Saudi Society of Agricultural Sciences* 20, 94–102.

FAOSTAT (2022) *Statistical Yearbook-World Food and Agriculture.* Available at: https://www.fao.org/3/cc2211en/cc2211en.pdf (last accessed October 2024).

Fenibo, E.O., Ijoma, G.N. and Matambo, T. (2021) Biopesticides in sustainable agriculture: a critical sustainable development driver governed by green chemistry principles. *Frontiers in Sustainable Food Systems* 5, 619058.

Filimon, M.N., Roman, D.L., Caraba, I.V. and Isvoran, A. (2021) Assessment of the effect of application of the herbicide S-metolachlor on the activity of some enzymes found in soil. *Agriculture* 11, 469.

Fouad, M.R., Shamsan, A.Q.S. and Abdel-Raheem, S.A.A. (2023) Toxicity of atrazine and metribuzin herbicides on earthworms (*Aporrectodea caliginosa*) by filter paper contact and soil mixing techniques. *Current Chemistry Letters* 12, 185–192.

Franco-Andreu, L., Gómez, I., Parrado, J., García, C., Hernández, T. *et al.* (2016) Behavior of two pesticides in a soil subjected to severe drought. Effects on soil biology. *Applied Soil Ecology* 105, 17–24.

Geiger, F., Bengtsson, J., Berendse, F., Weisser, W.W., Emmerson, M. *et al.* (2010) Persistent negative effects of pesticides on biodiversity and biological control potential on European farmland. *Basic and Applied Ecology* 11, 97–105.

Gkimprixi, E., Lagos, S., Nikolaou, C.N., Karpouzas, D.G. and Tsikou, D. (2023) Veterinary drug albendazole inhibits root colonization and symbiotic function of the arbuscular mycorrhizal fungus *Rhizophagus irregularis. FEMS Microbiology Ecology* 99, fiad048.

Gonzalez-Lopez, J., Martinez-Toledo, M.V., Rodelas, B. and Salmeron, V. (1992) Studies on the effects of the insecticides phorate and malathion on soil microorganisms. *Environmental Toxicology and Chemistry* 12, 1209.

Goswami, M., Pati, U.K., Chowdhury, A. and Mukhopadhyay, A. (2013) Studies on the effect of cypermethrin on soil microbial biomass and its activity in an alluvial soil. *Semantics Scholar.* Available at: https://www.semanticscholar.org/paper/Studies-on-the-effect-of-cypermethrin-on-soil-and-Goswami-Pati/a6b8587112c83ac54b6f0be5d58160987fd4404b (last accessed October 2024).

Gunstone, T., Cornelisse, T., Klein, K., Dubey, A. and Donley, N. (2021) Pesticides and soil invertebrates: a hazard assessment. *Frontiers in Environmental Science* 9, 643847.

Gureev, A.P., Sitnikov, V.V., Pogorelov, D.I., Vitkalova, I.Y., Igamberdiev, A.U. *et al.* (2022) The effect of pesticides on the NADH-supported mitochondrial respiration of permeabilized potato mitochondria. *Pesticide Biochemistry and Physiology* 183, 105056.

Hashimi, M.H., Hashimi, R. and Ryan, Q. (2020) Toxic effects of pesticides on humans, plants, animals, pollinators and beneficial organisms. *Asian Plant Research Journal* 5, 37–47.

Hatamleh, A.A., Danish, M., Al-Dosary, M.A., El-Zaidy, M. and Ali, S. (2022) Physiological and oxidative stress responses of *Solanum lycopersicum* (L.) (tomato) when exposed to different chemical pesticides. *RSC Advances* 12, 7237–7252.

Heděnec, P., Jiménez, J.J., Moradi, J., Domene, X., Hackenberger, D. *et al.* (2022) Global distribution of soil fauna functional groups and their estimated litter consumption across biomes. *Scientific Reports* 12, 17362.

Homayoonzadeh, M., Moeini, P., Talebi, K., Roessner, U. and Hosseininaveh, V. (2020) Antioxidant system status of cucumber plants under pesticides treatment. *Acta Physiologiae Plantarum* 42, 161.

Hong, Z., Faqir, Y., Kalhoro, G.M., Kalhoro, M.T., Kaleri, A.R. *et al.* (2023) Hazardous impacts of clopyralid, benazolin, and clethodim on seed germination and seedling of oil rapeseed (*Brassica napus*). *Polish Journal of Environmental Studies* 32, 2623–2636.

Huang, H. and Xiong, Z.-T. (2009) Toxic effects of cadmium, acetochlor and bensulfuron-methyl on nitrogen metabolism and plant growth in rice seedlings. *Pesticide Biochemistry and Physiology* 94, 64–67.

Huang, Y., Xiao, L., Li, F., Xiao, M., Lin, D. *et al.* (2018) Microbial degradation of pesticide residues and an emphasis on the degradation of cypermethrin and 3-phenoxy benzoic acid: a review. *Molecules* 23, 2313.

Hussain, S., Siddique, T., Saleem, M., Arshad, M. and Khalid, A. (2009) Chapter 5 Impact of pesticides on soil microbial diversity, enzymes, and biochemical reactions. *Advances in Agronomy* 102, 159–200.

Isenring, R. (2010) *Pesticides and the Loss of Biodiversity.* Pesticide Action Network Europe, London.

Jacquet, F., Jeuffroy, M.-H., Jouan, J., Le Cadre, E., Litrico, I. *et al.* (2022) Pesticide-free agriculture as a new paradigm for research. *Agronomy for Sustainable Development* 42, 8.

Jiang, L., Wang, H., Zong, X., Wang, X. and Wu, C. (2022) Effects of soil treated fungicide fluopimomide on tomato (*Solanum lycopersicum* L.) disease control and plant growth. *Open Life Sciences* 17, 800–810.

Johnsen, K., Jacobsen, C., Torsvik, V. and Sørensen, J. (2001) Pesticide effects on bacterial diversity in agricultural soils – a review. *Biology and Fertility of Soils* 33, 443–453.

Jyot, G., Mandal, K. and Singh, B. (2015) Effect of dehydrogenase, phosphatase and urease activity in cotton soil after applying thiamethoxam as seed treatment. *Environmental Monitoring and Assessment* 187, 298.

Kanungo, P.K., Adhya, T.K. and Rao, V.R. (1995) Influence of repeated applications of carbofuran on nitrogenase activity and nitrogen-fixing bacteria associated with rhizosphere of tropical rice. *Chemosphere* 31, 3249–3257.

Karpouzas, D.G., Karanasios, E., Giannakou, I.O., Georgiadou, A. and Menkissoglu-Spiroudi, U. (2005) The effect of soil fumigants methyl bromide and metham sodium on the microbial degradation of the nematicide Cadusafos. *Soil Biology and Biochemistry* 37, 541–550.

Khan, N., Ray, R.L., Sargani, G.R., Ihtisham, M., Khayyam, M. *et al.* (2021) Current progress and future prospects of agriculture technology: gateway to sustainable agriculture. *Sustainability* 13, 4883.

Khmelevtsova, L., Konstantinova, E., Karchava, S., Klimova, M., Azhogina, T. *et al.* (2023) Influence of pesticides and mineral fertilizers on the bacterial community of arable soils under pea and chickpea crops. *Agronomy* 13, 750.

Kumar, J., Patel, A., Tiwari, S., Tiwari, S., Srivastava, P.K. *et al.* (2018) Pretilachlor toxicity is decided by discrete photo-acclimatizing conditions: physiological and biochemical evidence from Anabaena sp. and Nostoc muscorum. *Ecotoxicology and Environmental Safety* 156, 344–353.

Kumar, U., Berliner, J., Adak, T., Rath, P.C., Dey, A. *et al.* (2017) Non-target effect of continuous application of chlorpyrifos on soil microbes, nematodes and its persistence under sub-humid tropical rice-rice cropping system. *Ecotoxicology and Environmental Safety* 135, 225–235.

Kumari, J.A., Rao, P. and Madhavi, M. (2020) Effects of herbicides on soil enzyme L-asparaginase activity. *Bangladesh Journal of Botany* 49, 1177–1183.

Lampiri, E. and Athanassiou, C.G. (2021) Insecticidal effect of phosphine on eggs of the khapra beetle (Coleoptera: Dermestidae). *Journal of Economic Entomology* 114, 1389–1400.

Lastdrager, J., Hanson, J. and Smeekens, S. (2014) Sugar signals and the control of plant growth and development. *Journal of Experimental Botany* 65, 799–807.

Lee, J.-W., Mwamula, A.O., Choi, J.-H., Lee, H.-W., Lee, Y.S. *et al.* (2023) The potency of abamectin formulations against the pine wood nematode, *Bursaphelenchus xylophilus. Plant Pathology Journal* 39, 290–302.

Li, J., Jin, Q., Li, S., Wang, Y., Yuan, S. *et al.* (2023a) Effects of profenofos on the growth, reproduction, behavior, and gene transcription of daphnia magna. *Environmental Science and Pollution Research* 30, 74928–74938.

Li, X., Zhang, M., Li, Y., Yu, X. and Nie, J. (2021) Effect of neonicotinoid dinotefuran on root exudates of *Brassica rapa* var. *chinensis. Chemosphere* 266, 129020.

Li, Z., Wang, Y., Qin, Q., Chen, L., Dang, X. *et al.* (2023b) Imidacloprid disrupts larval molting regulation and nutrient energy metabolism, causing developmental delay in honey bee *Apis mellifera*. *eLife* 12, R88772.

Liu, H., Fu, G., Li, W., Liu, B., Ji, X. *et al.* (2023) Oxidative stress and mitochondrial damage induced by a novel pesticide fluopimomide in *Caenorhabditis elegans*. *Environmental Science and Pollution Research* 30, 91794–91802.

Lo, C.-C. (2010) Effect of pesticides on soil microbial community. *Journal of Environmental Science and Health Part B* 45, 348–359.

Lykogianni, M., Bempelou, E., Karamaouna, F. and Aliferis, K.A. (2021) Do pesticides promote or hinder sustainability in agriculture? The challenge of sustainable use of pesticides in modern agriculture. *Science of the Total Environment* 795, 148625.

Mahapatra, B., Adak, T., Patil, N.K.B., Pandi G, G.P., Gowda, G.B. *et al.* (2017) Imidacloprid application changes microbial dynamics and enzymes in rice soil. *Ecotoxicology and Environmental Safety* 144, 123–130.

Marioara Nicoleta, F., Voia, O., Popescu, R., Dumitrescu, G., Ciochina, L. *et al.* (2015) The effect of some insecticides on soil microorganisms based on enzymatic and bacteriological analyses. *Romanian Biotechnological Letters* 20, 10439–10447.

Martinez-Toledo, M.V., Salmeron, V. and Gonzalez-Lopez, J. (1992) Effect of the insecticides methylpyrimifos and chlorpyrifos on soil microflora in an agricultural loam. *Plant and Soil* 147, 25–30.

Meena, R., Kumar, S., Datta, R., Lal, R., Vijayakumar, V. *et al.* (2020) Impact of agrochemicals on soil microbiota and management: a review. *Land* 9, 34.

Meghana, D., Bharathi, C. and Rangaswamy, V. (2023) Interaction between selected pesticides and urease enzyme activity in ground nut (*Arachis hypogaea* L.) cultivated red soil and black soil. *International Journal of Current Science* 13, 981–988.

Mishra, C.S.K., Samal, S. and Samal, R.R. (2022) Evaluating earthworms as candidates for remediating pesticide contaminated agricultural soil: a review. *Frontiers in Environmental Science* 10, 924480.

Monkiedje, A. and Spiteller, M. (2002) Effects of the phenylamide fungicides, mefenoxam and metalaxyl, on the microbiological properties of a sandy loam and a sandy clay soil. *Biology and Fertility of Soils* 35, 393–398.

Moustafa, M.A.M., Fouad, E.A., Ibrahim, E., Erdei, A.L., Kárpáti, Z. *et al.* (2023) The comparative toxicity, biochemical and physiological impacts of chlorantraniliprole and indoxacarb on *Mamestra brassicae* (Lepidoptera: Noctuidae). *Toxics* 11, 212.

Nannipieri, P., Ascher, J., Ceccherini, M.T., Landi, L., Pietramellara, G. *et al.* (2003) Microbial diversity and soil functions. *European Journal of Soil Science* 54, 655–670.

Ngowi, A., Mrema, E. and Kishinhi, S. (2016) Pesticide health and safety challenges facing informal sector workers: a case of small-scale agricultural workers in Tanzania. *New Solutions* 26, 220–240.

Niewiadomska, A. and Klama, J. (2005) Pesticide side effect on the symbiotic efficiency and nitrogenase activity of Rhizobiaceae bacteria family. *Polish Journal of Microbiology* 54, 43–48.

Nistala, S. and Kumar, A. (2023) Effect of toxicological interaction of chlorpyrifos, cypermethrin, and arsenic on soil dehydrogenase activity in the terrestrial environment. *Ecotoxicology* 32, 606–617.

Pan, H.-Q., Zhou, H., Miao, S., Guo, D.-A., Zhang, X.-L. *et al.* (2021) Plant metabolomics for studying the effect of two insecticides on comprehensive constituents of Lonicerae Japonicae Flos. *Chinese Journal of Natural Medicines* 19, 70–80.

Paniagua-López, M., Jiménez-Pelayo, C., Gómez-Fernández, G.O., Herrera-Cervera, J.A. and López-Gómez, M. (2023) Reduction in the use of some herbicides favors nitrogen fixation efficiency in *Phaseolus vulgaris* and *Medicago sativa*. *Plants* 12, 1608.

Parrales-Macias, V., Michel, P.P., Tourville, A., Raisman-Vozari, R., Haïk, S. *et al.* (2023) The pesticide chlordecone promotes parkinsonism-like neurodegeneration with tau lesions in midbrain cultures and C. elegans worms. *Cells* 12, 1336.

Parween, T., Jan, S., Mahmooduzzafar, S., Fatma, T. and Siddiqui, Z.H. (2016) Selective effect of pesticides on plant – a review. *Critical Reviews in Food Science and Nutrition* 56, 160–179.

Pathak, V.M., Verma, V.K., Rawat, B.S., Kaur, B., Babu, N. *et al.* (2022) Current status of pesticide effects on environment, human health and its eco-friendly management as bioremediation: a comprehensive review. *Frontiers in Microbiology* 13, 962619.

Pereira, P.C.G., Soares, L.O.S., Júnior, S.F.S., Saggioro, E.M. and Correia, F.V. (2020) Sub-lethal effects of the pesticide imazalil on the earthworm *Eisenia andrei*: reproduction, cytotoxicity, and oxidative stress. *Environmental Science and Pollution Research* 27, 33474–33485.

Potera, C. (2007) Agriculture: pesticides disrupt nitrogen fixation. *Environmental Health Perspectives* 115, A579.

Pozo, C., Salmeron, V., Rodelas, B., Martinez-Toledo, M.V. and Gonzalez-Lopez, J. (1994) Effects of the herbicide alachlor on soil microbial activities. *Ecotoxicology* 3, 4–10.

Prasad, R., Bhattacharyya, A. and Nguyen, Q.D. (2017) Nanotechnology in sustainable agriculture: recent developments, challenges, and perspectives. *Frontiers in Microbiology* 8, 1014.

Rasool, N., Reshi, Z.A. and Shah, M.A. (2014) Effect of butachlor (G) on soil enzyme activity. *European Journal of Soil Biology* 61, 94–100.

Rasool, S., Rasool, T. and Gani, K.M. (2022) A review of interactions of pesticides within various interfaces of intrinsic and organic residue amended soil environment. *Chemical Engineering Journal Advances* 11, 100301.

Reuhs, B.L., Geller, D.P., Kim, J.S., Fox, J.E., Kolli, V.S.K. *et al.* (1998) *Sinorhizobium fredii* and *Sinorhizobium meliloti* produce structurally conserved lipopolysaccharides and strain-specific K antigens. *Applied and Environmental Microbiology* 64, 4930–4938.

Riah, W., Laval, K., Laroche-Ajzenberg, E., Mougin, C., Latour, X. *et al.* (2014) Effects of pesticides on soil enzymes: a review. *Environmental Chemistry Letters* 12, 257–273.

Rillig, M.C., van der Heijden, M.G.A., Berdugo, M., Liu, Y.-R., Riedo, J. *et al.* (2023) Increasing the number of stressors reduces soil ecosystem services worldwide. *Nature Climate Change* 13, 478–483.

Roman, D.L., Voiculescu, D.I., Filip, M., Ostafe, V. and Isvoran, A. (2021) Effects of triazole fungicides on soil microbiota and on the activities of enzymes found in soil: a review. *Agriculture* 11, 893.

Saengsanga, T. and Phakratok, N. (2023) Biodegradation of chlorpyrifos by soil bacteria and their effects on growth of rice seedlings under pesticide-contaminated soil. *Plant, Soil and Environment* 69, 210–220.

Sahoo, S., Adak, T., Bagchi, T.B., Kumar, U., Munda, S. *et al.* (2017) Effect of pretilachlor on soil enzyme activities in tropical rice soil. *Bulletin of Environmental Contamination and Toxicology* 98, 439–445.

Sankoh, A.I., Whittle, R., Semple, K.T., Jones, K.C. and Sweetman, A.J. (2016) An assessment of the impacts of pesticide use on the environment and health of rice farmers in Sierra Leone. *Environment International* 94, 458–466.

Sannino, F. and Gianfreda, L. (2001) Pesticide influence on soil enzymatic activities. *Chemosphere* 45, 417–425.

Sarkar, C., Chatterjee, A., Barik, A. and Saha, N.C. (2022) Study on the effects of organophosphate insecticide triazophos, biopesticide spinosad and a pyrethroid insecticide cypermethrin on oxidative stress biomarkers of *Branchiura sowerbyi* (Beddard, 1892). *Nature Environment and Pollution Technology* 21, 787–793.

Savary, S., Willocquet, L., Pethybridge, S.J., Esker, P., McRoberts, N. *et al.* (2019) The global burden of pathogens and pests on major food crops. *Nature Ecology and Evolution* 3, 430–439.

Schmidt, R., Spangl, B., Gruber, E., Takács, E., Mörtl, M. *et al.* (2023) Glyphosate effects on earthworms: active ingredients vs. commercial herbicides at different temperature and soil organic matter levels. *Agrochemicals* 2, 1–16.

Schneider, K., Turrion, M.-B., Grierson, P. and Gallardo, J. (2001) Phosphatase activity, microbial phosphorus, and fine root growth in forest soils in the Sierra de Gata, western central Spain. *Biology and Fertility of Soils* 34, 151–155.

Schutter, O.D. and Vanloqueren, G. (2011) The new green revolution: how twenty-first-century science can feed the world. *Solutions* 2, 33–44.

Serrão, J.E., Plata-Rueda, A., Martínez, L.C. and Zanuncio, J.C. (2022) Side-effects of pesticides on non-target insects in agriculture: a mini-review. *Science of Nature* 109, 17.

Sharma, A., Kumar, V., Thukral, A.K. and Bhardwaj, R. (2017) Responses of plants to pesticide toxicity: an overview. *Planta Daninha* 37, e019184291.

Sharma, A., Kumar, V., Shahzad, B., Tanveer, M., Sidhu, G.P.S. *et al.* (2019) Worldwide pesticide usage and its impacts on ecosystem. *SN Applied Sciences* 1, 1446.

Shivani Grewal, S.K., Gill, R.K., Virk, H.K. and Bhardwaj, R.D. (2023) Effect of herbicide stress on synchronization of carbon and nitrogen metabolism in lentil (*Lens culinaris* Medik.). *Plant Physiology and Biochemistry* 196, 402–414.

Siddiqui, Z.S. and Ahmed, S. (2006) Combined effects of pesticide on growth and nutritive composition of soybean plants. *Pakistan Journal of Botany* 38, 721.

Sim, J.X.F., Drigo, B., Doolette, C.L., Vasileiadis, S., Karpouzas, D.G. *et al.* (2022) Impact of twenty pesticides on soil carbon microbial functions and community composition. *Chemosphere* 307, 135820.

Singh, B.L. and Singh, R. (2023) Effects of pesticides on enzyme activity of dehydrogenase and urease enzymes in soil of Aligarh region (UP) India. *Journal of Survey in Fisheries Sciences* 10, 1038–1045.

Soni, R. and Verma, S.K. (2020) Impact of herbicide pretilachlor on reproductive physiology of walking catfish, *Clarias batrachus* (Linnaeus). *Fish Physiology and Biochemistry* 46, 2065–2072.

Storck, V., Nikolaki, S., Perruchon, C., Chabanis, C., Sacchi, A. *et al.* (2018) Lab to field assessment of the ecotoxicological impact of chlorpyrifos, isoproturon, or tebuconazole on the diversity and composition of the soil bacterial community. *Frontiers in Microbiology* 9, 1412.

Sujeeth, N.K., Aravinth, R., Thandeeswaran, M., Angayarkanni, J., Rajasekar, A. *et al.* (2023) Toxicity analysis and biomarker response of quinalphos organophosphate insecticide (QOI) on eco-friendly exotic *Eudrilus eugeniae* earthworm. *Environmental Monitoring and Assessment* 195, 274.

Sun, Y.B., Wang, L., Xu, Y.M., Liang, X.F. and Zheng, S.N. (2020) Ecoltoxological effect of mesotrione on enzyme activity and microbial community in agicultural soils. *Applied Ecology and Environmental Research* 18, 3525–3541.

Supreeth, M., Chandrashekar, M.A., Sachin, N. and Raju, N.S. (2016) Effect of chlorpyrifos on soil microbial diversity and its biotransformation by Streptomyces sp. HP-11. *3 Biotech* 6, 147.

Tang, F.H.M., Lenzen, M., McBratney, A. and Maggi, F. (2021) Risk of pesticide pollution at the global scale. *Nature Geoscience* 14, 206–210.

Tedersoo, L., Bahram, M. and Zobel, M. (2020) How mycorrhizal associations drive plant population and community biology. *Science* 367, eaba1223.

Tejada, M., Morillo, E., Gómez, I., Madrid, F. and Undabeytia, T. (2017) Effect of controlled release formulations of diuron and alachlor herbicides on the biochemical activity of agricultural soils. *Journal of Hazardous Materials* 322, 334–347.

Tudi, M., Daniel Ruan, H., Wang, L., Lyu, J., Sadler, R. *et al.* (2021) Agriculture development, pesticide application and its impact on the environment. *International Journal of Environmental Research and Public Health* 18, 1112.

Tyler, H.L. (2022) Impact of 2,4-D and glyphosate on soil enzyme activities in a resistant maize cropping system. *Agronomy* 12, 2747.

UN (United Nations) (2017) World population projected to reach 9.8 billion in 2050, and 11.2 billion in 2100. Available at: www.un.org/en/desa/world-population-projected-reach-98-billion-2050-and-112-billion-2100 (last accessed October 2024).

Van Loon, S., Vicente, V.B. and van Gestel, C.A.M. (2022) Long-term effects of imidacloprid, thiacloprid, and clothianidin on the growth and development of *Eisenia andrei*. *Environmental Toxicology and Chemistry* 41, 1686–1695.

Wang, K., Chen, H., Fan, R.-L., Lin, Z.-G., Niu, Q.-S. *et al.* (2022) Effect of carbendazim on honey bee health: assessment of survival, pollen consumption, and gut microbiome composition. *Ecotoxicology and Environmental Safety* 239, 113648.

Wang, X., Lu, Z., Miller, H., Liu, J., Hou, Z. *et al.* (2020) Fungicide azoxystrobin induced changes on the soil microbiome. *Applied Soil Ecology* 145, 103343.

Wang, X., Wang, Y., Ma, X., Saleem, M., Yang, Y. *et al.* (2023) Ecotoxicity of herbicide diuron on the earthworm *Eisenia fetida*: oxidative stress, histopathology, and DNA damage. *International Journal of Environmental Science and Technology* 20, 6175–6184.

Wrońska-Pilarek, D., Maciejewska-Rutkowska, I., Lechowicz, K., Bocianowski, J., Hauke-Kowalska, M. *et al.* (2023) The effect of herbicides on morphological features of pollen grains in *Prunus serotina* Ehrh. in the context of elimination of this invasive species from European forests. *Scientific Reports* 13, 4657.

Xiao, Z., Hou, K., Zhou, T., Zhang, J., Li, B. *et al.* (2023) Effects of the fungicide trifloxystrobin on the structure and function of soil bacterial community. *Environmental Toxicology and Pharmacology* 99, 104104.

Xing, Q., Kim, Y.W., Park, J.-S., Han, Y.-S., Yarish, C. *et al.* (2023) Effects of triazine herbicide terbutryn on physiological responses and gene expression in *Alexandrium catenella*. *Journal of Applied Phycology* 35, 1663–1671.

Yadav, A., Yadav, K. and Abd-Elsalam, K.A. (2023) Exploring the potential of nanofertilizers for a sustainable agriculture. *Plant Nano Biology* 5, 100044.

Yadav, I.C. and Devi, N.L. (2017) Pesticides classification and its impact on human and environment. *Environmental Science and Engineering* 6, 140–158.

Yan, H., Wang, D., Dong, B., Tang, F., Wang, B. *et al.* (2011) Dissipation of carbendazim and chloramphenicol alone and in combination and their effects on soil fungal: bacterial ratios and soil enzyme activities. *Chemosphere* 84, 634–641.

Yang, C., Hamel, C., Vujanovic, V. and Gan, Y. (2011) Fungicide: modes of action and possible impact on nontarget microorganisms. *ISRN Ecology* 2011, 1–8.

Yousaf, S., Khan, S. and Aslam, M.T. (2013) Effect of pesticides on the soil microbial activity. *Pakistan Journal of Zoology* 45, 1063–1067.

Zhai, W., Zhang, L., Liu, H., Zhang, C., Liu, D. *et al.* (2022) Enantioselective degradation of prothioconazole in soil and the impacts on the enzymes and microbial community. *Science of the Total Environment* 824, 153658.

Zhang, J., Zhang, B., Zhu, F. and Fu, Y. (2021a) Baseline sensitivity and fungicidal action of propiconazole against penicillium digitatum. *Pesticide Biochemistry and Physiology* 172, 104752.

Zhang, Y., Zhang, J., Shi, B., Li, B., Du, Z. *et al.* (2021b) Effects of cloransulam-methyl and diclosulam on soil nitrogen and carbon cycle-related microorganisms. *Journal of Hazardous Materials* 418, 126395.

Zhang, Y., Guan, T., Wang, L., Ma, X., Zhu, C. *et al.* (2023) Metamifop as an estrogen-like chemical affects the pituitary-hypothalamic-gonadal (HPG) axis of female rice field eels (*Monopterus albus*). *Frontiers in Physiology* 14, 1088880.

4 CRISPR-based Genome Engineering for Biofortification of Crops

Ayushi Singh, Sonal Chaudhary and Shalini Porwal*

Amity Institute of Microbial Technology, Amity University, Noida, Uttar Pradesh, India

Abstract

Agricultural output is restricted by climate change and adverse abiotic and biotic stresses, making it increasingly challenging for crop scientists to ensure food sustainability worldwide. Current biotechnological innovations have incorporated biofortification into several food crops to prevent malnutrition. Precision breeding for crop enhancement is now more viable because of recent developments in genome engineering technologies, notably CRISPR/Cas technology, which makes tailored genetic alteration of crops more practical. CRISPR/Cas targets a specific gene, thus making it a top-tier technique free of ethical problems. This genome engineering technique has a simple design, low technique costs, remarkable efficiency, outstanding reproducibility and a fast cycle. Many cereal and vegetable crops, including barley, wheat, rice, maize, potato and tomato, have been biofortified utilizing CRISPR/Cas. This chapter focuses on enhancing crop quality using CRISPR/Cas9 technologies. It entails adjusting the appearance, flavour, nutritional value and other desired characteristics of various crops.

Keywords: Biofortification, CRISPR/Cas, crop, genome engineering

4.1 Introduction

The primary objective of crop development is to increase crop output, increase abiotic and biotic stress resistance, elevate quality and nutritional value, and ensure sustainable crop production systems. Over the past several decades, modern agricultural technologies have significantly increased crop productivity. Because crops include numerous elements, including minerals, fibre, vitamins, proteins and bioactive chemicals directly impacting human health, consumers are more interested in crop quality (Liu *et al.*, 2021). It is well documented that researchers and breeders have changed their approach from elevating productivity to enhancing crop quality. Genetic variation is the foundation of agricultural progress, and plant genetics aims to generate and utilize these biological variations. Throughout the extensive timeline of plant breeding, four primary methods were employed: traditional crossbreeding, transgenic and mutation breeding and genomic editing for breeding purposes (Chen *et al.*, 2019) (Fig. 4.1).

A variety of techniques, as well as traditional crossbreeding, radiation and chemical-induced molecular marker-aided breeding, mutation breeding and genome editing breeding,

*Corresponding author: sporwal@amity.edu

© CAB International 2025. *Soil Health and Nutrition Management*
(eds N.C. Joshi, T. Leustek and P.K. Singh)
DOI: 10.1079/9781800624597.0004

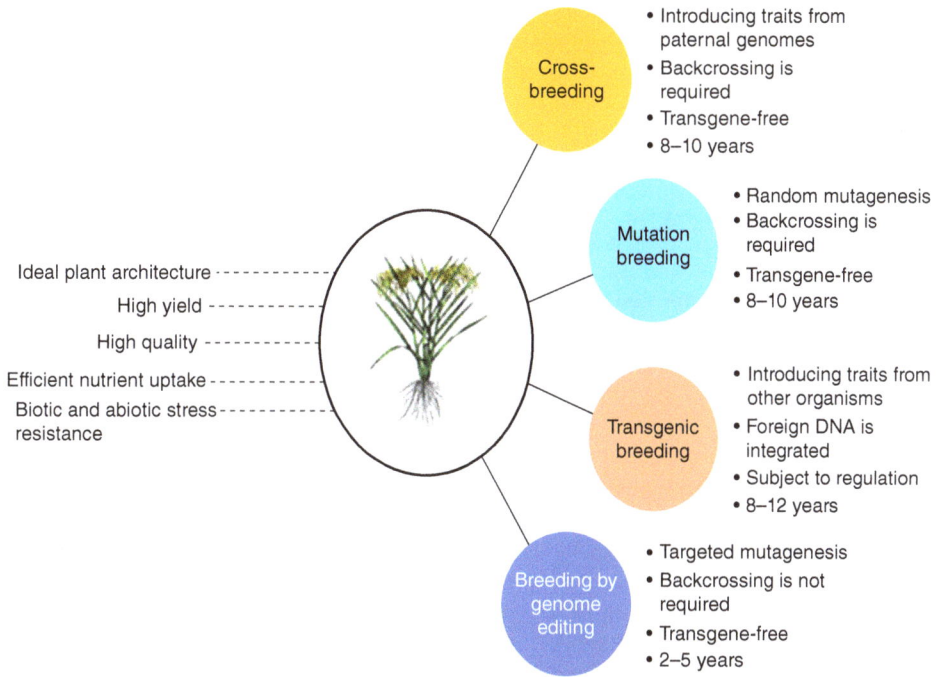

Fig. 4.1. Plant breeding has evolved through four key techniques: cross-breeding for trait introduction, mutation breeding for genome-wide mutations, transgenic breeding for foreign trait integration (with a lengthy regulatory process) and genome editing for precise trait modification without introducing foreign DNA. These precise techniques characterize next-generation plant breeding.

have been successfully used to improve various critical agricultural features (Chaudhary *et al.*, 2019). However, conventional mutagenesis-based breeding techniques require a lot of time and work, especially when developing polyploid crops (Parry *et al.*, 2009). Recently, crop development has shown significant benefits from genome editing (GE), which precisely and predictably modifies plant genome information (Gaj *et al.*, 2013). When some regions of the genome are altered, predictable and inheritable mutations can arise with few unexpected consequences and no convergence of foreign gene sequences. The DNA alterations introduced by GE include single-nucleotide substitutions (SNPs), big fragment replacements, insertions and deletions. The nucleotide excision approach uses four site-directed nuclease (SDN) families: homing endonucleases (HEs) or meganucleases, zinc-finger nucleases (ZFNs), CRISPR (clustered regularly interspaced short palindromic repeat)-associated protein (Cas) and transcription activator-like

effector nucleases (TALENs) (Christian *et al.*, 2010). Plants' endogenous repair systems correct double-stranded breaks (DSBs) in one of two ways. The two techniques for repairing DNA damage are homologous directed recombination (HDR) and non-homologous end joining (NHEJ). Due to large insertions or fragment replacement, a homologous flanking sequence or an exogenous rebuild is used to repair the break blueprint, in contrast to the error-prone NHEJ, which causes insertion and deletion mutation at the breakage site. The zinc finger DNA-binding site was combined with the FokI endonuclease site to create first-generation genome-editing nucleases (Kim *et al.*, 1996). To produce TALENs, TALE proteins require a specific DNA-binding and FokI cleavage site. TALENs technology outperforms ZFNs technology regarding target-binding selectivity and off-target probability (Joung and Sander, 2013). Studies have demonstrated the widespread utilization of gene-editing techniques in tomato, rice, maize and wheat (Zhang,

2013; Wang *et al.*, 2014; Čermák *et al.*, 2015). But even so, both require complicated construction methods, limiting their applicability to large-scale plants. CRISPR was discovered in *Escherichia coli* in 1987 and was initially described as an immune system that combats invading viruses and plasmid DNA (Ishino *et al.*, 1987). The CRISPR/Cas system, which has grown in popularity over the past few years, is the most widely used GE technology. The CRISPR/Cas system is less complicated and more efficient for GE than other SDNs because the selectivity of editing is regulated by the complementary nucleotides of the guide RNA to a specific region rather than by complicated protein engineering. Therefore, several scientists have employed CRISPR/Cas approaches to examine gene function (Lino *et al.*, 2018). When used in crop development, GE can significantly speed up the insertion of necessary attributes while reducing labour and other costs. The number of agricultural enhancement incidents due to GE has also increased. Among many target features, crop quality is one of the most crucial goals for crop improvement. This chapter reviews recent developments in CRISPR/Cas9-mediated improved crop quality and expands the debate on GE applications.

4.2 Utilization of Multiple Genome Editing Techniques for Improving Crops and Operating Mechanisms

Mega-nucleases can target sites as large as 18 bp. ZFNs are produced when many zinc-finger DNA-binding domains join the FokI endonuclease non-specific domain to identify the 3 bp module (Chandrasegaran and Carroll, 2016). Several transcription activator-like effector domains link with a FokI endonuclease domain to generate TALENs that can detect single base pairs (Sprink *et al.*, 2015). In comparison with ZFNs, TALENs have superior target-binding selectivity and produce fewer off-targets, which has led to their widespread application in crop genome editing (Fig. 4.2). Clustered frequently in archaea and bacteria, an immune system (adaptive) is made up of interspaced palindromic repeats (IPR-CRISPR) and Cas proteins (Barrangou, 2015). Four distinct phases make up the CRISPR/Cas immune system: (i) adaptation, in

which brief spacers from invasive virus and plasmid are identified, acquired, processed and incorporated into the CRISPR locus; (ii) the CRISPR locus is processed and combined with brief DNA fragments from plasmid and viral invasions during integration; (iii) during transcription, the CRISPR locus is translated into a long pre-crRNA and the pre-CRISPR RNA matures into crRNA; and (iv) when interference occurs, Cas effector nuclease uses guided RNA to detect the complementary target DNA sequences (Hsu *et al.*, 2014). A double-stranded DNA break results from Cas effectors binding to recognized targeted DNA. The CRISPR/Cas system is encompassed of six distinct categories. The type I, III and IV effector complexes are multi-subunit, whereas the type II, V and VI effector complexes are single subunits (Molina *et al.*, 2020). The *Streptococcus pyogenes* class 2, type II CRISPR/Cas9 system relies on RNA-guided designed nucleases (Mehta and Merkel, 2020).

ZFNs, megonucleases and TALENs recognize target sequences via DNA/protein interactions (Raza *et al.*, 2022). CRISPR/Cas9 uses guided RNA (sgRNA) (Blin *et al.*, 2016). These sgRNAs target a particular genomic sequence with short (20 nt) nucleotide sequences with a predetermined sequence. Cas9 nucleases subsequently divide the resulting RNA/DNA system. This results in the production of a DSB with a conserved protospacer adjacent motif (PAM) at the targeted location (Karvelis *et al.*, 2017). Frameshift or gene knockdown results from indels inserted by NHEJ repair into protein-coding regions (Auer and Del Bene, 2014). CRISPR/Cas9 reagents for genome editing have been utilized extensively because of the ease with which DNA can be targeted via base pairing. SpCas9, which is the most generic form of Cas9 and requires the 'NGG' PAM sequence in the target DNA sequence, is derived from *Streptococcus pyogenes* (Haeussler and Concordet, 2016). Plant genomes were modified using Cas9 variants with different PAM specifications.

CRISPR tools were later expanded to include Cas12 nucleases from class 2 and type V CRISPR systems (Koonin, 2015). Cas12 nucleases are characterized by a single 42 nt crRNA, which is the main difference between Cas12 and Cas9 nucleases. The most commonly used Cas12 variant in plant gene editing is LbCas12a, which recognizes a T-rich PAM ('TTTV') (Zhang,

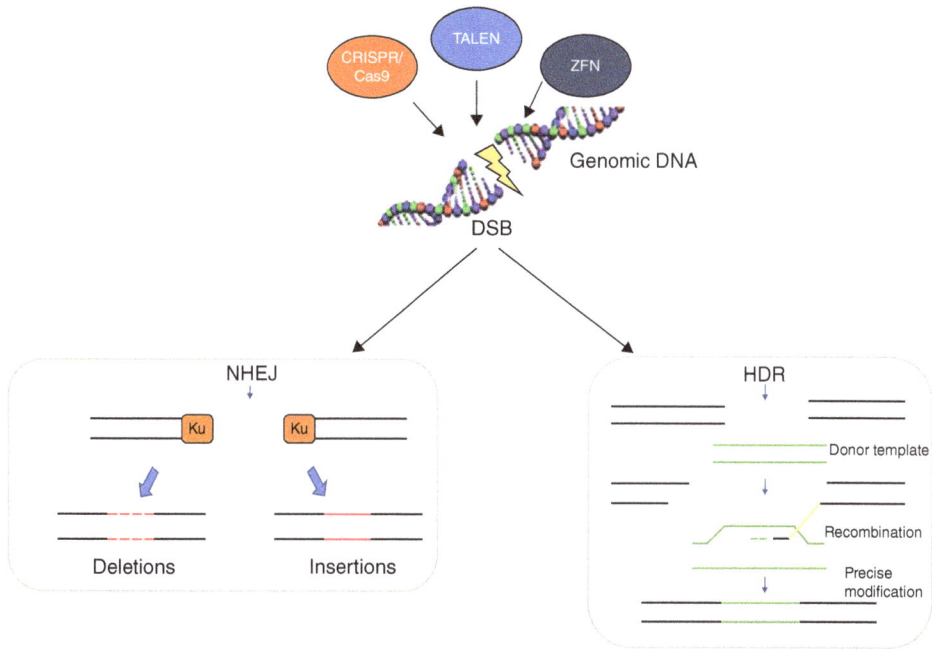

Fig. 4.2. Double-stranded breaks (DSBs) caused by site-specific nuclease (SSN) can be rectified using four primary techniques: zinc-finger nuclease (ZFN), transcription activator-like effector nuclease (TALEN), CRISPR/Cas9 and meganucleases. Where a template homologous to the DSB location is absent, DSBs are repaired using the non-homologous end joining (NHEJ) pathway. Ku, a dimeric protein complex, repairs damaged chromosomal termini by binding to DNA DSBs, resulting in either small deletions or insertions. DSBs can also be rectified through the homology-directed repair (HDR) pathway when a template with homology to the DSB site is available. In the process of HDR, the target locus is reconjoined with the donor template, leading to precise modifications at the specific site. This method presents a lower risk of errors than NHEJ and enables the introduction of customized changes with high accuracy.

2023). Cas12a is unique in that it can process mature crRNA without the need for tracrRNA, making it useful for base editing, multiplex gene editing, transcription and epigenetic modification (Aquino-Jarquin, 2023).

CRISPR/Cas8 is the most recent type V system to be added to the CRISPR arsenal. It consists of a single 70 kDa Cas8 protein. Embracing a staggered cut with 5′-overhangs and PAM 5′-TBN-3′ (where B represents G, T or C), Cas8 and Cas12a differ from Cas9 and Cas12a in not requiring tracrRNA. A reported 0.85% editing efficiency for CRISPR/Cas8-mediated genome editing in *Arabidopsis* underscores its potential in genetic manipulation.

The exact modification of DNA and RNA at the single-base level has recently been accomplished via base editing (Bharat *et al.*, 2020).

Transcriptomes and nuclear and organellar genomes (DNA BEs) may be precisely modified using base editors (BEs) in both dividing and non-dividing cells (RNA BEs) (Nerkar *et al.*, 2022). BEs comprise a DNA repair protein, nucleotide deaminase and catalytically ineffective Cas nuclease (dCas9: D10A and H840A). Individual nicks produced by BE are repaired more accurately than SpCas9-generated DSBs, which are fixed via the error-prone base excision repair pathway (NHEJ), minimizing undesirable side effects of gene editing (Kumar *et al.*, 2022). Various forms of DNA BEs include organelles, cytosine, adenine, cytosine-to-guanine, dual-base editors and adenine-to-guanine. CBEs have been used for GE in many crops, including cotton, soybean, rapeseed, tomato, rice, wheat, maize and *Arabidopsis*. However, base editors cannot produce additional base transversions, insertions or deletions because they can only produce base transitions without DNA donors (Hodges and Conlon, 2019).

4.3 Crop Improvement Breeding Approaches

Traditionally, breeding procedures have significantly contributed to the genetic improvement of today's premier crop types (Shiferaw *et al.*, 2013). Modern improvements to conventional breeding techniques have been made, such as introgression from wild crops through hybrid breeding, breeding via mutation, wide crosses, tissue culturing including embryo and ovule release, and fusion of protoplasts. The following are some common crop improvement breeding approaches.

1. **Conventional breeding:**
   - Mass selection: identifying and selecting plants with desirable traits from a population and using their seeds for the next generation.
   - Pure line selection: repeated self-pollination of selected plants to create genetically uniform, stable lines with consistent traits (Jain *et al.*, 2023).
2. **Hybridization:**
   - Cross-breeding: crossing two genetically distinct plants to combine desirable traits from each parent. F1 hybrids often exhibit improved vigour and yield compared with their parents.
   - Double cross hybrid: breeding involves two crosses to create a final hybrid with the desired characteristics.
3. **Mutation breeding:** inducing genetic mutations through radiation, chemicals or other methods to generate novel genetic variations. Mutants with desirable traits are selected for further breeding (Singer *et al.*, 2021).
4. **Marker-assisted selection (MAS):** using molecular markers to specific genes to select plants with desired characteristics more efficiently. This approach accelerates breeding by admitting early selection of desirable traits without waiting for phenotypic expression (Roychowdhury *et al.*, 2023).
5. **Genomic selection:** predicting the breeding value of plants based on genomic information. This approach utilizes large-scale genomic data to estimate plants' genetic potential, improving selection accuracy and efficiency.
6. **Transgenic (genetic engineering):** introducing genes from other organisms into the crop to confer specific traits, such as pest resistance, herbicide tolerance or improved nutritional content. Commonly known as genetically modified organisms (GMOs) (Khan *et al.*, 2019).
7. **Genome editing:** minimum modification of targeted genes in plant genomes using tools like CRISPR/Cas9. This technique allows for targeted changes without introducing foreign genes, providing a more precise and controlled form of genetic modification.
8. **Participatory plant breeding:** involving farmers, end users and other stakeholders in the breeding process to ensure that the developed varieties meet local needs and preferences (Roychowdhury *et al.*, 2023).

However, the most challenging plant transformation issues continue to involve the cost, time-consuming processes and difficulty transforming obstinate crops. Genetic transformation, which includes the arbitrary incorporation of transgenes into the nuclear genome, frequently results in transgene silence (George *et al.*, 2020). Precision genome editing is necessary to overcome this obstacle.

4.4 Genome Editing in the Development of Crops with Valuable Agronomic Traits

Genome editing, which has long been the goal of plant breeders worldwide, has transformed crop growth by producing precise adjustments in plant genomes. It has been reported in several food crops, including vegetable and fruit crops, since the initial report on the practice in rice. In the following sections, we examine the use of CRISPR/Cas9 to produce crops with valuable agronomic characteristics.

4.4.1 CRISPR/Cas9-induced disease resistance genome editing

Numerous crops, including tomato, wheat, rice, banana, citrus, cucumber, grapes and cassava, have been subjected to CRISPR-mediated modification for disease resistance (Mushtaq *et al.*, 2019). Owing to the ability of loci to impart resistance to a variety of pathogen species or races, wide-ranging resistance is a beneficial strategy for managing crop diseases. Zhou *et al.* (2018) discovered in rice the *bsr-k1* allele, as well as the *bsr-k1* mutant, which boasts a wide range of resistance to *Xanthomonas oryzae* sp. and *Magnaporthe oryzae* without sacrificing significant agronomic traits. *X. oryzae* is responsible for causing blight, leading to substantial loss in rice yield. Additionally, *SWEET1*, -*3* and -*14* expression results in disease vulnerability (these are sucrose transporter genes). Along with the two major types, *IR64* and *Ciherang-Sub1*, Oliva *et al.* (2019) produced wide-range resistance in Kitaake (the rice line).

4.4.2 CRISPR/Cas9-driven crop development for abiotic stress tolerance and high yield

Abiotic stress poses a severe threat to agricultural productivity and yield. To create mutant crops that are high yielding and resistant to abiotic stress, CRISPR/Cas has been quickly embraced for agricultural genome editing (Nerkar *et al.*, 2022). Three genes, *GS3*, *OsMYB30* and *OsPIN5b*, were altered concurrently by CRISPR/Cas9 technology to create many unique rice mutants with high yields and exceptionally low-temperature tolerance that were also persistent in T2 generations as well (Zeng *et al.*, 2020). By controlling *Feronia* (*Fer*) homologues in tomatoes, upregulated brassinosteroid regulator (BZR) caused the tomatoes to become thermotolerant (Saini *et al.*, 2015). Major crops, including rice, wheat, tomato and *Brassica napus*, have enhanced drought tolerance by altering crucial transcription factors (Joshi *et al.*, 2016). A transcription factor for rice that responds to drought, *OsNAC14*, has also been functionally characterized (Shim *et al.*, 2018). *OsNAC14* was overexpressed throughout the vegetative growth stage in rice mutants, which conferred drought tolerance. *OsNAC14*-overexpressing transgenic rice lines performed better in the field during drought conditions than non-transgenic plants in terms

of panicle production and filling rate. Wheat with *TaDREB2* (*dehydration response element binding protein 2*) and *TaERF3* (*ethylene-responsive factor 3*), whose genomes have been edited by CRISPR/Cas9, has shown improved drought tolerance (Trono and Pecchioni, 2022). Li *et al.* (2019) discovered that tomato lines upregulated for the *SlNPR1* gene have substantially lower drought abiotic tolerance. They isolated *SlNPR1* from tomatoes and utilized the CRISPR/Cas9 system to produce *slnpr1* mutants. This work provides insight into *NPR1*'s function in plant response. The utilization of CRISPR/Cas9 editing in rapeseed (*B. napus*) has enabled a deeper understanding of the functions of DELLA proteins in this plant's drought tolerance (Wu *et al.*, 2020). Specifically, editing the *rga-D*, *bnarga* and *bnaa6* genes has revealed the enhancement of drought tolerance in *bnaa6* and *rga-D* mutants and the physical interaction between *BnaRGAs* and *BnaA10* in providing abscisic acid (ABA) signalling. The use of CRISPR/Cas9 in editing the *bnarga*, *bnaa6* and *rga-D* genes has facilitated the elucidation of the DELLA proteins' role in *B. napus* drought tolerance. *BnaA6* mutants with *rga-D* had increased drought tolerance, and *BnaRGAs* physically interacted with *BnaA10*. Comparing the resultant *OsPQT3* knockout mutants (*ospqt3*) with the wild type, they showed better agronomic performance with greater yield under salinity stress in field conditions and the greenhouse. Li *et al.* (2018) used multiplex CRISPR/Cas9 in four wild tomatoes (abiotic stress tolerance) allied with morphological characteristics, early and better flowering and fruiting development, and ascorbic acid synthesis. These genotypes exhibit a 3–7% increase in grain yield compared with the wild type and are the result of modifying the *GS3* and *Gn1a* genes, which play a role in determining the size and number of grains. Hao *et al.* (2019) found that mutants with changed genomes that were produced by changing the *OsGRF3* and *GL2/OsGRF4* genes, which are important for grain yield and size, had larger grains, respectively. *DEP1*, *Gn1a*, *IPA1* and *GS3* genomes were edited utilizing the CRISPR/Cas9 system, and the results included more grains, dense erect panicles, larger grains and a range in the number of tillers in the T2 generation (Hao *et al.*, 2019).

4.4.3 CRISPR/Cas9 biofortification of crops

Enhancing grain nutrition is one of the top priorities of breeders to reduce problems linked to nutrient deficiencies (Gaikwad *et al.*, 2020). The genes *DHPS* (dapA) and *AK* (lysC), which encode crucial enzymes in the biosynthesis of lysine, were altered, increasing the amount of lysine in rice by up to 25-fold (Yang *et al.*, 2016). These lines with higher levels of lysine also showed improved physicochemical properties without altering the starch composition. There were only slight variations in plant height and grain colour during field testing; otherwise, the plants thrived. The genomes of *CrtI* and *PSY* were modified to produce marker-free mutants with elevated levels of carotenoids, allowing for the biofortification of carotenes in rice (Stra *et al.*, 2023). Biofortified tomatoes contain a range of nutrients, including γ-aminobutyric acid (GABA) which is reported to play a role in controlling anxiety and hypertension through its function as a neurotransmitter (Heli *et al.*, 2022). To increase the levels of GABA in these tomatoes, the glutamate decarboxylase, an enzyme involved in the production of GABA, has a C-terminal autoinhibitory domain, eliminated using CRISPR/Cas9 gene editing. This resulted in the development of mutant tomatoes with elevated GABA levels.

CRISPR/Cas9 gene editing was also used to develop yellow-seeded rapeseed mutants, resulting in a 9.47% increase in oil content in generation T2. These double mutants, which can produce a large amount of oil with altered fatty acid compositions and improved nutritional qualities, may prove to be valuable in the breeding of rapeseed. In tetraploid potatoes, Tuncel *et al.* (2019) developed a potato line with elevated levels of resistant starch using Cas9-driven mutagenesis of starch branching enzymes, resulting in better control of insulin levels. These results indicate that Cas9-driven mutagenesis has the potential to improve yields suitable for commercialization. Gene-edited crops using the CRISPR/Cas9 system are listed in Table 4.1.

4.4.4 Reduction of anti-nutritional factors in crops modified through CRISPR/Cas9 technology

Reducing the presence of anti-nutrients in crops is a crucial breeding strategy that greatly enhances the overall quality of the harvest. Various breeding methods have utilized induced and

Table 4.1. CRISPR/Cas9 system for gene-edited crops.

Crops	Species
Fibre crops	*Gossypium herbaceum*
Feed crops	*Medicago sativa*
Crops for industrial use	*Cichorium intybus, Coffea, Taraxcum officinale, Jatropha curcas, Pennisetum glaucum, Papaver somniferum, Parasponia andersonii, Sorghum biocolor, Saccharum officinarum, Panicum virgatum, Targopogon dubis Scop, Tripterygium wilfordii*
Edible crops	*Musa, Ocimum basilicum, Vaccinium* sect. *cyanococcus, Brassica oleracea, Daucus carota, Manihot esculenta, Cicer arietinum, Citrus, Cocos nucifer* L., *Cucumis sativus, Citrus* x *paradisi, Malus pumila, Vitis vinifera, Brassica oleracea, Actinidia deliciosa, Lactuca sativa, Citrus* x *lemon, Hordeum vulgare, Litchi chinensis, Zea mays* L., *Cucumis melo, Avena sativa, Citrus* x *sinensis, Carica papaya, Pyrus communis* L., *Piper nigrum, Solanum tuberosum, Cucurbita, Oryza sativa, Crocus sativus, Fragaria* x *ananassa, Beta vulgaris, Ipomoea batatas, Solanum lycopersicum, Citrullus lanatus, Triticum aesticum, Dioscorea*
Ornamental crops	*Lilium, Nelumbo nucifera, Petunia, Populus, Rosa rubiginosa, Sedum, Antirrhinum majus, Torenia fournieri*
Oil crops	*Brassica napus, Linum usitatissimum, Elaeis guineensis, Glycine max, Helianthus annuus*

natural genetic resources, including mutation, selection, backcrossing, population enhancement and hybridization. The efforts to diminish anti-nutrients in crops, notably with glandless cotton, began in the early 1960s. More recently, advanced techniques such as gene editing and silencing have been applied to generate crop lines with lower levels of anti-nutrients (Duraiswamy *et al.*, 2023).

A product of the *ITPK* gene lowers the amount of phytic acid in rapeseeds by catalysing the penultimate stage of phytate synthesis (Liu *et al.*, 2021). Without affecting plant performance, *ITPK* gene knockout using CRISPR/Cas9 in rapeseed decreased phytic acid content by 35% (Tian *et al.*, 2022). For those who are gluten intolerant, wheat gluten protein is another significant anti-nutritional component that may cause coeliac disease. Using traditional breeding methods to reduce gluten is challenging because the wheat genome has more than 100 loci that encode this protein. CRISPR/Cas9 produced wheat lines with low levels of gluten, and those without transgenes target a conserved region of the gliadin genes specifically (Liu *et al.*, 2020). An outstanding application of CRISPR/Cas9 technology in rice breeding is developing cultivars resistant to heavy metal contamination (Tang *et al.*, 2017). When administered in significant doses, the human carcinogen cadmium can potentially result in kidney failure. By manipulating *OsNramp5*, Tang *et al.* (2017) engineered unique indica rice varieties with reduced cadmium uptake in the roots, thereby controlling its concentration in the grain. *OsNramp5* mutant field performance showed that high cadmium conditions had an insignificant impact on agronomic traits or grain output.

4.5   Editing the Genome to Create Resistance Under Varying Climatic Conditions

Crop genome editing is now exposing genomic functions and regulatory mechanisms. Climate change is the biggest obstacle to further agricultural development (Altpeter *et al.*, 2016). Therefore, the most crucial challenge for breeders is to boost crop productivity in insufficient environments. The factors impacting agricultural output

were investigated by Bailey-Serres *et al.* (2019), who also suggested breeding techniques to increase crop productivity under challenging conditions. It is important to note that genome editing plays a major role in our understanding of gene activity during stress responses and the adaptive mechanisms plants have evolved to deal with challenging environmental conditions. Sequence-dependent breeding approaches have been developed by genomic techniques and next-generation sequencing (NGS), which has sped up identifying and replicating genetic loci crucial for environmental stress tolerance. Additionally, this creates previously unheard-of opportunities to use the genetic variety found in crops' wild cousins, expanding the pool of genetic resources available to breeders (Fahmideh *et al.*, 2023) (Fig. 4.3).

4.5.1   CRISPR/Cas-edited crop regulation

The management of genetically modified crops remains a contentious issue on a worldwide scale. GE is a new technique in plant breeding that involves making specific changes to a plant's genome without using transgenic sequences (Yin *et al.*, 2017). According to researchers, GE results in a few genetic alterations that are common in nature. Gene editing enables scientists to modify the genome more precisely than earlier methods (Adli, 2018). Genome-edited crops must be regulated to enhance crops that produce fuel, food and fibre for the world's growing people in the context of climate change (Turnbull *et al.*, 2021). On the one hand, even though this technology has proven adaptable in numerous significant crops, including cereals crops, fruits, etc., international discussions are seeking clarity and transparency in legal regulations concerning the approval of genome-edited products and their derivatives. A soybean variety that produces an oil with an extensive shelf life was used by Calyxt of Roseville, Minnesota, to market the first genetically modified food item in the USA in 2019 (Ricroch *et al.*, 2022). A gene-edited tomato with more GABA was released in Japan in 2024 for field testing of gene-edited plants; the UK has announced its intention to loosen the regulations governing field research on genetically modified crops (Gramazio *et al.*, 2020). The government of India's Ministry of Environment,

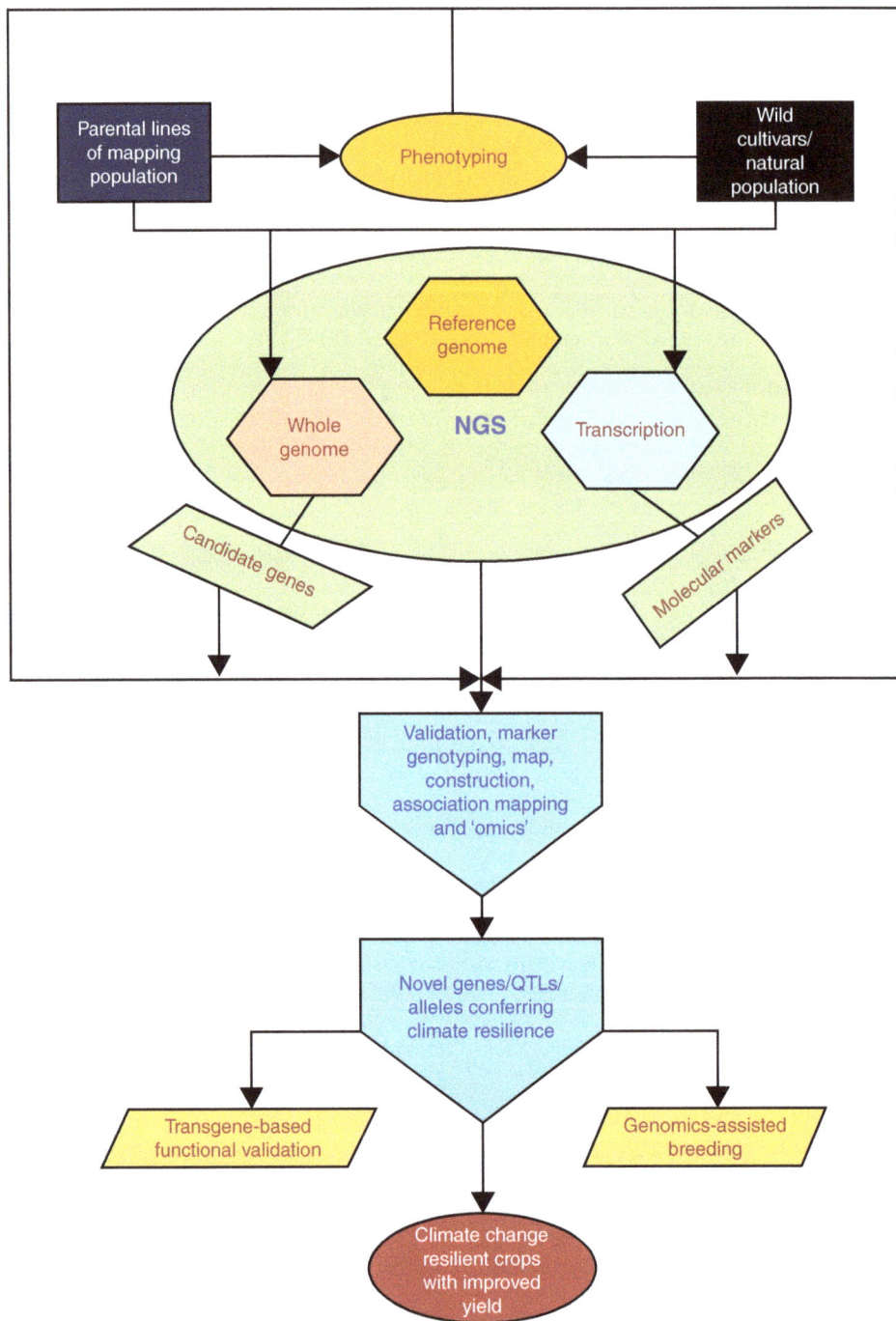

Fig. 4.3. Procedure for developing crops resilient to climate change through genomics and next-generation sequencing (NGS) technology. QTLs, quantitative trait loci.

Forests and Climate Change released a notification in 2022 recognizing that the SDN1 and SDN2 plant categories are transgene-free and exempting them from Rules 7 and 11 (Ali *et al.*, 2018). SDN1 incorporates modifications to the plant genome by minute deletions and insertions, whereas SDN2 creates the required alteration in the plant genome using a relatively small blueprint. Regulator exemption is still needed for plants with altered genomes in most nations.

However, this will quicken India's progress toward genome-edited crops. Even though several nations worldwide are currently waiting for unregulated products made from genome-edited materials, scientists think GE contains effective methods for ensuring future food security and should be approved rather than delayed. Since breeding programmes do not utilize mutations that produce detrimental characteristics (Carroll *et al.*, 2016) and given that specific naturally occurring alterations (such as a protein that may cause allergies) can arise without human intervention, it is unclear why genetic engineering – which creates specific genetic alterations – should be more strictly controlled. In contrast, point mutations, which introduce random variations into each genome, are not subject to the same level of regulation, despite producing results comparable to natural genetic mutations in the genomes of food.

The alteration of agricultural organisms' genomes through GE can be executed precisely, with no introduction of foreign DNA (Metje-Sprink *et al.*, 2019). Thus, the same level of scrutiny applied to other food items should be applied to the results of GE rather than the method used to make them. Public funds have largely financed the development of this technology, and its judicious application would benefit the broader population (Wang *et al.*, 2020). While claims are being made regarding the potential of these technologies to address all issues, it is important to recognize that many of the challenges they pose are social and require changes in attitudes and behaviour to be resolved. The decision-making process regarding these tools and their implementation should be entrusted to well-informed stakeholders, including consumers and farmers.

Crops genetically engineered and/or with altered genomes cannot substitute for sustainable agricultural methods such as cover crops,

crop rotation and crop diversity. As one of the many instruments accessible to farmers at all stages of production to adapt to local conditions and difficulties, they should be used in conjunction with sustainable practices. Regulation of gene-edited crops can be categorized into two main frameworks (Fig. 4.4). The first model is that used in India, New Zealand, Europe and Australia. Under this approach, the outcomes of novel breeding techniques are regulated by the methods used to produce them. Crops genetically modified by GE are subject to the same stringent regulatory framework as GMOs. In the second, less stringent regulatory paradigm, the focus is on the uniqueness of the trait rather than the method used to obtain it. This approach is employed in Argentina, the USA and Canada.

4.6 CRISPR Applications for Biofortification

Many crops have benefited from the effective application of CRISPR-based genome modification for biofortification, targeting different micronutrients and beneficial compounds. Some examples are given here.

4.6.1 Provitamin A biofortification

Insufficient vitamin A can cause blindness, weakened immune systems and higher death rates. This is a severe worldwide health concern, especially in impoverished nations and especially among children and pregnant women. Enhancing staple crops with provitamin A (β-carotene) through biofortification has been a crucial strategy to address this deficiency (Nkhata *et al.*, 2020).

Using CRISPR/Cas9, provitamin A levels in rice, maize and sweet potatoes have been raised. Researchers have disrupted genes producing β-carotene hydroxylases, which degrade provitamin A, leading to higher quantities of β-carotene in grains or tubers. In the extensively studied case of golden rice, the phytoene synthase gene from maize and the carotene desaturase gene from *Erwinia* were introduced. The *OsBCH2* gene, which encodes a β-carotene hydroxylase, was also disrupted using CRISPR/Cas9 (Bassolino *et al.*, 2022). This

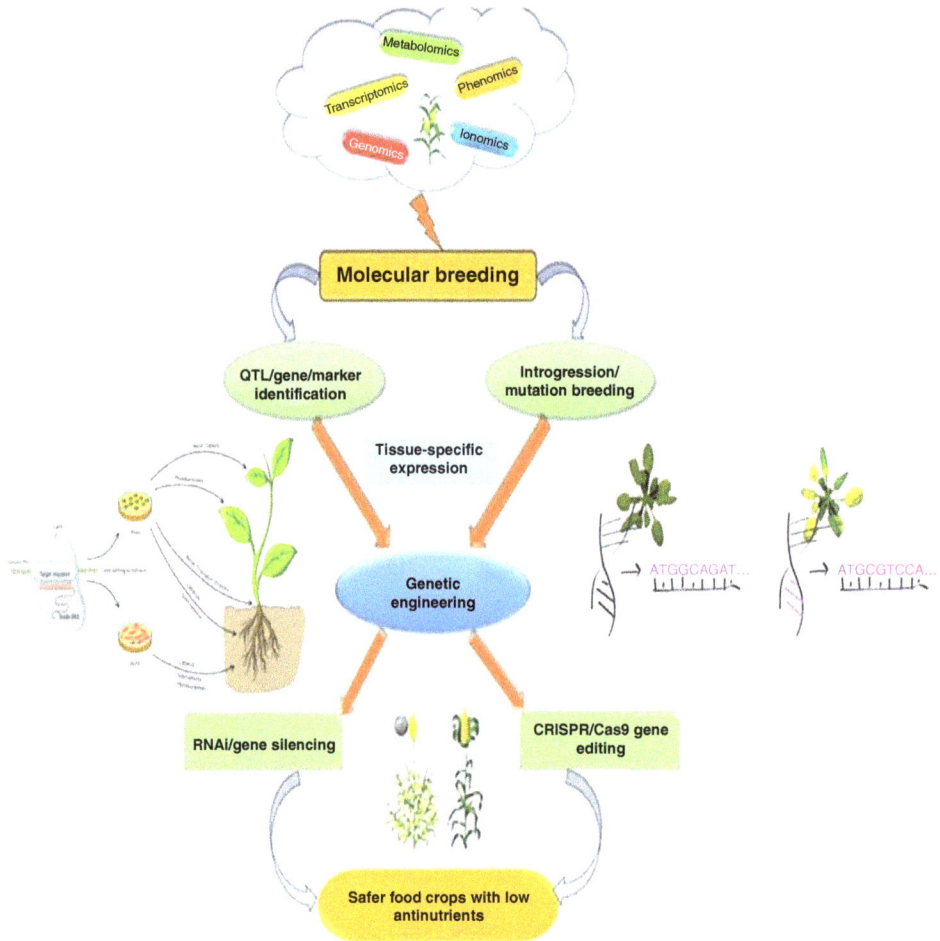

Fig. 4.4. Future opportunities to improve food crop quality. QTLs, quantitative trait loci.

combinatorial strategy significantly increased the rice endosperm's provitamin A levels.

4.6.2   Iron and zinc biofortification

Among the most common micronutrient deficits worldwide are those involving iron and zinc, affecting over 2 billion people. These deficiencies can lead to anaemia, impaired cognitive development and weakened immune systems, particularly in children and pregnant women.

It has proven possible to increase the concentrations of these vital minerals in crops, including sorghum, wheat and rice, using CRISPR/Cas9. Strategies include knocking out genes that encode chelators or transporters that sequester or mobilize iron and zinc away from the edible parts of the plant or overexpressing genes involved in mineral uptake and accumulation (Stangoulis and Knez, 2022). To disrupt the *OsNRAMP*5 gene in rice, for instance, researchers have utilized CRISPR/Cas9, which encodes a transporter that sequesters iron in the root vacuoles, leading to increased iron levels in the grains (Songmei *et al.*, 2019). In wheat, disruption of the *TaNRAMP*6 gene led to increased amounts of iron and zinc in the grain.

4.6.3 Folate biofortification

Folate, or vitamin B9, is essential for cell division, DNA synthesis and a properly developing fetus. A lack of folate increases the risk of cardiovascular disease and several types of cancer, as well as causing neural tube abnormalities in neonates (Liu *et al.*, 2021). By focusing on genes involved in folate production and transport, CRISPR/Cas9 has been used to increase the amounts of folate in crops such as tomatoes and rice. Researchers have disrupted the *OsSPDSCL1* gene in rice, which encodes an enzyme that degrades folate, resulting in higher folate accumulation in the grains. The disruption of the *FTRC* gene by CRISPR/Cas9 with the overexpression of the *GTPCHI* gene in tomatoes resulted in a significant rise in the fruit's folate content (Yadava *et al.*, 2018).

4.6.4 Amino acid biofortification

The primary dietary protein sources for many communities globally are cereal grains, frequently deficient in essential amino acids, including lysine, methionine and tryptophan. Deficiencies in these amino acids can lead to protein malnutrition and compromised growth and development, especially in children. By either preventing the degradation of these vital amino acids' genes or improving their manufacturing pathways, CRISPR/Cas9 has been used to increase the amounts of these amino acids. For instance, in maize, scientists have disrupted the *LKR/SDH* gene, which codes for an enzyme involved in the catabolism of lysine, using CRISPR/Cas9 (Kumar *et al.*, 2022).

4.6.5 Enhancing beneficial compounds

Other advantageous substances that may be added to crops using CRISPR include antioxidants, phytochemicals and vital amino acids. For instance, to increase the nutritional value of fruits and vegetables, researchers have targeted genes involved in the manufacture of anthocyanins or antioxidants or altered the expression of storage proteins in grains.

CRISPR-based genome engineering offers several advantages for biofortification compared with conventional breeding methods.

1. Precision and efficiency: With the use of CRISPR, accurate and targeted genome alterations are possible, eliminating the need for significant backcrossing or the transmission of unwanted characteristics and facilitating the introduction or improvement of desired features.

2. Time saving: Compared with traditional breeding procedures, which might take years or even decades to produce new biofortified crop types, the CRISPR process can be finished quickly.

3. Versatility: With CRISPR, particular nutritional demands can be tailored to various crops by targeting different micronutrients or advantageous chemicals (Strobbe and Van Der Straeten, 2017).

The application of CRISPR for biofortification is not without its difficulties, though.

1. Regulatory challenges: The regulatory landscape for genetically modified crops varies across different countries and regions, and there is an ongoing debate regarding the classification and regulation of CRISPR-edited crops.

2. Public acceptance: Genetically modified organisms can face considerable obstacles in public image and adoption; thus, it is important to address concerns and advance knowledge through effective communication and education initiatives.

3. Off-target effects: Even though CRISPR is often regarded as precise, there is a chance of unintentional off-target alterations, which need to be carefully assessed and reduced by thorough characterization and screening.

4. Complex trait regulation: Some traits, such as micronutrient accumulation, may be regulated by complex genetic networks, making it challenging to achieve substantial improvements through single-gene modifications.

4.7 Future Perspectives

Plant research, both basic and applied, has been significantly impacted by the application of GE technologies. Several crop selections with elevated nutritional value and improved agronomic traits have been created utilizing precise GE techniques based on CRISPR. We anticipate that CRISPR/Cas technologies will contribute to the achievement of Sustainable Development Goal 2 (SDG2: end hunger) by 2030 because of their widespread use in agriculture and their significant potential for crop modification. By improving photosynthetic efficiency and plant architecture, the CRISPR/Cas system has been used to enhance the quality of grain, crop tolerance to biotic and abiotic problems and the ideo-typing of crop kinds to end hunger and eliminate malnutrition. Furthermore, in addition to nuclear genes, organelles like mitochondria and chloroplasts inhibit several essential genes that adversely regulate plant respiration and photosynthesis. Since these organelles are not simple candidates for genetic alteration, their genomes present a substantial opportunity for creating more productive food crop varieties. Selected crops may benefit from genetic editing using CRISPR to enhance their photosynthetic systems, resulting in increased agricultural yield and biomass output. Notably, wheat, rice and barley – which are C3 food crops – may eventually be transformed into C4-like plants with greater net primary productivity and water consumption competence, as well as high photosynthesis rates, using the CRISPR/Cas technology.

4.8 Conclusion

Using innovative technologies in breeding techniques may offer a solution to global food security issues in the future. The combination of genetic resources and cutting-edge technologies, such as genome editing, is important for developing crops with vital agronomic traits that can improve food security worldwide and lessen the environmental impact of agriculture. The most popular crop breeding method in the last 10 years is CRISPR/Cas9, which has significantly advanced crop production of disease-resistant, abiotic stress-tolerant crops with higher yields, nutrient content and storage stability. The devotion of GE expertise in crop growth can expand our understanding of new gene activities and the governing procedures of genes that impact key agronomical traits in plants. This knowledge can be harnessed to discover and modify genes important for stress tolerance and yield enhancement, thus creating crops that are resilient to climate change. The regulatory framework is already in place, even though the transfer of genome-amended agricultural investigation of the field is still ongoing.

References

Adli, M. (2018) The CRISPR tool kit for genome editing and beyond. *Nature Communications* 9, 1911.

Ali, M., Anwar, S., Shuja, M.N., Tripathi, R.K. and Singh, J. (2018) The genus Luteovirus from infection to disease. *European Journal of Plant Pathology* 151, 841–860.

Altpeter, F., Springer, N.M., Bartley, L.E., Blechl, A., Brutnell, T.P. *et al.* (2016) Advancing crop transformation in the era of genome editing. *Plant Cell* 28, 1510–1520.

Aquino-Jarquin, G. (2023) Genome and transcriptome engineering by compact and versatile CRISPR-Cas systems. *Drug Discovery Today* 28, 103793.

Auer, T.O. and Del Bene, F. (2014) CRISPR/cas9 and TALEN-mediated knock-in approaches in zebrafish. *Methods* 69, 142–150.

Bailey-Serres, J., Parker, J.E., Ainsworth, E.A., Oldroyd, G.E.D. and Schroeder, J.I. (2019) Genetic strategies for improving crop yields. *Nature* 575, 109–118.

Barrangou, R. (2015) The roles of CRISPR–Cas systems in adaptive immunity and beyond. *Current Opinion in Immunology* 32, 36–41.

Bassolino, L., Petroni, K., Polito, A., Marinelli, A., Azzini, E. *et al.* (2022) Does plant breeding for antioxidant-rich foods have an impact on human health? *Antioxidants* 11, 794.

Bharat, S.S., Li, S., Li, J., Yan, L. and Xia, L. (2020) Base editing in plants: current status and challenges. *Crop Journal* 8, 384–395.

Blin, K., Pedersen, L.E., Weber, T. and Lee, S.Y. (2016) CRISPy-web: an online resource to design sgRNAs for CRISPR applications. *Synthetic and Systems Biotechnology* 1, 118–121.

Carroll, D., Van Eenennaam, A.L., Taylor, J.F., Seger, J. and Voytas, D.F. (2016) Regulate genome-edited products, not genome editing itself. *Nature Biotechnology* 34, 477–479.

Čermák, T., Baltes, N.J., Čegan, R., Zhang, Y. and Voytas, D.F. (2015) High-frequency, precise modification of the tomato genome. *Genome Biology* 16, 1–5.

Chandrasegaran, S. and Carroll, D. (2016) Origins of programmable nucleases for genome engineering. *Journal of Molecular Biology* 428, 963–989.

Chaudhary, J., Alisha, A., Bhatt, V., Chandanshive, S., Kumar, N. *et al.* (2019) Mutation breeding in tomato: advances, applicability and challenges. *Plants* 8, 128.

Chen, K., Wang, Y., Zhang, R., Zhang, H. and Gao, C. (2019) CRISPR/Cas genome editing and precision plant breeding in agriculture. *Annual Review of Plant Biology* 70, 667–697.

Christian, M., Cermak, T., Doyle, E.L., Schmidt, C., Zhang, F. *et al.* (2010) Targeting DNA double-strand breaks with TAL effector nucleases. *Genetics* 186, 757–761.

Duraiswamy, A., Sneha A., N.M., Jebakani K., S., Selvaraj, S., Pramitha J., L. *et al.* (2023) Genetic manipulation of anti-nutritional factors in major crops for a sustainable diet in future. *Frontiers in Plant Science* 13, 1070398.

Fahmideh, L., Khodadadi, E., Khodadadi, E., Zeinalzadeh, E., Dao, S. *et al.* (2023) Transcriptome analysis methods: from the serial analysis of gene expression and microarray to sequencing new generation methods. *Biointerface Research in Applied Chemistry* 13, 543.

Gaikwad, K.B., Rani, S., Kumar, M., Gupta, V., Babu, P.H. *et al.* (2020) Enhancing the nutritional quality of major food crops through conventional and genomics-assisted breeding. *Frontiers in Nutrition* 7, 533453.

Gaj, T., Gersbach, C.A. and Barbas, C.F. (2013) ZFN, TALEN, and CRISPR/cas-based methods for genome engineering. *Trends in Biotechnology* 31, 397–405.

George, J., Kahlke, T., Abbriano, R.M., Kuzhiumparambil, U., Ralph, P.J. *et al.* (2020) Metabolic engineering strategies in diatoms reveal unique phenotypes and genetic configurations with implications for algal genetics and synthetic biology. *Frontiers in Bioengineering and Biotechnology* 8, 513.

Gramazio, P., Takayama, M. and Ezura, H. (2020) Challenges and prospects of new plant breeding techniques for GABA improvement in crops: tomato as an example. *Frontiers in Plant Science* 11, 577980.

Haeussler, M. and Concordet, J.P. (2016) Genome editing with CRISPR-cas9: can it get any better? *Journal of Genetics and Genomics* 43, 239–250.

Hao, L., Ruiying, Q., Xiaoshuang, L., Shengxiang, L., Rongfang, X. *et al.* (2019) CRISPR/Cas9-mediated adenine base editing in rice genome. *Rice Science* 26, 125–128.

Heli, Z., Hongyu, C., Dapeng, B., Yee Shin, T., Yejun, Z. *et al.* (2022) Recent advances of γ-aminobutyric acid: physiological and immunity function, enrichment, and metabolic pathway. *Frontiers in Nutrition* 9, 1076223.

Hodges, C.A. and Conlon, R.A. (2019) Delivering on the promise of gene editing for cystic fibrosis. *Genes and Diseases* 6, 97–108.

Hsu, P.D., Lander, E.S. and Zhang, F. (2014) Development and applications of CRISPR-Cas9 for genome engineering. *Cell* 157, 1262–1278.

Ishino, Y., Shinagawa, H., Makino, K., Amemura, M. and Nakata, A. (1987) Nucleotide sequence of the iap gene, responsible for alkaline phosphatase isozyme conversion in Escherichia coli, and identification of the gene product. *Journal of Bacteriology* 169, 5429–5433.

Jain, S.K., Wettberg, E.J. von, Punia, S.S., Parihar, A.K., Lamichaney, A. *et al.* (2023) Genomic-mediated breeding strategies for global warming in chickpeas (*Cicer arietinum* l.). *Agriculture* 13, 1721.

Joshi, R., Wani, S.H., Singh, B., Bohra, A., Dar, Z.A. *et al.* (2016) Transcription factors and plants response to drought stress: current understanding and future directions. *Frontiers in Plant Science* 7, 1029.

Joung, J.K. and Sander, J.D. (2013) TALENs: a widely applicable technology for targeted genome editing. *Nature Reviews Molecular Cell Biology* 14, 49–55.

Karvelis, T., Gasiunas, G. and Siksnys, V. (2017) Methods for decoding cas9 protospacer adjacent motif (PAM) sequences: a brief overview. *Methods* 121–122, 3–8.

Khan, S., Anwar, S., Yu, S., Sun, M., Yang, Z. *et al.* (2019) Development of drought-tolerant transgenic wheat: achievements and limitations. *International Journal of Molecular Sciences* 20, 3350.

Kim, Y.-G., Cha, J. and Chandrasegaran, S. (1996) Hybrid restriction enzymes: zinc finger fusions to Fok I cleavage domain. *Proceedings of the National Academy of Sciences USA* 93, 1156–1160.

Koonin, E.V. (2015) Origin of eukaryotes from within archaea, archaeal eukaryome and bursts of gene gain: eukaryogenesis just made easier? *Philosophical Transactions of the Royal Society Series B, Biological Sciences* 370, 20140333.

Kumar, D., Yadav, A., Ahmad, R., Dwivedi, U.N. and Yadav, K. (2022) CRISPR-based genome editing for nutrient enrichment in crops: a promising approach toward global food security. *Frontiers in Genetics* 13, 932859.

Li, R., Liu, C., Zhao, R., Wang, L., Chen, L. et al. (2019) CRISPR/Cas9-mediated slnpr1 mutagenesis reduces tomato plant drought tolerance. *BMC Plant Biology* 19, 38.

Li, T., Yang, X., Yu, Y., Si, X., Zhai, X. et al. (2018) Domestication of wild tomato is accelerated by genome editing. *Nature Biotechnology* 36, 1160–1163.

Lino, C.A., Harper, J.C., Carney, J.P. and Timlin, J.A. (2018) Delivering CRISPR: a review of the challenges and approaches. *Drug Delivery* 25, 1234–1257.

Liu, H., Wang, K., Tang, H., Gong, Q., Du, L. et al. (2020) CRISPR/Cas9 editing of wheat TaQ genes alters spike morphogenesis and grain threshability. *Journal of Genetics and Genomics* 47, 563–575.

Liu, Q., Yang, F., Zhang, J., Liu, H., Rahman, S. et al. (2021) Application of CRISPR/Cas9 in crop quality improvement. *International Journal of Molecular Sciences* 22, 4206.

Mehta, A. and Merkel, O.M. (2020) Immunogenicity of Cas9 protein. *Journal of Pharmaceutical Sciences* 109, 62–67.

Metje-Sprink, J., Menz, J., Modrzejewski, D. and Sprink, T. (2018) DNA-free genome editing: past, present and future. *Frontiers in Plant Science* 9, 1957.

Molina, R., Sofos, N. and Montoya, G. (2020) Structural basis of CRISPR-Cas type III prokaryotic defence systems. *Current Opinion in Structural Biology* 65, 119–129.

Mushtaq, M., Sakina, A., Wani, S.H., Shikari, A.B., Tripathi, P. et al. (2019) Harnessing genome editing techniques to engineer disease resistance in plants. *Frontiers in Plant Science* 10, 550.

Nerkar, G., Devarumath, S., Purankar, M., Kumar, A., Valarmathi, R. et al. (2022) Advances in crop breeding through precision genome editing. *Frontiers in Genetics* 13, 880195.

Nkhata, S.G., Chilungo, S., Memba, A. and Mponela, P. (2020) Biofortification of maize and sweetpotatoes with provitamin A carotenoids and implication on eradicating vitamin A deficiency in developing countries. *Journal of Agriculture and Food Research* 2, 100068.

Oliva, R., Ji, C., Atienza-Grande, G., Huguet-Tapia, J.C., Perez-Quintero, A. et al. (2019) Broad-spectrum resistance to bacterial blight in rice using genome editing. *Nature Biotechnology* 37, 1344–1350.

Parry, M.A.J., Madgwick, P.J., Bayon, C., Tearall, K., Hernandez-Lopez, A. et al. (2009) Mutation discovery for crop improvement. *Journal of Experimental Botany* 60, 2817–2825.

Raza, S.H.A., Hassanin, A.A., Pant, S.D., Bing, S., Sitohy, M.Z. et al. (2022) Potentials, prospects and applications of genome editing technologies in livestock production. *Saudi Journal of Biological Sciences* 29, 1928–1935.

Ricroch, A.E., Martin-Laffon, J., Rault, B., Pallares, V.C. and Kuntz, M. (2022) Next biotechnological plants for addressing global challenges: the contribution of transgenesis and new breeding techniques. *New Biotechnology* 66, 25–35.

Roychowdhury, R., Das, S.P., Gupta, A., Parihar, P., Chandrasekhar, K. et al. (2023) Multi-omics pipeline and omics-integration approach to decipher plant's abiotic stress tolerance responses. *Genes* 14, 1281.

Saini, S., Sharma, I. and Pati, P.K. (2015) Versatile roles of brassinosteroid in plants in the context of its homoeostasis, signaling and crosstalks. *Frontiers in Plant Science* 6, 950.

Shiferaw, B., Smale, M., Braun, H.J., Duveiller, E., Reynolds, M. et al. (2013) Crops that feed the world 10. Past successes and future challenges to the role played by wheat in global food security. *Food Security* 5, 291–317.

Shim, J.S., Oh, N., Chung, P.J., Kim, Y.S., Choi, Y.D. et al. (2018) Overexpression of *OsNAC14* improves drought tolerance in rice. *Frontiers in Plant Science* 9, 310.

Singer, S.D., Laurie, J.D., Bilichak, A., Kumar, S. and Singh, J. (2021) Genetic variation and unintended risk in the context of old and new breeding techniques. *Critical Reviews in Plant Sciences* 40, 68–108.

Songmei, L., Jie, J., Yang, L., Jun, M., Shouling, X. et al. (2019) Characterization and evaluation of *OsLCT1* and *OsNramp5* mutants generated through CRISPR/Cas9-mediated mutagenesis for breeding low Cd rice. *Rice Science* 26, 88–97.

Sprink, T., Metje, J. and Hartung, F. (2015) Plant genome editing by novel tools: TALEN and other sequence specific nucleases. *Current Opinion in Biotechnology* 32, 47–53.

Stangoulis, J.C.R. and Knez, M. (2022) Biofortification of major crop plants with iron and zinc – achievements and future directions. *Plant and Soil* 474, 57–76.

Stra, A., Almarwaey, L.O., Alagoz, Y., Moreno, J.C. and Al-Babili, S. (2022) Carotenoid metabolism: new insights and synthetic approaches. *Frontiers in Plant Science* 13, 1072061.

Strobbe, S. and Van Der Straeten, D. (2017) Folate biofortification in food crops. *Current Opinion in Biotechnology* 44, 202–211.

Tang, L., Mao, B., Li, Y., Lv, Q., Zhang, L. *et al.* (2017) Knockout of *OsNramp5* using the CRISPR/cas9 system produces low cd-accumulating *indica* rice without compromising yield. *Scientific Reports* 7, 14438.

Tian, Q., Li, B., Feng, Y., Zhao, W., Huang, J. *et al.* (2022) Application of CRISPR/Cas9 in rapeseed for gene function research and genetic improvement. *Agronomy* 12, 824.

Trono, D. and Pecchioni, N. (2022) Candidate genes associated with abiotic stress response in plants as tools to engineer tolerance to drought, salinity and extreme temperatures in wheat: an overview. *Plants* 11, 3358.

Tuncel, A., Corbin, K.R., Ahn-Jarvis, J., Harris, S., Hawkins, E. *et al.* (2019) Cas9-mediated mutagenesis of potato starch-branching enzymes generates a range of tuber starch phenotypes. *Plant Biotechnology Journal* 17, 2259–2271.

Turnbull, C., Lillemo, M. and Hvoslef-Eide, T.A.K. (2021) Global regulation of genetically modified crops amid the gene edited crop boom – a review. *Frontiers in Plant Science* 12, 630396.

Wang, D., Vasconcelos, N.P. de, Poirier, M.J., Chieffi, A., Mônaco, C. *et al.* (2020) Health technology assessment and judicial deference to priority-setting decisions in healthcare: quasi-experimental analysis of right-to-health litigation in Brazil. *Social Science and Medicine* 265, 113401.

Wang, L., Li, J., Zhou, Q., Yang, G., Ding, X.L. *et al.* (2014) Rare earth elements activate endocytosis in plant cells. *Proceedings of the National Academy of Sciences USA* 111, 12936–12941.

Wu, J., Yan, G., Duan, Z., Wang, Z., Kang, C. *et al.* (2020) Roles of the *Brassica napus* DELLA protein BnaA6.RGA, in modulating drought tolerance by interacting with the ABA signaling component BnaA10.ABF2. *Frontiers in Plant Science* 11, 577.

Yadava, D.K., Hossain, F. and Mohapatra, T. (2018) Nutritional security through crop biofortification in India: status and future prospects. *Indian Journal of Medical Research* 148, 621–631.

Yang, Q., Zhang, C., Chan, M., Zhao, D., Chen, J. *et al.* (2016) Biofortification of rice with the essential amino acid lysine: molecular characterization, nutritional evaluation, and field performance. *Journal of Experimental Botany* 67, 4285–4296.

Yin, K., Gao, C. and Qiu, J.L. (2017) Progress and prospects in plant genome editing. *Nature Plants* 3, 17107.

Zeng, Y., Wen, J., Zhao, W., Wang, Q. and Huang, W. (2020) Rational improvement of rice yield and cold tolerance by editing the three genes *OsPIN5b*, *GS3*, and *OsMYB30* with the CRISPR–Cas9 system. *Frontiers in Plant Science* 10, 1663.

Zhang, Z.Q. (2013) Animal biodiversity: an update of classification and diversity in 2013. In: Zhang, Z.Q. (ed.) Animal biodiversity: an outline of higher-level classification and survey of taxonomic richness (addenda 2013). *Zootaxa* 3703, 5–11.

Zhou, X., Liao, H., Chern, M., Yin, J., Chen, Y. *et al.* (2018) Loss of function of a rice TPR-domain RNA-binding protein confers broad-spectrum disease resistance. *Proceedings of the National Academy of Sciences USA* 115, 3174–3179.

5 Role of Arbuscular Mycorrhizal Fungi in Nutrient Attainment and Plant Growth Development

Knight Nthebere[1], Manikyala Bhargava Narasimha Yadav[2], Rojalin Hota[3], Rahul Kumar[4] and Jaagriti Tyagi[5,6]*

[1]*Department of Soil Science and Agricultural Chemistry, Jayashankar Telangana State Agricultural University, Telangana, Hyderabad, India;* [2]*Department of Soil Science and Agricultural Chemistry, University of Agricultural Sciences, Dharwad, Karnataka, India;* [3]*Department of Soil Science and Agricultural Chemistry, MITS Institute of Professional Studies, Rayagada, Odisha, India;* [4]*Department of Soil Science and Agricultural Chemistry, Jharkhand RAI University, Ranchi, Odisha, India;* [5]*Amity Institute of Microbial Technology, Amity University, Noida, Uttar Pradesh, India;* [6]*Biotechnology Department, Dr. MPS Group of Institutions, Agra, Uttar Pradesh, India*

Abstract

Conventional agriculture relies extensively on synthetic chemical fertilizers, depleting vital soil nutrients essential for optimal plant nutrition, which may result in substantial land degradation. Plant growth-promoting microorganisms (PGPMs) inhabit the rhizosphere, converting several essential nutrients into available forms for plant uptake and utilization. PGPM-secreted hormones interact with beneficial or pathogenic counterparts in the rhizosphere, enhancing soil composition, fertility and functionality. This directly or indirectly supports plant growth, particularly in nutrient-deficient conditions. Soil microbiomes are crucial for maintaining essential biogeochemical nutrient cycling and transformation processes vital for plant growth. The microbiome's responses to different types of stress can serve as practical indicators for effective management strategies. Mycorrhizal fungi are the most ubiquitous among microbes, forming a primitive and extensive mutualistic association with tracheophytes. Within this category, arbuscular mycorrhizal fungi (AMF) are crucial for field crops. The mutualism between AMF and tracheophytes involves a network of extraradical mycelium (ERM) that extends into the rhizosphere. These ERM networks increase soil utilization and nutrient absorption, especially nitrogen and phosphorus, which are immobile. Increased biotic and abiotic stress tolerance, gene expression alteration, host defence system regulation and hormonal balancing promote host plant growth. The microbiological activities in the rhizosphere are altered in the presence of AMF due to reduced root exudation resulting from AMF activity, leading to qualitative modifications. In this chapter, we discuss various components of rhizosphere engineering, the role of AMF and mechanisms for alleviating nutrient stress for plant development in various field crops.

Keywords: Arbuscular mycorrhizal fungi, plant growth-promoting microorganisms, rhizosphere engineering components, nutrient stress, soil microbiome, plant growth development

*Corresponding author: jaagriti.tyagi13@gmail.com; jtyagi@amity.edu

© CAB International 2025. *Soil Health and Nutrition Management*
(eds N.C. Joshi, T. Leustek and P.K. Singh)
DOI: 10.1079/9781800624597.0005

5.1   Introduction

The swift rise in the global human population, climate change with both biotic and abiotic stressors and land shortages have negatively impacted global food production. With projections showing a further increase in the world's population, surpassing 9 billion by 2050, food insecurity is ongoing and anticipated to escalate (Kumar and Dubey, 2020). To maximize production from the remaining agricultural fields, the pressure from population growth has resulted in an intensive, excessive utilization of synthetic fertilizers and agrochemicals. Plants with poor nutrient absorption rates take up approximately 20–30% of the applied fertilizer, and over half of the chemical fertilizers used are lost to the environment because of inefficiencies in nutrient utilization and soil dynamics in agriculture (Fageria, 2014). Additionally, the yields of numerous plant varieties that emerged during the Green Revolution have reached a plateau, with diminished responsiveness to fertilizers. Soil-applied fertilizers, such as nitrogen and phosphorus, are rapidly volatilized, easily washed away and gradually converted into unavailable forms due to natural processes, posing a greater risk to the ecosystem and biosphere for future generations. One promising approach to enhance the efficiency of scarce non-renewable fertilizers/resources is the application of rhizosphere symbionts known as mycorrhizal fungi, which have garnered significant attention as a potential low-input solution to increase the efficiency of crop hosts. This has the potential to boost plant productivity and alleviate environmental pressures simultaneously.

Arbuscular mycorrhizal fungi (AMF) form the most extensive symbiotic associations between plants and fungi, significantly enhancing soil quality across its physical, chemical and biological dimensions. This is achieved through the extension of AMF hyphae into the rhizosphere, thereby improving nutrient absorption capacity, particularly of phosphorus and micronutrients (Mohandas et al., 2013; Akhzari et al., 2015; Ebbisa, 2022) (Table 5.1). The induction of mycorrhizal connections elicits physiological alterations in host plants. Like other soil microorganisms, AMF act as ecosystem engineers on plants' roots and root surfaces. These remarkable characteristics of AMF are attributed mainly to their mycelium. These mycelia, or hyphae, absorb nutrients through an osmotrophic process and explore a larger external area compared with non-mycorrhizal roots. As soil microorganisms, AMF contribute to forming soil aggregates due to the symbiotic interactions that profoundly alter root function (Rillig, 2004). Additionally, they produce substances such as glomalin and organic acids, which directly contribute to maintaining soil structure (Hashem et al., 2018). Indeed, the hyphae of AMF can unite soil particles, thereby improving soil structure, augmenting moisture retention, encouraging the formation of soil macropores and enhancing water-holding capacity and porosity (Akhzari et al., 2015).

Many research studies have recorded the significant influence of soil biology, especially mycorrhizal fungi, on soil fertility and overall quality. Mycorrhizal roots and their secretions can acquire additional nutrients from nutrient-deficient soils compared with non-mycorrhizal roots, enabling them to access more soil resources (Hashem et al., 2018). Additionally, AMF plays a major part in various functions of the soil, which entails carbon (C), nitrogen (N) and phosphorus (P) cycling for supporting the growth of the plant and nutrition in agroecosystems. Plants inoculated with mycorrhizae exhibit enhanced competitiveness in P-limited soil conditions. Several investigations have reported that mycorrhizal associations lead to improved soil structure and increased nutrient and water uptake due to the hyphae of AMF expansion within the rhizosphere soil (Karandashov and Bucher, 2005; Smith and Read, 2008). Medina and Azcón (2010) advise introducing mycorrhizal fungus to the soil to improve plant growth in P-deficient soils in Mediterranean climates as a useful biostrategy. Mycorrhizal inoculation has also been found to reduce soil salinity levels and electrical conductivity (Li et al., 2012).

Furthermore, AMF colonization can enhance a plant's ability to withstand stress conditions such as drought and salinity. In a study conducted with pepper plants grown in soil with various salt concentrations, Çekiç et al. (2012) found that mycorrhizal inoculation enhanced relative water content, P content, total chlorophyll levels and carotenoid content. The chapter explores the practical utilization of AMF in augmenting soil fertility, with a particular emphasis on their functions in enhancing the soil's biochemical,

Table 5.1. Roles of arbuscular mycorrhiza fungi (AMF) in nutrient acquisition in plants.

Fungi	Nutrients	Plant	Response	Reference
Rhizophagus irregularis	Zinc (Zn), phosphorus (P)	*Medicago truncatula*	Increased tolerance to P and Zn soil deficiency, *PT* genes, and *MtZIP5* and *MtZIP2* genes induced in inoculated seedlings	Nguyen et al., 2019
Funneliformi and *Glomus* sp.	Zn, P	*Pistacia vera*	Zn and P uptake was increased in AMF seedlings, improved plant growth, biomass and photosynthetic parameters	Paymaneh et al., 2023
Funneliformis mosseae	N, P, potassium (K)	*Prunus maritima*	Mitigation of salt stress and increased content of N and P in AMF-inoculated roots. Increased net photosynthetic rate (P_n), stomatal conductance (G_s), transpiration rate (E) and intercellular CO_2 concentration (C_i) value in inoculated plants. Higher F_v/F_m, qP and Φ_{PSII} values (indicating high electron transport rate, severe water deficit stress and low light intensity, respectively)	Zai et al., 2021
Rhizophagus intraradices	K, P	*Iris wilsonii*	Increased activity of antioxidant enzymes, proline and malondialdehyde. Improved K and P concentration in treated seedlings under chromium (Cr) stress	Huang et al., 2021
25 species of glomales	P, K, sodium (Na), calcium (Ca)	*Ceratonia siliqua*	Increased plant growth, nutrient content, water content, organic solutes and stomatal conductance under drought stress in AMF seedlings. H_2O_2 and monodialdehyde (MDA) content was reduced in AMF seedlings	Boutasknit et al., 2020
Glomus spp.	Zn	*Zea mays*	Higher stomatal conductance and photosynthetic rate, increased antioxidant enzyme activity and Zn content in AMF plants	Saboor et al., 2021
Glomus versiforme	N, P, K	*Nicotiana tabacum* L.	Increased chlorophyll and carotenoid content, photosynthesis and PSII efficiency. Reduction of electrolyte leakage and lipid peroxidation. Enhanced plant growth hormones (abscisic acid (ABA) and indole-3-acetic acid (IAA)) and secondary metabolite (phenyl and flavonoid) concentration in AMF treatments under drought stress	Begum et al., 2022
Rhizophagus irregularis	Zn, K, N	*Eucalyptus grandis*	Ten genes were upregulated, whereas 19 genes were downregulated during the stress. High Zn stress tends to upregulate three P-related transporters, six phosphate transporters (PHTs) and two nitrate transporters (NRTs), whereas P-related transporters, four PHTs and four nitrogen-related transporters were downregulated during the high Zn stress	Wang et al., 2022a, 2022b
Glomus sp.	N, P	*Iris tectorum*	Increased the biomass and nutrient levels, upregulated heavy metal transporter genes (*P-atPase*, *MIT*, *CDF* and *ABC*) under Cr stress	Zhao et al., 2023
Funneliformis mosseae and *Diversispora tortuosa*	Na, P, K, magnesium (Mg)	*Zelkova serrata* (Thunb.) Makino	Improved plant biomass by reducing Na^+ content and increasing P, K^+ and Mg^{2+} content. Increased leaf photosynthetic activity, photosynthetic pigments and stomatal conductance	Wang et al., 2019

Fungi	Nutrient	Plant	Effect	Reference
Rhizoglomus irregulare	Na, K	Zea mays	Increased level of K^+ escorted by an effective decline in Na^+ ions in plant tissues under saline stress. Improved plant biomass, nutrient status, plant growth and development	Moreira et al., 2020
Ambispora leptoticha	N, P, K	Glycine max MAUS 2 and MAUS 212	Increased total biomass, biovolume index, root volume, seed yield and leaf growth parameters	Ashwin et al., 2022
Rhizoglomus intraradices, Claroideoglomus etunicatum, C. claroideum and Funneliformis mosseae	N, P, carbon (C)	Cajanus cajan (L.) Millsp.	Increased soil enzyme activity, reciprocally increased the bioavailability of N, P and C content	Bisht and Garg, 2022
Rhizophagus clarus	N, K	Glycine max L.	Increased photosynthesis rate and chlorophyll fluorescence. Improved physiological, morphological and nutritional traits were improved in AMF seedlings under drought stress	Oliveira et al., 2022
Glomus mossae, G. hoi and Rhizophagus intraradices	N, P, K, Zn, copper (Cu), manganese (Mn)	Glycine max	Increased seed oil content, P concentration of grain, Zn, leaf K and Cu and stem Mn	Ezzati Lotfabadi et al., 2022
AMF consortium 'Rhizolive consortium' 26 species	P, Ca, K, Na	Olea europaea	Enhanced plant biomass and nutrient content under biotic stress	Boutaj et al., 2020
Rhizophagus intraradices	N, P	Catalpa bungei C.A.Mey	Improved leaf area, gas exchange and photosynthetic parameters. Increased concentration of plant growth hormone like IAA, gibberellin, ABA and zeatin. Improved absorption of phosphorus and P, also quality of the soil was enhanced by accumulation of glomalin-related soil protein (GRSP) and the ratio of macroaggregates in the surrounding soil	Chen et al., 2020
Glomus intraradices	Nickel (Ni)	Triticum aestivum	Reduced Ni stress, increased chlorophyll, stomatal conductance, soluble sugar content and leaf transpiration rate. Reduction in MDA, H_2O_2 content and electrolyte leakage in AMF seedlings	Rehman et al., 2022
Funneliformis mosseae and Corymbiglomus tortuosum	N, P	Gleditsia sinensis Lam	Increased activity of enzymes like nitrate reductase, niacinamide adenine dinucleotide oxidase and other soil enzymes like alkali hydrolysable nitrogen (AN), total P and P under NaCl stress	Ma et al., 2022a, 2020b

Continued

Table 5.1. Continued.

Fungi	Nutrients	Plant	Response	Reference
Funneliformis mosseae BGC XJ01	N, P, K	*Nicotiana tabacum* L.	Root growth parameters (root biomass, root/shoot ratio, root architecture) were improved, accelerated the capture and conversion of solar energy (photosynthetic rate, Φ_{PSII}), and increased nutrient (N, P, K) uptake. Increased antioxidant enzyme activity and photosynthetic activity while decline in MDA and H_2O_2 content was observed	Liu *et al.*, 2021
Claroideoglomus etunicatum	K, P, Ca, Mg	*Zea mays*	Enhanced shoot and root fresh and dry weight, increased uptake of K, P, Ca and Mg. Other bacterial and fungal species regulated in soil including a decline in lanthanum stress in soil	Hao *et al.*, 2021
Funneliformis mosseae	N, P, K, Ca, Mg	*Zelkova serrata*	Enhanced nutrient uptake in leaves and roots. Photosynthetic parameters were increased. Antioxidant enzymes and antioxidant metabolites increased in AMF seedlings. Depletion in H_2O_2 and MDA content under salt stress	Wahab *et al.*, 2023
Rhizophagus irregularis	N, P	*Lycopersicon esculentum* Mill.	Increased plant growth parameters, N and P content in AMF seedlings	Roussis *et al.*, 2022
Claroideoglomus etunicatum BEG168	Molybdenum (Mo), N, P, sulfur (S)	*Sorghum bicolor*	Improved plant growth, biomass (shoot dry weight and root dry weight), improved photosystem (PS) II photochemical efficiency	Shi *et al.*, 2020
Funneliformis mossae	Na, K	*Leymus chinensis*	Increased plant growth, soluble sugar, proline and Na$^+$ and K$^+$ content. Decreased MDA content in AMF seedlings	Wang *et al.*, 2022a, 2022b
Rhizophagus irregularis	N, P	*Lycopersicon esculentum* Mill.	Enhanced nutrient uptake and plant growth. Total dry weight of seeds was increased in AMF seedlings	Roussis *et al.*, 2022
Funneliformis mosseae	N, P, K	*Malus domestica* Borkh	Increased average diameter, length and number of roots and root forks, leaf surface area, soluble carbohydrates and photosynthetic rate. Decline in relative electrolyte leakage and MDA content. Upregulation of *MdTYDC* promoted AMF symbiosis genes	Gao *et al.*, 2020
Funneliformis mosseae	N, P, K	*Euonymus maackii*	Increased plant growth, photosynthetic efficiency and antioxidant enzyme activity	Li *et al.*, 2020
Rhizophagus irregularis	C, N	*Medicago truncatula*	Enhancement of plant growth and root growth parameters. Increased activity of acid invertase, sucrose synthetase and sucrose phosphate synthase activity. Increment of carbohydrate and amino acid content. Overexpression of sugar transportation genes *MtSUT1-1* and *MtSUT4-1* genes. Increased cysteine and sucrose content	Zhang *et al.*, 2020
Funneliformis mosseae and *Claroideoglomus etunicatum*	C, N, P	*Cinnamomum migao* (*C. migao*)	Enhanced plant growth and biomass. Improved water and nutrient (C, N, P) content under drought stress	Xiao *et al.*, 2023
Glomus spp.	P, K, Mg, Ca, aluminium (Al), iron (Fe)	*Sorghum bicolor*	Increased uptake of P and inhibition of Fe and K uptake from root to shoot. Increased plant growth and biomass	Zhen *et al.*, 2023

AMF	Nutrients	Plant species	Effect	References
Glomus spp.	Zn, Fe, Cu, Mn, Na, K	Triticum aestivum	Increased chlorophyll a, b, total, carotenoids and micronutrient (Zn, Fe, Cu, Mn) content. Increased uptake of K ions while Na was decreased. Increment in antioxidant enzyme actime activity while MDA and H_2O_2 content was decreased. Other plant growth parameters were increased in inoculated plants	Huang et al., 2023
AMF	Ca	Malus robusta	Upregulated expression of auxin synthesis genes, organic acid secretion genes, calcium transporters and channels (e.g. MdAux/IAAs, MdGH3, MdSAUR), TCA cycles (MdCS, MdMDH, MdACO), phosphate transporters (MdPHT1;1, MdPHT1;10, MdPHT1;3) and Ca^{2+} signal transduction pathways (MdCa^{2+}/ATPase, MdCML, MdTPC1, MdCDPK). Increased translocation of Ca from roots to shoot tissue	Wu et al., 2023
Funneliformis and Glomus sp.	Zn, Cu, Fe, Mn	Pistacia vera	Increment in plant growth parameters like root length, shoot length, root fresh weight, root dry weight, shoot fresh weight and shoot dry weight. Increased P and Zn uptake in AMF seedlings	Paymaneh et al., 2023
Funneliformis mosseae and F. geosporum	P, K, Na	Sorghum bicolor	Increased activity of dehydrogenase and alkaline phosphatase enzymes. Increased plant growth parameters and plant biomass. Increased population of other plant growth-promoting microorganisms like bacteria, actinobacteria and fungi	Chandra et al., 2022
Glomus mosseae	N, P, K, Fe, Mn	Salvia rosmarinus, Amaranthus sp. and Brassica oleracea	Increased nutrient uptake, decreased cadmium (Cd) content in root cells. Improved plant growth parameters like shoot, root length, shoot root fresh and dry weight	Nasiri et al., 2022
Claroideoglomus etunicatum	P, K, Ca, Mg, Fe, Mn, Zn	Camellia sinensis	Increased plant height, biomass, root volume, number of lateral roots and plant growth hormones like ABA, brassinosteroids, gibberelins and IAA. Improved uptake of P, K, Ca, Mg, Fe, Mn and Zn nutrients	Liu et al., 2023
Funneliformis mosseae	S, Ca, NO_3, PO_4	Triticum durum	Increased plant growth and relative water content. Increased expression of drought stress responsive genes TdSHN1 and TdDRF1, sulfur homeostasis genes TdSULTR1.1, TdSULTR1, TdOASTL1 (C) and TdSAT1 and mycorrhizal colonization genes FM18S and TdPT11	Fiorilli et al., 2022
Glomus sp. and Gigaspora	N, P, K	Barley (Hordeum vulgare L.)	Improved plant height, biomass, spike length and number, grain weight and yield. Increased uptake of N, P and K in grains and straw and increased tolerance to salt stress	Masrahi et al., 2023
Funneliformis mosseae and Corymbiglomus tortuosum	N	Gleditsia sinensis Lam. and Zelkova serrata	Increased plant growth, chlorophyll content, photosynthetic nitrogen use efficiency, nitrate reductase activity and leaf nitrogen content	Ma et al., 2022a, 2022b

physical and organic attributes. Additionally, it highlights the concept of rhizosphere engineering as a strategy to ameliorate nutrient stress for optimal plant growth.

5.2 Exchange of Nutrient Mechanisms by Arbuscular Mycorrhizas

Mycorrhizal fungi, particularly AMF, are acknowledged as the primary microbial partners for most crops, holding particular significance in the mobilization and transportation of P in soils (Read and Perez-Moreno, 2003). These fungi's hyphae have a far higher surface-to-volume ratio than root hairs, and they can spread up to 8 cm beyond nutrient-depleted zones surrounding the roots (Millner and Wright, 2002). This unique feature enables AMF to retrieve even highly immobile nutrients, such as phosphate, efficiently. The hyphae of AMF have specialized structures named arbuscules that are formed inside the root cells of the host plant (Fig. 5.1). These arbuscules are where the actual uptake of P takes place. The hyphal membranes found in arbuscules include transport proteins, such as phosphate transporters, which aid in the absorption of inorganic phosphate ions ($H_2PO_4^-$) from the soil, as highlighted in studies by Bolan (1991) and George *et al.* (1992, 1995). It serves as a phosphorus activator, expediting the process of transforming P into readily available forms through chemical reactions and biological interactions (Zhu *et al.*, 2018). In the case of P deficiency, plants frequently get the advantage from AMF plant symbiosis, and AMF rapidly promotes plant development. For sustainable agriculture, adding AMF is an essential management practice when the native mycorrhizal potential improves the quantity and quality of soil (Luginbuehl *et al.*, 2017).

AMF rely on the photosynthetic carbon from host plants for their life cycle and, in exchange, provide nutrients, including P, N and zinc (Zn) to the plant, with the arbuscular serving as the leading site for nutrient exchange (Balestrini and Bonfante, 2005). Arbuscules' molecular and structural configurations play a crucial role in making nutrient exchange easier and more stable, and studies also show that extraradical hyphae of AMF take in things like ammonium, NO_3^- and amino acids (Chalot *et al.*, 2006; Ellerbeck *et al.*, 2013). A significant quantity of N is believed to be absorbed as

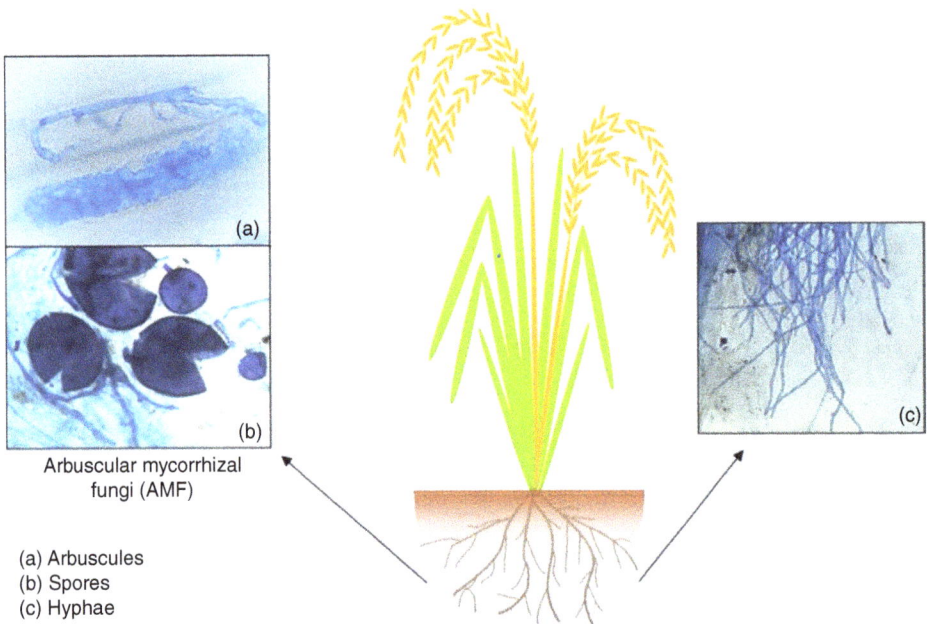

Arbuscular mycorrhizal
fungi (AMF)

(a) Arbuscules
(b) Spores
(c) Hyphae

Fig. 5.1. Characteristic features of arbuscular mycorrhiza fungi in nutrient acquisition.

ammonia through fungal-encoded AMT1 family transporters found in *Glomus* sp. (Rui *et al.*, 2022). However, there is no evidence to support the movement of ammonium or nitrate ions by the fungus; instead, N is proposed to be transported in the form of the amino acid arginine (Govindarajulu *et al.*, 2005). It is possible to supply amino acids straight to the interfacial apoplast so that the plant can absorb them. It was discovered that AMF reduced nitrate leaching 40 times more than tomato mutant plants that were not mycorrhizal. Moreover, the considerable biomass and high N requirement of AMF mean that they symbolize a global N pool to fine roots and play a substantial and overlooked part in the N cycle (Hodge and Fitter, 2010).

AMF have a complex and fascinating impact on plants' Zn metabolism. AMF are highly efficient in the sequestration of Zn from soils with low Zn concentrations, making them available to the plant. Additionally, AMF also have a protective role by limiting the build up of Zn to harmful levels in plants cultivated in soils with high Zn content, as highlighted by Cavagnaro *et al.* (2010) and Watts-Williams and Cavagnaro (2012). Research conducted by Bürkert and Robson (1994) and Jansa *et al.* (2003) documented the Zn absorption by AMF at distances ranging from 40 to 50 mm away from the root's surface. AMF can promote the uptake of Zn in the soil by raising the density of hyphal length in Zn patches.

Additionally, it can also increase the expression levels of genes responsible for Zn uptake (Maldonado-Mendoza *et al.*, 2001). The processes involved in transporting Zn over long distances from the extraradical hyphae of AMF in the soil to the intracellular symbiotic interface within the root cortex are not entirely understood. Irrespective of the mechanism by which Zn is carried in the extraradical hyphae of AMF, any Zn that reaches the interface between the plant and the fungus must be released into the interfacial apoplastic space and subsequently absorbed by the plant (Cavagnaro, 2008).

5.3   Impact of Arbuscular Mycorrhizal Fungi on Soil Physical Attributes

The dynamics of aggregate soil characteristics are often associated with changes in management practices and the environment. Generally, soils in harsh environments are characterized by poor and compacted soil structures, low water-holding capacity and soil organic matter (SOM) content, and widespread nutrient deficiency. According to Kavdir and Smucker (2005), the structure of the soil has a profound impact on various critical soil functions, including but not limited to soil productivity, biological activity, root growth, soil stability and nutrient cycling. The presence of AMF contributes to the enhancement of soil aggregate stability, attributed to the generation of extraradical hyphae and the secretion of a protein called glomalin (De Novais *et al.*, 2019). Glomalin is insoluble in water and stable at high temperatures (>120°C). Extraction from hyphae or soil requires citrate at neutral to alkaline pH. Glomalin, a glycoprotein with N-linked oligosaccharides, has some characteristics of hydrophobins, which contribute to the hydrophobicity of soil particles to allow for air penetration and water drainage (Hashem *et al.*, 2018). It remains in the soil after hyphae die (Driver *et al.*, 2005) (Fig. 5.2). In several soil types where organic matter is the main binder, glomalin and water-stable aggregates are highly correlated (Rillig *et al.*, 2003; Rillig, 2004). Wright and Upadhyaya (1996) noted that heightened aggregate stability, promoting improved soil structure, is pivotal in mitigating soil erosion, thereby fostering enhanced plant production (Lehmann *et al.*, 2020). Glomalin production from AMF also contributes significantly to stabilizing the soil condition and retaining water. Furthermore, it has been indicated that higher levels of glomalin increase water infiltration, improve porosity and air permeability, result in better root development, improve microbial efficiency and increase resistance to surface sealing and erosion (Bedini *et al.*, 2009). Driver *et al.* (2005) indicated that glomalin is thought to be deposited in soil by the degradation of extraradical hyphae of AMF. A study conducted by Bedini *et al.* (2009) demonstrated a positive correlation between the total hyphae length and hyphae density figures of AMF and mean weight diameter (MWD) figures of soil aggregates. In addition, they demonstrated that mycorrhizal-inoculated soils had a substantially greater MWD of macroaggregates than non-mycorrhizal soils.

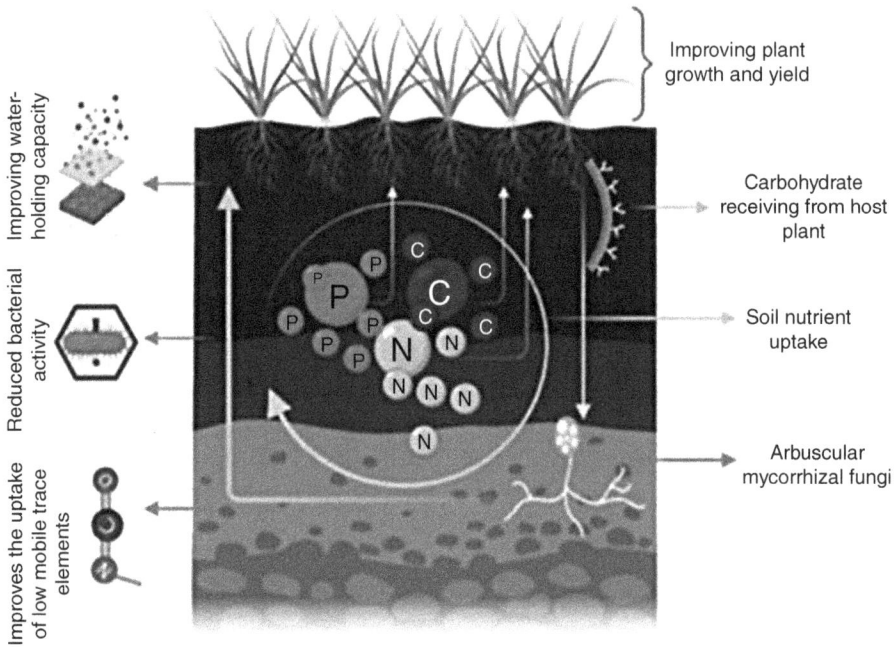

Fig. 5.2. Roles of arbuscular mycorrhiza fungi on chemical attributes of the soil.

5.4 Significance of Vesicular Arbuscular Mycorrhiza on Storage of Organic Carbon in Soil

Arbuscular mycorrhizal symbioses play a role in carbon fluxes between plants and the atmosphere through various pathways. Several research investigations, including those by Boldt *et al.* (2011), Parihar *et al.* (2020) and Hannula *et al.* (2020), have consistently shown that mycorrhizal symbiosis significantly enhances overall carbon assimilation by plants. As per Grimoldi *et al.* (2006), this symbiotic association can produce an additional carbon flux of 3–8% of gross photosynthesis into the soil. AMF consistently receive recently produced plant photoassimilates, which they utilize to construct their extensive extraradical mycelium (ERM) network. Soil organic matter pools may be balanced by the mycorrhizal hyphae network turnover at a rate greater than that of plant litter inputs (Godbold *et al.*, 2006). Zhu and Miller (2003) pointed out that variables like the amount of hyphal biomass produced, the turnover time

of accumulated hyphal biomass and the role of these fungi in stabilizing the formation of soil aggregates could determine the overall impact of AMF on soil carbon sequestration. Soil extraradical hyphae comprise a significant amount of microbial biomass, with a typical dry weight ranging from 0.03 to 0.5 mg/g (Olsson, 1999). The study by Zhu and Miller (2003) found that soil organic carbon (SOC) generated directly from AMF ranges from 54 to 900 kg/ha for a soil depth of 30 cm, a bulk density of 1.2 g/cm^3 and a 50% carbon content of dry hyphae. The fact that extraradical hyphae can vary in number and type indicates that soil hyphae retain measurable amounts of carbon, even though live hyphae turnover is fast. This highlights the practical significance of a stable hyphal network for subsurface carbon sequestration (Solaiman, 2014).

The hyphae of AMF are involved in carbon cycling in tandem with root exudates. According to Steinberg and Rillig (2003), these hyphae contribute to the production of glomalin, a glycoprotein-like material discovered by Wright and Upadhyaya (1998), which is very stable in soils. According to radiocarbon dating, the operationally defined glomalin extract surpasses the reported residence period for AMF hyphae in soils, with a residence time ranging from 6 to 42 years. Research in tropical forest soils has shown that glomalin carbon makes up as much as 5% of the total soil carbon, significantly higher than the proportion of carbon in soil microbe biomass. According to Wright and Upadhyaya (1996), there is a substantial link between soil glomalin levels, hyphal length and aggregate stability. This suggests that glomalin may indirectly affect soil carbon storage by stabilizing aggregates. Still, much is unclear about the connection between hyphal turnover and glomalin inputs, and additional measurement is needed to determine glomalin's exact contribution to carbon cycling.

5.5 Heavy Metal Toxicity Attenuation by Arbuscular Mycorrhizal Fungi

These fungi inhabit every soil rich in different metals, and their diverseness enables them to reduce the toxicity of heavy metals (HMs) present in these soils. AMFs are of great importance in boosting the tolerance capability of plants in accumulating HMs, particularly in soils contaminated with these metals, ultramafic soils and mining-degraded areas. Various techniques that exclude HMs by AMF have been proposed, including chelation by extracellular binding of the cell wall and the accumulation of HMs in the extraradical mycelium (Colpaert et al., 2011). By discharging complex agents into the soil solution phase, AMF can neutralize the toxicity of heavy metal ions. Mycorrhizal fungi are known to produce citric, malic and oxalic acids, which can either deploy or demobilize metals through processes such as complexation, which in turn rely upon certain factors such as soil pH, as highlighted by Meharg (2003) and Ahonen-Jonnarth et al. (2000). As seen in the research of Cabala et al. (2009), metals in the rhizosphere soil developed during mutual interaction of mycorrhiza and bacteria in different AMF and ectomycorrhizal plants. These compositions enhance the attachment of metals like zinc, lead and manganese in the soil rhizosphere. Research findings have indicated that glomalin, an abundant

glycoprotein released by AMF into the soil, contributes significantly to maintaining the stable aggregates in the soil, which may be involved in the deactivation of toxicity of HMs in soil (Malekzadeh et al., 2016; Zhou et al., 2023). Glomalin taken from soil regarded as polluted and/or hyphae of mycorrhiza irrevocably sequestrate metals like copper, cadmium, zinc and arsenic (Gonzalez-Chavez et al., 2002; Cornejo et al., 2008; Vodnik et al., 2008). According to Bellion et al. (2006), certain metal ions that mycorrhizal-infected plants detoxify are retained by fungal walls, and different attachments of cell wall molecules were seen, including glucan, chitin, galactosamine polymers, lesser peptides and proteins, together with potential attachment sites like free carboxyl, amino, hydroxyl, phosphate and mercapto compounds. Analysis of elemental dissemination in plants with mycorrhiza of the nickel hyperaccumulator Berkheya coddii found that the extraradical mycelium had a good attachment capacity for elements such as zinc, copper and nickel (Orłowska et al., 2008). On the other hand, González-Guerrero et al. (2008) demonstrated that at toxic levels of concentration, copper, zinc and cadmium were partially localized in the cell walls of fungi.

Following the passage of HMs through the fungal wall, additional mechanisms of avoidance may be triggered, including the modification of processes related to the influx of HM transporters as well as an augmentation efflux of HMs via the cell components, particularly the cell membrane (Meharg, 2003; Ouziad et al., 2005). Strategies used to separate the intracellular component and inactivate the absorbed segment of HMs are well established and documented. These HMs are transported into the vacuoles of fungi and sequestered a distance from the cytosol. According to González-Guerrero et al. (2008), the greatest metal concentration is located in spores. Ferrol et al. (2009) reported the development of AMF mycelium in a copper-enriched medium. They also observed the presence of spores in clusters containing only a small amount of copper, thus shielding the remaining colonies of fungi. Orłowska et al. (2008) demonstrated that the intraradical mycelium vesicles can also act as storage sites for HMs. When HM cations are unneutralized with the techniques above, these may exhibit oxidation-reduction activity, generating redox imbalance. This imbalance induces the production of highly reactive radical hydroxyl and superoxide groups, resulting in changes in cellular reactions. Although the adjustment techniques for this imbalance are not fully understood in mycorrhiza, it has been proposed that these may involve non-enzymatic antioxidant systems such as glutathione and vitamins C, E and B6, as well as enzymatic mechanisms such as catalases, superoxide dismutases (SOD), thioredoxins and glutaredoxins (Ferrol *et al.*, 2009). A slow down in SOD, particularly in the shoots of the plant, has been discovered to have an association with plants' inoculation with AMF (Neagoe *et al.*, 2013), which was due to lower oxidative imbalance in plants inoculated with AMF.

5.6   Arbuscular Mycorrhizal Fungi as Remediators for Salt-affected Soils (Saline Soils)

Arbuscular mycorrhizal fungi can mitigate the stress brought about by salinity, thus improving the development and performance of soils. Several techniques by which AMF tolerate salinity have been suggested, as well as some benefits:

1. Improvement of plant nutrient uptake and/or elevated leaf sequestration of salts like chlorides.
2. Modification of osmotic balance in the plants, leading to a decrease in water imbalance in the host plants and also reducing the concentration of toxic elements like sodium and chloride ions (Gupta and Mukerji, 2000).
3. Compartmentalization of these elemental ions (Jennings and Burke, 1990).
4. Osmotic adaptation is achieved by producing osmo-protectants including soluble carbohydrates, amino acids and nitrogen compounds (Hampp and Schaeffer, 1995).

The mechanisms through which AMF confer salt tolerance include improving plant nutrition, particularly phosphorus, or improving the acquisition of nutrients with low mobility (Ruiz-Lozano and Azcon, 2000; Al-Karaki, 2001). The preservation of a high potassium-to-sodium ratio, particularly in the shoots, has been seen as a potential technique to improve toleration of salinity (Cakmak, 2005). The increased potassium content in the mycorrhizal plants may not directly result from

improved phosphorus plant nutrition (Poss *et al.*, 1985). Among other elemental ions, potassium has been discovered to have the potential to alleviate the adverse impacts of salinity for plants (Cakmak, 2005). At the cell structural level, potassium requirements could impact salt-induced oxidative imbalance and associated destruction of the cell.

Consequently, the supply of more potassium to plants growing in salt-affected soils may mitigate the issue of cell destruction (Cakmak, 2005). Thus, enhancing the potassium concentration or reducing sodium content for mycorrhizal plants can also be a determinant factor for enhancing plant salinity tolerance by increasing the internal potassium-to-sodium ratio (Rinaldelli and Mancuso, 1996). Some of the probable techniques that can be employed for improving tolerance to salinity by AMF include their impact on slowing down the imbalance of water to the plants and a reduction in the concentration of elemental ions that are toxic like sodium and chloride (Gupta and Mukerji, 2000; Al-Karaki, 2001; Augé, 2001).

Osmotic adaptation is seen in the ability of the internal solute content to reduce outward water potential, especially that triggered by increased salt stress (Jennings and Burke, 1990). This represents a well-documented plant response to salt stress, involving the accumulation of soluble materials with less molecular mass (e.g. proline and betaine) (Ben Khaled *et al.*, 2003). Osmotic adjustment through compatible solutes helps salinized plants preserve the turgidity of their leaves as well as boosting various physiological functions such as photosynthesis, transpiration, conductance and water-use efficiency (Augé, 2001; Hu *et al.*, 2020). Likewise, some strategies through which AFM adapt to an increase in salt contents involve adaptation to osmotic environments through the composition of interlinkable solutes like polyols and the build up of betaine and proline. These intertwined solutes enhance the content of the structural cells of numerous AMFs when responding to salt stress (Hampp and Schaeffer, 1995; Naidu, 1998).

5.7   Effect of Mycorrhizae on Soil Quality

The importance of soil quality is paramount for sustainable development, and this significance is increasing rapidly due to swift urbanization and industrialization. The physical and chemical properties of soil matter and the diversity and health of soil biota also play a role in determining soil quality (Doran and Linn, 1994). AMFs are an active and major component of the soil microbial community and are important in soil quality (Bowen and Rovira, 1999). Barea *et al.* (2011) indicated that symbiotic mycorrhizal associations are fundamental in optimizing plant fitness and soil quality. Plants with a symbiotic relationship with AMF may withstand heavy metals and prevent water stress by influencing the rhizosphere's physical, chemical and biological activity (Smith and Read, 2008). AMF are anticipated to yield synergistic benefits for crops by enhancing soil quality. The potential impact of AMF on soil quality is associated with improvements in soil physical properties (Ortas *et al.*, 2013). Without passing through decomposition, the carbon from AMF plants can flow straight into the soil via the extraradical hyphae, adding to the carbon reservoirs of the soil. As a result, this increase in carbon intake seems to specifically boost the quantity and effectiveness of some soil organisms, especially those that can counteract soil-borne disease pathogens (Linderman, 2000).

AMF improve soil quality by binding microaggregates and primary soil particles together, resulting in: (i) improved soil structure; (ii) increased water infiltration and retention; and (iii) reduced soil erosion by minimizing leaching and runoff while making the nutrients available to plants. Another crucial role of mycorrhizal fungal mycelium is its contribution to creating water-stable soil aggregates (Miller and Jastrow, 2000; Barbosa *et al.*, 2019). Glomalin is a stable hydrophobic glycoprotein produced by AMF that helps improve soil architecture and properties. Numerous studies have concluded that glycoprotein serves as a binding protein when placed on the exterior walls of the extraradical mycelium (Wright and Upadhyaya, 1998, 1999). Hence, the extraradical hyphae, in collaboration with fibrous roots, can create a 'viscous-string container that aids in the interweaving of soil particles to produce macroaggregates' (Miller and Jastrow, 2000), which is a crucial component of soil composition. The significance of soil aggregates in the sequestration and stabilization of SOC is essential, as Smith (2008) described. As a result, AMFs are essential for maintaining soil structure in agricultural soils.

AMF inoculation often increases the success of ecological restoration (Jeffries *et al.*, 2003). For their role in maintaining soil structure and influencing soil ecological interactions, AMFs were singled out by Rillig (2004) as essential components of high-quality soil. However, the influence of AMF contributions to soil physical development is contingent on various factors and substantially impacts ecosystem services. The ecosystem services facilitated by AMF are based on the modification of root morphology and the development of an intricate, branched mycelial network in the soil. These interactions enhance the adherence between plants and soil and contribute to soil stability, including the binding action and improvement of soil structure. This, in turn, strengthens plants' uptake of microelements and water. Numerous studies have demonstrated the binding action of the mycorrhizal mycelial network on soil and its positive impact on soil structure (Caravaca *et al.*, 2006; Rillig *et al.*, 2010). A significant role of AMF in soils is their contribution to maintaining soil structure, which is vital in preserving soil quality. Mycorrhizal roots release AMF into the soil, creating a sophisticated web of hyphae that can reach concentrations of 30 m per gram of soil (Cavagnaro *et al.*, 2005; Wilson *et al.*, 2009).

5.8 Conclusion

In conclusion, AMF offer a number of benefits to the host plant through significant mechanisms and functional roles in various aspects of soil health, plant growth and development and environmental sustainability. These fungi have demonstrated their significance in enhancing nutrient acquisition, improving soil structure, sequestering carbon, alleviating heavy metal toxicity and mitigating salt stress. AMF contribute significantly to plant nutrition, especially in absorbing immobile nutrient elemental ions such as phosphates. Their extensive hyphal network and specialized structures like arbuscules enable efficient nutrient absorption, benefiting plant growth and development. AMF enhances phosphorus uptake and contributes to nitrogen acquisition by facilitating the uptake of ammonium and amino acids. This can slow down nitrate leaching and improve nitrogen's overall use efficiency. AMF have a role in plants increasing their zinc uptake through increased hyphal length, density and gene expression related to zinc uptake. Mycorrhizal fungi promote soil aggregate stability by producing extraradical hyphae and glomalin, a stable glycoprotein. This improves soil structure, reduces erosion and enhances water retention and infiltration. AMF contribute to soil carbon sequestration by funnelling plant-derived carbon into the soil as extraradical hyphae. This helps build up soil organic carbon, benefiting soil health and mitigating climate change. Mycorrhizal fungi are of utmost importance in assisting the plant in tolerating heavy metal pollution through exclusion, chelation and binding of heavy metals, preventing their harmful impacts on plants and promoting the growth and development of plants in contaminated soils. AMF assist in salt stress tolerance by improving plant nutrition,

altering water balance and aiding in osmotic adjustment. They help plants cope with salinity and maintain better growth under saline conditions. Mycorrhizal associations help improve soil quality by enhancing soil structure, nutrient availability and microbial activity. They have a positive impact on ecosystem services and are beneficial for ecological restoration efforts. Mycorrhizal fungi are essential allies in sustainable agriculture and environmental management. Their multifaceted contributions to soil health, plant growth and ecosystem resilience highlight their importance in promoting a more sustainable and productive future for agriculture and beyond. Recognizing and harnessing the potential of these fungi can have far-reaching benefits for food security, environmental conservation and the overall well-being of our planet.

References

Ahonen-Jonnarth, U., Van Hees, P.A.W., Lundström, U.S. and Finlay, R.D. (2000) Organic acids produced by mycorrhizal *Pinus sylvestris* exposed to elevated aluminium and heavy metal concentrations. *New Phytologist* 146, 557–567.

Akhzari, D., Attaeian, B., Arami, A., Mahmoodi, F. and Aslani, F. (2015) Effects of vermicompost and arbuscular mycorrhizal fungi on soil properties and growth of *Medicago polymorpha* L. *Compost Science and Utilization* 23, 142–153.

Al-Karaki, G.N. (2001) Salt stress response of salt-sensitive and tolerant durum wheat cultivars inoculated with mycorrhizal fungi. *Acta Agronomica Hungarica* 49, 25–34.

Ashwin, R., Bagyaraj, D.J. and Mohan Raju, B. (2022) Dual inoculation with rhizobia and arbuscular mycorrhizal fungus improves water stress tolerance and productivity in soybean. *Plant Stress* 4, 100084.

Augé, R.M. (2001) Water relations, drought and vesicular-arbuscular mycorrhizal symbiosis. *Mycorrhiza* 11, 3–42.

Balestrini, R. and Bonfante, P. (2005) The interface compartment in arbuscular mycorrhizae: a special type of plant cell wall? *Plant Biosystems* 139, 8–15.

Barbosa, M.V., Pedroso, D. de F., Curi, N. and Carneiro, M.A.C. (2019) Do different arbuscular mycorrhizal fungi affect the formation and stability of soil aggregates? *Ciência e Agrotecnologia* 43.

Barea, J.M., Palenzuela, J., Cornejo, P., Sánchez-Castro, I., Navarro-Fernández, C. *et al.* (2011) Ecological and functional roles of mycorrhizas in semi-arid ecosystems of southeast Spain. *Journal of Arid Environments* 75, 1292–1301.

Bedini, S., Pellegrino, E., Avio, L., Pellegrini, S., Bazzoffi, P. *et al.* (2009) Changes in soil aggregation and glomalin-related soil protein content as affected by the arbuscular mycorrhizal fungal species *Glomus mosseae* and *Glomus intraradices*. *Soil Biology and Biochemistry* 41, 1491–1496.

Begum, N., Wang, L., Ahmad, H., Akhtar, K., Roy, R. *et al.* (2022) Co-inoculation of arbuscular mycorrhizal fungi and the plant growth-promoting rhizobacteria improve growth and photosynthesis in tobacco under drought stress by up-regulating antioxidant and mineral nutrition metabolism. *Microbial Ecology* 83, 971–988.

Bellion, M., Courbot, M., Jacob, C., Blaudez, D. and Chalot, M. (2006) Extracellular and cellular mechanisms sustaining metal tolerance in ectomycorrhizal fungi. *FEMS Microbiology Letters* 254, 173–181.

Ben Khaled, L., Gomez, A.M., Ouarraqi, E.M. and Oihabi, A. (2003) Physiological and biochemical responses to salt stress of mycorrhized and/or nodulated clover seeding (*Trifolium alexandrinum* L). *Agronomie* 23, 571–580.

Bisht, A. and Garg, N. (2022) AMF species improve yielding potential of Cd stressed pigeonpea plants by modulating sucrose-starch metabolism, nutrients acquisition and soil microbial enzymatic activities. *Plant Growth Regulation* 96, 409–430.

Bolan, N.S. (1991) A critical review on the role of mycorrhizal fungi in the uptake of phosphorus by plants. *Plant and Soil* 134, 189–207.

Boldt, K., Pörs, Y., Haupt, B., Bitterlich, M. and Kühn, C. (2011) Photochemical processes, carbon assimilation and RNA accumulation of sucrose transporter genes in tomato arbuscular mycorrhiza. *Journal of Plant Physiology* 168, 1256–1263.

Boutaj, H., Meddich, A., Chakhchar, A., Wahbi, S., El Alaoui-Talibi, Z. *et al.* (2020) Arbuscular mycorrhizal fungi improve mineral nutrition and tolerance of olive tree to Verticillium wilt. *Archives of Phytopathology and Plant Protection* 53, 673–689.

Boutasknit, A., Baslam, M., Ait-El-Mokhtar, M., Anli, M., Ben-Laouane, R. *et al.* (2020) Arbuscular mycorrhizal fungi mediate drought tolerance and recovery in two contrasting carob (*Ceratonia siliqua* L.) ecotypes by regulating stomatal, water relations, and (in) organic adjustments. *Plants* 9, 80.

Bowen, G.D. and Rovira, A.D. (1999) The rhizosphere and its management to improve plant growth. *Advances in Agronomy* 66, 1–102.

Bürkert, B. and Robson, A. (1994) 65Zn uptake in subterranean clover (*Trifolium subterraneum* L.) by three vesicular-arbuscular mycorrhizal fungi in a root-free sandy soil. *Soil Biology and Biochemistry* 26, 1117–1124.

Cabala, J., Krupa, P. and Misz-Kennan, M. (2009) Heavy metals in mycorrhizal rhizospheres contaminated by Zn–Pb mining and smelting around Olkusz in southern Poland. *Water, Air and Soil Pollution* 199, 139–149.

Cakmak, I. (2005) The role of potassium in alleviating detrimental effects of abiotic stresses in plants. *Journal of Plant Nutrition and Soil Science* 168, 521–530.

Caravaca, F., Alguacil, M.M., Azcón, R. and Roldán, A. (2006) Formation of stable aggregates in rhizosphere soil of *Juniperus oxycedrus*: effect of AM fungi and organic amendments. *Applied Soil Ecology* 33, 30–38.

Cavagnaro, T.R. (2008) The role of arbuscular mycorrhizas in improving plant zinc nutrition under low soil zinc concentrations: a review. *Plant and Soil* 304, 315–325.

Cavagnaro, T.R., Smith, F.A., Smith, S.E. and Jakobsen, I. (2005) Functional diversity in arbuscular mycorrhizas: exploitation of soil patches with different phosphate enrichment differs among fungal species. *Plant, Cell and Environment* 28, 642–650.

Cavagnaro, T.R., Dickson, S. and Smith, F.A. (2010) Arbuscular mycorrhizas modify plant responses to soil zinc addition. *Plant and Soil* 329, 307–313.

Çekiç, F.Ö., Ünyayar, S. and Ortaş, İ. (2012) Effects of arbuscular mycorrhizal inoculation on biochemical parameters in *Capsicum annuum* grown under long term salt stress. *Turkish Journal of Botany* 36, 63–72.

Chalot, M., Blaudez, D. and Brun, A. (2006) Ammonia: a candidate for nitrogen transfer at the mycorrhizal interface. *Trends in Plant Science* 11, 263–266.

Chandra, P., Singh, A., Prajapat, K., Rai, A.K. and Yadav, R.K. (2022) Native arbuscular mycorrhizal fungi improve growth, biomass yield, and phosphorus nutrition of sorghum in saline and sodic soils of the semi-arid region. *Environmental and Experimental Botany* 201, 104982.

Chen, W., Meng, P., Feng, H. and Wang, C. (2020) Effects of arbuscular mycorrhizal fungi on growth and physiological performance of *Catalpa bungei* C.A.Mey. under drought stress. *Forests* 11, 1117.

Colpaert, J.V., Wevers, J.H.L., Krznaric, E. and Adriaensen, K. (2011) How metal-tolerant ecotypes of ectomycorrhizal fungi protect plants from heavy metal pollution. *Annals of Forest Science* 68, 17–24.

Cornejo, P., Meier, S., Borie, G., Rillig, M.C. and Borie, F. (2008) Glomalin-related soil protein in a Mediterranean ecosystem affected by a copper smelter and its contribution to Cu and Zn sequestration. *Science of the Total Environment* 406, 154–160.

De Novais, C.B., Avio, L., Giovannetti, M., de Faria, S.M., Siqueira, J.O. *et al.* (2019) Interconnectedness, length and viability of arbuscular mycorrhizal mycelium as affected by selected herbicides and fungicides. *Applied Soil Ecology* 143, 144–152.

Doran, J.W. and Linn, D.M. (1994) Microbial ecology of conservation management systems. In: Hatfield, J.L. and Stewart, B.A. (eds) *Soil Biology: Effects on Soil Quality (Advances in Soil Science)*. Lewis, Boca Raton, Florida, pp. 1–27.

Driver, J.D., Holben, W.E. and Rillig, M.C. (2005) Characterization of glomalin as a hyphal wall component of arbuscular mycorrhizal fungi. *Soil Biology and Biochemistry* 37, 101–106.

Ebbisa, A. (2022) Arbuscular mycorrhizal fungi (AMF) in optimizing nutrient bioavailability and reducing agrochemicals for maintaining sustainable agroecosystems In: de Sousa, R.N. (ed.) *Arbuscular Mycorrhizal Fungi in Agriculture – New Insights*. IntechOpen.

Ellerbeck, M., Schüßler, A., Brucker, D., Dafinger, C. and Loos, F. (2013) Characterization of three ammonium transporters of the glomeromycotan fungus *Geosiphon pyriformis*. *Eukaryotic Cell* 12, 1554–1562.

Ezzati Lotfabadi, Z., Weisany, W., Abdul-Razzak Tahir, N. and Mohammadi Torkashvand, A. (2022) Arbuscular mycorrhizal fungi species improve the fatty acids profile and nutrients status of soybean cultivars grown under drought stress. *Journal of Applied Microbiology* 132, 2177–2188.

Fageria, N.K. (2014) Yield and yield components and phosphorus use efficiency of lowland rice genotypes. *Journal of Plant Nutrition* 37, 979–989.

Ferrol, N., González-Guerrero, M., Valderas, A., Benabdellah, K. and Azcón-Aguilar, C. (2009) Survival strategies of arbuscular mycorrhizal fungi in Cu-polluted environments. *Phytochemistry Reviews* 8, 551–559.

Fiorilli, V., Maghrebi, M., Novero, M., Votta, C., Mazzarella, T. *et al.* (2022) Arbuscular mycorrhizal symbiosis differentially affects the nutritional status of two durum wheat genotypes under drought conditions. *Plants* 11, 804.

Gao, T., Liu, X., Shan, L., Wu, Q., Liu, Y. *et al.* (2020) Dopamine and arbuscular mycorrhizal fungi act synergistically to promote apple growth under salt stress. *Environmental and Experimental Botany* 178, 104159.

George, E., Haussler, K., Kothari, S.K., Ki, X.-L., Marschner, H. *et al.* (1992) Contribution of mycorrhizal hyphae to nutrient and water uptake by plants. In: Read, D.J. (ed.) *Mycorrhizas in Ecosystems*. CAB International, Wallingford, UK.

George, E., Marschner, H. and Jakobsen, I. (1995) Role of arbuscular mycorrhizal fungi in uptake of phosphorus and nitrogen from soil. *Critical Reviews in Biotechnology* 15, 257–270.

Godbold, D.L., Hoosbeek, M.R., Lukac, M., Cotrufo, M.F., Janssens, I.A. *et al.* (2006) Mycorrhizal hyphal turnover as a dominant process for carbon input into soil organic matter. *Plant and Soil* 281, 15–24.

Gonzalez-Chavez, C., D'Haen, J., Vangronsveld, J. and Dodd, J.C. (2002) Copper sorption and accumulation by the extraradical mycelium of different *Glomus* spp. (arbuscular mycorrhizal fungi) isolated from the same polluted soil. *Plant and Soil* 240, 287–297.

González-Guerrero, M., Melville, L.H., Ferrol, N., Lott, J.N.A., Azcón-Aguilar, C. *et al.* (2008) Ultrastructural localization of heavy metals in the extraradical mycelium and spores of the arbuscular mycorrhizal fungus *Glomus intraradices*. *Canadian Journal of Microbiology* 54, 103–110.

Govindarajulu, M., Pfeffer, P.E., Jin, H., Abubaker, J., Douds, D.D. *et al.* (2005) Nitrogen transfer in the arbuscular mycorrhizal symbiosis. *Nature* 435, 819–823.

Grimoldi, A.A., Kavanová, M., Lattanzi, F.A., Schäufele, R. and Schnyder, H. (2006) Arbuscular mycorrhizal colonization on carbon economy in perennial ryegrass: quantification by $^{13}CO_2/^{12}CO_2$ steady-state labelling and gas exchange. *New Phytologist* 172, 544–553.

Gupta, R. and Mukerji, K.G. (2000) The growth of VAM fungi under stress conditions. In: Mukerji, K.G., Chamola, B.P. and Singh, J. (eds) *Mycorrhizal Biology*. Kluwer, New York, pp. 57–62.

Hampp, R. and Schaeffer, C. (1995) Mycorrhiza-carbohydrate and energy metabolism. In: Varma, A. and Hock, B. (eds) *Mycorrhiza*. Springer, Berlin, pp. 267–296.

Hannula, S.E., Morriën, E., van der Putten, W.H. and de Boer, W. (2020) Rhizosphere fungi actively assimilating plant-derived carbon in a grassland soil. *Fungal Ecology* 48, 100988.

Hao, L., Zhang, Z., Hao, B., Diao, F., Zhang, J. *et al.* (2021) Arbuscular mycorrhizal fungi alter microbiome structure of rhizosphere soil to enhance maize tolerance to La. *Ecotoxicology and Environmental Safety* 212, 111996.

Hashem, A., Abd- Allah, E.F., Alqarawi, A.A., Radhakrishnan, R. and Kumar, A. (2018) Plant defense approach of *Bacillus subtilis* (BERA 71) against *Macrophomina phaseolina* (Tassi) Goid in mung bean. *Journal of Plant Interaction* 12, 390401.

Hodge, A. and Fitter, A.H. (2010) Substantial nitrogen acquisition by arbuscular mycorrhizal fungi from organic material has implications for N cycling. *Proceedings of the National Academy of Sciences USA* 107, 13754–13759.

Hu, S., Chen, Z., Vosátka, M. and Vymazal, J. (2020) Arbuscular mycorrhizal fungi colonization and physiological functions toward wetland plants under different water regimes. *Science of the Total Environment* 716, 137040.

Huang, G.M., Srivastava, A.K., Zou, Y.N., Wu, Q.S. and Kuča, K. (2021) Exploring arbuscular mycorrhizal symbiosis in wetland plants with a focus on human impacts. *Symbiosis* 84, 311–320.

Huang, S., Gill, S., Ramzan, M., Ahmad, M.Z., Danish, S. *et al.* (2023) Uncovering the impact of AM fungi on wheat nutrient uptake, ion homeostasis, oxidative stress, and antioxidant defense under salinity stress. *Scientific Reports* 13, 8249.

Jansa, J., Mozafar, A. and Frossard, E. (2003) Long-distance transport of P and Zn through the hyphae of an arbuscular mycorrhizal fungus in symbiosis with maize. *Agronomie* 23, 481–488.

Jeffries, P., Gianinazzi, S., Perotto, S., Turnau, K. and Barea, J.-M. (2003) The contribution of arbuscular mycorrhizal fungi in sustainable maintenance of plant health and soil fertility. *Biology and Fertility of Soils* 37, 1–16.

Jennings, D.H. and Burke, R.M. (1990) Compatible solutes – the mycological dimension and their role as physiological buffering agents. *New Phytologist* 116, 277–283.

Karandashov, V. and Bucher, M. (2005) Symbiotic phosphate transport in arbuscular mycorrhizas. *Trends in Plant Science* 10, 22–29.

Kavdır, Y. and Smucker, A.J.M. (2005) Soil aggregate sequestration of cover crop root and shoot-derived nitrogen. *Plant and Soil* 272, 263–276.

Kumar, A. and Dubey, A. (2020) Rhizosphere microbiome: engineering bacterial competitiveness for enhancing crop production. *Journal of Advanced Research* 24, 337–352.

Lehmann, J., Hansel, C.M., Kaiser, C., Kleber, M., Maher, K. *et al.* (2020) Persistence of soil organic carbon caused by functional complexity. *Nature Geoscience* 13, 529–534.

Li, Y., Chen, Y.L., Li, M., Lin, X.-G. and Liu, R.-J. (2012) Effects of arbuscular mycorrhizal fungi communities on soil quality and the growth of cucumber seedlings in a greenhouse soil of continuously planting cucumber in North China. *Pedosphere* 22, 79–87.

Li, Z., Wu, N., Meng, S., Wu, F. and Liu, T. (2020) Arbuscular mycorrhizal fungi (AMF) enhance the tolerance of *Euonymus maackii* Rupr. at a moderate level of salinity. *PLoS One* 15, e0231497.

Linderman, R.G. (2000) Effects of mycorrhizas on plant tolerance to diseases. In: Kapulnik, Y. and Douds, D.D., Jr (eds) *Arbuscular Mycorrhizas: Physiology and Function.* Kluwer Academic Publishers, Dordrecht, the Netherlands, pp. 345–365.

Liu, C.-Y., Hao, Y., Wu, X.-L., Dai, F.-J., Abd-Allah, E.F. *et al.* (2023) Arbuscular mycorrhizal fungi improve drought tolerance of tea plants via modulating root architecture and hormones. *Plant Growth Regulation* 102, 13–22.

Liu, L., Li, D., Ma, Y., Shen, H., Zhao, S. *et al.* (2021) Combined application of arbuscular mycorrhizal fungi and exogenous melatonin alleviates drought stress and improves plant growth in tobacco seedlings. *Journal of Plant Growth Regulation* 40, 1074–1087.

Luginbuehl, L.H., Menard, G.N., Kurup, S., Van Erp, H., Radhakrishnan, G.V. *et al.* (2017) Fatty acids in arbuscular mycorrhizal fungi are synthesized by the host plant. *Science* 356, 1175–1178.

Maldonado-Mendoza, I.E., Dewbre, G.R. and Harrison, M.J. (2001) A phosphate transporter gene from the extra-radical mycelium of an arbuscular mycorrhizal fungus *Glomus intraradices* is regulated in response to phosphate in the environment. *Molecular Plant-Microbe Interactions* 14, 1140–1148.

Malekzadeh, E., Aliasgharzad, N., Majidi, J., Abdolalizadeh, J. and Aghebati-Maleki, L. (2016) Contribution of glomalin to Pb sequestration by arbuscular mycorrhizal fungus in a sand culture system with clover plant. *European Journal of Soil Biology* 74, 45–51.

Ma, S., Yue, J., Wang, J., Jia, Z., Li, C. *et al.* (2022a) Arbuscular mycorrhizal fungi alleviate salt stress damage by coordinating nitrogen utilization in leaves of different species. *Forests* 13, 1568.

Ma, S., Zhu, L., Wang, J., Liu, X., Jia, Z. *et al.* (2022b) Arbuscular mycorrhizal fungi promote *Gleditsia sinensis* root growth under salt stress by regulating nutrient uptake and physiology. *Forests* 13, 688.

Masrahi, A.S., Alasmari, A., Shahin, M.G., Qumsani, A.T. and Oraby, H.F. (2023) Role of arbuscular mycorrhizal fungi and phosphate solubilizing bacteria in improving yield, yield components, and nutrients uptake of barley under salinity soil. *Agriculture* 13, 537.

Medina, A. and Azcón, R. (2010) Effectiveness of the application of arbuscular mycorrhiza fungi and organic amendments to improve soil quality and plant performance under stress conditions. *Journal of Soil Science and Plant Nutrition* 10, 354–372.

Meharg, A.A. (2003) The mechanistic basis of interactions between mycorrhizal associations and toxic metal cations. *Mycological Research* 107, 1253–1265.

Miller, R.M. and Jastrow, J.D. (2000) Mycorrhizal fungi influence soil structure. In: Kapulnik, Y. and Douds, D.D., Jr (eds) *Arbuscular Mycorrhizas: Physiology and Functions.* Kluwer Academic Publishers, Dordrecht, the Netherlands, pp. 3–18.

Millner, P.D. and Wright, S. (2002) Tools for support of ecological research on arbuscular mycorrhizal fungi. *Symbiosis* 33, 101.

Mohandas, S., Poovarasan, S., Panneerselvam, P., Saritha, B., Upreti, K.K. *et al.* (2013) Guava (*Psidium guajava* L.) rhizosphere *Glomus mosseae* spores harbor actinomycetes with growth promoting and antifungal attributes. *Scientia Horticulturae* 150, 371–376.

Moreira, H., Pereira, S.I.A., Vega, A., Castro, P.M.L. and Marques, A.P.G.C. (2020) Synergistic effects of arbuscular mycorrhizal fungi and plant growth-promoting bacteria benefit maize growth under increasing soil salinity. *Journal of Environmental Management* 257, 109982.

Naidu, B.P. (1998) Separation of sugars, polyols, proline analogues, and betaines in stressed plant extracts by high performance liquid chromatography and quantification by ultra violet detection. *Functional Plant Biology* 25, 793.

Nasiri, K., Babaeinejad, T., Ghanavati, N. and Mohsenifar, K. (2022) Arbuscular mycorrhizal fungi affecting the growth, nutrient uptake and phytoremediation potential of different plants in a cadmium-polluted soil. *Biometals* 35, 1243–1253.

Neagoe, A., Iordache, V., Bergmann, H. and Kothe, E. (2013) Patterns of effects of arbuscular mycorrhizal fungi on plants grown in contaminated soil. *Journal of Plant Nutrition and Soil Science* 176, 273–286.

Nguyen, T.D., Cavagnaro, T.R. and Watts-Williams, S.J. (2019) The effects of soil phosphorus and zinc availability on plant responses to mycorrhizal fungi: a physiological and molecular assessment. *Scientific Reports* 9, 14880.

Oliveira, T.C., Cabral, J.S.R., Santana, L.R., Tavares, G.G., Santos, L.D.S. *et al.* (2022) The arbuscular mycorrhizal fungus *Rhizophagus clarus* improves physiological tolerance to drought stress in soybean plants. *Scientific Reports* 12, 9044.

Olsson, P.A. (1999) Signature fatty acids provide tools for determination of the distribution and interactions of mycorrhizal fungi in soil. *FEMS Microbiology Ecology* 29, 303–310.

Orłowska, E., Mesjasz-Przybyłowicz, J., Przybyłowicz, W. and Turnau, K. (2008) Nuclear microprobe studies of elemental distribution in mycorrhizal and non-mycorrhizal roots of Ni-hyperaccumulator *Berkheya coddii*. *X-ray Spectrometry* 37, 129–132.

Ortas, I., Akpinar, C. and Lal, R. (2013) Long-term impacts of organic and inorganic fertilizers on carbon sequestration in aggregates of an entisol in Mediterranean Turkey. *Soil Science* 178, 12–23.

Ouziad, F., Hildlebrandt, U., Schmelzer, E. and Bothe, H. (2005) Differential gene expressions in arbuscular mycorrhizal-colonized tomato grown under heavy metal stress. *Journal of Plant Physiology* 162, 634–649.

Parihar, M., Rakshit, A., Meena, V.S., Gupta, V.K., Rana, K. *et al.* (2020) The potential of arbuscular mycorrhizal fungi in C cycling: a review. *Archives of Microbiology* 202, 1581–1596.

Paymaneh, Z., Sarcheshmehpour, M., Mohammadi, H. and Askari Hesni, M. (2023) Vermicompost and/or compost and arbuscular mycorrhizal fungi are conducive to improving the growth of pistachio seedlings to drought stress. *Applied Soil Ecology* 182, 104717.

Poss, J.A., Pond, E., Menge, J.A. and Jarrell, W.M. (1985) Effect of salinity on mycorrhizal onion and tomato in soil with and without additional phosphate. *Plant and Soil* 88, 307–319.

Read, D.J. and Perez-Moreno, J. (2003) Mycorrhizas and nutrient cycling in ecosystems – a journey towards relevance? *New Phytologist* 157, 475–492.

Rehman, S., Mansoora, N., Al-Dhumri, S.A., Amjad, S.F., Al-Shammari, W.B. *et al.* (2022) Associative effects of activated carbon biochar and arbuscular mycorrhizal fungi on wheat for reducing nickel food chain bioavailability. *Environmental Technology and Innovation* 26, 102539.

Rillig, M.C. (2004) Arbuscular mycorrhizae, glomalin, and soil aggregation. *Canadian Journal of Soil Science* 84, 355–363.

Rillig, M.C., Ramsey, P.W., Morris, S. and Paul, E.A. (2003) Glomalin, an arbuscular-mycorrhizal fungal soil protein, responds to land-use change. *Plant and Soil* 253, 293–299.

Rillig, M.C., Wagner, M., Salem, M., Antunes, P.M., George, C. *et al.* (2010) Material derived from hydrothermal carbonization: effects on plant growth and arbuscular mycorrhiza. *Applied Soil Ecology* 45, 238–242.

Rinaldelli, E. and Mancuso, S. (1996) Response of young mycorrhizal and non-mycorrhizal plants of olive tree (*Olea europaea* L.) to saline conditions. I. Short-term electrophysiological and longterm vegetative salt effects. *Advances in Horticultural Science* 10, 126–134.

Roussis, I., Beslemes, D., Kosma, C., Triantafyllidis, V., Zotos, A. *et al.* (2022) The influence of arbuscular mycorrhizal fungus *Rhizophagus irregularis* on the growth and quality of processing tomato (*Lycopersicon esculentum* Mill.) seedlings. *Sustainability* 14, 9001.

Rui, W., Mao, Z. and Li, Z. (2022) The roles of phosphorus and nitrogen nutrient transporters in the arbuscular mycorrhizal symbiosis. *International Journal of Molecular Sciences* 23, 11027.

Ruiz-Lozano, J.M. and Azcón, R. (2000) Symbiotic efficiency and infectivity of an autochthonous arbuscular mycorrhizal *Glomus* sp. from saline soils and *Glomus deserticola* under salinity. *Mycorrhiza* 10, 137–143.

Saboor, A., Ali, M.A., Danish, S., Ahmed, N., Fahad, S. *et al.* (2021) Effect of arbuscular mycorrhizal fungi on the physiological functioning of maize under zinc-deficient soils. *Scientific Reports* 11, 18468.

Shi, Z., Zhang, J., Lu, S., Li, Y. and Wang, F. (2020) Arbuscular mycorrhizal fungi improve the performance of sweet sorghum grown in a Mo-contaminated soil. *Journal of Fungi* 6, 44.

Smith, P. (2008) Land use change and soil organic carbon dynamics. *Nutrient Cycling in Agroecosystems* 81, 169–178.

Smith, S.E. and Read, D.J. (2008) *Mycorrhizal Symbiosis*. Academic Press, San Diego, California.

Solaiman, Z.M. (2014) Contribution of arbuscular mycorrhizal fungi to soil carbon sequestration. In: Solaiman, Z.M., Abbott, L.K. and Varma, A. (eds) *Mycorrhizal Fungi: Use in Sustainable Agriculture and Land Restoration*. Springer, Berlin, Heidelberg, pp. 287–296.

Steinberg, P.D. and Rillig, M.C. (2003) Differential decomposition of arbuscular mycorrhizal fungal hyphae and glomalin. *Soil Biology and Biochemistry* 35, 191–194.

Vodnik, D., Grcman, H., Macek, I., van Elteren, J.T. and Kovacevic, M. (2008) The contribution of glomalin-related soil protein to Pb and Zn sequestration in polluted soil. *Science of the Total Environment* 392, 130–136.

Wahab, A., Muhammad, M., Munir, A., Abdi, G. and Zaman, W. (2023) Role of arbuscular mycorrhizal fungi in regulating growth, enhancing productivity, and potentially influencing ecosystems under abiotic and biotic stresses. *Plants* 12, 3102.

Wang, J., Fu, Z., Ren, Q., Zhu, L., Lin, J. *et al.* (2019) Effects of arbuscular mycorrhizal fungi on growth, photosynthesis, and nutrient uptake of *Zelkova serrata* (Thunb.) Makino seedlings under salt stress. *Forests* 10, 186.

Wang, X., Liang, J., Liu, Z., Kuang, Y., Han, L. *et al.* (2022a) Transcriptional regulation of metal metabolism- and nutrient absorption-related genes in *Eucalyptus grandis* by arbuscular mycorrhizal fungi at different zinc concentrations. *BMC Plant Biology* 22, 76.

Wang, Y., Lin, J., Yang, F., Tao, S., Yan, X. *et al.* (2022b) Arbuscular mycorrhizal fungi improve the growth and performance in the seedlings of *Leymus chinensis* under alkali and drought stresses. *PeerJ* 10, e12890.

Watts-Williams, S.J. and Cavagnaro, T.R. (2012) Arbuscular mycorrhizas modify tomato responses to soil zinc and phosphorus addition. *Biology and Fertility of Soils* 48, 285–294.

Wilson, G.W.T., Rice, C.W., Rillig, M.C., Springer, A. and Hartnett, D.C. (2009) Soil aggregation and carbon sequestration are tightly correlated with the abundance of arbuscular mycorrhizal fungi: results from long-term field experiments. *Ecology Letters* 12, 452–461.

Wright, S.F. and Upadhyaya, A. (1996) Extraction of an abundant and unusual protein from soil and comparison with hyphal protein of arbuscular mycorrhizal fungi. *Soil Science* 161, 575–586.

Wright, S.F. and Upadhyaya, A. (1998) A survey of soils for aggregate stability and glomalin, a glycoprotein produced by hyphae of arbuscular mycorrhizal fungi. *Plant and Soil* 198, 97–107.

Wright, S.F. and Upadhyaya, A. (1999) Quantification of arbuscular mycorrhizal fungi activity by the glomalin concentration on hyphal traps. *Mycorrhiza* 8, 283–285.

Wu, S.W., Shi, Z.Y., Huang, M., Yang, S., Yang, W.Y. and Li, Y.J. (2023) Influence of mycorrhiza on C:N:P stoichiometry in senesced leaves. *Journal of Fungi* 9, 588.

Xiao, X., Liao, X., Yan, Q., Xie, Y., Chen, J. *et al.* (2023) Arbuscular mycorrhizal fungi improve the growth, water status, and nutrient uptake of *Cinnamomum migao* and the soil nutrient stoichiometry under drought stress and recovery. *Journal of Fungi* 9, 321.

Zai, X.-M., Fan, J.-J., Hao, Z.-P., Liu, X.-M. and Zhang, W.-X. (2021) Effect of co-inoculation with arbuscular mycorrhizal fungi and phosphate solubilizing fungi on nutrient uptake and photosynthesis of beach palm under salt stress environment. *Scientific Reports* 11, 5761.

Zhang, X., Zhang, H., Zhang, Y., Liu, Y., Zhang, H. *et al.* (2020) Arbuscular mycorrhizal fungi alter carbohydrate distribution and amino acid accumulation in *Medicago truncatula* under lead stress. *Environmental and Experimental Botany* 171, 103950.

Zhao, W., Chen, Z., Yang, X., Sheng, L., Mao, H. *et al.* (2023) Metagenomics reveal arbuscular mycorrhizal fungi altering functional gene expression of rhizosphere microbial community to enhance *Iris tectorum*'s resistance to Cr stress. *Science of the Total Environment* 895, 164970.

Zhen, L.I., Songlin, W.U., Yunjia, L.I.U., Qing, Y.I., Merinda, H.A.L.L. *et al.* (2023) Arbuscular mycorrhizal fungi regulate plant mineral nutrient uptake and partitioning in iron ore tailings undergoing eco-engineered pedogenesis. *Pedosphere* 34, 385–398.

Zhou, X., Wang, T., Wang, J., Chen, S. and Ling, W. (2023) Research progress and prospect of glomalin-related soil protein in the remediation of slightly contaminated soil. *Chemosphere* 344, 140394.

Zhu, J., Li, M. and Whelan, M. (2018) Phosphorus activators contribute to legacy phosphorus availability in agricultural soils: a review. *Science of the Total Environment* 612, 522–537.

Zhu, Y.G. and Miller, R.M. (2003) Carbon cycling by arbuscular mycorrhizal fungi in soil–plant systems. *Trends in Plant Science* 8, 407–409.

6 Impact of Climate Change on Soil Health and Nutrient Cycling

Hansa Sehgal, Era Sharma, Chandrakant Pant, Sushant Malhotra and Mukul Joshi*

Department of Biological Sciences, Birla Institute of Technology and Science (BITS), Pilani Campus, Vidya Vihar, Pilani, Rajasthan, India

Abstract

The soil environment is influenced by different parameters and processes including topography, climate and parent material throughout the landscape. Major soil properties including, (i) soil moisture which plays a fundamental role in interactions between the land and the atmosphere, (ii) soil respiration which is attributed by the soil carbon pool to the atmosphere, (iii) soil organic matter (SOM), and (iv) soil nutrient cycles which go hand-in-hand with ecological restoration practices, are affected by the soil ecosystem. Plant species play a crucial role in nutrient cycling in natural ecosystems, with plant growth, litter quality and herbivory affecting the rates of nutrient cycling. The diversity and activity of soil microorganisms are essential for sustainable agriculture, and organic farming and tillage can improve soil health. Cycling nutrients, including nitrogen and phosphorus, is a vital ecosystem service that incorporates reusing agricultural and municipal organic residues. Soil quality is not a constant value for nutrient cycling or other soil functions, as soil properties can simultaneously enhance or weaken the performance of one or more functions depending on prevailing climatic conditions, and evaluations must be site-specific. Soil health is crucial in delivering various ecosystem services, such as sustaining water quality and plant productivity, controlling soil nutrient recycling and decomposition, and reducing greenhouse gases from the atmosphere. This chapter focuses on the effects of climate change on soil health and nutrient cycling, leading to variations in plant productivity.

Keywords: Climate change, nutrient cycling, plant productivity, soil properties, soil health

6.1 Introduction

The impacts of climate change are now discernible in soil properties and biodiversity. The World Meteorological Organization indicated that changes in the climate pattern are one of the causes of disastrous heat waves and floods (Kawamiya *et al.*, 2020). In recent years, extreme climate change has intensified concern, prompting an urgent need to take action to mitigate its devastating impact on the ecosystem. An increase in carbon dioxide, methane and nitrous oxide gases and depletion of water resources causing drought situations are paramount because of anthropogenic activities that have driven climate change (Bogati and Walczak, 2022). It is predicted that with climate change, temperate parts of the northern hemisphere will face increased flooding situations during the autumn/winter season and extended dry spells during summer.

Climate warming has hastened the rate of mineralization of organic carbon and caused a spate of physical, biological and chemical changes in soil,

*Corresponding author: mukul.joshi@pilani.bits-pilani.ac.in

© CAB International 2025. *Soil Health and Nutrition Management*
(eds N.C. Joshi, T. Leustek and P.K. Singh)
DOI: 10.1079/9781800624597.0006

which has further accelerated the decomposition of organic matter via microbial communities, disturbed the flow of nutrients and led to the deterioration of soil quality and decline in the fertility of the soil (Guoju *et al.*, 2012). By 2050, climate change will negatively impact cultivated areas, and reducing the production of economically important crops will threaten food security and biodiversity (Raza *et al.*, 2019). One prominent reason for the change in climate patterns is global warming. An exponential increase in the atmospheric concentration of greenhouse gases (GHG) leads to global warming. According to historical data and climate trends, climate change will lead to warmer winters in colder regions, as observed. The Earth's temperature is increasing by an average of 0.18°C per decade, and the soil changes by 0.47°C per decade (Kehler *et al.*, 2021). From 1959 to 2019, the concentration of CO_2 increased from 315.98 to 411.43 ppm. CO_2 (76%) constitutes the major proportion of GHG, followed by methane (CH_4: 16%), nitrous oxide (N_2O: 6%) and others (2%) (Malhi *et al.*, 2020). Agriculture practices are the most significant contributors (about 14%) to greenhouse gas emissions. However, agriculture can also solve the problem if appropriately managed and with effective policies adopted (IPCC, 2024). Another event occurring due to climate change is increased rainfall, which leads to increased soil erosion rates, alteration in soil moisture and nutrient cycling (Hamidov *et al.*, 2018). The frequency and intensity of rainfall are important factors because faster-moving water and raindrops can easily move larger particles and aggregates. Therefore, fine-textured soils are prone to more erosion (Wasan and Wasan, 2023). A lack of rainfall precipitation causes a reduction in the water status of the soil profile. Over the last three decades, there has been a significant increase in drought in Europe; some lasted more than 3 months. Geographically, semi-arid regions are more prone to drought (Furtak and Wolińska, 2023), and according to the United Nations Convention to Combat Desertification report, Africa has been most affected with 300 droughts situations in the last 100 years (UN-ESCAP, 2020).

In terms of population, Asian countries are severely affected by droughts, which have affected over 66 million people in South-East Asia over the past 30 years (UN-ESCAP, 2020). The prediction reveals that many more dry years will occur in the future with a shift and expansion of drought-affected areas – moreover, microbial community size and activity will change during drought. Increasing drought resilience will require much better forecasting and more efficient responses at both national and regional levels. So far, many studies have been conducted to understand the effect of climate change on soil health and nutrient cycling. In this chapter, we will highlight some important parameters that help us understand the gravity of the topic.

Soil health is crucial for maintaining plant and animal fitness, ecological biodiversity, primary productivity and environmental quality. A robust framework is needed to monitor physical, chemical and biological indicators to maintain soil health (Maaroufi and De Long, 2020). Hatten and Liles (2019) discussed soil health impacted by direct and indirect disturbances, monitoring and management practices, and identified soil vulnerability to degradation and change. Soil properties matter, but they should be considered in the context of climate and management because their impacts vary with the circumstances and how they interact with other soil properties. These traits may amplify specific processes while attenuating others; this complexity has also been observed in other soil functions. As a result, site-specific assessments and management, which consider local suitability and difficulties, are essential (Schröder *et al.*, 2016). Soil health indicators are a comprehensive array of measurable physical, chemical and biological attributes directly correlating with essential soil processes. These indicators serve as a means to assess the soil health status (Raghavendra *et al.*, 2020). In the context of climate change, the definition of soil health is intricately linked to the anticipated impacts of various global change drivers. These drivers encompass rising atmospheric carbon dioxide levels, increased temperatures, modifications in precipitation patterns (rainfall) and the decomposition of atmospheric nitrogen. There is a need to understand how these climate change drivers affect soil chemical, physical and biological functions when defining and evaluating soil health in the context of evolving environmental conditions (French *et al.*, 2009) (Fig. 6.1). In the following sections, we highlight the biological,

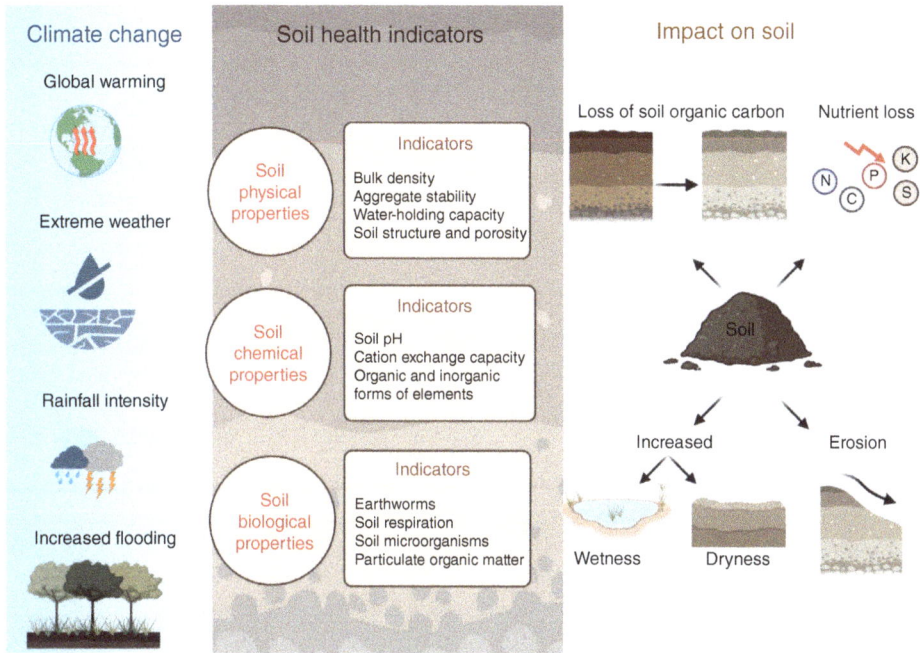

Fig. 6.1. The change in climatic patterns leads to global warming, droughts, floods and changes in rainfall intensity among different global parameters. These changes prominently affect the soil health indicators (physical, chemical and biological), which can ultimately make soil unfit for the ecosystem in various ways. Created with BioRender.com.

physical and chemical indicators and their role in maintaining soil quality.

6.2 Biological Indicators

This refers to a species or a group of species representing the state of the environment's abiotic or biotic stress. The effect of climate change on a habitat, community or ecosystem can have a positive or negative impact (Parmar *et al.*, 2016). Numerous organisms are sensitive to environmental changes that may interrupt their metabolism, growth and reproduction. Chowdhury *et al.* (2023) reviewed different categories of bioindicators that play a role during environmental stress conditions. Microbial indicators in particular exhibit high sensitivity toward the slightest change in the environment. Apart from microbes, insects can also act as bioindicators of climate change. Soil extracellular enzymes produced by microbial communities play an important role in decomposing soil organic matter and nutrient cycling. These enzymes are categorized into two groups: hydrolytic and oxidative (Singh *et al.*, 2020). The soil microbial community drives the nitrogen (N), carbon (C), phosphorus (P) and sulfur (S) cycles. Hence, it is essential to understand the activities of certain enzymes across the regions such as β-glucosidase, glucosaminidase, alkaline or acid phosphatase and arylsulfatase, as they are part of the nutrient cycles (Pérez-Guzmán *et al.*, 2021). There are *c.*10^{29} microorganisms present in terrestrial environments (Cavicchioli *et al.*, 2019); microorganisms in loamy and sandy soils are present within six phyla, mainly *Actinobacteria*, constituting >95% of the total abundance of bacteria. During drought conditions, the C/N ratio changes the loss of certain microbes, which are reported as sensitive to nitrogen ratios (Bogati and Walczak, 2022).

Looking into the distribution of soil microorganisms, soil aggregates create a physical environment for microorganisms. In heavy, loamy, calcareous soil, bacteria, fungi and actinomycetes were higher in aggregates of 1–3 mm than 5–7 mm (Tahat *et al.*, 2020). Along with microbes, nematodes and abundant metazoans are key contributors to soil food webs and bioindicators of soil health. It is suggested that beneficial nematodes be recognized as plant-promoting agents to harness their potential for improved soil health, decomposition services, plant performance and carbon sequestration. The goal is to contribute to a well-balanced and well-managed system, enhancing productivity, ensuring food security and reducing the environmental footprint (Pires *et al.*, 2023). The different types of microbes react to changes in different physical and chemical environmental factors (Table 6.1).

N deposition impacts soil nematodes in the context of N enrichment. Xing *et al.* (2022) conducted a meta-analysis of 66 N addition experiments, revealing that N inputs influence the abundance of nematodes, suppress their growth in the colder regions but promote it in warm regions. The study suggests that a future increase in N deposition may significantly reduce nematode communities, impacting plant and microbial activities and simplifying soil food webs, particularly in cold, dry areas (Xing *et al.*, 2022). However, their responses are generally slower than microbiological and biochemical properties such as soil enzymes, soil respiration, mycorrhiza, lipid profiling and the presence of earthworms. Normally, changes observed in these biological indicators make them especially effective for detecting perturbation from various agricultural management practices. The comprehensive use of these indicators allows for a nuanced understanding of soil health, facilitating timely and responsive management strategies to promote sustainable soil use and overall environmental well-being (Raghavendra *et al.*, 2020).

6.3 Physical and Chemical Indicators

Physical indicators like bulk density, soil aggregate stability and water-holding capacity emerge as precious metrics due to their sensitivity to change induced by natural or anthropogenic actions. Chemical indicators, including pH, electrical conductivity (EC), soil organic carbon and nutrient status, are well-established components of soil health evaluation. In the following sections we will discuss the effect of climate on physical and chemical indicators of soil in detail.

6.3.1 Climate effect on soil respiration

Soil respiration (SR) is an essential component in the global carbon cycle, contributing *c.*78–95 Pg of atmospheric carbon annually. SR is sensitive to environmental changes, and the slightest changes in SR can significantly impact the dynamics of the C cycle (Rodtassana *et al.*, 2021). Heterotrophic respiration (decomposition of soil organic matter) and autotrophic respiration (plant root respiration) result in SR (Hashimoto *et al.*, 2023). Since the 1980s, the soil to atmosphere flux of CO_2 has been extensively perceived. Due to the limited data availability, SR estimation was sparse. To fill the gap, researchers developed the Soil Respiration Database (SRDB), which will accelerate the studies related to SR (Hashimoto *et al.*, 2015). In various ecosystems, the relationship between the response of SR and decreasing precipitation has been reported.

In semi-arid and arid regions, a decline in precipitation suppresses SR, but contrasting results have been reported in relation to tropical rainforests. Such contradictory results lead to different microbial communities and differences in the type of vegetation (Shen *et al.*, 2021). In South-East Asian tropical forests, it is crucial to understand carbon dioxide emissions. Sensitive to environmental changes, SR was measured in three successional forests during wet and dry periods. Older forests and wet conditions showed higher SR. Younger forests had no response to soil temperature (ST), while old-growth forests did, likely due to varied forest structure. Soil moisture (SM) limited SR in the wet period, while SR varied with organic matter content in the dry period. Results highlight temporal and successional variations in SR responses to environmental factors, underscoring the need for more detailed investigations in South-East Asian tropical forests (Rodtassana *et al.*, 2021). Dhital *et al.* (2022) investigated soil CO_2 efflux, a process releasing carbon from soil to the atmosphere, in a subtropical mixed forest in central Nepal. Conducted over 2 years, the study focused on ecological parameters influencing soil CO_2 efflux, including ST, water content and litter. A positive correlation was reported between soil CO_2 efflux and temperature, with a temperature sensitivity (Q10 value) ranging from 3.2 to 3.6. Soil water content also positively affects soil CO_2 efflux. The study highlights the vulnerability of subtropical forests to climate change, emphasizing the dynamic nature of the forest carbon cycle in tropical regions. Another study was conducted over 3 years in a seasonally wet tropical forest in central Panama, and it monitored soil CO_2 effluxes through automated and manual measurements. The 2-year mean total annual soil CO_2 efflux for the forest is estimated at 904.76 g C m^{-2} year^{-1}. A substantial spatial and temporal variability was reported, and temporal fluctuations were predominantly linked to surface soil water dynamics, which influence seasonal and diurnal cycles, rain-induced pulses and interannual variability. Not entirely explained, it showed the influence of forest structure, ST and topography. Mean annual soil CO_2 effluxes exhibited an increasing trend, associated with an El Niño/Southern Oscillation cycle, particularly during the intense 2015/16 event, attributed to a mild wet season with less persistent soil saturation (Rubio et al., 2017).

The intricate relationship between land, soil and climate emphasizes their pivotal role in the global ecosystem. Agriculture and forestry contribute significantly to greenhouse gas emissions; Leal Filho et al. assessed the impact of climate change and land use changes on soil biodiversity and related services. The findings underscore the importance of well-managed soils for resilient production systems, advocate for integrated agricultural approaches as climate-resilient systems, and propose agricultural zoning as a tool to mitigate climate change effects. Continuous environmental monitoring and intersectoral collaboration are emphasized, highlighting the need for proactive measures in addressing climate change and extreme events (Leal Filho et al., 2023).

The impact of fast-growing non-native species like *Eucalyptus* and *Acacia* exhibited significant productivity and reduction of SR, and their effects on ecosystem services in degraded tropical forestland. Despite being used for quick rehabilitation, non-native species, particularly in north-east Thailand, showed limitations in mitigating extreme soil conditions, impacting microbial decomposition and reducing SR. This denotes the importance of careful tree species selection to optimize carbon sequestration, storage and nutrient cycling for effective ecosystem services (Ontong et al., 2023).

6.3.2 Climate effect on soil temperature

Global variations in ST are caused by various factors, including changing weather patterns, the ratio of soil to air temperature and others, in addition to climate change. It varies geographically and is associated with the operation of numerous biogeochemical cycles. ST is influenced by several physical, chemical and microbial alterations that impact staple crops, nutrient recycling and other areas of agriculture (Hamidov et al., 2018; Dorau et al., 2022). The long-term disparity between historical and projected climatic conditions is best illustrated by the rise in ST between the historical and 2020–2050 timeframes and the subsequent significant increase by 2070–2100. Over 90% of climate models agreed that yearly ST and air temperature will rise throughout the dryland domain. Long-term effects of air temperature and precipitation patterns also affect soil temperatures. The increased respiration rates will lower overall ecosystem carbon stocks. Furthermore, these rising temperatures highlight the increased risk of hot droughts in these already water-limited ecosystems, and warmer ST may also impact ecosystem carbon fluxes (Bradford et al., 2019). Different soil temperatures in European Russia were evaluated in response to climate change using data from four meteorological stations. Studies demonstrated that throughout the past few decades, a consistent trend toward warming has been seen, with the rate of warming declining from north to south. The average ST rose by 0.7–0.9°C at 20 cm depth and by 0.5–0.8°C at 160 cm. The soil profile shows warming in warm and cold seasons (Reshotkin and Khudyakov, 2019). The daily surface ST data was recorded in China and gathered from 360 weather stations between 1962 and 2011. Based on the data, the rise in ST during the day and night was not equal. The daily minimum and maximum surface ST increased throughout 50 years at rates of 0.055 and 0.031°C/year, respectively, while the average rate of decline in the soil diurnal temperature range at most stations was –0.025°C/year (Zhang et al., 2019a). The positive correlation between solar duration and the uneven temperature increase between the air and soil likely caused this decline. The biggest drop occurs in the winter (–0.08°C/year). The temperature differential between the soil and the air at night decreases, and perennials in regions close to the zero contour line would benefit from this for their wintering (Zhang et al., 2019a). Global warming of the soil was somewhat slower than that of the air above it. For representative concentration pathway (RCP) values 4.5 and 8.5, respectively, the model ensemble projected a worldwide mean soil warming of 2.3 ± 0.7 and 4.5 ± 1.1°C at 100 cm depth by the end of the 21st century. The rate of soil warming at 100 cm was nearly identical to that of near-surface (c.1 cm) soils. In tropical and arid locations, soil warming was maintained with regional air warming, but it trailed behind warming in colder regions. For this reason, in frigid climates where snow and ice prevent sensible heat from moving directly from the atmosphere to the soil, air warming may not always be an effective stand-in for soil warming. Despite this impact, high-latitude soils were still predicted to warm more quickly than other places but more slowly than the surface air above them (Soong et al., 2020). In the Shaanxi Province of China, four rebuilt soils in arid gravel land had respiration and hydrothermal properties observed before, during and following two precipitation episodes. Both precipitation episodes considerably lowered the ST, although large temperature swings followed the second precipitation event. ST had a greater effect when the water content remained constant. Temperature and SM impacted soil respiration before, during and after the precipitation event (Lei and Han, 2020). Gains and losses in surface radiation, evaporation, heat conduction through the soil profile and convective transfer through the movement of gas and water influence the temperature regime of the soil. A warming trend in ST over 50 years from 1958 to 2008 in Canada was correlated with changes in air temperatures and snow cover

depth, and the patterns in ST are linked to climate change (Qian et al., 2011). Additionally, a noteworthy downward trend in the snow cover depth during the winter and spring has been linked to rising air temperatures. Like SM, ST is a major factor in most soil processes. A warmer soil will hasten several processes, including a quick breakdown of organic matter, boosting microbiological activity, releasing nutrients more quickly, raising the nitrification rate and generally emphasizing the chemical weathering of minerals. However, the kind of vegetation growing on the surface of the soil will also impact its temperature, and this vegetation may be altered due to climate change or management of adaptation (Karmakar et al., 2016). The minimum and maximum temperatures may be affected differently by the warming climate. The diurnal range may shift as a result of these asymmetrical changes. It was found that the diurnal range of ST increases dramatically during the spring while it decreases during the other seasons. There are now considerably fewer cold days and a greater number of warmer days. Furthermore, the rise in scorching temperatures over recent years is especially noteworthy (Zhang et al., 2019a).

6.3.3 Climate effect on soil pH

Soil pH is one of the essential regulators of soil and is ineluctably regulated by different factors. Soils present in different regions have distinct pH levels. Among soil chemical properties, soil acidity is a key feature and can regulate the ecosystem functions and processes. In the majority of ecosystems, the increase in the emission of sulfur dioxide and nitrogen oxide is one of the major causes of high soil acidity (Wei et al., 2019). Arid climate soil is usually alkaline, whereas humid climate soil is acidic with a low pH (Zhang et al., 2019b). The reason for soil alkalinity in semi-arid and arid regions is the reduced availability of micro- and macronutrients (Abd El-Mageed et al., 2021). Significant impacts of climate change on soil pH are being observed due to increased mean annual air temperature and precipitation. It is observed that there is a negative correlation between pH, temperature and precipitation (Chytrý et al., 2007; Cheng-Jim et al., 2014). These changes are expected to enhance the chemical weathering of soil minerals, increasing base cation weathering rates and raising soil pH. Increased soil pH could improve the acid–base status and fertility of soils in eastern Canada, suggesting a potential amelioration of soil conditions despite climate change challenges (Houle *et al.*, 2020). Another study was conducted on the Tibetan plateau to study the effect of climate on soil pH at three different depths. At 0–10 cm, climate change caused alkalization, and at 10–30 cm depth caused acidification (Sun *et al.*, 2023). Wei *et al.* (2019) mentioned that different phases may exist to buffer acid inputs in soil, and soil buffering capacity varies from soil type across the ecosystems. For example, carbonated soils range from 27.2 to 188.5 mmol kg^{-1} pH $unit^{-1}$, and non-carbonated soils range from 10.4 to 58.7 mmol kg^{-1} pH $unit^{-1}$. Soil organic matter and cation exchange capacity affect the immensity of soil acid-buffering capacity. In Ethiopia's highlands, acidity has become a severe problem, as 47% of the total land area is affected by acidity. A total of 32% of arable land is acidic soil around the globe, and the biological and chemical characteristics of acidic soil are poor (Belay *et al.*, 2023).

Under drought conditions, the pH of the soil can become more acidic due to decreased leaching of base cations and increased concentration of CO_2 in soil water, which forms carbonic acid when combined with water (Hue, 2022). This acidification process can affect nutrient availability, microbial activity and soil health. Drought reduces SM, which is crucial for the dissolution and transport of nutrients, leading to changes in soil chemistry that can negatively impact plant growth and soil fertility. The effect of drought on soil acidification varies due to many factors. First, less water means less removal of acidic ions like aluminium and hydrogen, lowering pH and reducing leaching (Gelybó *et al.*, 2018). Second, as water evaporates, salts become more concentrated and focused, making the soil acidic (Duan *et al.*, 2022).

Rainfall can lead to higher soil pH levels by enhancing the leaching of acidic components and promoting the dissolution and transportation of base cations. This process can mitigate soil acidity, improving nutrient availability and conditions for plant growth. However, excessive rainfall might also wash away essential nutrients,

negatively affecting soil fertility. The overall effect on soil pH depends on the balance between these processes and the initial soil conditions (Rengel, 2011).

The influence of rainfall on soil pH is a complex phenomenon. While drought can lead to an increase in soil pH, high rainfall areas usually experience acidification. However, the specific effects may vary based on a range of factors.

6.3.4 Climate effect on soil moisture

Soil moisture is the water content in the soil and it can be determined in several ways, for example W (column of water in the given depth of the soil), P (porosity), FC (field capacity), W0 (total water holding capacity), etc. In the environment, SM regulates the various processes and life cycles, like the pH of the soil, diffusion rate of gases and solvents, nutrient availability, dynamics of microorganisms and the mineralization rate. The physical properties of soil, like aeration and the stability of the colloid, are affected by changes in the water status of the soil (Furtak and Wolińska, 2023). Changes in rainfall patterns affect the SM content, and every type of soil has a water threshold; an increase in instances of short-term rainfall will increase the microbial mass present in the soil. However, excessive rainfall inhibits the growth of aerobic microorganisms and increases the activity of anaerobic microorganisms, and sometimes disturbs the stability of soil organic carbon (Wang *et al.*, 2023). During drought conditions, the soil gets dried, leading to air saturation and affecting the diffusion and distribution of gases, nutrients and soil extracellular enzymes.

Moreover, due to the accumulation of solvent compounds, soil osmotic pressure increases, leading to a hypertonic solution, causing dehydration of the microbial community. In case of flooding, the soil is compacted with water, reducing the exchange of gases like oxygen and nitrogen. Limitation of oxygen in the soil causes inhibition of growth and activity of microbes. During such a situation, microbes start utilizing alternative substrates, and the dynamics of nutrient cycles get disturbed. The major changes occurring in soil due to flood situations include: (i) an increase in the concentration of NH_4^+ ions to the level where it becomes toxic for the development of roots; (ii) leaching of nutrients and humus; and (iii) change in pH of the soil (Furtak and Wolińska, 2023).

6.3.5   Effect of climate on the nutrient cycle

Soil facilitates the conversion and recycling of organic waste into nutrients. This involves four processes: potential for nutrient uptake, nutrient bioavailability, enabling nutrient uptake by plants and harvest of nutrient-enriched plants (Schröder *et al.*, 2016). Biochemical cycles like C, P and N play significant roles in ecosystem functioning and structure. Their availability and concentration in soils indicate soil health (Shen *et al.*, 2022). Soil is a huge reservoir of C and N. The amount of organic and inorganic forms of C and N present in the upper meter of mineral soil is *c.*2200 and 135 Pg, respectively (Certini and Scalenghe, 2023).

Carbon cycle

Among different geochemical cycles, the C cycle is integral to biological activities; hence, it has received the most attention (Fig. 6.2a). The global C cycle is divided into fast and slow domains (Jones *et al.*, 2016). The largest reservoir of C is soil, which stores *c.*1505 Pg carbon in the upper 1 m layer. In terrestrial soil, the majority of the C pool is present in the form of soil organic carbon (SOC), nearly 60%, which regulates the global C cycle and functioning of the ecosystem. Researchers explore how changes in climate and the amount of water in soil affect the SOC dynamics and significantly change the concentration of atmospheric CO_2. The soil C pool can function as a source or sink depending on the soil water content (SWC) and climate warming and can directly or indirectly affect the decomposition of SOC (Zhao *et al.*, 2021). Cheng *et al.* (2023) conducted a study investigating the impact of temperature and precipitation on SOC and soil alkali-hydrolysed nitrogen (SAN) content. They found that the response of SOC is not directly linked to climate, especially during winter and spring. An increase in precipitation stimulates the accumulation of SOC. In the terrestrial C cycle, it is presumed that an increase in temperature will hasten SOC mineralization and

elevate CO_2 levels, further aggravating climate warming (Hari and Tyagi, 2022). However, the effect of climate will depend on the type of SOC affected. SOC is divided into three groups according to the turnover rate: annual cycling C (0–5%), decadally cycling C (60–85%) and millennial cycling C. The global C cycle is most influenced by decadally cycling C. Hence, it is essential during warming to understand the response of decadally cycling C (Liu *et al.*, 2023a).

Moreover, climate change and other factors influence the soil C pool, including soil management practices and land use change. Xu *et al.* (2020) performed a meta-analysis of soil data from 2001 to 2019 and discussed the underlying factors responsible for the loss of the soil C pool. Among other factors, the feedback mechanism between climate change and the C cycle depends upon geographical locations, too.

A significant cause of climate change is anthropogenic human activities, causing prodigious plant and animal extinctions, which are well researched and documented. However, the context of the effect of climate change on microorganisms should be discussed further. The global C cycle is dependent on microorganisms. Microbial enzymes are regarded as crucial engines that drive biogeochemical cycles. In the C cycle, plants eliminate CO_2 from the atmosphere via photosynthesis and produce organic matter, whereas autotrophic and heterotrophic respiration add CO_2. Temperature influences the dynamics between the two processes (Mangodo *et al.*, 2020).

EFFECT OF ABIOTIC STRESS.   Excessive carbon emission harms the C cycle, resulting in stunted vegetation growth, death and increased wildfires. The relationship between the carbon and water cycles in an ecosystem is measured using water use efficiency (WUE). During droughts, the WUE tends to increase, but the extent of the increase varies depending on the vegetation type. A study conducted using data from ERA5 and global land surface satellites from 1982 to 2018 discussed the impact of soil moisture drought on WUE. The effects of soil moisture drought on WUE can be negative or positive (Ji *et al.*, 2021). During droughts, water availability in the soil decreases, inhibiting the decomposition process and increasing the physical protection of soil organic matter

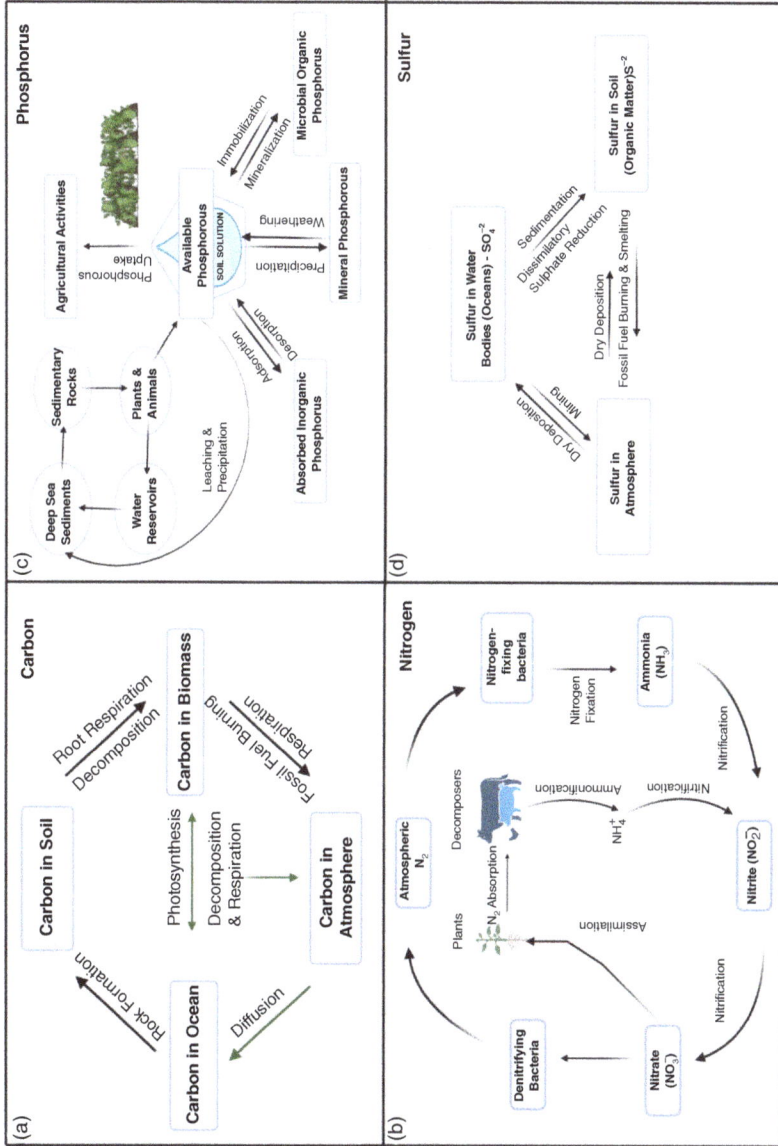

Fig. 6.2. Four major biogeochemical/nutrient cycles in the ecosystem. (a) Carbon cycle: carbon is present in four major ecosystem components, i.e. soil, biomass in living forms, atmosphere in gaseous form and oceans or other water beds. Different forms of carbon participate according to the environment to maintain carbon flux in the ecosystem. (b) Nitrogen cycle: nitrogen exists in various forms, and its conversions occur via different processes. Each form has a significant role, mainly occurring in the terrestrial ecosystem. (c) Phosphorus cycle: phosphorus is a major part of aquatic ecosystems. The connecting link between terrestrial and aquatic cycles is via precipitation and leaching of phosphorus in water beds, mainly of oceans (i.e. deep-sea sediments). (d) Sulfur cycle: the sulfur element is mainly sourced from the atmosphere, soil, organic matter and oceans. Anthropogenic activities such as mining also play a role in its cycling. The dissimilatory sulfate reduction process helps change the sulfate form to reduced sulfur. Created with BioRender.com.

(SOM). The frequency, duration and intensity of heat waves are expected to increase with global warming, which will reduce the C sink of an ecosystem and escalate the risk of fires. Numerous studies have shown the detrimental effect of heat waves on C sinks (Qu *et al.*, 2024).

According to reports, precipitation regimes will decline in subtropical regions and increase at higher altitudes (Feng *et al.*, 2018). Altered precipitation patterns affect the abiotic processes of soil and biotic ones, including microbial communities, vegetation and the activity of soil enzymes. Due to a change in the pattern of precipitation, water availability in the soil will change, leading to escalated C inputs and decomposition of SOM (Navarro-Pedreño *et al.*, 2021). The effect of alteration in precipitation patterns has been observed in arid regions (mean annual precipitation <500 mm) and humid regions (mean annual precipitation ≥500 mm). In arid areas, adding water significantly increases the microbial biomass; however, humid areas will fill the soil spaces and reduce the oxygen content, impeding aerobic microbes' growth. A comprehensive meta-analysis reported that increased precipitation positively affected the C cycle in arid areas. Conversely, the decline in rainfall in humid regions had a negative effect (Wang *et al.*, 2021).

Nitrogen cycle

For biogeochemistry on Earth, the N cycle is fundamental, and biological N fixation fixes most atmospheric N in terrestrial ecosystems (Fowler *et al.*, 2013). In the N cycle, unreactive N_2 is initially converted and reduced into ammonium compounds. Further, ammonium compounds are converted into other oxidized forms or amino acids by the action of microorganisms and utilized by plants (Fig. 6.2b). To maintain the balance between the forms of N, N_2 is returned to the atmosphere via microbial denitrification processes. The functioning of the plant and animal kingdoms is highly dependent on the availability of reactive nitrogen (Nr). Lightning, biomass burning and the N fixation of N_2 form Nr (Erisman *et al.*, 2013). Nr is present in ammonia, ammonium, nitric oxide, nitric acid, nitrous acid, nitrous oxide and nitrogen dioxide. Over time, the N cycle has changed considerably. An important mediator of N cycling is a particular group of microbial communities, and new microorganisms have been identified in recent times with an improvement in sequencing technologies, for example *Nitrosopumilus maritimus*, a chemoautotrophic ammonia-oxidizing archaeon (Sanjuan *et al.*, 2020; Aryal *et al.*, 2022). The inadequate fixation of natural N causes the ecosystem to shift and adapt to low rates of Nr supply. However, the Nr required to produce food is approximately tenfold higher than its consumption (Erisman *et al.*, 2013).

In the ecosystem, the N and C cycles are tightly regulated, and litter decomposition is a fundamental process in these cycles. The decomposition rate depends on climate, nutrient availability, litter quality and diversity, and abundance of microbes present in the soil. Usually, the decomposition rate is stimulated during high temperatures and precipitation, but when substrate moisture lies <30% and >80%, an average temperature below 10°C will inhibit the process. On a global scale, climate can lead to approximately 68% of the variability in the litter decomposition rate (Kwon *et al.*, 2021). N is abundantly present in the atmosphere in the form of N_2 gas, which accounts for 78% of gases. However, the N_2 form of nitrogen is not accessible to microorganisms. Therefore, distinct biochemical phenomena stimulate the emergence of multiple chemical and oxidation states of N (Aryal *et al.*, 2022).

EFFECT OF ABIOTIC STRESS. Nitrogen dynamics depend on mineralization, N_2 fixation and immobilization in soil. So far, research conducted to study the effect of drought shows contrasting results; in one study, it is seen that even prolonged drought has no negative impact on N cycling, whereas another study indicates that total N concentration under drought decreases. However, the presence of dissolved organic N and minerals increases. Therefore, the effect depends on the type of ecosystem and soil (Liu *et al.*, 2023b). After analysing the topsoil data of 80 sites across different grasslands for N and P, it was suggested that an increase in the mean annual temperature increases the availability of N. At the same time, P is reduced, resulting in the shift of the N/P ratio. Under warmer conditions, the N/P ratio decoupling is more pronounced in alpine grasslands (Geng *et al.*, 2017). The amount of rainfall determines the dynamics of soil N pools. Studies have shown that NH_4^+-N is primarily present at low amounts of rainfall, whereas the concentration of NO_3-N increases at high amounts. Excessive rain can lead to the leaching of nitrates (Zhang *et al.*, 2020a).

Phosphorus cycle

Phosphorus is one of the essential macronutrients required for the functioning of all organisms. P is a crucial structural component of nucleic acids, lipids, proteins and enzymes and drives metabolic processes. The solubility of naturally occurring P-containing compounds is low, and its cycle is relatively slow. These are two significant constraints on the availability and efficiency of P in ecology. Geographically, P is limited in the weathered soils of tropical regions (Fig. 6.2c). Other factors like topography, plate tectonics, rock type, dust transport and biological responses diminish the P limitation. In nutrient cycling, P input compared with C and N is low. Hence, the slightest change in an influx of P will have consequences over an extended period. In terrestrial ecosystems, a significant influx of P occurs by atmospheric deposition and weathering of rocks (Menge *et al.*, 2023). P exists in different forms in soil, and orthophosphate concentrations in a managed soil system range between 500 and 800 mg/kg of dry soil (Kehler *et al.*, 2021). Less than 5% of total P is readily available for plants, and the rest present in the soil is not accessible. Most P is associated with soil as a primary mineral, secondary mineral, organic constituent or in occluded form, regulating P availability (Hou *et al.*, 2018). The effect of biotic or abiotic drivers on P cycling remains unexplored due to scanty research and limited understanding. Soil P interacts with microbes, plants, minerals and dead organic matter, which are influenced by climate, making it challenging to understand its cycle. Geochemical and biological factors regulate the bioavailability and dynamics of organic and inorganic forms of P. Soluble phosphate may precipitate with factors including soil pH, and can form insoluble minerals with aluminium, calcium or iron, further adsorbed by sesquioxides or assimilated into SOM and other colloids (García-Velázquez *et al.*, 2020). In the normal state, P is mainly concentrated in topsoil, and climate change affects P availability by direct or indirect processes. Climate primarily affects the availability of P in the soil via three factors: (i) the form of P present in the soil; (ii) the properties of the soil; and (iii) the uptake and return of P by plants and microbes (Hou *et al.*, 2018).

Kehler *et al.* (2021) discussed the source and sink of P, considering the impact of climate change on P cycles, highlighting the reduced form of P in the nutrient cycle as climate change is likely to favour the abundance of reduced P and free phosphatase in temperate soils. Across the globe, warming has stimulated mineralization, and mobilization of P is increased. Regarding the change in forest soil P pools due to soil warming, long-term soil warming can reduce bioavailable P (Tian *et al.*, 2023). This leads to substantial losses of topsoil and increased inorganic sorption and accumulation of P. An increase in temperature leads to an increased occluded form of P through its absorption on secondary minerals, and P availability reduces as drier conditions affect biological activity. At the same time, precipitation influences the loss of P via the chemical weathering of primary P (García-Velázquez *et al.*, 2020). Abrupt drought–flood alternation increases the soil P loss by 4.6 times. However, during prolonged drought stress, the concentration of P available in soil increases, which may be because of a decline in the population of P-accretion enzymes (Bi *et al.*, 2023).

There are three key points regarding the increase of soil P during drought: (i) the enhanced physical weathering process; (ii) the secretion of soluble P by damaged or dead microbial cells; and (iii) the degradation of SOM (Turner and Haygarth, 2003). Microbes regulate the P cycle via the mineralization of organic P, immobilization of the inorganic form of P, synthesizing the new organic form of P and affecting the solubility of P. To achieve the following tasks, three major gene groups are involved for the uptake and transport of P, mineralization and solubilization, and response during starvation of P (Li *et al.*, 2022). Tian *et al.* (2021) focused on the role of phosphate-solubilizing microorganisms in P biogeochemical cycles. The effect of climate change on these groups of bacteria needs to be well explored, and detailed research is required.

Sulfur cycle

One of the main elemental cycles is the S cycle. S has a wide range of oxidation states, ranging from sulfides in the II state to sulfates in the VI state (Fig. 6.2d). Many sulfides of metals or sulfates, like those of calcium (gypsum) and/or barium (barite), are commonly found as minerals (Brimblecombe, 2015). Soil solution containing sulfate is one direct source of S that plants can use. The amounts of S that plants can uptake in the growth season can be impacted by sulfates present at a depth in the soil profile. One significant method of restocking accessible S is mineralization, which converts organic S – such as humus and crop residues – to sulfate. Sulfates are converted to sulfide by microorganisms (e.g. *Desulfovibrio* and *Desulfotomaculum*) under reducing conditions, where the redox potential of S is reduced due to inadequate aeration, flooding and high oxygen consumption after other electron acceptors, such as oxygen and nitrate, are exhausted (Schoenau and Malhi, 2008).

The sediment–water interface is the primary location for sulfate reduction in the majority of totally convective lakes (Luo, 2018). The mineralization of plant-based organic matter doubles with every 8–9°C increase in the mean annual air temperature. Since the mineralization of organic materials occurs concurrently with the breakdown of sulfate, this may significantly impact the S cycle (Luo, 2018; Grzyb *et al.*, 2020). It seems that temperature is a key factor in altering the architectures of populations that reduce sulfate. Other than bacterial genera like *Desulfovibrio* and *Desulfotomaculum*, archaeal sequences (similarity 94%) are connected to the genus *Archaeoglobus*, which might have a role in S cycling. To reduce sulfate at temperatures close to their respective optimal conditions for microbial activity, some specific sulfate reducers were activated, which also caused hindrances in the growth of other microbial communities and had a significant role in S cycling (Cheng *et al.*, 2014). With increasing temperature, the relative abundance of the *dsrB* (*dissimilatory sulfite reductase beta subunit*) gene (which catalyses the reduction of sulfite to sulfide in the reductive sulfate assimilation pathway) increased. At the same time, its operational taxonomical unit (OTU) level diversity (cutoff 97%) decreased, suggesting that temperature has a major impact on the *dsrB* gene's distribution (Ma *et al.*, 2021).

These biogeochemical/nutrient cycles in soil are associated with microbes, and changes in any nutrient element leads to changes in the concentration of different microbes (Table 6.1).

6.4   Future Perspectives

Soil health status concerning climate change is crucial and can be explored based on biological, physical and chemical indicators. Extreme climate change in recent decades is a new reality, and anthropogenic interference with nature has intensified concerns regarding its negative impact on the ecosystem. Various global change drivers, including rising atmospheric CO_2 levels, high temperatures, rain patterns, decomposition of atmospheric N, soil acidification, etc., are directly connected with soil health (Fig. 6.1). Plastic and chemical pesticide pollution in soils and water bodies are major challenges to soil health and fertility that are directly proportional to food security and human health (Stubenrauch and Ekardt, 2020). Biological indicators (i.e. microbial communities and nematodes) interrupt nutrient cycling through multiple factors in climate change.

Different physical and chemical parameters, including soil respiration, temperature, pH, moisture and nutrient cycling, discussed in detail here, are decisive indicators of climate change. The major nutrient elements C, N, P and S are available in different forms in soil, and plants absorb these in a few specific forms only (Fig. 6.2). Climate change may drastically change the composition of these bioavailable nutrients in soil. The cumulative understanding of these indicators will also allow us to understand soil health, helping to promote sustainable soil use practices for the benefit of the ecosystem (Raghavendra *et al.*, 2020).

The growing global population and the challenge to support people, necessitates a simultaneous increase in food production and mitigation of climate change impacts (Tumwesigye *et al.*, 2019). This objective can be achieved by adopting climate-smart agriculture (CSA) practices and technologies worldwide. The shift from traditional agricultural methods to CSA is crucial in the 21st century, aligning with the UN 2015–2030 sustainable development goals (SDGs) and fostering sustainable development. While some developing nations have embraced CSA, broader awareness and implementation efforts are vital for universal adoption. Ensuring localized initiatives can contribute to global scalability and benefit both populations

Table 6.1. Summary of how different microbe phyla/class/family/species react to changes in environmental factors like air or soil temperature, drought and precipitation.*

Microbe (air/soil temperature)	Soil type	Concentration	Biogeochemical cycle involved	Geography	Reference
Examples of microbes affected by air/soil temperature					
Planktothrix	Tidal mud flats (mesocosm experiment)	Decreased with temperature increase	Nitrogen (N), phosphorus (P)	Scotland, UK	Hicks et al., 2018; Peck, 2020
Lutibacter species	Tidal mud flats (mesocosm experiment)	Decreased with temperature increase	Carbon (C)	Scotland, UK	Hicks et al., 2018
Proteobacteria	Tengchong National Field Station	Increased with temperature decrease	C	China	Li et al., 2015
Firmicutes	Tengchong National Field Station	Increased with temperature increase	C	China	Li et al., 2015
Chloroflexi	Mesocosm experiment	Increased with temperature increase	C, N	China	Pan et al., 2021
Actinobacter	Subalpine tussock grassland	Increased with temperature increase	C, N	New Zealand	Adair et al., 2019
Sphingomonas	Subalpine tussock grassland	Decreased with temperature increase	C, N	New Zealand	Fang et al., 2021
Nitrosospira	Alpine Research Station	Increased with temperature increase	N	Eastern Tibetan Plateau	Zhang et al., 2020b
Thauera	Tengchong National Field Station	Decreased with temperature increase	P, S	China	Li et al., 2015
Chlorobi	Two arctic-alpine mountain regions	Increased with temperature increase	S	Norway	Frindte et al., 2019
Examples of microbes affected by drought					
Acidobacteria	Grassland soil	Increased with water content decrease	N	California, USA	Siebielec et al., 2020
Actinobacteria	Silty loam soil with low pH (5.2)	Increased under drought stress	C, N, P, potassium (K), etc.	Albany, California, USA	Naylor et al., 2017; Naylor and Coleman-Derr, 2018
Ensifer	Thar desert	Increased in drought stress	N	Rajasthan, India	Ardley, 2017
Streptomyces	Field	Increased in drought stress	C	Parlier, California, USA	Xu et al., 2018
Sphingomonas	Sand and silt loam	Decreased in drought stress	N	Pulawy, Poland	Siebielec et al., 2020
Chloroflexi	Greenhouse	Increased in drought stress	C, N, iron (Fe)	California, USA	Santos-Medellin et al., 2017; Narsing Rao et al., 2022
Nitrosomonas	Grassland (near central Alps)	Decreased in drought stress	N	Austria	Bogati and Walczak, 2022

Continued

Table 6.1. Continued.

Microbe (air/soil temperature)	Soil type	Concentration	Biogeochemical cycle involved	Geography	Reference
Nitrosospira	Grassland (near central Alps)	Decreased in drought stress	N	Austria	Bogati and Walczak, 2022
Bacilli	Greenhouse	Increased in drought stress	N, P	California, USA	Santos-Medellín *et al.*, 2017; Hashem *et al.*, 2019
Clostridia	Greenhouse	Decreased in drought stress	C	California, USA	Zhao *et al.*, 2014; Santos-Medellín *et al.*, 2017
Examples of microbes affected by precipitation					
Acidobacteria	Grassland soil	Decreased with dry-down	N	California, USA	Barnard *et al.*, 2013
Actinobacteria (*Gaiella*)	Grassland soil	Increased with dry-down	C, N, P, K, etc.	California, USA	Schimel, 2018; Siebielec *et al.*, 2020
Firmicutes	Grassland soil	Increased with rewetting	C, N, P, K, etc.	California, USA	Schimel, 2018
Chloroflexi	Hot spring	First increased and then decreased with dry-down	C, N, Fe	Maharashtra, India	Preece *et al.*, 2019; Narsing Rao *et al.*, 2022
Flavobacterium	Lake Cajititlán	Abundant in the rainy season	C, S, N	Mexico	Díaz-Torres *et al.*, 2022; Choi *et al.*, 2023
Pseudomonas	Lake Cajititlán	Abundant in rain-fed soil	N	Mexico	Díaz-Torres *et al.*, 2022; Zhu *et al.*, 2022
Cyanobacteria	Paddy soil	Increased in rain-fed soil	C	Ang Thong Province, central Thailand	Reim *et al.*, 2017; Sánchez-Baracaldo *et al.*, 2022
Planctomycetes	Paddy soil	Increased in rain-fed soil	S	Ang Thong Province, central Thailand	Reim *et al.*, 2017; Storesund *et al.*, 2020
Methanogens	Paddy soil	Some increased and some decreased upon desiccation and rewetting	C	Ang Thong Province, central Thailand	Reim *et al.*, 2017; Buan, 2018
Gemmatimonadetes	—	Decreased with precipitation increase	C	China	Zhou *et al.*, 2018; Mujakić *et al.*, 2022

*This table compiles soil data from various places such as Scotland, China, New Zealand and the USA. Understanding these responses helps us grasp how ecosystems cope with environmental shifts and anticipate microbial reactions to ongoing changes. Biogeochemical cycles associated with microbes are also mentioned as the fluctuation in microbial activity might impact the nutrient cycle.

and the environment. By harmonizing farming techniques, climate-resilient practices can be executed while concurrently reducing greenhouse gas emissions. Updated ecosystem-based management and governance practices, necessitating multistakeholder collaboration to effectively address current challenges in food security and climate change, are a way forward.

Acknowledgements

HS and ES are thankful to BITS Pilani for the Institute fellowship. CP is thankful to the Anusandhan National Research Foundation, India (Erstwhile Science and Engineering Research Board) for the project fellowship under grant number SRG/2021/002390.

References

Abd El-Mageed, T.A., Belal, E.E., Rady, M.O.A., Abd El-Mageed, S.A., Mansour, E. *et al.* (2021) Acidified biochar as a soil amendment to drought stressed (*Vicia faba* L.) plants: influences on growth and productivity, nutrient status, and water use efficiency. *Agronomy* 11, 1290.

Adair, K.L., Lindgreen, S., Poole, A.M., Young, L.M., Bernard-Verdier, M. *et al.* (2019) Above and belowground community strategies respond to different global change drivers. *Scientific Reports* 9, 2540.

Ardley, J. (2017) Legumes of the Thar desert and their nitrogen fixing Ensifer symbionts. *Plant and Soil* 410, 517–520.

Aryal, B., Gurung, R., Camargo, A.F., Fongaro, G., Treichel, H. *et al.* (2022) Nitrous oxide emission in altered nitrogen cycle and implications for climate change. *Environmental Pollution* 314, 120272.

Barnard, R.L., Osborne, C.A. and Firestone, M.K. (2013) Responses of soil bacterial and fungal communities to extreme desiccation and rewetting. *ISME Journal* 7, 2229–2241.

Belay, A.M., Selassie, Y.G., Tsegaye, E.A., Meshesha, D.T. and Addis, H.K. (2023) Soil pH mapping as a function of land use, elevation, and rainfall in the Lake Tana basin, northwestern of Ethiopia. *Agrosystems, Geosciences and Environment* 6, e20420.

Bi, W., Zhang, D., Weng, B., Dong, Z., Wang, J. *et al.* (2023) Research progress on the effects of droughts and floods on phosphorus in soil-plant ecosystems based on knowledge graph. *HydroResearch* 6, 29–35.

Bogati, K. and Walczak, M. (2022) The impact of drought stress on soil microbial community, enzyme activities and plants. *Agronomy* 12, 189.

Bradford, J.B., Schlaepfer, D.R., Lauenroth, W.K., Palmquist, K.A., Chambers, J.C. *et al.* (2019) Climate-driven shifts in soil temperature and moisture regimes suggest opportunities to enhance assessments of dryland resilience and resistance. *Frontiers in Ecology and Evolution* 7.

Brimblecombe, P. (2015) Biogeochemical cycles: sulfur cycle. In: North, G.R., Pyle, J.A. and Zhang, F. (eds) *Encyclopedia of Atmospheric Sciences,* 2nd edn. Elsevier, pp. 187–193.

Buan, N.R. (2018) Methanogens: pushing the boundaries of biology. *Emerging Topics in Life Sciences* 2, 629–646.

Cavicchioli, R., Ripple, W.J., Timmis, K.N., Azam, F., Bakken, L.R. *et al.* (2019) Scientists' warning to humanity: microorganisms and climate change. *Nature Reviews Microbiology* 17, 569–586.

Certini, G. and Scalenghe, R. (2023) The crucial interactions between climate and soil. *Science of the Total Environment* 856, 159169.

Cheng, T.W., Lin, L.H., Lin, Y.T., Song, S.R. and Wang, P.L. (2014) Temperature-dependent variations in sulfate-reducing communities associated with a terrestrial hydrocarbon seep. *Microbes and Environments* 29, 377–387.

Cheng, X., Zhou, T., Liu, S., Sun, X., Zhou, Y. *et al.* (2023) Effects of climate on variation of soil organic carbon and alkali-hydrolyzed nitrogen in subtropical forests: a case study of Zhejiang province, China. *Forests* 14, 914.

Cheng-Jim, J.I., Yuan-He, Y.A.N.G., Wen-Xuan, H.A.N., Yan-Fang, H.E., Smith, J. and Smith, P. (2014) Climatic and edaphic controls on soil pH in alpine grasslands on the Tibetan Plateau, China: a quantitative analysis. *Pedosphere* 24, 39–44.

Choi, A., Cha, I.T., Lee, K.E., Son, Y.K., Yu, J. *et al.* (2023) The role of *Flavobacterium enshiense* R6S-5-6 in the wetland ecosystem revealed by whole-genome analysis. *Current Microbiology* 80, 83.

Chowdhury, S., Dubey, V.K., Choudhury, S., Das, A., Jeengar, D. *et al.* (2023) Insects as bioindicator: a hidden gem for environmental monitoring. *Frontiers in Environmental Science* 11, 1146052.

Chytrý, M., Danihelka, J., Ermakov, N., Hájek, M., Hájková, P. *et al.* (2007) Plant species richness in continental southern Siberia: effects of pH and climate in the context of the species pool hypothesis. *Global Ecology and Biogeography* 16, 668–678.

Dhital, D., Manandhar, R., Manandhar, P. and Maharjan, S.R. (2022) Soil CO_2 efflux dynamics and its relationship with the environmental variables in a sub-tropical mixed forest. *Open Journal of Forestry* 12, 312–336.

Díaz-Torres, O., Lugo-Melchor, O.Y., de Anda, J., Pacheco, A., Yebra-Montes, C. *et al.* (2022) Bacterial dynamics and their influence on the biogeochemical cycles in a subtropical hypereutrophic lake during the rainy season. *Frontiers in Microbiology* 13, 832477.

Dorau, K., Bamminger, C., Koch, D. and Mansfeldt, T. (2022) Evidences of soil warming from long-term trends (1951–2018) in North Rhine-Westphalia, Germany. *Climatic Change* 170, 170.

Duan, D., Jiang, F., Lin, W., Tian, Z., Wu, N. *et al.* (2022) Effects of drought on the growth of *Lespedeza davurica* through the alteration of soil microbial communities and nutrient availability. *Journal of Fungi* 8, 384.

Erisman, J.W., Galloway, J.N., Seitzinger, S., Bleeker, A., Dise, N.B. *et al.* (2013) Consequences of human modification of the global nitrogen cycle. *Philosophical Transactions of the Royal Society Series B, Biological Sciences* 368, 20130116.

Fang, J., Wei, S., Shi, G., Cheng, Y., Zhang, X. *et al.* (2021) Potential effects of temperature levels on soil bacterial community structure. *E3S Web of Conferences* 292, 01008.

Feng, X., Liu, C., Xie, F., Lu, J., Chiu, L.S. *et al.* (2018) Precipitation characteristic changes due to global warming in a high-resolution (16 km) ECMWF simulation. *Quarterly Journal of the Royal Meteorological Society* 145, 303–317.

Fowler, D., Coyle, M., Skiba, U., Sutton, M.A., Cape, J.N. *et al.* (2013) The global nitrogen cycle in the twenty-first century. *Philosophical Transactions of the Royal Society of London Series B, Biological Sciences* 368, 20130164.

French, S., Levy-Booth, D., Samarajeewa, A., Shannon, K.E., Smith, J. *et al.* (2009) Elevated temperatures and carbon dioxide concentrations: effects on selected microbial activities in temperate agricultural soils. *World Journal of Microbiology and Biotechnology* 25, 1887–1900.

Frindte, K., Pape, R., Werner, K., Löffler, J. and Knief, C. (2019) Temperature and soil moisture control microbial community composition in an arctic-alpine ecosystem along elevational and micro-topographic gradients. *ISME Journal* 13, 2031–2043.

Furtak, K. and Wolińska, A. (2023) The impact of extreme weather events as a consequence of climate change on the soil moisture and on the quality of the soil environment and agriculture – a review. *CATENA* 231, 107378.

García-Velázquez, L., Rodríguez, A., Gallardo, A., Maestre, F.T., Dos Santos, E. *et al.* (2020) Climate and soil micro-organisms drive soil phosphorus fractions in coastal dune systems. *Functional Ecology* 34, 1690–1701.

Gelybó, G., Tóth, E., Farkas, C., Horel, Á., Kása, I. *et al.* (2018) Potential impacts of climate change on soil properties. *Agrokémia és Talajtan* 67, 121–141.

Geng, Y., Baumann, F., Song, C., Zhang, M., Shi, Y. *et al.* (2017) Increasing temperature reduces the coupling between available nitrogen and phosphorus in soils of Chinese grasslands. *Scientific Reports* 7, 43524.

Grzyb, A., Wolna-Maruwka, A. and Niewiadomska, A. (2020) Environmental factors affecting the mineralization of crop residues. *Agronomy* 10, 1951.

Guoju, X., Qiang, Z., Jiangtao, B., Fengju, Z. and Chengke, L. (2012) The relationship between winter temperature rise and soil fertility properties. *Air, Soil and Water Research* 5, 15–22.

Hamidov, A., Helming, K., Bellocchi, G., Bojar, W., Dalgaard, T. *et al.* (2018) Impacts of climate change adaptation options on soil functions: a review of European case-studies. *Land Degradation and Development* 29, 2378–2389.

Hari, M. and Tyagi, B. (2022) Terrestrial carbon cycle: tipping edge of climate change between the atmosphere and biosphere ecosystems. *Environmental Science* 2, 867–890.

Hashem, A., Tabassum, B. and Fathi Abd_Allah, E. (2019) Bacillus subtilis: a plant-growth promoting rhizobacterium that also impacts biotic stress. *Saudi Journal of Biological Sciences* 26, 1291–1297.

Hashimoto, S., Carvalhais, N., Ito, A., Migliavacca, M., Nishina, K. *et al.* (2015) Global spatiotemporal distribution of soil respiration modeled using a global database. *Biogeosciences* 12, 4121–4132.

Hashimoto, S., Ito, A. and Nishina, K. (2023) Divergent data-driven estimates of global soil respiration. *Communications Earth and Environment* 4, 1–8.

Hatten, J. and Liles, G. (2019) A 'healthy' balance – the role of physical and chemical properties in maintaining forest soil function in a changing world. *Developments in Soil Science* 36, 373–396.

Hicks, N., Liu, X., Gregory, R., Kenny, J., Lucaci, A. *et al.* (2018) Temperature driven changes in benthic bacterial diversity influences biogeochemical cycling in coastal sediments. *Frontiers in Microbiology* 9, 1730.

Hou, E., Chen, C., Luo, Y., Zhou, G., Kuang, Y. *et al.* (2018) Effects of climate on soil phosphorus cycle and availability in natural terrestrial ecosystems. *Global Change Biology* 24, 3344–3356.

Houle, D., Marty, C., Augustin, F., Dermont, G. and Gagnon, C. (2020) Impact of climate change on soil hydro-climatic conditions and base cations weathering rates in forested watersheds in eastern Canada. *Frontiers in Forests and Global Change* 3, 535397.

Hue, N. (2022) Soil acidity: development, impacts, and management. In: *Structure and Functions of Pedosphere*. Springer Nature Singapore, Singapore, pp. 103–131.

IPCC (2024). *Climate Change and Land*. Available at: https://www.ipcc.ch/srccl/ (last accessed October 2024).

Ji, Y., Li, Y., Yao, N., Biswas, A., Zou, Y. *et al.* (2021) The lagged effect and impact of soil moisture drought on terrestrial ecosystem water use efficiency. *Ecological Indicators* 133, 108349.

Jones, M.T., Jerram, D.A., Svensen, H.H. and Grove, C. (2016) The effects of large igneous provinces on the global carbon and sulphur cycles. *Palaeogeography, Palaeoclimatology, Palaeoecology* 441, 4–21.

Karmakar, R., Das, I., Dutta, D. and Rakshit, A. (2016) Potential effects of climate change on soil properties: a review. *Science International* 4, 51–73.

Kawamiya, M., Hajima, T., Tachiiri, K., Watanabe, S. and Yokohata, T. (2020) Two decades of Earth system modeling with an emphasis on Model for Interdisciplinary Research on Climate (MIROC). *Progress in Earth and Planetary Science* 7, 64.

Kehler, A., Haygarth, P., Tamburini, F. and Blackwell, M. (2021) Cycling of reduced phosphorus compounds in soil and potential impacts of climate change. *European Journal of Soil Science* 72, 2517–2537.

Kwon, T., Shibata, H., Kepfer-Rojas, S., Schmidt, I.K., Larsen, K.S. *et al.* (2021) Effects of climate and atmospheric nitrogen deposition on early to mid-term stage litter decomposition across biomes. *Frontiers in Forests and Global Change* 4, 1–18.

Leal Filho, W., Nagy, G.J., Setti, A.F.F., Sharifi, A., Donkor, F.K. *et al.* (2023) Handling the impacts of climate change on soil biodiversity. *Science of the Total Environment* 869, 161671.

Lei, N. and Han, J. (2020) Effect of precipitation on respiration of different reconstructed soils. *Scientific Reports* 10, 7328.

Li, H., Yang, Q., Li, J., Gao, H., Li, P. *et al.* (2015) The impact of temperature on microbial diversity and AOA activity in the Tengchong Geothermal Field, China. *Scientific Reports* 5, 17056.

Li, Y., Wang, J., He, L., Xu, X., Wang, J. *et al.* (2022) Different mechanisms driving increasing abundance of microbial phosphorus cycling gene groups along an elevational gradient. *iScience* 25, 105170.

Liu, C., Siri, M., Li, H., Ren, C., Huang, J. *et al.* (2023a) Drought is threatening plant growth and soil nutrients of grassland ecosystems: a meta-analysis. *Ecology and Evolution* 13, e10092.

Liu, D., Zhang, W., Xiong, C. and Nie, Q. (2023b) Warming increases the relative change in the turnover rate of decadally cycling soil carbon in microbial biomass carbon and soil respiration. *Frontiers in Earth Science* 10, 1089544.

Luo, Y. (2018) Geochemical cycle and environmental effects of sulfur in lakes. *IOP Conference Series* 394, 052039.

Ma, L., She, W., Wu, G., Yang, J., Phurbu, D. *et al.* (2021) Influence of temperature and sulfate concentration on the sulfate/sulfite reduction prokaryotic communities in the Tibetan hot springs. *Microorganisms* 9, 1–15.

Maaroufi, N.I. and De Long, J.R. (2020) Global change impacts on forest soils: linkage between soil biota and carbon-nitrogen-phosphorus stoichiometry. *Frontiers in Forests and Global Change* 3, 1–8.

Malhi, G.S., Kaur, M. and Kaushik, P. (2020) Impact of climate change on agriculture and its mitigation strategies: a review. *Sustainability* 13, 1318.

Mangodo, C., Adeyemi, T.O.A., Bakpolor, V.R. and Adegboyega, D.A. (2020) Impact of microorganisms on climate change: a review. *World News of Natural Sciences* 31, 36–47.

Menge, D.N.L., Kou-Giesbrecht, S., Taylor, B.N., Akana, P.R., Butler, A. *et al.* (2023) Terrestrial phosphorus cycling: responses to climatic change. *Annual Review of Ecology, Evolution and Systematics* 54, 429–449.

Mujakić, I., Piwosz, K. and Koblížek, M. (2022) Phylum gemmatimonadota and its role in the environment. *Microorganisms* 10, 151.

Narsing Rao, M.P., Luo, Z.-H., Dong, Z.-Y., Li, Q., Liu, B.-B. *et al.* (2022) Metagenomic analysis further extends the role of Chloroflexi in fundamental biogeochemical cycles. *Environmental Research* 209, 112888.

Navarro-Pedreño, J., Almendro-Candel, M.B. and Zorpas, A.A. (2021) The increase of soil organic matter reduces global warming, myth or reality? *Sci* 3, 18.

Naylor, D. and Coleman-Derr, D. (2018) Drought stress and root-associated bacterial communities. *Frontiers in Plant Science* 8, 2223.

Naylor, D., DeGraaf, S., Purdom, E. and Coleman-Derr, D. (2017) Drought and host selection influence bacterial community dynamics in the grass root microbiome. *ISME Journal* 11, 2691–2704.

Ontong, N., Poolsiri, R., Diloksumpun, S., Staporn, D. and Jenke, M. (2023) Effects of tree functional traits on soil respiration in tropical forest plantations. *Forests* 14, 715.

Pan, M., Wang, T., Hu, B., Shi, P., Xu, J. *et al.* (2021) Mesocosm experiments reveal global warming accelerates macrophytes litter decomposition and alters decomposition-related bacteria community structure. *Water* 13, 1940.

Parmar, T.K., Rawtani, D. and Agrawal, Y.K. (2016) Bioindicators: the natural indicator of environmental pollution. *Frontiers in Life Science* 9, 110–118.

Peck, D.H. (2020) The role of nitrogen availability on the dominance of *Planktothrix agardhii* in Sandusky Bay, Lake Erie. MSc thesis, Bowling Green State University, Ohio.

Pérez-Guzmán, L., Phillips, L.A., Seuradge, B.J., Agomoh, I., Drury, C.F. *et al.* (2021) An evaluation of biological soil health indicators in four long-term continuous agroecosystems in Canada. *Agrosystems, Geosciences and Environment* 4, e20164.

Pires, D., Orlando, V., Collett, R.L., Moreira, D., Costa, S.R. *et al.* (2023) Linking nematode communities and soil health under climate change. *Sustainability* 15, 11747.

Preece, C., Verbruggen, E., Liu, L., Weedon, J.T. and Peñuelas, J. (2019) Effects of past and current drought on the composition and diversity of soil microbial communities. *Soil Biology and Biochemistry* 131, 28–39.

Qian, B., Gregorich, E.G., Gameda, S., Hopkins, D.W. and Wang, X.L. (2011) Observed soil temperature trends associated with climate change in Canada. *Journal of Geophysical Research* 116, D02106.

Qu, L.-P., Chen, J., Xiao, J., De Boeck, H.J., Dong, G. *et al.* (2024) The complexity of heatwaves impact on terrestrial ecosystem carbon fluxes: factors, mechanisms and a multi-stage analytical approach. *Environmental Research* 240, 117495.

Raghavendra, M., Sharma, M.P., Ramesh, A., Richa, A., Billore, S.D. *et al.* (2020) Soil health indicators: methods and applications. In: Rakshit, A., Ghosh, S., Chakraborty, S., Philip, V., Datta, A. (eds *Soil Analysis: Recent Trends and Applications*. Springer, Singapore, pp. 221–253.

Raza, A., Razzaq, A., Mehmood, S.S., Zou, X., Zhang, X. *et al.* (2019) Impact of climate change on crops adaptation and strategies to tackle its outcome: a review. *Plants* 8, 34.

Reim, A., Hernández, M., Klose, M., Chidthaisong, A., Yuttitham, M. *et al.* (2017) Response of methanogenic microbial communities to desiccation stress in flooded and rain-fed paddy soil from Thailand. *Frontiers in Microbiology* 8, 785.

Rengel, Z. (2011) Soil pH, soil health and climate change. In: Singh, B.P., Cowie, A.L. and Chan, K.Y. (eds) *Soil Health and Climate Change*. Springer, Berlin, Heidelberg, pp. 69–85.

Reshotkin, O.V. and Khudyakov, O.I. (2019) Soil temperature response to modern climate change at four sites of different latitude in the European part of Russia. *IOP Conference Series* 368, 012040.

Rodtassana, C., Unawong, W., Yaemphum, S., Chanthorn, W., Chawchai, S. *et al.* (2021) Different responses of soil respiration to environmental factors across forest stages in a southeast Asian forest. *Ecology and Evolution* 11, 15430–15443.

Rubio, V.E., Detto, M. and Vanessa Rubio, C.E. (2017) Spatiotemporal variability of soil respiration in a seasonal tropical forest. *Ecology and Evolution* 7, 7104–7116.

Sánchez-Baracaldo, P., Bianchini, G., Wilson, J.D. and Knoll, A.H. (2022) Cyanobacteria and biogeochemical cycles through earth history. *Trends in Microbiology* 30, 143–157.

Sanjuan, J., Delgado, M.J. and Girard, L. (2020) Editorial: Microbial control of the nitrogen cycle. *Frontiers in Microbiology* 11, 950.

Santos-Medellín, C., Edwards, J., Liechty, Z., Nguyen, B. and Sundaresan, V. (2017) Drought stress results in a compartment-specific restructuring of the rice root-associated microbiomes. *mBio* 8, e00764-17.

Schimel, J.P. (2018) Life in dry soils: effects of drought on soil microbial communities and processes. *Annual Review of Ecology, Evolution and Systematics* 49, 409–432.

Schoenau, J.J. and Malhi, S.S. (2008) Sulfur forms and cycling processes in soil and their relationship to sulfur fertility. In: Jez, J. (ed.) *Sulfur: A Missing Link Between Soils, Crops, and Nutrition*. Agronomy Monograph No. 50. American Society of Agronomy, pp. 1–10.

Schröder, J.J., Schulte, R.P.O., Creamer, R.E., Delgado, A., van Leeuwen, J. *et al.* (2016) The elusive role of soil quality in nutrient cycling: a review. *Soil Use and Management* 32, 476–486.

Shen, H., Zhang, L., Meng, H., Zheng, Z., Zhao, Y. *et al.* (2021) Response of soil respiration and its components to precipitation exclusion in *Vitex negundo* var. *Heterophylla* shrubland of the middle Taihang mountain in north China. *Frontiers in Environmental Science* 9, 712301.

Shen, X., Ma, J., Li, Y., Li, Y. and Xia, X. (2022) The effects of multiple global change factors on soil nutrients across China: a meta-analysis. *International Journal of Environmental Research and Public Health* 19, 15230.

Siebielec, S., Siebielec, G., Klimkowicz-Pawlas, A., Gałązka, A., Grządziel, J. *et al.* (2020) Impact of water stress on microbial community and activity in sandy and loamy soils. *Agronomy* 10, 1429.

Singh, A.K., Jiang, X.-J., Yang, B., Wu, J., Rai, A. *et al.* (2020) Biological indicators affected by land use change, soil resource availability and seasonality in dry tropics. *Ecological Indicators* 115, 106369.

Soong, J.L., Phillips, C.L., Ledna, C., Koven, C.D. and Torn, M.S. (2020) CMIP5 models predict rapid and deep soil warming over the 21st century. *Journal of Geophysical Research:Biogeoscience* 125, e2019JG005266.

Storesund, J.E., Lanzèn, A., Nordmann, E.-L., Armo, H.R., Lage, O.M. *et al.* (2020) *Planctomycetes* as a vital constituent of the microbial communities inhabiting different layers of the meromictic lake sælenvannet (Norway). *Microorganisms* 8, 1–18.

Stubenrauch, J. and Ekardt, F. (2020) Plastic pollution in soils: governance approaches to foster soil health and closed nutrient cycles. *Environments* 7, 38.

Sun, W., Li, S., Zhang, G., Fu, G., Qi, H. *et al.* (2023) Effects of climate change and anthropogenic activities on soil pH in grassland regions on the Tibetan plateau. *Global Ecology and Conservation* 45, e02532.

Tahat, M.M., Alananbeh, K.M., Othman, Y.A. and Leskovar, D.I. (2020) Soil health and sustainable agriculture. *Sustainability* 12, 4859.

Tian, J., Ge, F., Zhang, D., Deng, S. and Liu, X. (2021) Roles of phosphate solubilizing microorganisms from managing soil phosphorus deficiency to mediating biogeochemical P cycle. *Biology* 10, 1–19.

Tian, Y., Shi, C., Malo, C.U., Kwatcho Kengdo, S., Heinzle, J. *et al.* (2023) Long-term soil warming decreases microbial phosphorus utilization by increasing abiotic phosphorus sorption and phosphorus losses. *Nature Communications* 14, 864.

Tumwesigye, W., Aschalew, A., Wilber, W., Atwongyire, D., Nagawa, G.M. and Ndizihiwe, D. (2019) Climate-Smart Agriculture for improving crop production and biodiversity conservation: opportunities and challenges in the 21st century – a narrative review. *Journal of Water Resources and Ocean Science* 8, 56–62.

Turner, B.L. and Haygarth, P.M. (2003) Changes in bicarbonate-extractable inorganic and organic phosphorus by drying pasture soils. *Soil Science Society of America Journal* 67, 344–350.

UN-ESCAP (United Nations – The Economic and Social Commission for Asia and the Pacific) (2020). *Ready for the Dry Years: Building Resilience to Drought in South-East Asia.* Available at: https://www.unescap.org/sites/default/files/publications/Ready%20for%20the%20Dry%20Years.pdf (last accessed October 2024).

Wang, B., Chen, Y., Li, Y., Zhang, H., Yue, K. *et al.* (2021) Differential effects of altered precipitation regimes on soil carbon cycles in arid versus humid terrestrial ecosystems. *Global Change Biology* 27, 6348–6362.

Wang, H., Wu, J., Li, G., Yan, L. and Liu, S. (2023) Effects of extreme rainfall frequency on soil organic carbon fractions and carbon pool in a wet meadow on the Qinghai-Tibet Plateau. *Ecological Indicators* 146, 109853.

Wasan, J.P.M. and Wasan, K.M. (2023) Effects of climate change on soil health resulting in an increased global spread of neglected tropical diseases. *PLoS Neglected Tropical Diseases* 17, e0011378.

Wei, H., Liu, Y., Xiang, H., Zhang, J., Li, S. *et al.* (2019) Soil pH responses to simulated acid rain leaching in three agricultural soils. *Sustainability* 12, 280.

Xing, W., Lu, X., Niu, S., Chen, D., Wang, J. *et al.* (2022) Global patterns and drivers of soil nematodes in response to nitrogen enrichment. *CATENA* 213, 106235.

Xu, L., Naylor, D., Dong, Z., Simmons, T., Pierroz, G. *et al.* (2018) Drought delays development of the sorghum root microbiome and enriches for monoderm bacteria. *Proceedings of the National Academy of Sciences USA* 115, E4284–E4293.

Xu, S., Sheng, C. and Tian, C. (2020) Changing soil carbon: Influencing factors, sequestration strategy and research direction. *Carbon Balance and Management* 15, 2.

Zhang, H., Liu, B., Zhou, D., Wu, Z. and Wang, T. (2019a) Asymmetric soil warming under global climate change. *International Journal of Environmental Research and Public Health* 16, 1504.

Zhang, X.-Y., Li, Q.-W., Gao, J.-Q., Hu, Y.-H., Song, M.-H. *et al.* (2020a) Effects of rainfall amount and frequency on soil nitrogen mineralization in Zoigê alpine wetland. *European Journal of Soil Biology* 97,
   103170.
Zhang, Y., Zhang, N., Yin, J., Zhao, Y., Yang, F. *et al.* (2020b) Simulated warming enhances the responses
   of microbial N transformations to reactive N input in a Tibetan alpine meadow. *Environment International*
   141, 105795.
Zhang, Y.-Y., Wu, W. and Liu, H. (2019b) Factors affecting variations of soil pH in different horizons in hilly
   regions. *PLoS One* 14, e0218563.
Zhao, F., Wu, Y., Hui, J., Sivakumar, B., Meng, X. *et al.* (2021) Projected soil organic carbon loss in response
   to climate warming and soil water content in a loess watershed. *Carbon Balance and Management*
   16, 24.
Zhao, M., Xue, K., Wang, F., Liu, S., Bai, S. *et al.* (2014) Microbial mediation of biogeochemical cycles
   revealed by simulation of global changes with soil transplant and cropping. *ISME Journal* 8, 2045–
   2055.
Zhou, Z., Wang, C. and Luo, Y. (2018) Response of soil microbial communities to altered precipitation: a
   global synthesis. *Global Ecology and Biogeography* 27, 1121–1136.
Zhu, H.Z., Jiang, C.Y. and Liu, S.J. (2022) Microbial roles in cave biogeochemical cycling. *Frontiers in
   Microbiology* 13, 950005.

7 Impact of Rising Atmospheric Carbon Dioxide on Soil Health and Plant Nutrition

Roshni Patel[1], Deviprasad Samantaray[2], Arti Hansda[1], K. Santosh Kumar[3] and Swati Mohapatra[1]*

[1]*Department of Life Science, School of Science, GSFC University, Vadodara, Gujarat, India;* [2]*Department of Microbiology, Orissa University of Agriculture and Technology, Bhubaneswar, India;* [3]*Department of Chemistry, School of Science, GSFC University, Vadodara, Gujarat, India*

Abstract

Elevated atmospheric CO_2 has been reported to affect soil structure, soil organic matter and microbial communities through increasing microbial biomass, modifying microbial community structure and stimulating the growth of specific microbial groups. Soil organic matter accumulation is also influenced by elevated CO_2 levels, increased organic matter and higher decomposition rates, which may lead to soil carbon sequestration. Nutrient availability is another important factor influenced by elevated CO_2, with studies showing increased nitrogen and phosphorus uptake by plants. However, this effect may vary depending on soil nutrient status and plant species. Understanding the impact of rising atmospheric CO_2 on soil health and plant nutrition is very important for maintaining ecosystems and food quality. Therefore, we discuss in this chapter the impact of elevated CO_2 on soil health and plant nutrition. The chapter provides readers with important insights into the intricate interactions between CO_2, soil ecosystems and the nutritional value of crops by examining the substantial effects of growing atmospheric CO_2 levels on soil health and plant nutrition. It is essential to understand these processes to tackle the issues brought on by climate change and guarantee sustainable agricultural practices for a future of food security.

Keywords: Elevated CO_2, soil health, plant nutrition, agriculture sustainability, microbial community, soil fertility, plant growth

7.1 Introduction

This modern era is inclining towards more industrialization, leading to climate change. This climate change, highly driven by human activities such as burning excess fossil fuels, deforestation and industrial processes, has accelerated atmospheric carbon dioxide (CO_2) concentration. The rising atmospheric CO_2 levels have multiple environmental consequences, including impacts on soil health and plant nutrition.

Over the past 150–200 years, there has been a consistent increase in atmospheric CO_2 concentrations. The significant increase in atmospheric CO_2 levels worldwide, attributed mainly to human activities such as the combustion of fossil fuels and deforestation, has emerged as a critical environmental issue with potential

---

*Corresponding author: swatimohapatraiitr@gmail.com

© CAB International 2025. *Soil Health and Nutrition Management*
(eds N.C. Joshi, T. Leustek and P.K. Singh)
DOI: 10.1079/9781800624597.0007

repercussions for both the well-being of humans and the environment. Rising atmospheric CO_2 acts as a greenhouse gas, contributing a lot to global warming and directly affecting the environment. The CO_2 concentration surpassed 400 ppm in 2013, the highest it had ever been in human history, and it is still rising, which is a severe issue for both people and the rest of the Earth (Williams *et al.*, 2018). Global change encompasses multiple components due to elevated CO_2, including climate changes (such as temperature and precipitation) and frequency of extreme conditions (such as heatwaves, hurricanes, droughts and floods), environmental modifications, land use changes (such as deforestation, wetland degradation and ploughing), soil degradation and declining fertility (such as erosion, salinization and nutrient losses) (Fig. 7.1).

Moreover, there is a rise in heavy metal contamination, urbanization, population expansion and exhaustion of non-renewable resources (such as agricultural land, freshwater, groundwater, fossil fuels and minerals) as well as chemical and organic pollution with pesticides, heavy metals, radionuclides, acid deposition and polyaromatic hydrocarbons as well as eutrophication. Additionally, the reduction in biodiversity and genetic diversity is an outcome of global changes, as highlighted by Williams *et al.* (2018). In agriculture, soil health and plant nutrition are critical components of agricultural productivity and ecosystem sustainability. Elevated atmospheric CO_2 mainly influence organic matter content, microbial communities, nutrient cycling and soil structure (Fig. 7.2).

Additionally, changes in CO_2 concentration can directly affect plant nutrition, leading to alterations in nutrient uptake, nutrient allocation and nutrient use efficiency. These changes can have positive and negative implications for agricultural production, ecosystem functioning and food quality. The major relationship between elevated atmospheric CO_2 and soil health involves complex interactions and feedback loops due to their intricate interplay. Recognizing the implications of the escalating atmospheric CO_2 levels for soil health and plant nutrition is essential in anticipating and controlling the repercussions of climate change on both agricultural systems and natural ecosystems (Elbasiouny *et al.*, 2022). Moreover, elevated CO_2 levels can significantly impact soil microbial communities by altering various soil factors, such as pH, carbon and nitrogen content, and the ratio of carbon to nitrogen. Understanding these complex interactions is essential for developing effective strategies to manage and maintain healthy soil ecosystems in the face of changing environmental conditions. This chapter therefore compiles the latest scientific findings on the effects of elevated CO_2 on soil health and plant nutrition (Elbasiouny *et al.*, 2022).

7.2 Impact of Elevated CO_2 on Soil Health, Plant Nutrition and Growth

7.2.1 Impact of elevated CO_2 on soil health

Soil health is a critical determinant of plant growth, and its development contributes significantly to

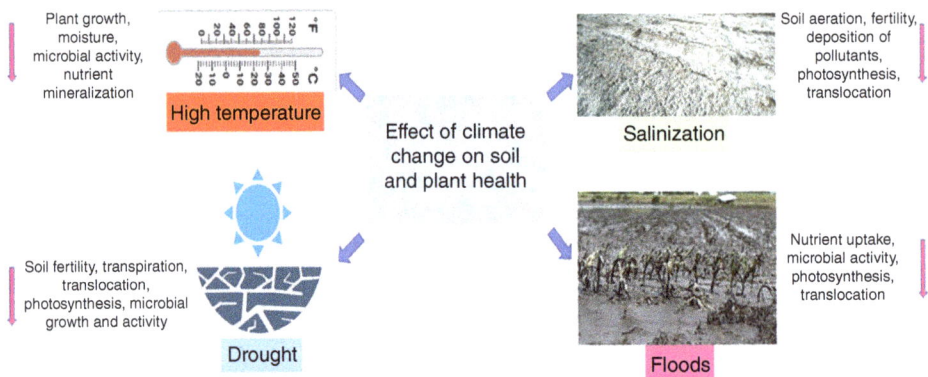

Fig. 7.1. The impact of climate change on the health of soil and plants.

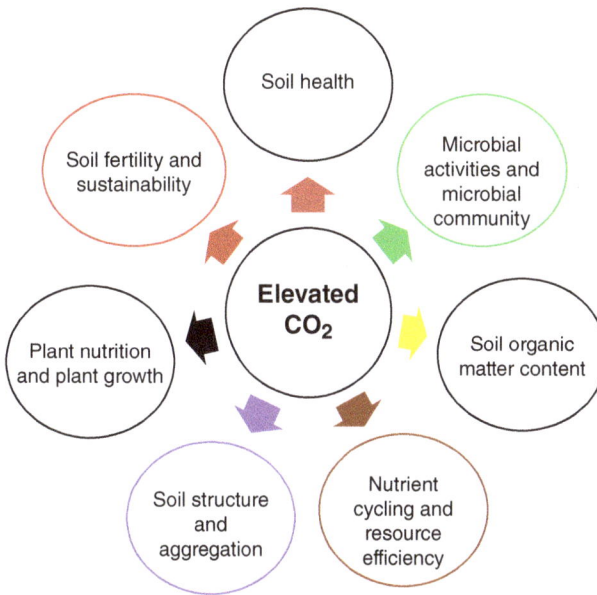

Fig. 7.2. The impact of increased carbon dioxide levels on plant nutrition and soil health.

ecosystem functioning. Elevated CO$_2$ levels can directly and indirectly impact soil health through various mechanisms. Firstly, increased CO$_2$ concentrations can alter plant physiology, leading to changes in plant growth, root exudation and litter decomposition rates. These changes can influence soil organic matter (SOM) dynamics, nutrient cycling and microbial activity and communities (Dutta and Dutta, 2016). Secondly, elevated CO$_2$ levels can indirectly affect soil health by altering plant–microbe interactions. For instance, changes in plant physiology and SOM dynamics can affect rhizosphere processes, including microbial colonization of plant roots, nutrient acquisition and disease suppression. Thirdly, heightened levels of CO$_2$ can impact soil physical characteristics like structure, porosity, water retention capacity and also increased soil inorganic carbon (Ferdush and Paul, 2021). It also raises the soil redox potential, which is crucial for the biogeochemical cycling of soil nutrients (Li *et al.*, 2024). Increased CO$_2$ raises the rate of soil nitrogen mineralization (Wu *et al.*, 2020). These changes can subsequently affect nutrient accessibility, soil fertility and the soil's overall health (Bijay-Singh, 2011).

7.2.2   Impact of elevated CO$_2$ on microbial activity and communities

The soil microbial community greatly aids the cycle of nutrients and the breakdown of organic matter. By providing bacteria, fungi and other soil microorganisms a direct carbon source in the form of exudates, increased CO$_2$ can affect their diversity, abundance and activity (Du *et al.*, 2017). This can result in changes in the competitive interactions between different microbial groups, potentially impacting nutrient cycling, organic matter decomposition and soil aggregation. The rhizosphere, the soil area around plant roots influenced by root exudates and root–microbe interactions, experiences changes in nutrient availability, decomposition rates and soil aggregation due to elevated CO$_2$. Moreover, elevated CO$_2$ can alter the composition and behaviour of microbial communities within the rhizosphere (Dutta and Dutta, 2016). For example, changes in root exudate composition under elevated CO$_2$ can alter the abundance and activity of specific microbial groups, such as nitrogen-fixing bacteria or mycorrhizal fungi, which can impact nutrient availability and cycling in the soil. Root exudates play a crucial role in shaping the composition and activity of soil microbial communities.

High CO_2 increase the release of sugars, amino acids, organic acids and other compounds in root exudates. These can result in the drainage of excess nutrients and liquids from the plants, affecting microbial communities. When exposed to rising CO_2, most plant species display enhanced photosynthesis, increased growth, lower water demand and decreased nitrogen and protein tissue concentrations. Moreover, it has an impact on the chemical composition of plant tissues. In the presence of elevated CO_2 under free-air CO_2 enrichment (FACE) conditions, there is an average increase of 30–40% in leaf non-structural carbohydrates (such as sugars and starches) per unit leaf area, attributed to heightened photosynthetic activity. This increase may influence their interactions with soil biota. Fast-growing plant species exhibit greater responsiveness to changes in CO_2 concentration compared with slow-growing species in the soil, as Singh *et al.* reported in 2019. For instance, some studies have reported that root exudates from plants grown under elevated CO_2 may have higher carbon-to-nitrogen (C/N) ratios, indicating a shift towards more carbon-rich exudates. This change in exudate quality can affect microbial processes in the soil, such as decomposition rates and nutrient cycling (Varma and Choudhary, 2019).

Additionally, changes in soil organic carbon and total nitrogen concentrations and a decrease in soil pH brought on by an increase in base nutrient intake to support greater biomass production can mediate the effects of rising CO_2 on microbial communities (Sun *et al.*, 2021). Elevated CO_2 levels can also increase the abundance and activity of saprotrophic fungi in soils. This is because many saprotrophic fungi are well adapted to utilizing the simple carbon compounds released by plant roots through exudation as a food source. With increased plant growth and root exudation stimulated by elevated CO_2, saprotrophic fungi can have more carbon substrates for growth and activity. The increased activity of saprotrophic fungi can also accelerate the decomposition of organic matter in the soil, leading to faster turnover of plant residues and litter. As a result, the availability of organic matter-derived carbon and nutrients in the soil can change, potentially affecting soil aggregation.

In some cases, increased decomposition of organic matter by saprotrophic fungi under elevated CO_2 can lead to decreased soil aggregation (Terrer *et al.*, 2021). This is because the breakdown of organic matter can release organic compounds that act as binding agents between soil particles, helping to stabilize soil aggregates. Nevertheless, it is important to note that no alterations in the microbial community or extracellular enzyme activities have been observed. Consequently, the responses of soil microbes to rising CO_2 levels can differ depending on soil type and environmental conditions. Additionally, increased CO_2 has been found to enhance soil microbial biomass and enzyme activity (Singh *et al.*, 2019).

7.2.3 Impact of elevated CO_2 on soil organic matter content, soil structure and aggregation

Elevated atmospheric CO_2 can affect the quantity and quality of plant residues that enter the soil through litterfall and root exudation. Increased plant productivity under elevated CO_2 may result in higher inputs of organic materials, leading to enhanced SOM content. However, changes in litter quality and decomposition rates can also alter the balance between carbon inputs and outputs, potentially affecting SOM or loss. Results of carbon isotope techniques showed that increasing atmospheric CO_2 (up to 60%) considerably boosted rhizosphere respiration, soluble carbon concentrations in the rhizosphere, dry mass accumulation and growth of wheat plants. Moreover, adding nitrogen increases SOM and CO_2 emissions (Cheng and Johnson, 1998). When organic matter decomposition is accelerated, as can happen under elevated CO_2, the availability of these binding agents may decrease, potentially leading to reduced soil aggregation. The wetlands are an enormous source of organic carbon and the environment with the fastest carbon accumulation rate. The equilibrium between organic matter production and decomposition in wetland soils is strongly related to the survival of this significant ecosystem and its ability to act as a carbon sink. In response to higher CO_2, increased decomposition will counteract increased productivity and soil carbon storage in wetlands, which leads to a negligible drop in soil organic carbon in wetlands (Kirwan and Blum, 2011).

The organization of soil particles and aggregates, known as soil structure, is pivotal in determining soil health, water infiltration and nutrient accessibility. Elevated CO_2 can influence soil structure and aggregation through changes in root growth, microbial activities and organic matter inputs. For example, increased root biomass under elevated CO_2 may enhance soil aggregation and stability, potentially improving soil structure. However, it is important to note that the impacts of elevated CO_2 on soil aggregation can be complex and may vary depending on other factors such as soil type, plant species and management practices. Changes in litter quality and decomposition rates can also affect soil aggregation, with implications for soil health and plant nutrition. Increased plant growth under elevated CO_2 can lead to higher water uptake and transpiration rates, influencing soil moisture dynamics and potentially affecting soil aggregation (He et al., 2019). In the croplands, higher CO_2 concentrations increased available potassium (AK) and total nitrogen (TN) (2.2% and 3.8%, respectively) while only slightly decreasing available nitrogen (AN) and available phosphorus (AP) (2.6% and 2.7%, respectively). Reductions in the AN and AP with elevated CO_2 concentrations encourage plant development and produce a significantly bigger absorption of nitrogen and phosphorus (Hoosbeek, 2016).

7.2.4 Impact of elevated CO_2 on nutrient cycling and resource efficiency

Soil health and plant nutrition are closely linked to nutrient cycling, which involves the uptake, cycling and recycling of nutrients by plants and soil microorganisms. Mineralization, immobilization and decomposition are three major processes in the soil that recycle nutrients from organic matter, crop leftovers and fertilizers, making them available for plant uptake (Singh et al., 2016). Efficient nutrient cycling in healthy soils can minimize nutrient losses through leaching or runoff and promote nutrient use efficiency, reducing the need for excessive fertilizer applications and improving resource management in agriculture. Elevated CO_2 can influence nutrient cycling in the soil through its effects on plant productivity and litter inputs and through direct effects on microbial activities. Increased plant productivity under elevated CO_2 can result

in higher nutrient demand and uptake, potentially leading to changes in nutrient cycling rates and patterns. The supply of micro- and macronutrients can influence changes in root exudation, microbial activities and organic matter decomposition, affecting the soil pH. More exudates can provide a larger energy source for soil microorganisms, potentially affecting their activities and interactions with the soil. It has also been shown that soil pH is the primary driver of total microbial biomass across different biomes. At the same time, the ratio of fungi to bacteria tends to increase with soil carbon to nitrogen ratios (Singh et al., 2016).

7.2.5 Impact of elevated CO_2 on soil fertility and sustainability

Soil health and plant nutrition are fundamental to maintaining soil fertility and long-term sustainability in agriculture. Healthy soils with balanced nutrient availability, optimal pH and good structure support sustainable crop production, reduce soil erosion, improve water infiltration and retention and promote beneficial soil microbial communities. These factors contribute to maintaining soil fertility, crucial for sustained agricultural productivity and resilience against environmental stresses, such as drought, disease and climate change (Kuzyakov et al., 2018). Accelerated carbon (C) turnover occurs due to heightened microbial activities like respiration, enzymatic actions and priming effects on soil organic matter, particularly in the presence of elevated C input from plants under rising atmospheric CO_2 levels. It can be concluded that elevated CO_2 is likely to have minor (or no) effects on C pools but will significantly increase C fluxes when comparing the impact of elevated CO_2 on changes in C pools with those effects (Kuzyakov et al., 2018). While the stable C pools may not be significantly affected, the intensified C fluxes will accelerate biogeochemical cycles under elevated CO_2 conditions.

7.2.6 Impact of elevated CO_2 on plant nutrition and plant growth

Plant nutrition is intricate and shaped by multiple factors, encompassing soil nutrient availability, plant nutrient uptake and nutrient use efficiency. Elevated CO_2 levels can impact plant nutrition by altering nutrient availability, influencing nutrient uptake and affecting nutrient use efficiency (Gavito *et al.*, 2001). Elevated CO_2 levels can influence nutrient availability by affecting SOM dynamics, mineralization and immobilization processes. For example, increased plant productivity under elevated CO_2 can result in higher litter inputs, altering SOM decomposition rates and nutrient release patterns. Elevated CO_2 can affect plant nutrient uptake by changing root morphology, exudation and mycorrhizal associations. It has also been shown that under elevated CO_2, plants tend to allocate more carbon to the roots, resulting in increased nutrient uptake, especially for nutrients like nitrogen (N) and phosphorus (P). Elevated CO_2 can also affect nutrient use efficiency by affecting plant nutrient utilization, assimilation and allocation. For instance, increased plant productivity and C assimilation under elevated CO_2 can

result in higher nutrient utilization efficiency, where plants use nutrients more effectively to produce biomass. Yet, the impact of elevated CO_2 on plant nutrition may vary among species and is contingent upon factors like nutrient availability, plant species and ecosystem type (Gavito *et al.*, 2001).

Elevated CO_2 levels, including increased root biomass and length, can stimulate plant growth (Table 7.1). The atmosphere's CO_2 directly affects plants, and too much CO_2 is predicted to promote plant growth and production (Singh *et al.*, 2019). Additionally, changes in plant root architecture, such as increased root branching and proliferation, can alter soil pore structure and impact soil aggregation. Evidence from prior environmental manipulation studies in tropical and temperate ecosystems suggests that plants can consistently sequester more C in both their above-ground and below-ground biomass when exposed to elevated CO_2 compared with ambient CO_2 levels (Singh *et al.*, 2019).

Table 7.1. The effect of elevated carbon dioxide on different plants.

Plant	Effect	Reference
Lettuce	Increased soluble sugar build-up by 27.1%, improving lettuce's taste	Pérez-López *et al.*, 2013; Becker and Kläring, 2016, Sgherri *et al.*, 2017
Tomato	Increased the amounts of fructose, glucose and total soluble sugars in tomatoes improving their quality and probably taste and yield	Zhang *et al.*, 2014; Pimenta *et al.*, 2023
Potato	Increased soluble sugar and starch concentrations while maintaining organic acid concentrations	Kumari and Agrawal, 2014
Barley	Higher iron build-up in soil-grown barley shoots	Haase *et al.*, 2008
Loblolly pine	Increased dissolved organic carbon from the loblolly pine roots (*Pinus taeda*)	Phillips *et al.*, 2011
Wheat	Photosynthesis and growth uplifted, enhancement of productivity observed, wateruse efficiency accelerated and also increase phosphorus availability	Ainsworth and Long, 2005; Jin *et al.*, 2022
Rice	Plant growth promoted, enhancement of grain yield and alteration in nutrients	Kimball, 2016; AbdElgawad *et al.*, 2023
Cotton	Enhancement of productivity growth and water use efficiency, and enhanced fibre quality	AbdElgawad *et al.*, 2015; Nguyen *et al.*, 2019
Tomato	Enhanced fruit yield and altered nutritional composition	Kimball *et al.*, 1994
Potato	Increased tuber yield and improved water use efficiency	Poorter *et al.*, 2012a
Barley	Increased photosynthesis, yield and improved water use efficiency	Ceulemans and Mousseau, 1994
Brassica crops	Improved biomass production and altered nutritional quality	Poorter *et al.*, 2012b

In the case of teak and *Butea* plants, after 46 months of treatment the plant height, dry leaf weight, steam weight and total above-ground biomass increased in elevated CO_2 compared with ambient CO_2 (Singh *et al.*, 2019).

Elevated CO_2 increases total non-structural carbohydrates (TNC; predominantly starch and sugars) but decreases protein and N concentrations in C3 plants (such as rice and wheat). The influence of elevated CO_2 on the ionome, encompassing the mineral and trace element composition of plants, however, has been obscured by inconsistent data. Consequently, CO_2-induced alterations in plant quality have not been adequately considered in estimating the potential impact of climate change on human well-being. Additionally, the study by Singh *et al.* (2019) shows that elevated CO_2 increases the ratio of TNC to minerals compared with the ratio of carbon to minerals in C3 plants, leading to an overall reduction in mineral concentrations.

The effects of CO_2 on certain elements were as follows: overall, elevated CO_2 decreased the concentrations of P, potassium (K), calcium (Ca), sulfur (S), magnesium (Mg), iron (Fe), zinc (Zn) and copper (Cu) by 6.5–10% ($P = 0.001$). Only manganese (Mn) did not change significantly and only C was increased. If the increase in carbohydrate content drives the shift in the plant ionome, then the dramatic contrast between C and mineral reactions to elevated CO_2 is to be expected. The presence of carbohydrates leads to a dilution effect, reducing the mineral content in most plant tissues, with a minimal impact on C (Loladze, 2002).

Elevated CO_2 results in a systemic reduction of N contents by 10–18% across various tissues, encompassing leaves, stems, roots, tubers and reproductive and edible components such as seeds and grains (Taub and Wang, 2008).

No loss in minerals was found in one study in rice grains taken from four FACE paddies in Japan, but there was a lower N concentration. Seneweera and Conroy (1997), pioneers in documenting decreased Fe and Zn in grains of rice grown under elevated CO_2 conditions, disagreed with the results. They cautioned that alterations in rice quality could significantly impact developing countries.

The general decline in plant mineral content has been hypothesized to be attributed to increased carbohydrate production and other impacts of elevated CO_2, such as reduced transpiration. Projected changes in mineral content are expected to vary across different minerals (Loladze, 2002). The researchers looked at the effects of rising CO_2 levels on the nutritious content of food plants using data from 130 different plant types. The dataset incorporates the results of 7761 observations conducted by academics worldwide over 30 years (Loladze, 2014). When CO_2 levels were raised, plants' average concentration of 25 important minerals, including Ca, K, Zn and Fe, fell by 8%. The researchers looked at the effects of rising CO_2 levels on the nutritious content of food plants using data from 130 different plant types (Loladze, 2014).

Additionally, Loladze discovered that these plants' proportion of carbohydrates to minerals rose with increased CO_2 exposure. The decrease in the nutritional content of plants poses potential implications for human health, as malnutrition can result from a diet lacking in minerals and other essential nutrients, even when caloric intake is sufficient (Loladze, 2002, 2014). At the US Zero Emission Research and Technology Center, natural vegetation exhibited chlorosis and discoloration just 4 days after exposure to elevated CO_2 levels. Parameters indicative of plant health, such as chlorophyll levels, photosynthesis rate, stomatal conductance and transpiration rate, were lower in soil with elevated CO_2 concentrations. Furthermore, compared with the CO_2 control group, morphological indicators like plant height, root length, leaf number, leaf area, seed number and pod number decreased in soils exposed to high CO_2 levels (Lakkaraju *et al.*, 2010). When plants are grown in an atmosphere with high CO_2 concentrations of 475–600 ppm, leaf photosynthetic rates rise by an average of 40%. Stomata, the pores through which plants exchange gas with their surroundings, are partly controlled by atmospheric CO_2. When stomata are open, CO_2 may seep into the leaf for photosynthesis, but water can also escape. To strike a balance between the need for high photosynthetic rates and minimal water loss, plants control the degree to which their stomata are open (a quantity known as stomatal conductance). Plants can maintain high photosynthetic rates despite keeping stomatal conductance low when CO_2 levels rise. It has been shown that the stomatal conductance of water is reduced by around 22% during growth under high CO_2 in a number of different studies. The overall impact of CO_2 will be contingent on its influence on other factors driving plant water consumption, including plant size, morphology and leaf temperature. However, a reasonable assumption is that it will lead to a reduction in plant water usage.

7.2.7 Impact of elevated CO_2 on environmental and water quality

Soil health and plant nutrition also significantly affect environmental and water quality. Proper nutrient management in agriculture helps prevent nutrient pollution such as excess N and P runoff into water bodies, which can cause eutrophication and harm aquatic ecosystems. Healthy soils with good structure and organic matter content can also reduce soil erosion and sediment runoff, protecting water quality and reducing the risk of soil and land degradation. Previous studies have consistently reported inhibited growth in plants exposed to high soil CO_2 levels, but the main driving factor behind these negative impacts remains unclear. Some researchers have observed reduced water absorption in plants, but they have not been able to distinguish whether this is caused by high soil CO_2 levels (c.100% at 20 cm depth), low oxygen levels or other trace gases such as H_2S and CH_4. Additionally, a few studies have suggested that reduced metabolism in bean and maize plants in CO_2-gassed plots may be due to decreased oxygen levels rather than changes in pH induced by high soil CO_2 concentrations (50–70% at 15–30 cm depth) (Patil et al., 2010; Kuzyakov et al., 2018).

7.3 Implications for Agriculture and Ecosystem Functioning

The impacts of rising atmospheric CO_2 on soil health and plant nutrition can significantly affect agriculture and ecosystem functioning. In agriculture, soil health and plant nutrition changes can affect crop productivity, nutrient management strategies and pest and disease dynamics. For example, SOM dynamics and nutrient availability changes can influence fertilizer

requirements, nutrient management practices and overall crop nutrient status. Changes in plant nutrition can also affect plant defence mechanisms, such as allelopathy and induced resistance, which can influence pest and disease dynamics. Moreover, alterations in soil physical characteristics, such as soil structure and water retention capacity, can influence the need for irrigation, water use efficiency and soil erosion rates. In natural ecosystems, soil health and plant nutrition changes can influence plant species composition, ecosystem productivity and nutrient cycling dynamics.

7.4 Sustainable Agriculture and Food Security

Sustainable agriculture practices, including maintaining soil health and optimizing plant nutrition, are essential for ensuring food security at a global scale. Healthy soils and balanced plant nutrition contribute to sustainable crop production, reducing the need for synthetic fertilizers and pesticides, improving soil resilience to climate variability and enhancing the nutritional quality of crops. Sustainable agriculture practices can also promote biodiversity, protect natural resources and support the livelihoods of small farmers, contributing to long-term food security and sustainable development.

7.5 Future Perspectives

While research in this field has made significant progress, there are still gaps in our understanding of the mechanisms driving the effects of elevated CO_2 on soil health and plant nutrition. Furthermore, the implications of these changes for agriculture and ecosystems are not yet fully understood, and there is a need for further investigation and policy considerations to ensure sustainable management practices.

7.6 Conclusion

Soil health and plant nutrition are of utmost importance in agriculture as they directly impact crop growth and yield, nutrient cycling and resource efficiency, soil fertility and sustainability, environmental and water quality and overall food security. Sustainable soil and nutrient management practices ensure long-term agricultural productivity, environmental sustainability and global food security. The increase in atmospheric CO$_2$ has led to a decline in the nutritional value of plants, affecting human health. Consequently, many individuals now consume a limited variety of staple meals and inadequate quantities of foods rich in minerals, including fruits, vegetables, dairy products and meats. Diets deficient in minerals, particularly zinc and iron, contribute to slower growth in children, reduced ability to combat infections and elevated rates of maternal and infant mortality.

This chapter aims to offer an overview of the current understanding of the influence of rising atmospheric CO$_2$ levels on soil health and plant nutrition. It summarized the direct and indirect consequences of increasing CO$_2$ levels on soil structure and characteristics, plant development and nutrition. Overall, understanding the complex relationships between rising atmospheric CO$_2$, soil health and plant nutrition is critical for creating methods to prevent climate change consequences on agriculture and ecosystems.

References

AbdElgawad, H., Farfan-Vignolo, E.R., de Vos, D. and Asard, H. (2015) Elevated CO$_2$ mitigates drought and temperature-induced oxidative stress differently in grasses and legumes. *Plant Science* 231, 1–10.

AbdElgawad, H., Mohammed, A.E., van Dijk, J.R., Beemster, G.T.S., Alotaibi, M.O. *et al.* (2023) The impact of chromium toxicity on the yield and quality of rice grains produced under ambient and elevated levels of CO$_2$. *Frontiers in Plant Science* 14, 1019859.

Ainsworth, E.A. and Long, S.P. (2005) What have we learned from 15 years of free-air CO$_2$ enrichment (FACE)? A meta-analytic review of the responses of photosynthesis, canopy properties and plant production to rising CO$_2$. *New Phytologist* 165, 351–371.

Beaubien, S., Ciotoli, G., Coombs, P., Dictor, M., Kruger, M. *et al.* (2008) The impact of a naturally occurring CO$_2$ gas vent on the shallow ecosystem and soil chemistry of a mediterranean pasture (Latera, Italy). *International Journal of Greenhouse Gas Control* 2, 373–387.

Becker, C. and Kläring, H.-P. (2016) CO$_2$ enrichment can produce high red leaf lettuce yield while increasing most flavonoid glycoside and some caffeic acid derivative concentrations. *Food Chemistry* 199, 736–745.

Bijay-Singh (2011) The nitrogen cycle: implications for management, soil health, and climate change. In: Singh, B., Cowie, A. and Chan, K. (eds) *Soil Health and Climate Change. Soil Biology*, vol. 29. Springer, Berlin, pp. 107–129.

Ceulemans, R. and Mousseau, M. (1994) Effects of elevated atmospheric CO$_2$ on woody plants. *New Phytologist* 127, 425–446.

Cheng, W. and Johnson, D.W. (1998) Elevated CO$_2$, rhizosphere processes, and soil organic matter decomposition. *Plant and Soil* 202, 167–174.

Du, W., Gardea-Torresdey, J.L., Xie, Y., Yin, Y., Zhu, J. *et al.* (2017) Elevated CO$_2$ levels modify TiO$_2$ nanoparticle effects on rice and soil microbial communities. *Science of the Total Environment* 578, 408–416.

Dutta, H. and Dutta, A. (2016) The microbial aspect of climate change. *Energy, Ecology and Environment* 1, 209–232.

Elbasiouny, H., El-Ramady, H., Elbehiry, F., Rajput, V.D., Minkina, T. *et al.* (2022) Plant nutrition under climate change and soil carbon sequestration. *Sustainability* 14, 914.

Ferdush, J. and Paul, V. (2021) A review on the possible factors influencing soil inorganic carbon under elevated CO$_2$. *CATENA* 204, 105434.

Gavito, M.E., Curtis, P.S., Mikkelsen, T.N. and Jakobsen, I. (2001) Interactive effects of soil temperature, atmospheric carbon dioxide and soil N on root development, biomass and nutrient uptake of winter wheat during vegetative growth. *Journal of Experimental Botany* 52, 1913–1923.

Haase, S., Rothe, A., Kania, A., Wasaki, J., Römheld, V. *et al.* (2008) Responses to iron limitation in Hordeum vulgare L. as affected by the atmospheric CO$_2$ concentration. *Journal of Environmental Quality* 37, 1254–1262.

He, W., Yoo, G., Moonis, M., Kim, Y. and Chen, X. (2019) Impact assessment of high soil CO$_2$ on plant growth and soil environment: a greenhouse study. *PeerJ* 7, e6311.

Hoosbeek, M.R. (2016) Elevated CO_2 increased phosphorous loss from decomposing litter and soil organic matter at two FACE experiments with trees. *Biogeochemistry* 127, 89–97.

Jin, J., Krohn, C., Franks, A.E., Wang, X., Wood, J.L. *et al.* (2022) Elevated atmospheric CO_2 alters the microbial community composition and metabolic potential to mineralize organic phosphorus in the rhizosphere of wheat. *Microbiome* 10, 12.

Kimball, B.A. (2016) Crop responses to elevated CO_2 and interactions with H_2O, N, and temperature. *Current Opinion in Plant Biology* 31, 36–43.

Kimball, B.A., LaMorte, R.L., Seay, R.S., Pinter, P.J., Jr, Rokey, R.R. *et al.* (1994) Effects of free-air CO_2 enrichment on energy balance and evapotranspiration of cotton. *Agricultural and Forest Meteorology* 70, 259–278.

Kirwan, M.L. and Blum, L.K. (2011) Enhanced decomposition offsets enhanced productivity and soil carbon accumulation in coastal wetlands responding to climate change. *Biogeosciences* 8, 987–993.

Kumari, S. and Agrawal, M. (2014) Growth, yield and quality attributes of a tropical potato variety (*Solanum tuberosum* L. cv Kufri chandramukhi) under ambient and elevated carbon dioxide and ozone and their interactions. *Ecotoxicology and Environmental Safety* 101, 146–156.

Kuzyakov, Y., Horwath, W.R., Dorodnikov, M. and Blagodatskaya, E. (2018) Effects of elevated CO_2 in the atmosphere on soil. *Developments in Soil Science* 35, 207–219.

Lakkaraju, V.R., Zhou, X., Apple, M.E., Cunningham, A., Dobeck, L.M. *et al.* (2010) Studying the vegetation response to simulated leakage of sequestered CO_2 using spectral vegetation indices. *Ecological Informatics* 5, 379–389.

Li, J., Zhang, H., Xie, W., Liu, C., Liu, X. *et al.* (2024) Elevated CO_2 increases soil redox potential by promoting root radial oxygen loss in paddy field. *Journal of Environmental Sciences* 136, 11–20.

Loladze, I. (2002) Rising atmospheric CO_2 and human nutrition: toward globally imbalanced plant stoichiometry? *Trends in Ecology and Evolution* 17, 457–461.

Loladze, I. (2014) Hidden shift of the ionome of plants exposed to elevated CO_2 depletes minerals at the base of human nutrition. *eLife* 3, e02245.

Nguyen, L.T.T., Broughton, K., Osanai, Y., Anderson, I.C., Bange, M.P. *et al.* (2019) Effects of elevated temperature and elevated CO_2 on soil nitrification and ammonia-oxidizing microbial communities in field-grown crop. *Science of the Total Environment* 675, 81–89.

Patil, R.H., Colls, J.J. and Steven, M.D. (2010) Effects of CO_2 gas as leaks from geological storage sites on agro-ecosystems. *Energy* 35, 4587–4591.

Pérez-López, U., Miranda-Apodaca, J., Muñoz-Rueda, A. and Mena-Petite, A. (2013) Lettuce production and antioxidant capacity are differentially modified by salt stress and light intensity under ambient and elevated CO_2. *Journal of Plant Physiology* 170, 1517–1525.

Phillips, R.P., Finzi, A.C. and Bernhardt, E.S. (2011) Enhanced root exudation induces microbial feedbacks to N cycling in a pine forest under long-term CO_2 fumigation. *Ecology Letters* 14, 187–194.

Pimenta, T.M., Souza, G.A., Brito, F.A.L., Teixeira, L.S., Arruda, R.S. *et al.* (2023) The impact of elevated CO_2 concentration on fruit size, quality, and mineral nutrient composition in tomato varies with temperature regimen during growing season. *Plant Growth Regulation* 100, 519–530.

Poorter, H., Bühler, J., van Dusschoten, D., Climent, J. and Postma, J.A. (2012a) Pot size matters: a meta-analysis of the effects of rooting volume on plant growth. *Functional Plant Biology* 39, 839.

Poorter, H., Niklas, K.J., Reich, P.B., Oleksyn, J., Poot, P. *et al.* (2012b) Biomass allocation to leaves, stems and roots: meta-analyses of interspecific variation and environmental control. *New Phytologist* 193, 30–50.

Seneweera, S.P. and Conroy, J.P. (1997) Growth, grain yield and quality of rice (*Oryza sativa* L.) in response to elevated CO_2 and phosphorus nutrition. *Soil Science and Plant Nutrition* 43, 1131–1136.

Sgherri, C., Pérez-López, U., Micaelli, F., Miranda-Apodaca, J., Mena-Petite, A. *et al.* (2017) Elevated CO_2 and salinity are responsible for phenolics-enrichment in two differently pigmented lettuces. *Plant Physiology and Biochemistry* 115, 269–278.

Singh, A.K., Rai, A., Kushwaha, M., Chauhan, P.S., Pandey, V. *et al.* (2019) Tree growth rate regulate the influence of elevated CO_2 on soil biochemical responses under tropical condition. *Journal of Environmental Management* 231, 1211–1221.

Singh, J.S., Koushal, S., Kumar, A., Vimal, S.R. and Gupta, V.K. (2016) Book review: Microbial inoculants in sustainable agricultural productivity – Vol. II: functional application. *Frontiers in Microbiology* 7, 2105.

Sun, Y., Wang, C., Yang, J., Liao, J., Chen, H.Y. and Ruan, H. (2021) Elevated CO_2 shifts soil microbial communities from K- to r-strategists. *Global Ecology and Biogeography* 30, 961–972.

Taub, D.R. and Wang, X. (2008) Why are nitrogen concentrations in plant tissues lower under elevated CO_2? A critical examination of the hypotheses. *Journal of Integrative Plant Biology* 50, 1365–1374.

Terrer, C., Phillips, R.P., Hungate, B.A., Rosende, J., Pett-Ridge, J. *et al.* (2021) A trade-off between plant and soil carbon storage under elevated CO_2. *Nature* 591, 599–603.

Varma, A. and Choudhary, D.K. (eds) (2019) *Mycorrhizosphere and Pedogenesis*. Springer Singapore.

Williams, A., Pétriacq, P., Beerling, D.J., Cotton, T.A. and Ton, J. (2018) Impacts of atmospheric CO_2 and soil nutritional value on plant responses to rhizosphere colonization by soil bacteria. *Frontiers in Plant Science* 9, 1493.

Wu, Q., Zhang, C., Liang, X., Zhu, C., Wang, T. *et al.* (2020) Elevated CO_2 improved soil nitrogen mineralization capacity of rice paddy. *Science of the Total Environment* 710, 136438.

Zhang, Z., Liu, L., Zhang, M., Zhang, Y. and Wang, Q. (2014) Effect of carbon dioxide enrichment on health-promoting compounds and organoleptic properties of tomato fruits grown in greenhouse. *Food Chemistry* 153, 157–163.

8 Role of Grain Legumes in Soil Conservation and Improving Soil Quality

Seema Pradhan[1]*, Chandra Kant[2] and Subodh Verma[3]

[1]*BRIC-Institute of Life Sciences, Bhubaneswar, Odisha, India;* [2]*Dharma Samaj College, Aligarh, Uttar Pradesh, India;* [3]*Central European Institute of Technology (CEITEC), Brno, Czech Republic*

Abstract

Good-quality soil is the foundation for agricultural productivity. Therefore, numerous measures with varying degrees of effectiveness have been implemented to improve soil fertility. The arrival of the Green Revolution brought a marked increase in the usage of chemical fertilizers, which continued to be the most widely applied method for replenishing soil fertility. However, these fertilizers tend to pose an environmental hazard, and soon farmers reverted to the traditional methods of soil enrichment, including compost, mulch and animal waste. Although viable, this may not be enough to mitigate the compound effects of a rapidly changing climate, which is projected only to worsen. Legumes can uniquely form symbiotic associations with specific rhizobacteria to form root nodules and fix atmospheric nitrogen in the soil. In addition, they are also a rich source of proteins and other macro- and micronutrients and substantially contribute to the economy. These qualities make them the second most important crop in India after cereals. Legumes are an integral part of the intercropping system to replenish soil nutrients. This chapter focuses on the various processes that harness grain legumes' potential for sustainably and cost-effectively improving soil fertility.

Keywords: Grain legumes, soil quality, soil conservation, atmospheric nitrogen, rhizobacteria

8.1 Introduction

According to the estimates of the United Nations (UN), the world's population is slated to reach more than 9.5 billion by 2050 (Godfray *et al.*, 2010). This implies that we need much more basic resources to ensure a certain quality of life for our citizens. One of the major requirements is food, which has already fallen short of feeding the entire population of India. This means that food production would have to grow exponentially to keep up with the demands of an ever-growing populace, or we will be heading into an impending global food crisis. In fact, based on a briefing report by the Food and Agriculture Organization of the United Nations (FAO, n.d.) it is estimated that food production would have to increase by as much as 70% to feed the world's population (Van Dijk *et al.*, 2021).

Despite such a bleak outlook, we must do what we can to mitigate this disaster and ensure

*Corresponding author: seema@ils.res.in

© CAB International 2025. *Soil Health and Nutrition Management*
(eds N.C. Joshi, T. Leustek and P.K. Singh)
DOI: 10.1079/9781800624597.0008

that the world's people are fed. Feeding the world is an ambitious project and requires judicious use of every resource available. World hunger is a multifaceted problem requiring specific solutions for each aspect of the issue. In addition to low productivity, there are also issues regarding marketing, farmer income disparity, non-uniform regulations and policies for different crops, etc. However, most of these issues result from lower yields in important crops. Therefore, it is necessary to analyse the problem at its root.

Soil quality is often overlooked as a major problem in the scheme of increasing crop productivity. India has one of the wealthiest biodiversities as a consequence of its varied geographical and climate zones. This also implies that the subcontinent has a diverse soil profile suited to specific crops. However, rapid climate change and human intervention have been instrumental in changing soil quality worldwide. In India, the relentless use of chemical fertilizers after the Green Revolution has been a significant factor in soil quality detriment. This chapter looks at the various aspects affecting soil quality and the role of legumes in restoring soil health.

8.2 Factors Affecting Soil Quality

The term 'soil quality' may have different meanings given the perspective. It is defined as 'the capacity of a soil to function within ecosystem and land-use boundaries to sustain biological productivity, maintain environmental quality, and promote plant and animal health' (Doran and Parkin, 1994, 1997). At its core, the quality of soil is determined by its capacity to support plant growth, protect watersheds or water tables and prevent water pollution by sequestering harmful chemicals in the form of chemical fertilizers, pesticides, industrial runoff, etc., in its layers (Sims *et al.*, 1997). Today, a number of factors threaten the integrity of this precious resource, and it is necessary to understand these problems before we explore the role of legumes in soil health.

8.2.1 Human activity

Overpopulation has become a major contributor to changes in soil quality. The growing population puts a burgeoning demand on limited land resources to fulfil multiple needs, including food and shelter. More and more land area is being used for developing housing and setting up industries, rapidly turning the world into a concrete jungle. Such rapid urbanization requires the development of other accessories like roadways and railways, effectively reducing the surface area exposed to the environment. This affects the watershed/groundwater levels, which depend on rainfall for replenishment. This directly affects soil quality as lower water content directly translates to lower soil water content. Unplanned expansion of cities also leads to the loss of soil wildlife, which lends each type of soil its characteristic properties. This unchecked development is a direct result of the Industrial Revolution, which, in addition to global economic growth, has also brought a set of problems that will only intensify in the absence of compensatory measures. According to a very informative report by Borrelli *et al.* (2020), soil erosion is a concerning outcome of anthropogenic activities. Such indiscriminate land use needs to be checked, and we need to become more resourceful to allow soil quality to replenish.

8.2.2 Poor farming practices

The report by Borrelli *et al.* (2020) also discusses the effects of modern-day agriculture on soil health. The Green Revolution did wonders for countries like India and brought the country back from the brink of a challenging food crisis. As a nation, India has come a long way from a food-deficient to a food-surplus nation (Ramachandra and Vidya, 2023). However, as is the case with any major developmental programme, the Green Revolution has also brought about a number of detrimental effects in its aftermath. For instance, the higher production of high-yielding varieties (HYVs) of rice and wheat created a demand for chemical fertilizers to help replenish the soil. This led to subsidizing fertilizers, which resulted in the often indiscriminate use of these components, leading to an imbalance in soil composition (Pradhan and Parida, 2023). As a result, large portions of arable lands are becoming barren due to changes in pH and loss of organic matter. This effect is compounded by overgrazing, poor intercropping systems and a host of other activities that put soil health at risk. In addition, the drive towards producing HYVs led to an alteration in farming practices where more and more farmers were incentivized to grow rice and wheat, irrespective of soil and climate conditions. This put undue pressure on the cultivated land and led to depletion of nutrients.

8.2.3   Climate change and poor farmer benefits

Although India has come a long way since its independence, several issues concerning the agricultural sector need a solution. A major issue is the income disparity between farmers who produce the food and marketers who sell the produce at a high profit margin. This leaves very little incentive for farmers to pursue good practices in a labour-intensive profession and at the mercy of a rapidly changing climate. Climate change has been a major concern since the last decade, with many researchers emphasizing the need for drastic measures to curb the lasting effects of this global devastation. Changes in rainfall patterns and higher temperatures affect vegetation and soil quality, and a very concerning consequence of climate change is the increasing salinity of soil worldwide (Singh, 2022). This only adds to the distress of smallholder farmers. The government of India has proposed a strategy to increase farmers' income to provide them with a better quality of life and, consequently, improve farming practices (Chand, 2017). This includes shifting farming practices to encourage more income-centric farming, promoting climate-resilient crops, using renewable sources of energy, especially solar energy, discouraging the use of chemical fertilizers and making policies that stabilize

agricultural markets and ensure better compensation to the farmers.

8.3   Role of Legumes in Soil Quality and Conservation

Grain legumes form a considerably large proportion of proteins in people's diets worldwide, and the demand is set to increase in the future, with more and more people opting for plant-based protein products (Nigam *et al.*, 2021). Also, legumes possess the unique ability of being able to fix atmospheric nitrogen in the soil. They replenish the nitrogen that is depleted during a harvest cycle and also have a major role in improving soil carbon sequestration (Graham and Vance, 2003). Therefore, their role in a dynamic and sustainable agricultural system cannot be overlooked. This section will discuss the various facets of how grain legumes provide a means for improving soil health.

8.3.1   Legumes of the world

Before detailing how grain legumes can contribute to preserving soil health, it would be good to have an overview of the important legumes grown in the world. According to Food and Agriculture Organization of the UN data (FAOSTAT), India is the world's largest producer of grain legumes, accounting for the highest production of legumes like chickpea, pigeonpea and dry beans (Parida *et al.*, 2023). The other high legume-producing countries include China, Myanmar, Canada, Australia, Brazil, Argentina, USA, Mexico and Russia (FAOSTAT). The grain legumes are grown on an area of about 81 million hectares, producing more than 90 million tonnes of crop product (Dutta *et al.*, 2022). The most widely cultivated and consumed grain legumes include chickpea (*Cicer arietinum* L.), pigeonpea (*Cajanus cajan* L.), cowpea (*Vigna unguiculata* L.), lentil (*Lens culinaris* L.), field pea (*Pisum sativum* L.), greengram (*V. radiate* L.), blackgram (*V. mungo* L. Hepper) and dry bean (*Phaseolus vulgaris* L.).

While India contributes about 28% of global grain legume output, it is also one of the

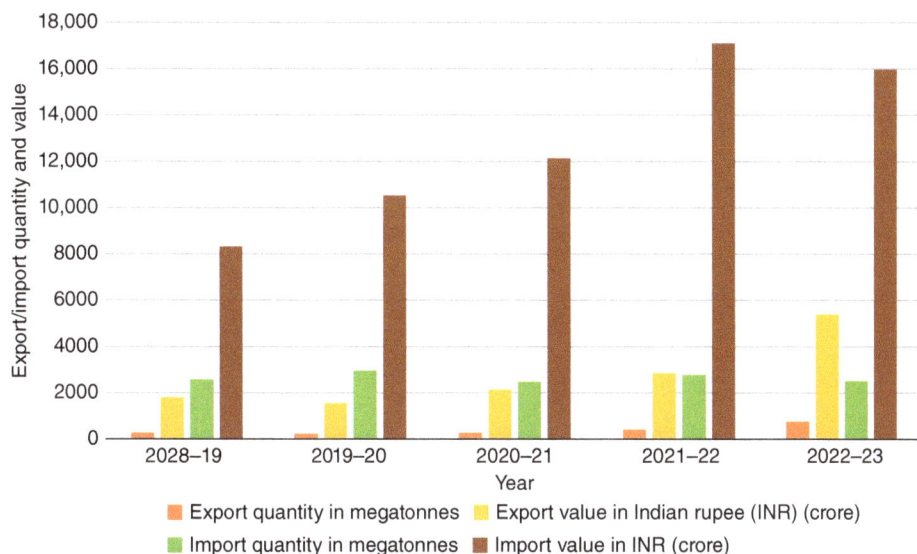

Fig. 8.1. Five-year trend of the pulses trade in India. There is an apparent excess in the import of pulses as compared with exports based on data provided by the Agricultural and Processed Food Products Export Development Authority (APEDA) (https://apeda.gov.in/apedawebsite/; last accessed October 2024).

largest importers of legumes, depending on external sources for fulfilling deficits at home (Fig. 8.1). This points to a gap in the country's supply and demand of these important crops. Steps are now being taken to diversify India's agricultural portfolio, and grain legumes are prominent on the list of crops projected to grow upwards in the global market. This increased demand for legumes has become an international trend wherein the combined production of major legumes has increased by 548.6% since the 1960s (Nigam *et al.*, 2021). The same study also reported that the highest annual growth rate was seen in the production of soybean, cowpea, lentil, beans and pigeonpea. It would not be an exaggeration to assume that, shortly, the world will rely on countries like India, Indonesia, China and South America to meet the demands for legume crops in the world market.

It is also notable that India's varied landscape and climate support a variety of minor legumes, such as moth bean and horse gram, which are more climate resilient and can contribute to the nation's food security. Moth bean (*Vigna aconitifolia*) is very nutritious and one of the hardiest legumes in that it grows in drought-prone areas of India like Rajasthan and can be a valuable genetic resource for legume improvement (Suranjika *et al.*, 2022). Similarly, horse gram is a marginally grown and consumed grain legume with medicinal properties in addition to its protein- and fibre-rich seeds (Aditya *et al.*, 2019). Apart from these, India produces soybeans, groundnuts, lentils and grass peas (Abate, 2020).

Ethiopia has been moving towards improving the production of grain legumes after partnering with the International Crops Research Institute for the Semi-Arid Tropics (ICRISAT) through the Tropical Legumes projects (Ojiewo *et al.*, 2020). This was a very fruitful collaboration that boosted chickpea production and was especially beneficial to the smallholder farmers. These efforts have been extended to other African countries where the production of groundnuts and cowpea has increased as a result of partnerships between farmers, governments and seed companies.

There have been similar efforts in Myanmar to double the production of chickpeas, which have positively impacted the agri-economy of the country (Ojiewo *et al.*, 2020). Along with chickpea, mung bean has also emerged as a front-runner in grain legumes of agroeconomic importance in Myanmar. The Department of Agricultural Research and the World Vegetable Center have made great strides in mung bean breeding and have contributed to the enhanced production and quality of mung bean varieties being grown in the country (Sequeros *et al.*, 2020).

Soybeans have occupied the top spot among the legumes cultivated in the USA for a long time. However, more recent studies have advocated a plant-based diet as a healthier alternative to animal proteins (Perera *et al.*, 2020; Semba *et al.*, 2021). A similar assessment has also been made for Sweden, where a group of researchers enumerate the benefits of switching from a diet comprised principally of meat to that with a higher proportion of legumes (Röös *et al.*, 2020). All data point to the fact that pulses will be a major part of the global economy and that they can be grown on various soils with varying properties such as moisture content, microbiome and composition, which implies that if employed judiciously, legumes can be an asset for soil conservation.

8.3.2   Nitrogen fixation by root nodulation in legumes

The ability to fix atmospheric nitrogen is perhaps the most valuable property of plants of the legume family, and directly impacts soil quality. Legumes such as *Medicago*, soybean, peas, chickpea and beans can make their nitrogen fertilizer by establishing a symbiotic relationship with nitrogen-fixing soil bacteria called rhizobia, a Gram-negative bacteria of the family *Rhizobiaceae*. In symbiotic nitrogen fixation, plants supply carbon sources to rhizobia for energy-dependent reduction of nitrogen and protect them from oxygen, and rhizobia work as mini-factories for nitrogen synthesis.

The development of nitrogen-fixing root nodules is a complex phenomenon that is governed by a host genetic programme and requires simultaneous coordination of two processes: the infection of rhizobia at the epidermis of the root and initiation of cell division and formation of nodule primordia in the cortical region. Understanding the process at a molecular level is important in devising methods for its application. The molecular exchange between plants and bacteria starts with the secretion of flavonoids from the plant, and legumes secrete different sets of phenolic compounds to the rhizosphere, which provide specificity in the interaction between the host and bacteria to some extent. The NodD proteins recognize the flavonoid signal from the host plant. NodD protein, when bound to the flavonoid, acts as a transcription regulator and activates the expression of an array of nodulation-related *NOD* genes. In response to flavonoid perception, the bacteria produce NOD factors, which some *NOD* genes synthesize. The NOD factors differ in the number of glucosamine residues, saturation level fatty acid residues and different types of modifications such as sulfuryl, methyl, carbamoyl, acetyl, fucosyl, arabinosyl and many other groups. These properties are cumulatively considered to account for the specificity of the host to the bacteria (Geurts and Bisseling, 2002). A consensus model developed by several studies suggested the presence of two receptor-like kinases (RLKs) on root epidermal cells that can perceive NOD factor (NF) signals. *MtLYK3/MtLYK4* and *MtNFP* in *Medicago truncatula*, *LjNFR1* and *LjNFR5* in *Lotus japonicus* and *GmNFR1α/β* and *GmNFR5α/β* in *Glycine max* are some of the well-studied NF receptors suggesting the validity of the model (Arrighi *et al.*, 2006; Indrasumunar, 2007; Indrasumunar *et al.*, 2010). When the legume root perceives the NF, two kinds of ionic changes occur in the root hair cell. First is an influx of Ca^{2+} followed by membrane depolarization, and second is oscillations in the concentration of cytoplasmic Ca^{2+} called calcium spiking. Within 1 min of NF application, a rapid influx of Ca^{2+} has been observed in root hair cells, followed by efflux of Cl and K+ (Felle *et al.*, 1999). The Ca^{2+} influx in root hair cells activates Cl and K+ efflux channels and helps initiate repolarization (Kurkdjian, 1995; Felle *et al.*, 1999). Approximately 10 min later, Ca^{2+} spiking is induced in the same cells, characterized by oscillation in cytosolic concentrations of Ca^{2+} (Wais *et al.*, 2000). Ca^{2+} spiking is a phenomenon in which a rapid initial increase in Ca^{2+} concentration around the nucleus occurs, followed by a gradual decline. Some ion-channel genes and nucleoporins involved in NF signal transduction have been characterized in model systems (Riely *et al.*, 2007; Saito *et al.*, 2007). Moreover, a calcium- and calmodulin-dependent protein kinase (CCaMK) has been found indispensable for rhizobial infection (Mitra *et al.*, 2004). The CCaMK perceives the calcium oscillation signals generated after NF application and has been shown to interact and phosphorylate the protein CYCLOPS (Yano *et al.*, 2008). Once CCaMK and CYCLOPS are activated, they activate an array of transcription factors downstream, such as nodulation signalling pathway 1 (NSP1), NSP2, ethylene responsive factor required for nodulation 1 (ERN 1) and nodule inception (NIN). These transcription factors translate the NF signal to regulate the expression of early nodulins, which initiate a cascade of events important for the initiation of nodule development (Andriankaja *et al.*, 2007; Marsh *et al.*, 2007). Initial NF signals activate a cascade of epidermal transcription factors that help initiate root nodules in the cortex. The initiation process is linked from the epidermis to the deep cortex through plant hormones that carry signals for modulation from the epidermis to the cortex. The key players in the phenomenon are the hormones cytokinin and auxin, which are important for maintaining a balance between cell division and cell differentiation in root cortical cells. The necessity of cytokinin in root nodule formation has been well established since the gain-of-function mutation in the cytokinin receptor gene *LHK1* of *L. japonicus* has been shown to trigger spontaneous nodule formation (Tirichine *et al.*, 2007). Expression of the transcription factor SHY2 is triggered by cytokinin application, which subsequently suppresses expression of the auxin (PIN) transporters (Ioio *et al.*, 2008), and the suppression of PIN auxin transporters creates a local low auxin region in the cortex (Pernisová *et al.*, 2009). Therefore, it is suggested that low auxin and high cytokinin levels are associated with nodule initiation. NSP1, NSP2 and NIN, which act downstream of

CCaMK in the epidermis after NF perception, are also indispensable for cell division in the root cortex (Heckmann *et al.*, 2006). NSP1 and NSP2 have been found to work downstream to the cytokinin receptor and CCaMK, as a lack of functional copies of NSP1 and NSP2 results in loss of spontaneous nodulation phenotypes even after overexpression of cytokinin and CCaMK (Gleason *et al.*, 2006; Oldroyd *et al.*, 2011). Upon perception of the NF, root hairs undergo significant morphological changes, such as swelling of the root hair tip and sometimes branching of the root hair (Esseling *et al.*, 2003). As a result, bacteria get entrapped in the modified structures called infection pockets, and the rhizobia continues to divide in the infection pocket (Geurts *et al.*, 2005). The entrapment of bacteria is required for localized plant cell wall degradation, which is necessary for infection thread initiation and subsequent infection (Turgeon and Bauer, 1985). The plant cell wall degradation enzymes can be of either bacterial or plant origin. Plant cell wall-degrading enzymes such as pectin methylesterase (Lievens *et al.*, 2002) and polygalacturonase (Muñoz *et al.*, 1998) have been found expressing on the infection locus. Bacterial cellulose has also been reported to be involved during the infection process (Robledo *et al.*, 2008).

Based on the available literature, a chickpea study identified several genes regulating root nodulation using transcriptomics, and the study reported a set of 5907 nodule-specific unigenes (Kant *et al.*, 2016).

8.3.3   Root architecture and microbiome associated with legume roots for soil health

Soil erosion is a concerning impact of sudden and erratic climate change and uncontrolled human activity, which adds to the precarious situation where the loss of a layer of soil harms agricultural production. Soil erosion occurs due to soil quality/composition and external environmental influences such as floods (Bryan, 2000). Plant roots provide higher tensile strength and effectively bind soil particles to retain the top layer of the soil and prevent it from eroding. In their article, Ola *et al.* (2015) detailed the various effects of roots on soil retention and how their architecture is affected by the presence or absence of certain nutrients. The simplistic idea is that the denser the roots, the more effective their ability to bind the soil and prevent its loss. Since most legumes grow in arid or semi-arid regions, they possess a naturally longer and denser root system. There have been numerous studies where researchers have attempted to understand the molecular mechanisms regulating legume root structure. This could help us develop customized systems for preventing soil erosion. One such report documents the quantitative trait loci (QTLs) associated with legume root architecture identified through genome-wide association studies (Ye *et al.*, 2018). Another study reported the key players that determine root architecture at the genic level and mentions the role of the gene *DRO1*, whose presence makes roots grow deeper in rice (Uga *et al.*, 2013; Lynch, 2022). However, similar studies that pinpoint the gene that could be responsible for root system architecture in legumes are yet to be reported but seem likely to be reported in the near future.

As mentioned, legume roots have the unique ability of fixing atmospheric nitrogen through their symbiotic association with rhizobia. The process is facilitated by certain phytochemicals secreted by the legume roots. A similar system exists in many plants where the root exudates, a mixture of multiple phytochemicals, may help form a zone of microbial associations (microbiome) around the roots (rhizosphere). One such association is formed between the fungi of the phylum *Glomeramycota*, forming arbuscular mycorrhizae (AM), and legume roots and is crucial for nutrient acquisition (Corradi and Bonfante, 2012). This symbiosis benefits phosphorus and nitrogen assimilation as AM fungi have a high affinity for inorganic phosphate (Harrison and von Buuren 1995). Researchers have studied this process in detail through RNA-Seq in the model legume *L. japonicus* and found 3641 genes to be differentially expressed during AM development in the roots. In comparison, 275 genes were coregulated in both AM and rhizobial symbioses (Handa *et al.*, 2015). Nitrogen and phosphorus are major

determinants of soil quality, so such studies using accurate sequencing platforms and analytical tools will allow the identification and characterization of this process at the molecular level and will allow us to manipulate root systems for soil sustainability by reducing the heavy usage of chemical fertilizers.

We have also mentioned the rising problem of soil salinity and its impact on the global economy. While it is challenging to remove excessive salts from the soil, the microbiome associated with halophytes provides a way to mitigate the effects of these salts on plant growth. Rhizobia, associated with nodule formation in legume plants, can also exist in non-legume plants as beneficial endophytes and aid in salinity tolerance (Tian *et al.*, 2017). Isolating and propagating the microbes present in the rhizosphere of salinity-tolerant legumes can benefit the growth of susceptible crops in saline soils. This, in turn, contributes to improving the soil quality through the plants' symbiotic associations.

8.4   Documented Cases of Using Legumes for Soil Improvement Across the World

8.4.1   Restoration of soil fertility in mine areas

India has rich resources of many ores and produces considerable amounts of coal. Coal mining using open-strip mining methods leads to heavy losses of the topsoil and has significant adverse effects on the ecosystem (Ahirwal and Maiti, 2018). Restoring the fertility of such barren lands is difficult and requires careful selection of vegetation that could help rehabilitate the soil. In one such coal mine in Jharkhand, India, a mixture of grass and legume was employed to improve the soil's nitrogen content (Kumari and Maiti, 2019). The study showed legumes are much more effective at biomass accumulation than grasses and thus provide higher amounts of organic matter. Using legumes in combination with grasses as initial colonizers in the degraded coal mine soils presents an effective and highly economical method for soil quality improvement and sustainability.

Muñoz-Rojas *et al.* (2016) conducted a study on disturbed soils with degraded quality in the Pilbara region of Western Australia. The region was an iron ore mine site with a semi-arid climate. The soil quality had deteriorated due to the mining activities and comprised mostly shrubland vegetation. They restored the soil quality of the region using a variety of methods, including planting native legumes. They concluded that several factors, such as soil microbiota, organic content, moisture and vegetation, contribute to the improvement of soil quality and must be used in appropriate combination to restore soil quality.

8.4.2 Intercropping with legumes in farming practices

Intercropping is one of the oldest and most widespread farming practices that allows optimum utilization of arable land. The system is prevalent due to its low input and feasibility for farmers of all socioeconomic strata. Since it involves growing multiple crop types in the same field, it requires careful selection of compatible crops for simultaneous growth – growing in similar climatic conditions, soil type, water content, etc. Grain legume–cereal intercropping is a widespread practice in agriculture. It is especially beneficial to areas that receive erratic rainfall as there are varieties of early-maturing legumes that can escape drought conditions.

Similarly, the rhizosphere of nodulating legumes helps soil enrichment. Some popular combinations of such intercropping are millet and green gram (Kaushik and Gautam, 1987), cowpea and sorghum (Reddy *et al.*, 1980). More examples of legume–cereal intercropping are documented by Chamkhi *et al.* (2022) in their review, where they mention various combinations such as maize–soybean/pigeon pea/groundnuts/cowpea, millets–groundnuts and rice–pulses among many others. The same review also reports the molecular basis of how such intercropping

improves the soil microbiome and abiotic stress response in plants.

Another good use of legumes in an intercropping system is their use as mulch. Mulch is any material that can be applied to the soil surface to allow moisture retention. Many leguminous cover crops can serve as living mulches and be intercropped with a main crop to suppress the growth of weeds, thereby reducing excessive soil tillage (Stein *et al.*, 2022). This model has been successfully implemented to improve the productivity of summer maize and wheat in the Himalayan regions of India (Sharma *et al.*, 2010; Das *et al.*, 2022).

The practice of intercropping with legumes is quite popular across the world. In their review, Kocira *et al.* (2020) documented the effects of various legumes such as *Pisum sativum*, *Medicago sativa* and *Trifolium pratense* on soil quality restoration. In the USA, an intercropping system involving chickpeas as one of the crops was found to be beneficial in restoring the nitrogen content of soils in drylands (Wieme *et al.*, 2020). Similar results were also found in cases of Mediterranean agroecosystems (Oliveira *et al.*, 2021) and these reports have been instrumental in projecting a simulation of cropping systems with legumes to combat the effects of adverse environments in Europe (Marteau-Bazouni *et al.*, 2024).

8.5 Conclusions and Future Perspectives

Agriculture accounts for a sizeable part of the Indian economy and employs a large sector of its population. However, the basic needs of the agrarian society have been largely ignored whilst they have been sustaining heavy losses due to climate change. Legumes are an integral part of diet and the economy, and should be explored for their role in maintaining soil health. We have the advantage of years of traditional knowledge of good farming practices and contemporary data from molecular characterization of important biological processes. A combination of these will provide a strong foundation for restoring and maintaining soil health.

References

Abate, T. (2020) *Grain Legumes of India. Technical Report*. Researchgate. Available at: https://www.researchgate.net/publication/342082115 (last accessed October 2024).

Aditya, J.P., Bhartiya, A., Chahota, R.K., Joshi, D., Chandra, N. *et al.* (2019) Ancient orphan legume horse gram: a potential food and forage crop of future. *Planta* 250, 891–909.

Ahirwal, J. and Maiti, S.K. (2018) Assessment of soil carbon pool, carbon sequestration and soil CO_2 flux in unreclaimed and reclaimed coal mine spoils. *Environmental Earth Sciences* 77, 1–13.

Andriankaja, A., Boisson-Dernier, A., Frances, L., Sauviac, L., Jauneau, A. *et al.* (2007) AP2-ERF transcription factors mediate Nod factor-dependent Mt *ENOD11* activation in root hairs via a novel cis-regulatory motif. *Plant Cell* 19, 2866–2885.

Arrighi, J.-F., Barre, A., Ben Amor, B., Bersoult, A., Soriano, L.C. *et al.* (2006) The *Medicago truncatula* lysin [corrected] motif-receptor-like kinase gene family includes NFP and new nodule-expressed genes. *Plant Physiology* 142, 265–279.

Borrelli, P., Robinson, D.A., Panagos, P., Lugato, E., Yang, J.E. *et al.* (2020) Land use and climate change impacts on global soil erosion by water (2015–2070). *Proceedings of the National Academy of Sciences USA* 117, 21994–22001.

Bryan, R.B. (2000) Soil erodibility and processes of water erosion on hillslope. *Geomorphology* 32, 385–415.

Chamkhi, I., Cheto, S., Geistlinger, J., Zeroual, Y., Kouisni, L. *et al.* (2022) Legume-based intercropping systems promote beneficial rhizobacterial community and crop yield under stressing conditions. *Industrial Crops and Products* 183, 114958.

Chand, R. (2017) Presidential address: Doubling farmers' income: strategy and prospects. *Indian Journal of Agricultural Economics* 71, 1–23.

Corradi, N. and Bonfante, P. (2012) The arbuscular mycorrhizal symbiosis: origin and evolution of a beneficial plant infection. *PLoS Pathogens* 8, e1002600.

Das, A., Babu, S., Singh, R., Kumar, S., Rathore, S.S. *et al.* (2022) Impact of live mulch-based conservation tillage on soil properties and productivity of summer maize in Indian Himalayas. *Sustainability* 14, 12078.

Doran, J.W. and Parkin, T.B. (1994) Defining and assessing soil quality. In: Doran, J.W., Coleman, D.C., Bezdicek, D.F. and Stewart, B.A. (eds) *Defining Soil Quality for a Sustainable Environment* Vol. 35. Soil Science Society of America, pp. 1–21.

Doran, J.W. and Parkin, T.B. (1997) Quantitative indicators of soil quality: a minimum data set. *Methods for Assessing Soil Quality* 49, 25–37.

Dutta, A., Trivedi, A., Nath, C.P., Gupta, D.S. and Hazra, K.K. (2022) A comprehensive review on grain legumes as climate-smart crops: challenges and prospects. *Environmental Challenges* 7, 100479.

Esseling, J.J., Lhuissier, F.G.P. and Emons, A.M.C. (2003) Nod factor-induced root hair curling: continuous polar growth towards the point of nod factor application. *Plant Physiology* 132, 1982–1988.

FAO (n.d.) *How to Feed the World in 2050*. Available at: https://www.fao.org/fileadmin/templates/wsfs/docs/expert_paper/How_to_Feed_the_World_in_2050.pdf (last accessed October 2024).

Felle, H.H., Kondorosi, É., Kondorosi, A. and Schultze, M. (1999) Elevation of the cytosolic free [Ca2+] is indispensable for the transduction of the nod factor signal in alfalfa. *Plant Physiology* 121, 273–280.

Geurts, R. and Bisseling, T. (2002) Rhizobium nod factor perception and signalling. *Plant Cell* 14 (Suppl. 1), S239–S249.

Geurts, R., Fedorova, E. and Bisseling, T. (2005) Nod factor signaling genes and their function in the early stages of rhizobium infection. *Current Opinion in Plant Biology* 8, 346–352.

Gleason, C., Chaudhuri, S., Yang, T., Muñoz, A., Poovaiah, B.W. *et al.* (2006) Nodulation independent of rhizobia induced by a calcium-activated kinase lacking autoinhibition. *Nature* 441, 1149–1152.

Godfray, H.C.J., Beddington, J.R., Crute, I.R., Haddad, L., Lawrence, D. *et al.* (2010) Food security: the challenge of feeding 9 billion people. *Science* 327, 812–818.

Graham, P.H. and Vance, C.P. (2003) Legumes: Importance and constraints to greater use. *Plant Physiology* 131, 872–877.

Handa, Y., Nishide, H., Takeda, N., Suzuki, Y., Kawaguchi, M. *et al.* (2015) RNA-seq transcriptional profiling of an arbuscular mycorrhiza provides insights into regulated and coordinated gene expression in *Lotus japonicus* and *Rhizophagus irregularis*. *Plant and Cell Physiology* 56, 1490–1511.

Harrison, M.J. and van Buuren, M.L. (1995) A phosphate transporter from the mycorrhizal fungus Glomus versiforme. *Nature* 378, 626–629.

Heckmann, A.B., Lombardo, F., Miwa, H., Perry, J.A., Bunnewell, S. *et al.* (2006) *Lotus japonicus* nodulation requires two GRAS domain regulators, one of which is functionally conserved in a non-legume. *Plant Physiology* 142, 1739–1750.

Indrasumunar, A. (2007) Molecular cloning and functional characterisation of soybean (Glycine max L.) nod factor receptor genes. PhD thesis, The University of Queensland, Brisbane, Australia.

Indrasumunar, A., Kereszt, A., Searle, I., Miyagi, M., Li, D. *et al.* (2010) Inactivation of duplicated nod factor receptor 5 (NFR5) genes in recessive loss-of-function non-nodulation mutants of allotetraploid soybean (Glycine max L. Merr.). *Plant and Cell Physiology* 51, 201–214.

Ioio, R.D., Nakamura, K., Moubayidin, L., Perilli, S., Taniguchi, M. *et al.* (2008) A genetic framework for the control of cell division and differentiation in the root meristem. *Science* 322, 1380–1384.

Kant, C., Pradhan, S. and Bhatia, S. (2016) Dissecting the root nodule transcriptome of chickpea (Cicer arietinum L.). *PloS One* 11, e0157908.

Kaushik, S.K. and Gautam, R.C. (1987) Effect of nitrogen and phosphorus on the production potential of pearl millet–cow pea or green gram intercropping systems under rainfed conditions. *Journal of Agricultural Science* 108, 361–364.

Kocira, A., Staniak, M., Tomaszewska, M., Kornas, R., Cymerman, J. *et al.* (2020) Legume cover crops as one of the elements of strategic weed management and soil quality improvement. A review. *Agriculture* 10, 394.

Kumari, S. and Maiti, S.K. (2019) Reclamation of coalmine spoils with topsoil, grass, and legume: a case study from India. *Environmental Earth Sciences* 78, 429.

Kurkdjian, A.C. (1995) Role of the differentiation of root epidermal cells in Nod factor (from Rhizobium meliloti)-induced root-hair depolarization of Medicago sativa. *Plant Physiology* 107, 783–790.

Lievens, S., Goormachtig, S., Herman, S. and Holsters, M. (2002) Patterns of pectin methylesterase transcripts in developing stem nodules of Sesbania rostrata. *Molecular Plant–Microbe Interactions* 15, 164–168.

Lynch, J.P. (2022) Harnessing root architecture to address global challenges. *Plant Journal* 109, 415–431.

Marsh, J.F., Rakocevic, A., Mitra, R.M., Brocard, L., Sun, J. *et al.* (2007) *Medicago truncatula NIN* is essential for rhizobial-independent nodule organogenesis induced by autoactive calcium/calmodulin-dependent protein kinase. *Plant Physiology* 144, 324–335.

Marteau-Bazouni, M., Jeuffroy, M.H. and Guilpart, N. (2024) Grain legume response to future climate and adaptation strategies in Europe: a review of simulation studies. *European Journal of Agronomy* 153, 127056.

Mitra, R.M., Gleason, C.A., Edwards, A., Hadfield, J., Downie, J.A. *et al.* (2004) A Ca^{2+}/calmodulin-dependent protein kinase required for symbiotic nodule development: gene identification by transcript-based cloning. *Proceedings of the National Academy of Sciences USA* 101, 4701–4705.

Muñoz, J.A., Coronado, C., Pérez-Hormaeche, J., Kondorosi, A., Ratet, P. *et al.* (1998) MsPG3, a Medicago sativa polygalacturonase gene expressed during the alfalfa–*Rhizobium meliloti* interaction. *Proceedings of the National Academy of Sciences USA* 95, 9687–9692.

Muñoz-Rojas, M., Erickson, T.E., Dixon, K.W. and Merritt, D.J. (2016) Soil quality indicators to assess functionality of restored soils in degraded semiarid ecosystems. *Restoration Ecology* 24 (Suppl. 2), S43–S52.

Nigam, S.N., Chaudhari, S., Deevi, K.C., Saxena, K.B. and Janila, P. (2021) Trends in legume production and future outlook. In: Saxena, K.B., Saxena, R.K. and Varshney, R.K. (eds) *Genetic Enhancement in Major Food Legumes: Advances in Major Food Legumes*. Springer, pp. 7–48.

Ojiewo, C.O., Omoigui, L.O., Pasupuleti, J. and Lenné, J.M. (2020) Grain legume seed systems for smallholder farmers: perspectives on successful innovations. *Outlook on Agriculture* 49, 286–292.

Ola, A., Dodd, I.C. and Quinton, J.N. (2015) Can we manipulate root system architecture to control soil erosion? *Soil* 1, 603–612.

Oldroyd, G.E., Murray, J.D., Poole, P.S. and Downie, J.A. (2011) The rules of engagement in the legume-rhizobial symbiosis. *Annual Review of Genetics* 45, 119–144.

Oliveira, M., Castro, C., Coutinho, J. and Trindade, H. (2021) Grain legume-based cropping systems can mitigate greenhouse gas emissions from cereal under mediterranean conditions. *Agriculture, Ecosystems and Environment* 313, 107406.

Parida, S.K., Mondal, N., Yadav, R., Vishwakarma, H., Rana, J.C. *et al.* (2023) Mining legume germplasm for genetic gains: an Indian perspective. *Frontiers in Genetics* 14, 996828.

Perera, T., Russo, C., Takata, Y. and Bobe, G. (2020) Legume consumption patterns in US adults: National Health and Nutrition Examination Survey (NHANES) 2011–2014 and Beans, Lentils, Peas (BLP) 2017 survey. *Nutrients* 12, 1237.

Pernisová, M., Klíma, P., Horák, J., Válková, M., Malbeck, J. *et al.* (2009) Cytokinins modulate auxin-induced organogenesis in plants via regulation of the auxin efflux. *Proceedings of the National Academy of Sciences USA* 106, 3609–3614.

Pradhan, S. and Parida, A. (2023) Perception of food crops developed by mutagenesis among various stakeholders. In: Bhattacharya, A., Parkhi, V. and Char, B. (eds) *TILLING and Eco-TILLING for Crop Improvement*. Springer Nature, Singapore, pp. 217–236.

Ramachandra, S. and Vidya, B. (2023) Agricultural revolutions which moved India from food deficit to food surplus since independence. *Delta National Journal of Multidisciplinary Research* 10, 110–112.

Reddy, M.S., Floyd, C.N. and Willey, R.W. (1980) Groundnut in intercropping systems. In: *Proceedings of the International Workshop on Groundnuts*, 13–17 October 1980, ICRISAT Center, Patancheru, India.

Riely, B.K., Lougnon, G., Ané, J.M. and Cook, D.R. (2007) The symbiotic ion channel homolog DMI1 is localized in the nuclear membrane of *Medicago truncatula* roots. *Plant Journal* 49, 208–216.

Robledo, M., Jiménez-Zurdo, J.I., Velázquez, E., Trujillo, M.E., Zurdo-Piñeiro, J.L. *et al.* (2008) *Rhizobium* cellulase CelC2 is essential for primary symbiotic infection of legume host roots. In: *Proceedings of the National Academy of Sciences USA* 105, 7064–7069.

Röös, E., Carlsson, G., Ferawati, F., Hefni, M., Stephan, A. *et al.* (2020) Less meat, more legumes: prospects and challenges in the transition toward sustainable diets in Sweden. *Renewable Agriculture and Food Systems* 35, 192–205.

Saito, K., Yoshikawa, M., Yano, K., Miwa, H., Uchida, H. *et al.* (2007) NUCLEOPORIN85 is required for calcium spiking, fungal and bacterial symbioses, and seed production in *Lotus japonicus*. *Plant Cell* 19, 610–624.

Semba, R.D., Ramsing, R., Rahman, N., Kraemer, K. and Bloem, M.W. (2021) Legumes as a sustainable source of protein in human diets. *Global Food Security* 28, 100520.

Sequeros, T., Schreinemachers, P., Depenbusch, L., Shwe, T. and Nair, R.M. (2020) Impact and returns on investment of mungbean research and development in Myanmar. *Agriculture and Food Security* 9, 1–9.

Sharma, A. R, Singh, R., Dhyani, S.K. and Dube, R. K (2010) Effect of live mulching with annual legumes on performance of maize (*Zea mays*) and residual effect on following wheat (*Triticum aestivum*). *Indian Journal of Agronomy* 55, 177–184.

Sims, J.T., Cunningham, S.D. and Sumner, M.E. (1997) Assessing soil quality for environmental purposes: roles and challenges for soil scientists. *Journal of Environmental Quality* 26, 20–25.

Singh, A. (2022) Soil salinity: a global threat to sustainable development. *Soil Use and Management* 38, 39–67.

Stein, S., Hartung, J., Möller, K. and Zikeli, S. (2022) The effects of leguminous living mulch intercropping and its growth management on organic cabbage yield and biological nitrogen fixation. *Agronomy* 12, 1009.

Suranjika, S., Pradhan, S., Nayak, S.S. and Parida, A. (2022) *De novo* transcriptome assembly and analysis of gene expression in different tissues of moth bean (*Vigna aconitifolia*) (Jacq.) Marechal. *BMC Plant Biology* 22, 1–15.

Tian, B., Zhang, C., Ye, Y., Wen, J., Wu, Y. *et al.* (2017) Beneficial traits of bacterial endophytes belonging to the core communities of the tomato root microbiome. *Agriculture, Ecosystems and Environment* 247, 149–156.

Tirichine, L., Sandal, N., Madsen, L.H., Radutoiu, S., Albrektsen, A.S. *et al.* (2007) A gain-of-function mutation in a cytokinin receptor triggers spontaneous root nodule organogenesis. *Science* 315, 104–107.

Turgeon, B.G. and Bauer, W.D. (1985) Ultrastructure of infection-thread development during the infection of soybean by *Rhizobium japonicum*. *Planta* 163, 328–349.

Uga, Y., Sugimoto, K., Ogawa, S., Rane, J., Ishitani, M. *et al.* (2013) Control of root system architecture by *DEEPER ROOTING 1* increases rice yield under drought conditions. *Nature Genetics* 45, 1097–1102.

Van Dijk, M., Morley, T., Rau, M.L., and Saghai, Y. (2021) A meta-analysis of projected global food demand and population at risk of hunger for the period 2010–2050. *Nature Food* 2(7), 494–501.

Wais, R.J., Galera, C., Oldroyd, G., Catoira, R., Penmetsa, R.V. *et al.* (2000) Genetic analysis of calcium spiking responses in nodulation mutants of *Medicago truncatula*. *Proceedings of the National Academy of Sciences USA* 97, 13407–13412.

Wieme, R.A., Reganold, J.P., Crowder, D.W., Murphy, K.M. and Carpenter-Boggs, L.A. (2020) Productivity and soil quality of organic forage, quinoa, and grain cropping systems in the dryland Pacific Northwest, USA. *Agriculture, Ecosystems and Environment* 293, 106838.

Yano, K., Yoshida, S., Müller, J., Singh, S., Banba, M. *et al.* (2008) CYCLOPS, a mediator of symbiotic intracellular accommodation. *Proceedings of the National Academy of Sciences USA* 105, 20540–20545.

Ye, H., Roorkiwal, M., Valliyodan, B., Zhou, L., Chen, P. *et al.* (2018) Genetic diversity of root system architecture in response to drought stress in grain legumes. *Journal of Experimental Botany* 69, 3267–3277.

9 Effects of Elevated Temperature on Rice Growth and Nutritional Value: Present Status and Future Directions

Priyanka Das[1]*, Sanghamitra Adak[1], Naveen Chandra Joshi[2], Prashant Kumar Singh[3] and Arun Lahiri Majumder[1]

[1]*Division of Plant Biology, Bose Institute, Kankurgachi, Kolkata, West Bengal, India;*
[2]*Amity Institute of Microbial Technology, Amity University, Noida, Uttar Pradesh, India;* [3]*Depatment of Biotechnology, Mizoram Central University, Mizoram, India*

Abstract

Rice grain is the principal component of the diet in more than 30 nations, providing essential nutrients, including dietary protein and fat. In the present climatic scenario, heat stress is the leading abiotic restraint for major crop productivity, including that of rice. Any additional upsurge in the intensity of heat stress, either independently or in combination with other abiotic stresses, would massively decrease grain yield, which is linked to nutritional efficiency and universal diet security. Heat stress affects rice growth at all developmental stages and the level of impairment during the reproductive/yield stage is high, ultimately affecting the nutritional quality of the grain. Heat stress considerably affects yield in rice too by reducing seed number and size, ultimately upsetting the saleable trait (i.e. 100-seed weight). Elevated temperatures can lessen the grain yield of rice by 50% and can result in no seed filling in rice by disturbing starch granule accumulation. Under normal conditions, nutrient accumulation in the grain/fruit results from complicated enzymatic processes. Furthermore, these processes are susceptible to elevated temperature stress because a group of transporters and enzymes are connected. Elevated temperature stress often reduces the total nutrient content of rice plants and can affect enzyme activities involved in nutrient uptake and metabolism. Here, we investigate the effects of oppressive heat on the nutritional composition of rice seeds at the cellular level. Furthermore, we present an update on the cultivation of upgraded rice nutritional features under elevated temperature/heat stress by adopting modern techniques such as mutagenesis and gene transformation technology.

Keywords: Heat stress, nutritional efficiency, food security, rice, plant growth

9.1 Introduction

The greater predictability of current international climate models indicates that a gradual increase in global temperature puts vegetation under stress (Battisti and Naylor, 2009). A temperature increase above a certain level causes heat stress to crop plants. Rice is one of the chief dietary products, serving nutriment for more than 50% of the global inhabitants (Seck *et al.,* 2012) and its yield is significantly impacted by considerable changes in climatic conditions (Jagadish *et al.,* 2015). Rice plants are vulnerable to heat, and the expected changes in present

*Corresponding author: prink.bot@gmail.com

© CAB International 2025. *Soil Health and Nutrition Management*
(eds N.C. Joshi, T. Leustek and P.K. Singh)
DOI: 10.1079/9781800624597.0009

and upcoming environmental situations will adversely affect worldwide rice production and seed quality (Jagadish *et al.*, 2010; Lin *et al.*, 2010; Teixeira *et al.*, 2013).

Each 1°C temperature increase in the growing season causes a 10% decrease in grain production (Peng *et al.*, 2004). Heat stress reduces both the economic and nutritional benefits of rice (Wang *et al.*, 2011; Lyman *et al.*, 2013). It is also clear that ecological factors such as CO_2 and temperature may act together under field circumstances, disturbing various growth phases of crops including those of rice (Mittler and Blumwald, 2010; Jagadish *et al.*, 2016). Increased environmental CO_2 is involved in decreasing the percentage of rice milled, increasing rice chalkiness, reducing protein levels in rice grains and decreasing nutrient and mineral concentrations in rice (Taub *et al.*, 2008; DaMatta *et al.*, 2010; Wang *et al.*, 2011; Myers *et al.*, 2014). The reproductive phase is critical for rice grain yield under elevated temperature stress (Yoshida, 1981). During the reproductive phase, grain quality and mineral content are affected by individual or communicating effects of temperature increase and/or environmental CO_2 levels (Terao *et al.*, 2005; Madan *et al.*, 2012; Kadam *et al.*, 2014; Usui *et al.*, 2014; Bahuguna *et al.*, 2015; Chaturvedi *et al.*, 2017). An intricate relationship of rice seed minerals with other quality traits – protein level, amylose level, chalkiness, gelatinization temperature and gel consistency – has been described under field environments as well as under elevated temperature stress (Zeng *et al.*, 2005; Jiang *et al.*, 2007; Gu *et al.*, 2015; Chaturvedi *et al.*, 2017).

Under elevated temperature stress conditions, rice panicles are essential for grain quality and yield maintenance. Panicle growth depends on humidity and air temperature (Weerakoon *et al.*, 2008; Tian *et al.*, 2010; Yoshimoto *et al.*, 2011; Matsui *et al.*, 2014). Dry and hot winds at the time of grain filling can quicken panicles' water loss and upturn the chalkiness of grains (Kang *et al.*, 2003; Hiroshi *et al.*, 2012).

Under heat stress and dry wind, rice endosperm cells depend on osmotic alteration to preserve their turgidity (Wada *et al.*, 2014, 2019). Midseason rice (indica variety) in central China frequently shows yield loss due to elevated temperature conditions (Tian *et al.*, 2010). Furthermore, some reports have also hypothesized that elevated temperature and decreased relative humidity worsen rice grain quality (Guo *et al.*, 2016; Tan and Shen, 2016; Liu *et al.*, 2019). Molecular and cellular alterations have been attempted to manage nutritional efficiency in rice. In this chapter, we highlight the consequences of elevated heat on rice plants, approaches for temperature tolerance and findings in rice crops, presenting how the nutritional properties of rice seeds are strongly impacted by elevated temperature stress. We also summarize the research on the ability of modern techniques such as mutagenesis and plant transformation to cultivate rice under elevated temperature conditions.

9.2   Rice Life Cycle at Elevated Temperature

Rice plants have different stages in their life cycle, such as the vegetative growth phase, initial reproductive growth stage and later reproductive growth period or maturity stage. Heat stress affects different phases of the life cycle of rice plants (Zhang *et al.*, 2018; Ren *et al.*, 2023). The flowering stage is more prone to elevated temperature stress than the other phases are (Chen *et al.*, 2020). Among the two major temperate region rice varieties, with varied morphological and physiological features, indica is more tolerant to heat than japonica (Lee *et al.*, 2017). Figure 9.1 shows some major influences of raised temperatures on different rice life cycle stages.

9.2.1   Effects of elevated temperature on the vegetative/somatic growth phase

Overall, the process of seed development, which includes seed latency, quality and propagation, is negatively influenced by elevated heat (Liu *et al.*, 2019; Ren *et al.*, 2023). Heat stress decreases the seed viability rate, resulting in poor germination (Fahad *et al.*, 2016). High-temperature stress is associated with a poor germination rate and poor seed vigour in rice because of a decrease in the thermostability of the cell membrane (Saidi *et al.*, 2010). Heat stress causes reduced seed size in terms of seed size, breadth and mass at the early stage of seed development

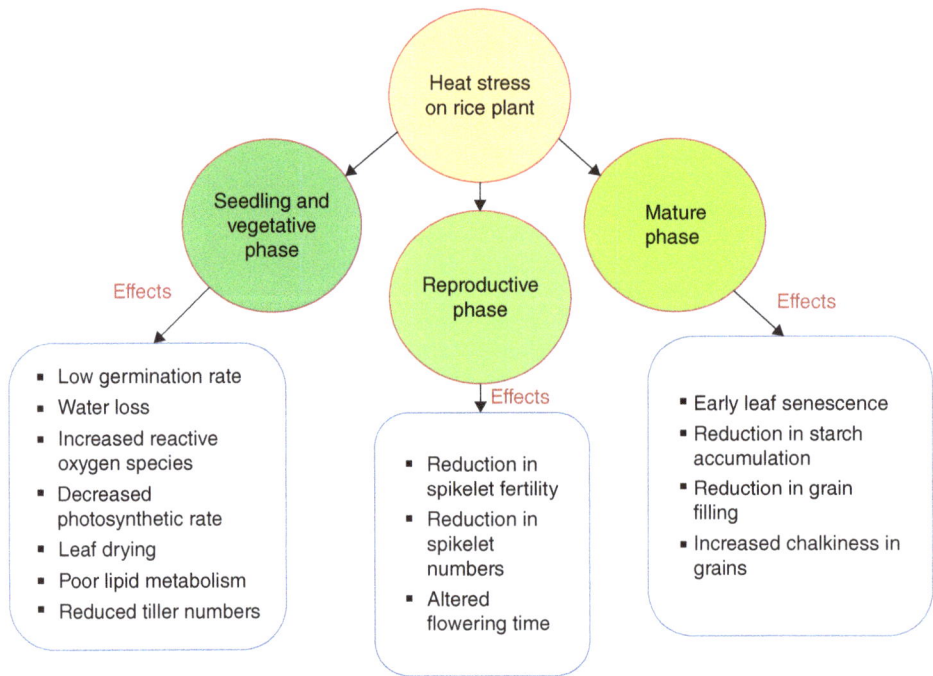

Fig. 9.1. Effects of heat stress on different phases of the life cycle of rice plants.

(Begcy *et al.*, 2018). The endosperm collapses when the temperature rises above 39°C (Begcy *et al.*, 2018). Elevated temperature also initiates leaf yellowing, leaf curling and decreased root growth at the plantlet stage (Das *et al.*, 2014, 2021). The tiller number and size in rice plants are also reduced under elevated temperature (Liu *et al.*, 2018; Xu *et al.*, 2020a).

9.2.2 Effects of elevated temperature on the early reproductive growth phase

Of all the phases of rice lifespan, the heading and flowering stages are the most delicate with regard to heat stress (Matsui *et al.*, 2014; Das *et al.*, 2021). An increase in temperature during the reproductive phase decreases the fertility rate and spikelet quantity, resulting in poor rice yield (Ren *et al.*, 2023). It has also been reported that high temperature stress (above 35°C) results in spikelet sterility and yield failure when applied for approximately 5 days (Satake and Yoshida, 1978). Preflowering heat stress exposure

is more harmful than postflowering heat stress, resulting in sterile spikelets (Jagadish *et al.*, 2007; Shi *et al.*, 2018). Stamens are more susceptible to elevated temperatures than pistils (Miura *et al.*, 2011; Arshad *et al.*, 2017; Hu *et al.*, 2021). Heat stress affects the process of meiosis during pollen formation to produce non-fertile pollen (Endo *et al.*, 2009). Increased temperatures also affect the process of pollination by altering the pollen grain moisture content in rice (Das *et al.*, 2014; Shrestha *et al.*, 2022). Furthermore, heat stress causes a reduction in the protein content present in the pollen and hence affects pollen tube growth, leading to sterile spikelets (Jagadish, 2020; Shrestha *et al.*, 2022).

9.2.3 Effects of elevated temperature on the late reproductive growth phase or maturity phase

During the late reproductive phase, grain filling occurs, which involves the alteration of sucrose to starch. Starch is the chief carbohydrate in rice grains and determines grain yield and quality (Tang *et al.*, 2018). Rice starch is a derivative of sugar that results after anthesis. Heat stress adversely affects anthesis, thus jeopardizing grain carbohydrate content and yield. High-temperature stress also affects the initiation of whole-plant senescence, which causes unsatisfactory seed filling (Plaut *et al.*, 2004; Yang and Zhang, 2006; Chen *et al.*, 2021a). Seed filling can reach 50% or no seed filling can happen due to elevated temperature at the time of the seed-filling stage (Sreenivasulu *et al.*, 2015; Chen *et al.*, 2017; Impa *et al.*, 2021). DNA methylation also occurs in the rice ABA promoter during the grain-filling phase due to elevated temperature stress, which can hinder the seed propagation rate (Suriyasak *et al.*, 2020). Heat stress can reduce the abundance of key genes involved in endosperm sugar to starch conversion and can decrease starch synthesis (Yamakawa and Hakata, 2010; Zhang *et al.*, 2021; Shirdelmoghanloo *et al.*, 2022). Heat stress hinders photosynthesis in somatic plant parts, decreasing the supply of fixed carbon to reproductive parts and affecting the rate of seed filling and yield (Yang and Zhang, 2010). He *et al.* (2018, 2021) showed that the overall thermotolerance of plants is linked to grain filling, and they also demonstrated that thermosusceptible lines showed reduced seed filling under heat stress.

Compared with wild-type seeds, the ERACTA mutants are heat sensitive and have smaller seeds (Wu *et al.*, 2022). Heat stress also affects endosperm quality, seed size and rice yield (Nevame *et al.*, 2018). The chalkiness of rice increases under elevated temperature stress, which causes a decrease in grain weight and quality, including a cooking taste (Wada *et al.*, 2019; Wang *et al.*, 2022a). Heat-induced grain weight loss is associated with decreased carbohydrate content, an immature vascular system and reduced glume mass (Zhang *et al.*, 2009; Cao *et al.*, 2016). It has also been reported that heat exposure at the panicle initiation phase decreases grain weight (Wu *et al.*, 2021).

9.3 Approaches for Heat Acceptance in Rice

Various procedures have been adopted to manage elevated temperatures in rice plants (Fig. 9.2), which can be summarized as follows.

9.3.1 Agronomic management and conventional breeding

Unalterable damage due to excessive heat stress has different harmful consequences on the

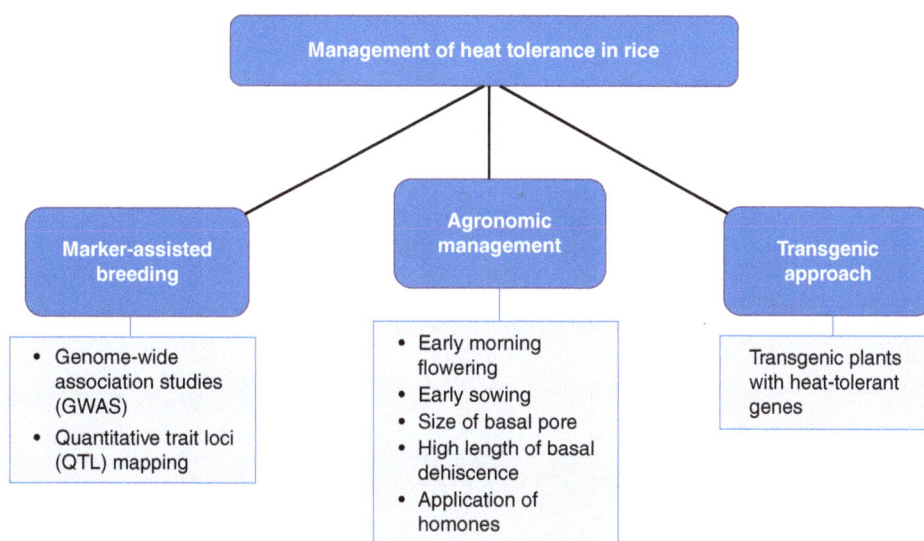

Fig. 9.2. Strategies for management of heat stress tolerance in rice.

procreative/reproductive stages of plants. The effects include unsuccessful pollination, enhanced sterility and reduced yield (Khan *et al.*, 2019). Various methods have been used to avoid, escape or tolerate high temperatures in rice (Khan *et al.*, 2019). Agronomic practices mainly include timely rice seeding, irrigation adaptation and site-specific cropping. Rice cultivars with early maturity during grain filling are important for agronomic management systems (O-he *et al.*, 2007; Krishnan *et al.*, 2011). Above 40°C, complete anther sterility is observed during the anthesis stage in many cultivars, including the tolerant cultivar N22 (Satake and Yoshida, 1978; Matsui *et al.*, 2001). Different rice genotypes have diverse floral opening times, fluctuating from sunrise to night. Excessive heat affects flowering time, ultimately reducing fertilization and flowering (Jagadish *et al.*, 2007; Djanaguiraman *et al.*, 2020). As a result, flowering at sunrise is chosen to avoid heat stress in rice. Early flowering can protect spikelets, preserving fruitfulness under high temperature stress in rice (Sheehy *et al.*, 2001; Jagadish *et al.*, 2008; Julia and Dingkuhn, 2012; Hirabayashi *et al.*, 2015; Jagadish, 2020; Ren *et al.*, 2023).

The size of basal pores in anthers and the length of anthers vary among rice cultivars. The selection of these traits helps rice cope with elevated thermal stress (Satake and Yoshida, 1978; Matsui and Kagata, 2003; Matsui *et al.*, 2005). The application of exogenous phytohormones such as salicylic acid (SA) to rice plants induces high temperature tolerance. Foliar SA spray (0.1 mM) enhanced heat stress by enhancing the quantities of sugars, proline and oxidation inhibitor enzymes (POD, APX and CAT). The concentration of growth regulators such as indole acetic acid (IAA), gibberellin 3 (GA3) and abscisic acid (ABA) also increased in spikelets, enhancing crop yield. Many reports have shown that the exogenous use of methyl jasmonate, ascorbic acid, brassinosteroids and α-tocopherol could escalate elevated temperature tolerance in rice through the enhancement of several endogenous hormones (Chang, 2007; Debolt *et al.*, 2007; Mohammed and Tarpley, 2009, 2011; Hayat *et al.*, 2010; Chandrakala *et al.*, 2013; Zhang *et al.*, 2017, 2018). It has been reported that pretreatment of rice plants with low nitric oxide or H_2O_2 levels can cause thermotolerance, as indicated by an elevated quantum yield of photosystem II (PSII) and decreased senescence. These signalling molecules activate antioxidant enzymes and enhance heat shock protein (HSP) and sucrose phosphate synthase transcript expression (Uchida *et al.*, 2002; Hasanuzzaman *et al.*, 2013). A recent study revealed that spray treatment of rice crop fields with mist enhanced heat tolerance by delaying leaf senescence and enhancing antioxidant enzymes (Jiang *et al.*, 2020).

The enhancement of thermotolerance by conventional breeding is a good way of lowering the detrimental influence of raised temperature on seed value and crop yield (Driedonks *et al.*, 2016). The first and crucial step is selecting elite thermotolerant cultivars for use as donors in conventional breeding. Some heat-tolerant breeding lines, such as N22, Giza178, HHT4, 996, IR2061, Habataki, EMF20 and Guodao 6, have been used in investigations. These rice genotypes showed better adaptability to high temperatures (Khan *et al.*, 2019). Some high-temperature lenient genotypes exhibit increased expression of enzymes associated with the Calvin cycle (i.e. phosphoribulokinase and Rubisco) under high temperatures (Makino *et al.*, 2007; Scafaro *et al.*, 2010). Rubisco activase from the high-temperature-tolerant rice *Oryza meridionalis* was found to be more lenient to heat (40°C) than that from the cultivated rice *O. sativa* (Scafaro *et al.*, 2010). Conventional breeders use backcrossed inbred, recombinant inbred and double haploid lines, which were acquired from successful crosses between tolerant and sensitive cultivars and screened based on their tolerance to high temperatures at various developmental stages. A successful introgression line was achieved from cross-pollination between *O. officinalis* (wild rice) and an indica rice cultivar, Koshihikari, which showed early flowering, greater spikelet fertility and better yield under elevated temperature (Ishimaru *et al.*, 2010).

9.3.2 Identification of elevated temperature-tolerant quantitative trait loci and marker-based breeding

Heat tolerance in rice can be reduced not only by conventional breeding but also by marker-supported breeding. Mapping of QTLs for heat acceptance has been performed for various rice cultivars, mainly during the floral stage (Cao et al., 2003; Chen et al., 2008; Zhang et al., 2008, 2009; Jagadish et al., 2010; Cheng et al., 2012; Ye et al., 2012, 2015). Sixty QTLs associated with heat confrontation were spotted during the flowering period. Among them, two main QTLs were documented on chromosome 1 (qHTSF1.1) and 4 (qHTSF4.1) from the N22/IR64 population (Ye et al., 2012). Zhao et al. (2016) recognised a QTL related to spikelet fertility through a cross between heat-susceptible japonica and heat-resistant indica cultivars. Several QTLs associated with spikelet fertility (qSFht4.2 and qSFht2) and flowering time (qDFT11, qDFT3, qDFT10.1 and qDFT8) under heat stress have been identified. Moreover, 12 QTLs associated with elevated temperature tolerance during booting have been identified (qHTB3-3 being the major QTL) (Jagadish et al., 2010; Zhu et al., 2017; Kilasi et al., 2018). Recently, the QTL qEMF3 was identified in wild rice (O. officinalis), reducing thermal stress damage by modifying the flowering opening time in the Nanjing 11 variety (Hirabayashi et al., 2015). Adding to the QTL qEMF3, another heat tolerance QTL, qHTSF4.1, was identified, and near-isogenic lines (NILs) were developed (Ye et al., 2022). The QTL TT1 was chosen for thermotolerance in African rice and used in molecular breeding (Li et al., 2015). Some QTLs, such as OsHTAS and SUS3, exhibit thermotolerance during the grain filling stages of rice (Wei, 2012; Liu et al., 2016; Ps et al., 2017). In a recent study, heat-tolerant rice QTLs positioned on chromosomes 1, 3, 4, 5, 7, 8, 9 and 10 were used for marker-assisted breeding (Buu et al., 2021).

9.3.3 Transgenic methods and gene editing technologies

Gene manipulation by genetic engineering is the most effective method for developing heat-tolerant rice. Multiple transcripts are associated in heat tolerance in rice, such as receptor-like kinase ERECTA, OsIF, OsMYB55, OsANN1, SNAC3, OsbZIP46CA1, SAPK6, OsRGB1, OsSIZ1, OsHIRP1, Rca and OsWRKY11 (Wu et al., 2009; El-Kereamy et al., 2012; Fang et al., 2015; Qiao et al., 2015; Shen et al., 2015; Driedonks et al., 2016; Arshad et al., 2017; Chang et al., 2017;

Mishra et al., 2018; Scafaro et al., 2018; Soda et al., 2018; Biswas et al., 2019; Kim et al., 2019). Overexpression of these genes improved thermotolerance in rice. Overexpressed lines of the kinase ERECTA displayed heat tolerance at 42°C (Shen et al., 2015). Recent research analysing differentially expressed genes (DEGs) revealed that OsNECED1-overexpressing rice lines improved their high temperature tolerance by increasing their antioxidant capacity (Zhou et al., 2022). HSPs are abundant in plant systems. Upon exposure to high temperature stress, HSPs are induced by stabilizing unfolded proteins. A considerable number of HSPs (HSP23.2, HSP26.7, HSP17.9A, HSP16.9A and HSP17.4) are upregulated under elevated temperature stress in different rice cultivars (Müller and Rieu, 2016). Furthermore, gene editing technology provides a new way to produce thermotolerant rice plants because it can mutate the target gene only, not the expression cassette (Gao, 2019). By means of the CRISPR/Cas9 system, the roles of AET1, OsNAC006, HSA1, OsCNGC14 and OsCNGC16 in thermotolerance in rice have been validated (Qiu et al., 2018; Chen et al., 2019; Cui et al., 2020; Wang et al., 2020). Some of the genes used in transgenic research and their effects on the thermal adaptation behaviour of rice plants are listed in Table 9.1.

9.4 Nutritional Efficiency in Rice Plants under Elevated Temperature Stress

The ideal temperature for rice photosynthesis is 28 ± 2°C during the light period and c.20°C during the dark period. The grain-filling phase is crucial for protein, carbohydrate and lipid synthesis in rice plants. Apart from yield, the poor quality of grain and the changed nutrient structure in grain could influence user acceptance (Lyman et al., 2013; Zhao and Fitzgerald, 2013). Raised temperature at the seed-filling phase disturbs the accretion of several nutritional elements, such as protein and starch, by restricting the biochemical or enzymatic reactions of protein and starch synthesis (Farooq et al., 2017).

Pravallika et al. (2020) described the nutritional superiority of three diverse rice types under non-stress and raised heat conditions at

Table 9.1 List of genes used in the development of heat-tolerant or heat-sensitive transgenic rice.

Gene	Expression/characteristic features compared with the wild type (WT)	Reference
GAD3	High-temperature tolerance	El-Kereamy et al., 2012
OsHYR	Increased total chlorophyll level in leaves under heat stress	Ambavaram et al., 2014
OsRBG1	The heat-tolerant phenotype, expressed in roots, shows that roots and young shoots are more extended than WT	Lo et al., 2020
OsNSUN2	Downregulated, temperature sensitive, expressed in roots, roots are shorter at the four-leaf stage	Zhu and Zhang, 2020
OsRCc3	Upregulated, temperature tolerant, expressed in roots, stronger roots	Li et al., 2018
OsZFP350	Upregulated, temperature tolerant, expressed in roots and leaves, higher volume and biomass of root	Kang et al., 2019
OsPL	Downregulated, temperature tolerant, expressed in roots, longer roots	Akhter et al., 2019
OsHYR	Upregulated, temperature tolerant, expressed in roots and leaves, longer and thicker roots, dark green and bright leaves	Ambavaram et al., 2014
OsPSL50	Downregulated, temperature sensitive, expressed in stem and leaves, shorter and prematurely senescent plants	He et al., 2021
OsGSK	Downregulated, temperature tolerant, expressed in stem, dwarf and short plants	Koh et al., 2007
OsPDT1	Downregulated, temperature sensitive, expressed in stem and dwarf plants	Deng et al., 2020
OsSPL7	Downregulated, expressed in stem	Park et al., 2020
OsFKBP20-1b	Downregulated, temperature sensitive, expressed in stem	Park et al., 2020
OsHSA1	Downregulated, temperature sensitive, expressed in stem and leaves, short and albino plants, chloroplast development is slower	Qiu et al., 2018
OsHTS1	Downregulated, temperature sensitive, expressed in stem and leaves, short plants with less chlorophyll content	Chen et al., 2021b
OsHSP40	Downregulated, temperature sensitive, expressed in leaves, leaves are smaller	Wang et al., 2022c
OsMDHAR4	Upregulated, temperature sensitive, expressed in leaves, under high temperature stress less percentage of closed stomata	Liu et al., 2018
OsqEMF3	Thermotolerant, expressed in flower, advanced flowering by approximately 2 h	Bheemanahalli et al., 2017
Rab7	Upregulated, expressed in leaves, heat tolerance with higher yield	El-Esawi and Alayafi, 2019
TT1	Heat tolerance by degrading ubiquitinated proteins	Li et al., 2015
ER (Erecta)	Heat tolerance (independent of water loss)	Shen et al., 2015
OsHTAS	Heat tolerance through reactive oxygen species (ROS) homeostasis functions at the seedling stage	Liu et al., 2016
TOGR1	rRNA homeostasis under heat stress functions at the seedling stage	Wang et al., 2016
Sus3	High-temperature tolerance, expressed at the ripening stage	Takehara et al., 2018
SLG1	High-temperature tolerance, expressed in green tissue at the reproductive stage	Xu et al., 2020b
HTH5	High-temperature tolerance by reducing ROS functions at the heading period.	Chen et al., 2021a, 2021b
MSD1	High-temperature tolerance by inducing the expression of ROS scavengers	Shiraya et al., 2015

Continued

Table 9.1. Continued.

Gene	Expression/characteristic features compared with the wild type (WT)	Reference
OsANN1	Modulates H_2O_2 production, high expression in seeds and panicle	Qiao et al., 2015
SNAC3	Modulates ROS functions at the seedling stage and is expressed ubiquitously	Fang et al., 2015
OsNTL3	Relay heat stress signals functions at the seedling stage	Liu et al., 2020
OsHIRP1	Acts as a regulator in response to heat, functions at the germination period and is highly expressed under heat stress	Kim et al., 2019
OsRab7	Modulates antioxidants, osmolytes and stress responsive genes, functions at seedling stage	El-Esawi and Alayafi, 2019
OsRGB1	Heat stress tolerance, functions at germination and seedling stage	Biswas et al., 2019
OsCNGC-14 and -16	Modulates calcium signals, expressed in most of the organs, functions at seedling stage	Cui et al., 2020
OsNSUN2	Helps to maintain chloroplast function	Tang et al., 2020
OsFBN1	Role in chloroplastic lipid metabolism, expressed in green tissue, functions at seedling stage	Li et al., 2019
Os-MDHAR4	H_2O_2-induced stomatal closure, ubiquitous expression, functions at seedling stage	Liu et al., 2018
OsUBP21	Plays negative role in basal thermotolerance in rice, expressed in embryo, shoot and inflorescence	Zhou et al., 2019

the milky and dough stages (Table 9.2). They reported that the total increase in reducing sugar content for all the varieties under heat stress compared with non-stressed plants was c.9.45%. The findings of Gill et al. (2001) also supported the sugar content patterns of Pravallika et al. (2020). A 20.10% increase in carbohydrate content was found in all the varieties, where the average carbohydrate content was 747.93 and 898.31 mg/g of fresh mass under non-stress and heat stress, respectively (Table 9.2). These results also agreed with the outcomes of Rowland-Bamford et al. (1996). Gunaratne et al. (2011) and Pravallika et al. (2020) reported increased starch levels in rice (c.58.10%) under elevated temperature conditions. Under increased heat conditions, the overall content of amylose decreased to 26.93% (Table 9.2). Jiang et al. (2003) also reported that the endospemic amylose content is decreased by 25% under elevated temperature stress conditions in indica rice. An overall decrease in the level of seed protein in rice varieties has also been observed under elevated temperature stress conditions (Triboi and Triboi-Blondel, 2002; Pravallika et al., 2020). The anthocyanin content decreases in rice varieties due to high temperature (Zaidi

et al., 2019; Pravallika et al., 2020). The flavonoid content of rice varieties improved due to elevated temperature (Zaidi et al., 2019; Pravallika et al., 2020).

The presence of amino acids is a vital dietary quality trait in rice. Huang et al. (2019) reported amino acid levels in the seeds of two rice varieties (Luliangyou 996 and Lingliangyou 268) under divergent temperature conditions during the early grain-filling phase under field conditions in 2016 and 2017. The accretion and nitrogen levels of the grains were markedly lower due to high temperature stress. This report also suggested that heat stress can reduce nitrogen accumulation in rice seeds and, therefore, reduce the amino acid level in the seeds. Heat stress and elevated atmospheric CO_2 are reported to negatively affect rice seed quality (Wang et al., 2011; Usui et al., 2014). In the future, rice plants are predicted to face heat stress along with further increases in CO_2 (Mittler and Blumwald, 2010). Chaturvedi et al. (2017) explored the influences of high CO_2 with temperature stress on rice (NL-44 and Pusa 1121) seed quality and mineral nutrients. They found lower grain production under heat combined with elevated CO_2 in comparison with plants subjected

Table 9.2. Nutrient content in rice at the dough stage.

Varieties	Control	Heat stress
Reducing sugar (mg/g)		
Hraswa	15.54*	16.71*
Prathyasa	14.29*	15.61*
Manuratha	14.30*	15.71*
Total carbohydrate (mg/g)		
Hraswa	745.12*	889.89*
Prathyasa	753.97*	903.38*
Manuratha	748.72*	898.40*
Starch content (mg/g)		
Hraswa	6.97*	9.06*
Prathyasa	11.94*	26.53*
Manuratha	14.38*	18.13*
Amylose content (%)		
Hraswa	27.43*	20.10*
Prathyasa	25.35*	17.33*
Manuratha	25.79*	16.11*

*Indicates the statistical significance of the result obtained by Pravallika *et al.* (2020), where a minimum of three replicates were taken to generate the data.

to only elevated CO_2. Similar studies also revealed that rice yield decreased under elevated CO_2 along with heat stress (Madan *et al.*, 2012; Roy *et al.*, 2012; Usui *et al.*, 2014). The expression of starch synthesis enzymes was downregulated, and the expression of starch-degrading enzymes increased under heat (Mitsui *et al.*, 2016). Increased amylose content under elevated CO_2 is evident (Conroy *et al.*, 1994; Seneweera and Conroy, 1997). Chaturvedi *et al.* (2017) reported that combined elevated temperature and CO_2 offset the elevated CO_2-induced accretion of amylose.

A reduction in nitrogen levels due to a high photosynthetic rate and carbohydrate accretion under elevated CO_2 has been connected with decreased protein content in the seed (Uprety *et al.*, 2010; Panozzo *et al.*, 2014). Furthermore, heat stress affects carbohydrate and protein accretion in rice (Lin *et al.*, 2010; Shi *et al.*, 2013). Chaturvedi *et al.* (2017) reported further adverse outcomes of heat and CO_2 on the protein content of Pusa 1121 grains. Heat stress and elevated CO_2 are reported to disturb macro- and micromineral element build-up and composition in rice seeds (Seneweera, 2011; Usui *et al.*, 2014; Chaturvedi *et al.*, 2017). Wang *et al.* (2016) evaluated the impact of heat on rice chalkiness, seed mass and nutrient content.

According to their studies, rice seeds/grains were categorized into four sets based on the chalkiness of the scanned images of the seeds. The mineral nutrient, protein and amylose levels of each chalkiness group were assessed. Heat stress during seed filling reduced the seed mass and level of amylose in rice but improved the concentrations of most minerals, protein content and chalky areas in the grain. Their results showed that upsurges in chalky grain regions due to heat during seed filling also increase the mineral content of the grain.

9.5   Conclusion and Future Perspectives

The global population has gradually increased demand resulting in increased rice production and other agricultural output (Ray *et al.*, 2013). Universal temperatures have also increased over the last few decades, challenging the world economy and food safety. As rice plants are susceptible to heat stress, they have evolved to some extent naturally or manually to regulate their development and adapt to changing environmental temperatures.

This chapter offers a significant speculative base for understanding the impact of elevated temperature stress on rice at different growth phases and the methods adopted to generate heat-tolerant rice plants. Furthermore, the nutritional efficiency of rice under heat stress has been analysed. At times of excess environmental temperatures, all the stages of rice plants are affected, and the plants are under elevated temperature stress. This stress can result in late germination, decreased pollen viability/efficiency, irregular development of the ovary and unsatisfactory nutrient content (Chen *et al.*, 2017; Zhao *et al.*, 2018; Afzal *et al.*, 2019; Liu *et al.*, 2019). Several advanced techniques, such as transgenic generation or gene editing methods, have been adopted to generate heat-tolerant rice plants that can withstand temperatures above the optimal level, increase yield and preserve or increase nutrient levels in the grains under elevated temperature conditions. However, there is still a long way to go in understanding the exact heat tolerance mechanisms in rice plants at different developmental phases and under different extents of heat stress. Elevated temperature proteins must be thoroughly studied to better understand the biochemical reactions involved during high-temperature periods. Research on heat stress proteins and heat stress-related transcription factors is crucial for understanding the tolerance of rice to elevated temperature stress (Afzal *et al.*, 2019; Wang *et al.*, 2022b). Phenotypic studies have also made a significant contribution to the identification of thermotolerance in rice as a preliminary index.

Furthermore, more phenotypic investigations must be performed under diverse increased temperature stress conditions, in various plant parts and at different growth phases, to map the behaviour of plants in temperature variations. Complete 'omics' studies are also required to detect the genes, proteins or transcription factors associated with the stress acceptance machinery in rice plants. Through these future studies and understanding, rice researchers will be able to generate heat stress-tolerant, high-yielding and nutrient-efficient rice to fulfil food demand to a large extent.

References

Afzal, S., Sirohi, P., Yadav, A.K., Singh, M.P., Kumar, A. et al. (2019) A comparative screening of abiotic stress tolerance in early flowering rice mutants. *Journal of Biotechnology* 302, 112–122.

Akhter, D., Qin, R., Nath, U.K., Eshag, J., Jin, X. et al. (2019) The rice gene OsPL, encoding a MYB family of transcription factors, is involved in anthocyanin synthesis, heat stress response and hormonal signalling. *Gene* 699, 62–72.

Ambavaram, M.M., Basu, S., Krishnan, A., Ramegowda, V., Batlang, U. et al. (2014) Coordinated regulation of photosynthesis in rice increases yield and tolerance to environmental stress. *Nature Communications* 5, 5302.

Arshad, M.S., Farooq, M., Asch, F., Krishna, J.S.V., Prasad, P.V.V. et al. (2017) Thermal stress impacts reproductive development and grain yield in rice. *Plant Physiology and Biochemistry* 115, 57–72.

Bahuguna, R.N., Jha, J., Pal, M., Shah, D., Lawas, L.M.F. et al. (2015) Physiological and biochemical characterization of NERICA-L-44: a novel source of heat tolerance at the vegetative and reproductive stages in rice. *Physiologia Plantarum* 154, 543–559.

Battisti, D.S. and Naylor, R.L. (2009) Historical warnings of future food insecurity with unprecedented seasonal heat. *Science* 323, 240–244.

Begcy, K., Sandhu, J. and Walia, H. (2018) Transient heat stress during early seed development primes germination and seedling establishment in rice. *Frontiers in Plant Science* 9, 1768.

Bheemanahalli, R., Sathishraj, R., Manoharan, M., Sumanth, H.N., Muthurajan, R. et al. (2017) Is early morning flowering an effective trait to minimize heat stress damage during flowering in rice? *Field Crops Research* 203, 238–242.

Biswas, S., Islam, M.N., Sarker, S., Tuteja, N. and Seraj, Z.I. (2019) Overexpression of heterotrimeric G protein beta subunit gene (OsRGB1) confers both heat and salinity stress tolerance in rice. *Plant Physiology and Biochemistry* 144, 334–344.

Buu, B.C., Chan, C.Y. and Lang, N.T. (2021) Molecular breeding for improving heat stress tolerance in rice: recent progress and future perspectives. In: Hossain, M.A., Hassan, L., Ifterkharuddaula, K.M., Kumar, A. and Henry, R. (eds) *Molecular Breeding for Rice Abiotic Stress Tolerance and Nutritional Quality.* Wiley, pp. 92–119.

Cao, W.H., Dong, Y., Zhang, J.S. and Chen, S.Y. (2003) Characterization of an ethylene receptor homolog gene from rice. *Science in China Series C Life Sciences* 46, 370–378.

Cao, Y.Y., Chen, Y.H., Chen, M.X., Wang, Z.Q., Wu, C.F., Bian, X.C., Yang, J.C. and Zhang, J.H. (2016) Growth characteristics and endosperm structure of superior and inferior spikelets of indica rice under high-temperature stress. *Biologia Plantarum* 60(3), 532–542.

Chandrakala, J.U., Chaturvedi, A.K., Ramesh, K.V., Rai, P., Khetarpal, S. et al. (2013) Acclimation response of signalling molecules for high temperature stress on photosynthetic characteristics in rice genotypes. *Indian Journal of Plant Physiology* 18, 142–150.

Chang, P.F.L., Jinn, T.L., Huang, W.K., Chen, Y., Chang, H.M. et al. (2007) Induction of a cDNA clone from rice encoding a class II small heat shock protein by heat stress, mechanical injury, and salicylic acid. *Plant Science* 172, 64–75.

Chang, Y., Nguyen, B.H., Xie, Y., Xiao, B., Tang, N. *et al.* (2017) Co-overexpression of the constitutively active form of OsbZIP46 and ABA-activated protein kinase SAPK6 improves drought and temperature stress resistance in rice. *Frontiers in Plant Science* 8, 1102.

Chaturvedi, A.K., Bahuguna, R.N., Pal, M., Shah, D., Maurya, S. *et al.* (2017) Elevated CO_2 and heat stress interactions affect grain yield, quality and mineral nutrient composition in rice under field conditions. *Field Crops Research* 206, 149–157.

Chen, F., Dong, G., Wang, F., Shi, Y., Zhu, J. *et al.* (2021a) A β-ketoacyl carrier protein reductase confers heat tolerance via the regulation of fatty acid biosynthesis and stress signaling in rice. *New Phytologist* 232, 655–672.

Chen, J., Tang, L., Shi, P., Yang, B., Sun, T. *et al.* (2017) Effects of short-term high temperature on grain quality and starch granules of rice (*Oryza sativa* L.) at post-anthesis stage. *Protoplasma* 254, 935–943.

Chen, J., Xu, Y., Fei, K., Wang, R., He, J. *et al.* (2020) Physiological mechanism underlying the effect of high temperature during anthesis on spikelet-opening of photo-thermo-sensitive genic male sterile rice lines. *Scientific Reports* 10, 2210.

Chen, K., Guo, T., Li, X.M., Zhang, Y.M., Yang, Y.B. *et al.* (2019) Translational regulation of plant response to high temperature by a dual-function tRNA[his] guanylyltransferase in rice. *Molecular Plant* 12, 1123–1142.

Chen, M., Fu, Y., Mou, Q., An, J., Zhu, X. *et al.* (2021b) Spermidine induces expression of stress associated proteins (SAPs) genes and protects rice seed from heat stress-induced damage during grain-filling. *Antioxidants* 10, 1–14.

Chen, Q.Q., Yu, S.B. and Li, C.H. (2008) Identification of QTLs for heat tolerance at flowering stage in rice. *Scientia Agricultura Sinica* 41, 315–321.

Cheng, L.R, Wang, J.M, Uzokwe, V., Meng, L.J, Wang, Y. *et al.* (2012) Genetic analysis of cold tolerance at seedling stage and heat tolerance at anthesis in rice (*Oryza sativa* L.). *Journal of Integrative Agriculture* 11, 359–367.

Conroy, J.P, Seneweera, S., Basra, A.S, Rogers, G. and Nissen-Wooller, B. (1994) Influence of rising atmospheric CO_2 concentrations and temperature on growth, yield and grain quality of cereal crops. *Functional Plant Biology* 21, 741.

Cui, Y., Lu, S., Li, Z., Cheng, J., Hu, P. *et al.* (2020) Cyclic nucleotide-gated ion channels 14 and 16 promote tolerance to heat and chilling in rice. *Plant Physiology* 183, 1794–1808.

DaMatta, F.M., Grandis, A., Arenque, B.C. and Buckeridge, M.S. (2010) Impacts of climate changes on crop physiology and food quality. *Food Research International* 43, 1814–1823.

Das, P., Bahuguna, R.N., Rathore, R.S., Abbat, S., Nongpiur, R.C. *et al.* (2021) Rice mutants with tolerance to multiple abiotic stresses show high constitutive abundance of stress-related transcripts and proteins. *Australian Journal of Crop Science* 15, 12–21.

Das, S., Krishnan, P., Nayak, M. and Ramakrishnan, B. (2014) High temperature stress effects on pollens of rice (*Oryza sativa* L.) genotypes. *Environmental and Experimental Botany* 101, 36–46.

Debolt, S., Melino, V. and Ford, C.M. (2007) Ascorbate as a biosynthetic precursor in plants. *Annals of Botany* 99, 3–8.

Deng, W., Li, R., Xu, Y., Mao, R., Chen, S. *et al.* (2020) A lipid transfer protein variant with a mutant eight-cysteine motif causes photoperiod- and thermo-sensitive dwarfism in rice. *Journal of Experimental Botany* 71, 1294–1305.

Djanaguiraman, M., Narayanan, S., Erdayani, E. and Prasad, P.V.V. (2020) Effects of high temperature stress during anthesis and grain filling periods on photosynthesis, lipids and grain yield in wheat. *BMC Plant Biology* 20, 268.

Driedonks, N., Rieu, I. and Vriezen, W.H. (2016) Breeding for plant heat tolerance at vegetative and reproductive stages. *Plant Reproduction* 29, 67–79.

El-Esawi, M.A. and Alayafi, A.A. (2019) Overexpression of rice Rab7 gene improves drought and heat tolerance and increases grain yield in rice (*Oryza sativa* L.). *Genes* 10, 56.

El-Kereamy, A., Bi, Y.M., Ranathunge, K., Beatty, P.H., Good, A.G. *et al.* (2012) The rice R2R3-MYB transcription factor OsMYB55 is involved in the tolerance to high temperature and modulates amino acid metabolism. *PLoS One* 7, e52030.

Endo, M., Tsuchiya, T., Hamada, K., Kawamura, S., Yano, K. *et al.* (2009) High temperatures cause male sterility in rice plants with transcriptional alterations during pollen development. *Plant and Cell Physiology* 50, 1911–1922.

Fahad, S., Hussain, S., Saud, S., Hassan, S., Ihsan, Z. *et al.* (2016) Exogenously applied plant growth regulators enhance the morpho-physiological growth and yield of rice under high temperature. *Frontiers in Plant Science* 7, 1250.

Fang, Y., Liao, K., Du, H., Xu, Y., Song, H. *et al.* (2015) A stress-responsive NAC transcription factor SNAC3 confers heat and drought tolerance through modulation of reactive oxygen species in rice. *Journal of Experimental Botany* 66, 6803–6817.

Farooq, M., Nadeem, F., Gogoi, N., Ullah, A., Alghamdi, S.S. *et al.* (2017) Heat stress in grain legumes during reproductive and grain-filling phases. *Crop and Pasture Science* 68, 985.

Gao, C. (2019) Precision plant breeding using genome editing technologies. *Transgenic Research* 28 (Suppl. 2), 53–55.

Gill, P.K., Sharma, A.D., Singh, P. and Bhullar, S.S. (2001) Effect of various abiotic stresses on the growth, soluble sugars and water relations of sorghum seedlings grown in light and darkness. *Bulgarian Journal of Plant Physiology* 27, 72–84.

Gu, J., Chen, J., Chen, L., Wang, Z., Zhang, H. *et al.* (2015) Grain quality changes and responses to nitrogen fertilizer of japonica rice cultivars released in the Yangtze River Basin from the 1950s to 2000s. *Crop Journal* 3, 285–297.

Gunaratne, A., Sirisena, N., Ratnayaka, U.K., Ratnayaka, J., Kong, X. *et al.* (2011) Erratum: Effect of fertilizer on functional properties of flour from four rice varieties grown in Sri Lanka. *Journal of the Science of Food and Agriculture* 91, 1728–1728.

Guo, W., Wang, X., Sun, J., Ding, A. and Zou, J. (2016) Comparison of land–atmosphere interaction at different surface types in the mid- to lower reaches of the Yangtze River Valley. *Atmospheric Chemistry and Physics* 16, 9875–9890.

Hasanuzzaman, M., Nahar, K., Alam, M.M., Roychowdhury, R. and Fujita, M. (2013) Physiological, biochemical, and molecular mechanisms of heat stress tolerance in plants. *International Journal of Molecular Sciences* 14, 9643–9684.

Hayat, Q., Hayat, S., Irfan, M. and Ahmad, A. (2010) Effect of exogenous salicylic acid under changing environment: a review. *Environmental and Experimental Botany* 68, 14–25.

He, Y., Li, L., Zhang, Z. and Wu, J.L. (2018) Identification and comparative analysis of premature senescence leaf mutants in rice (*Oryza sativa* L.). *International Journal of Molecular Sciences* 19, 140.

He, Y., Zhang, X., Shi, Y., Xu, X., Li, L. *et al.* (2021) PREMATURE SENESCENCE LEAF 50 promotes heat stress tolerance in rice (*Oryza sativa* L.). *Rice* 14, 53.

Hirabayashi, H., Sasaki, K., Kambe, T., Gannaban, R.B., Miras, M.A. *et al.* (2015) QEMF3, a novel QTL for the early-morning flowering trait from wild rice, *Oryza officinalis*, to mitigate heat stress damage at flowering in rice, *O. sativa*. *Journal of Experimental Botany* 66, 1227–1236.

Hiroshi, W., Hiroshi, N., Yabuoshi, Y., Fukuyo, T., Atsushi, M. *et al.* (2012) Mechanism for the formation of ring-shaped chalkiness in growing rice kernels under typhoon/foehn-induced dry wind condition and its practical applications. *Abstracts of Meeting of the CSSJ* 233, 190–190.

Hu, Q., Wang, W., Lu, Q., Huang, J., Peng, S. *et al.* (2021) Abnormal anther development leads to lower spikelet fertility in rice (*Oryza sativa* L.) under high temperature during the panicle initiation stage. *BMC Plant Biology* 21, 428.

Huang, M., Zhang, H., Zhao, C., Chen G. and Zou, Y. (2019) Amino acid content in rice grains is affected by high temperature during the early grain-filling period. *Science Reports* 9, 2700.

Hwang, O.J. and Back, K. (2019) Melatonin deficiency confers tolerance to multiple abiotic stresses in rice via decreased brassinosteroid levels. *International Journal of Molecular Sciences* 20, 5173.

Impa, S.M., Raju, B., Hein, N.T., Sandhu, J., Prasad, P.V.V. *et al.* (2021) High night temperature effects on wheat and rice: current status and way forward. *Plant, Cell and Environment* 44, 2049–2065.

Ishimaru, T., Hirabayashi, H., Ida, M., Takai, T., San-Oh, Y.A. *et al.* (2010) A genetic resource for early-morning flowering trait of wild rice *Oryza officinalis* to mitigate high temperature-induced spikelet sterility at anthesis. *Annals of Botany* 106, 515–520.

Jagadish, S.V.K. (2020) Heat stress during flowering in cereals – effects and adaptation strategies. *New Phytologist* 226, 1567–1572.

Jagadish, S.V.K., Craufurd, P.Q. and Wheeler, T.R. (2007) High temperature stress and spikelet fertility in rice (*Oryza sativa* L.). *Journal of Experimental Botany* 58, 1627–1635.

Jagadish, S.V.K., Craufurd, P.Q. and Wheeler, T.R. (2008) Phenotyping parents of mapping populations of rice for heat tolerance during anthesis. *Crop Science* 48, 1140–1146.

Jagadish, S.V.K., Cairns, J., Lafitte, R., Wheeler, T.R., Price, A.H. *et al.* (2010) Genetic analysis of heat tolerance at anthesis in rice. *Crop Science* 50, 1633–1641.

Jagadish, S.V.K., Murty, M.V.R. and Quick, W.P. (2015) Rice responses to rising temperatures – challenges, perspectives and future directions. *Plant, Cell and Environment* 38, 1686–1698.

Jagadish, S.V.K., Bahuguna, R.N., Djanaguiraman, M., Gamuyao, R., Prasad, P.V.V. *et al.* (2016) Implications of high temperature and elevated CO$_2$ on flowering time in plants. *Frontiers in Plant Science* 7, 913.

Jiang, H., Dian, W. and Wu, P. (2003) Effect of high temperature on fine structure of amylopectin in rice endosperm by reducing the activity of the starch branching enzyme. *Phytochemistry* 63, 53–59.

Jiang, S.L., Wu, J.G., Feng, Y., Yang, X.E. and Shi, C.H. (2007) Correlation analysis of mineral element contents and quality traits in milled rice (*Oryza stavia* L.). *Journal of Agricultural and Food Chemistry* 55, 9608–9613.

Jiang, X., Hua, M., Yang, X., Hu, N., Qiu, R. *et al.* (2020) Impacts of mist spray on rice field micrometeorology and rice yield under heat stress condition. *Scientific Reports* 10, 1579.

Julia, C. and Dingkuhn, M. (2012) Variation in time of day of anthesis in rice in different climatic environments. *European Journal of Agronomy* 43, 166–174.

Kadam, N.N., Jagadish, K.S., Xiao, G., Melgar, R.J., Bahuguna, R.N. *et al.* (2014) Agronomic and physiological responses to high temperature, drought, and elevated CO$_2$ interactions in cereals. *Advances in Agronomy* 127, 111–156.

Kang, D.-J., Yeo, U.-S., Oh, B.-G., Kang, J.-H., Yang, S.-J. *et al.* (2003) Physiological studies on the foehn tolerance of rice (*Oryza sativa* L.). *Japanese Journal of Crop Science* 72, 328–332.

Kang, Z., Qin, T. and Zhao, Z. (2019) Overexpression of the zinc finger protein gene OsZFP350 improves root development by increasing resistance to abiotic stress in rice. *Acta Biochimica Polonica* 66, 183–190.

Khan, S., Anwar, S., Ashraf, M.Y., Khaliq, B., Sun, M. *et al.* (2019) Mechanisms and adaptation strategies to improve heat tolerance in rice. A review. *Plants* 8, 508.

Kilasi, N.L., Singh, J., Vallejos, C.E., Ye, C., Jagadish, S.V.K. *et al.* (2018) Heat stress tolerance in rice (*Oryza sativa* L.): identification of quantitative trait loci and candidate genes for seedling growth under heat stress. *Frontiers in Plant Science* 9, 1578.

Kim, J.H., Lim, S.D. and Jang, C.S. (2019) *Oryza sativa* heat-induced RING finger protein 1 (OsHIRP1) positively regulates plant response to heat stress. *Plant Molecular Biology* 99, 545–559.

Koh, S., Lee, S.C., Kim, M.K., Koh, J.H., Lee, S. *et al.* (2007) T-DNA tagged knockout mutation of rice OsGSK1, an orthologue of Arabidopsis BIN2, with enhanced tolerance to various abiotic stresses. *Plant Molecular Biology* 65, 453–466.

Krishnan, P., Ramakrishnan, B., Reddy, K.R. and Reddy, V.R. (2011) High-temperature effects on rice growth, yield, and grain quality. *Advances in Agronomy* 111, 87–205.

Lee, S.Y., Kim, Y.H. and Lee, G.S. (2017) Mapping QTLs associated to germination stability following dry-heat treatment in rice seed. *3 Biotech* 7, 220.

Li, J., Yang, J., Zhu, B. and Xie, G. (2019) Overexpressing *OsFBN1* enhances plastoglobule formation, reduces grain-filling percent and jasmonate levels under heat stress in rice. *Plant Science* 285, 230–238.

Li, X., Chen, R., Chu, Y., Huang, J., Jin, L. *et al.* (2018) Overexpression of RCc3 improves root system architecture and enhances salt tolerance in rice. *Plant Physiology and Biochemistry* 130, 566–576.

Li, X.-M., Chao, D.-Y., Wu, Y., Huang, X., Chen, K. *et al.* (2015) Natural alleles of a proteasome α2 subunit gene contribute to thermotolerance and adaptation of African rice. *Nature Genetics* 47, 827–833.

Lin, Z., Zheng, D., Zhang, X., Wang, Z., Lei, J. *et al.* (2016) Chalky part differs in chemical composition from translucent part of japonica rice grains as revealed by a notched-belly mutant with white-belly. *Journal of the Science of Food and Agriculture* 96, 3937–3943.

Liu, J., Zhang, C., Wei, C., Liu, X., Wang, M. *et al.* (2016) The RING finger ubiquitin E3 ligase OsHTAS enhances heat tolerance by promoting H$_2$O$_2$-induced stomatal closure in rice. *Plant Physiology* 170, 429–443.

Liu, J., Sun, X., Xu, F., Zhang, Y., Zhang, Q. *et al.* (2018) Suppression of *OsMDHAR4* enhances heat tolerance by mediating H$_2$O$_2$-induced stomatal closure in rice plants. *Rice* 11, 38.

Liu, K., Deng, J., Lu, J., Wang, X., Lu, B. *et al.* (2019) High nitrogen levels alleviate yield loss of super hybrid rice caused by high temperatures during the flowering stage. *Frontiers in Plant Science* 10, 357.

Liu, X.H., Lyu, Y.S., Yang, W., Yang, Z.T., Lu, S.J. *et al.* (2020) A membrane-associated NAC transcription factor OsNTL3 is involved in thermotolerance in rice. *Plant Biotechnology Journal* 18, 1317–1329.

Lo, S.F., Cheng, M.L., Hsing, Y.I.C., Chen, Y.S., Lee, K.W. *et al.* (2020) Rice Big Grain 1 promotes cell division to enhance organ development, stress tolerance and grain yield. *Plant Biotechnology Journal* 18, 1969–1983.

Lyman, N.B., Jagadish, K.S.V., Nalley, L.L., Dixon, B.L. and Siebenmorgen, T. (2013) Neglecting rice milling yield and quality underestimates economic losses from high-temperature stress. *PLoS One* 8, e72157.

Madan, P., Jagadish, S.V.K., Craufurd, P.Q., Fitzgerald, M., Lafarge, T. *et al.* (2012) Effect of elevated CO_2 and high temperature on seed-set and grain quality of rice. *Journal of Experimental Botany* 63, 3843–3852.

Makino, T., Kamiya, T., Takano, H., Itou, T., Sekiya, N. *et al.* (2007) Remediation of cadmium-contaminated paddy soils by washing with calcium chloride: verification of on-site washing. *Environmental Pollution* 147, 112–119.

Matsui, T. and Kagata, H. (2003) Characteristics of floral organs related to reliable self pollination in rice (*Oryza sativa* L.). *Annals of Botany* 91, 473–477.

Matsui, T., Omasa, K. and Horie, T. (2001) The difference in sterility due to high temperatures during the flowering period among japonica-rice varieties. *Plant Production Science* 4, 90–93.

Matsui, T., Kobayasi, K., Kagata, H. and Horie, T. (2005) Correlation between viability of pollination and length of basal dehiscence of the theca in rice under a hot-and-humid condition. *Plant Production Science* 8, 109–114.

Matsui, T., Kobayasi, K., Nakagawa, H., Yoshimoto, M., Hasegawa, T. *et al.* (2014) Lower-than-expected floret sterility of rice under extremely hot conditions in a flood-irrigated field in New South Wales, Australia. *Plant Production Science* 17, 245–252.

Mishra, N., Srivastava, A.P., Esmaeili, N., Hu, W. and Shen, G. (2018) Overexpression of the rice gene os-siz1 in *Arabidopsis* improves drought-, heat-, and salt-tolerance simultaneously. *PLoS One* 13, e0201716.

Mitsui, T., Yamakawa, H. and Kobata, T. (2016) Molecular physiological aspects of chalking mechanism in rice grains under high-temperature stress. *Plant Production Science* 19, 22–29.

Mittler, R. and Blumwald, E. (2010) Genetic engineering for modern agriculture: challenges and perspectives. *Annual Review of Plant Biology* 61, 443–462.

Miura, K., Ashikari, M. and Matsuoka, M. (2011) The role of QTLs in the breeding of high-yielding rice. *Trends in Plant Science* 16, 319–326.

Mohammed, A.R. and Tarpley, L. (2009) Impact of high nighttime temperature on respiration, membrane stability, antioxidant capacity, and yield of rice plants. *Crop Science* 49, 313–322.

Mohammed, A.R. and Tarpley, L. (2011) Characterization of rice (*Oryza sativa* L.) physiological responses to α-tocopherol, glycine betaine or salicylic acid application. *Journal of Agricultural Science* 3, 3–13.

Müller, F. and Rieu, I. (2016) Acclimation to high temperature during pollen development. *Plant Reproduction* 29, 107–118.

Myers, S.S., Zanobetti, A., Kloog, I., Huybers, P., Leakey, A.D. *et al.* (2014) Increasing CO_2 threatens human nutrition. *Nature* 510, 139–142.

Nevame, A.Y.M., Emon, R.M., Malek, M.A., Hasan, M.M., Alam, M.A. *et al.* (2018) Relationship between high temperature and formation of chalkiness and their effects on quality of rice. *BioMed Research International* 2018, 1653721.

Oh-e, I., Saitoh, K. and Kuroda, T. (2007) Effects of high temperature on growth, yield and dry-matter production of rice grown in the paddy field. *Plant Production Science* 10, 412–422.

Panozzo, J.F., Walker, C.K., Partington, D.L., Neumann, N.C., Tausz, M. *et al.* (2014) Elevated carbon dioxide changes grain protein concentration and composition and compromises baking quality. A face study. *Journal of Cereal Science* 60, 461–470.

Park, H.J., You, Y.N., Lee, A., Jung, H., Jo, S.H. *et al.* (2020) OsFKBP20-1b interacts with the splicing factor OsSR45 and participates in the environmental stress response at the post-transcriptional level in rice. *Plant Journal* 102, 992–1007.

Peng, S., Huang, J., Sheehy, J.E., Laza, R.C., Visperas, R.M. *et al.* (2004) Rice yields decline with higher night temperature from global warming. *Proceedings of the National Academy of Sciences USA* 101, 9971–9975.

Plaut, Z., Butow, B.J., Blumenthal, C.S. and Wrigley, C.W. (2004) Transport of dry matter into developing wheat kernels and its contribution to grain yield under post-anthesis water deficit and elevated temperature. *Field Crops Research* 86, 185–198.

Pravallika, K., Arunkumar, C., Vijaykumar, A., Beena, R. and Jayalekshmi, V.G. (2020) Effect of high temperature stress on seed filling and nutritional quality of rice (*Oryza sativa* L.). *Journal of Crop and Weed* 16, 18–23.

Ps, S., Sv, A.M., Prakash, C., Mk, R., Tiwari, R. *et al.* (2017) High resolution mapping of QTLs for heat tolerance in rice using a 5K SNP array. *Rice* 10, 28.

Qiao, B., Zhang, Q., Liu, D., Wang, H., Yin, J. *et al.* (2015) A calcium-binding protein, rice annexin OsANN1, enhances heat stress tolerance by modulating the production of H_2O_2. *Journal of Experimental Botany* 66, 5853–5866.

Qiu, Z., Kang, S., He, L., Zhao, J., Zhang, S. *et al.* (2018) The newly identified heat-stress sensitive albino 1 gene affects chloroplast development in rice. *Plant Science* 267, 168–179.

Ray, D.K., Mueller, N.D., West, P.C. and Foley, J.A. (2013) Yield trends are insufficient to double global crop production by 2050. *PLoS One* 8, e66428.

Ren, H., Bao, J., Gao, Z., Sun, D., Zheng, S. *et al.* (2023) How rice adapts to high temperatures. *Frontiers in Plant Science* 14, 1–12.

Rowland-Bamford, A.J., Baker, J.T., Allen, L.H., Jr and Bowes, G. (1996) Interactions of CO_2 enrichment and temperature on carbohydrate accumulation and partitioning in rice. *Environmental and Experimental Botany* 36, 111–124.

Roy, K.S., Bhattacharyya, P., Neogi, S., Rao, K.S. and Adhya, T.K. (2012) Combined effect of elevated CO_2 and temperature on dry matter production, net assimilation rate, C and N allocations in tropical rice (*Oryza sativa* L.). *Field Crops Research* 139, 71–79.

Saidi, Y., Peter, M., Finka, A., Cicekli, C., Vigh, L. *et al.* (2010) Membrane lipid composition affects plant heat sensing and modulates Ca^{2+}-dependent heat shock response. *Plant Signaling and Behavior* 5, 1530–1533.

Satake, T. and Yoshida, S. (1978) High temperature-induced sterility in indica rices at flowering. *Japanese Journal of Crop Science* 47, 6–17.

Scafaro, A.P., Haynes, P.A. and Atwell, B.J. (2010) Physiological and molecular changes in *Oryza meridionalis* Ng., a heat-tolerant species of wild rice. *Journal of Experimental Botany* 61, 191–202.

Scafaro, A.P., Atwell, B.J., Muylaert, S., Reusel, B.V., Ruiz, G.A. *et al.* (2018) A thermotolerant variant of Rubisco activase from a wild relative improves growth and seed yield in rice under heat stress. *Frontiers in Plant Science* 9, 1663.

Seck, P.A., Diagne, A., Mohanty, S. and Wopereis, M.C.S. (2012) Crops that feed the world 7: Rice. *Food Security* 4, 7–24.

Seneweera, S. (2011) Effects of elevated CO_2 on plant growth and nutrient partitioning of rice (*Oryza sativa* L.) at rapid tillering and physiological maturity. *Journal of Plant Interaction* 6, 35–42.

Seneweera, S. and Conroy, J.P. (1997) Growth, grain yield and quality of rice (*Oryza sativa* L.) in response to elevated CO_2 and phosphorus nutrition. *Soil Science and Plant Nutrition* 43, 1131–1136.

Sheehy, J.E., Dionora, M.J.A. and Mitchell, P.L. (2001) Spikelet numbers, sink size and potential yield in rice. *Field Crops Research* 71, 77–85.

Shen, H., Zhong, X., Zhao, F., Wang, Y., Yan, B. *et al.* (2015) Overexpression of receptor-like kinase ERECTA improves thermotolerance in rice and tomato. *Nature Biotechnology* 33, 996–1003.

Shi, W., Muthurajan, R., Rahman, H., Selvam, J., Peng, S. *et al.* (2013) Source–sink dynamics and proteomic reprogramming under elevated night temperature and their impact on rice yield and grain quality. *New Phytologist* 197, 825–837.

Shi, W., Li, X., Schmidt, R.C., Struik, P.C., Yin, X. *et al.* (2018) Pollen germination and *in vivo* fertilization in response to high-temperature during flowering in hybrid and inbred rice. *Plant, Cell and Environment* 41, 1287–1297.

Shiraya, T., Mori, T., Maruyama, T., Sasaki, M., Takamatsu, T. *et al.* (2015) Golgi/plastid-type manganese superoxide dismutase involved in heat-stress tolerance during grain filling of rice. *Plant Biotechnology Journal* 13, 1251–1263.

Shirdelmoghanloo, H., Chen, K., Paynter, B.H., Angessa, T.T., Westcott, S. *et al.* (2022) Grain-filling rate improves physical grain quality in barley under heat stress conditions during the grain-filling period. *Frontiers in Plant Science* 13, 858652.

Shrestha, S., Mahat, J., Shrestha, J., Madhav, K.C. and Paudel, K. (2022) Influence of high-temperature stress on rice growth and development. a review. *Heliyon* 8, e12651.

Soda, N., Gupta, B.K., Anwar, K., Sharan, A., Govindjee, A.P. *et al.* (2018) Rice intermediate filament, OsIF, stabilizes photosynthetic machinery and yield under salinity and heat stress. *Scientific Reports* 8, 4072.

Sreenivasulu, N., Butardo, V.M., Jr, Misra, G., Cuevas, R.P., Anacleto, R. *et al.* (2015) Designing climate-resilient rice with ideal grain quality suited for high-temperature stress. *Journal of Experimental Botany* 66, 1737–1748.

Suriyasak, C., Oyama, Y., Ishida, T., Mashiguchi, K., Yamaguchi, S. *et al.* (2020) Mechanism of delayed seed germination caused by high temperature during grain filling in rice (*Oryza sativa* L.). *Scientific Reports* 10, 17378.

Takehara, K., Murata, K., Yamaguchi, T., Yamaguchi, K., Chaya, G. *et al.* (2018) Thermo-responsive allele of *sucrose synthase 3* (*Sus3*) provides high-temperature tolerance during the ripening stage in rice (*Oryza sativa* L.). *Breeding Science* 68, 336–342.

Tan, S. and Shen, S. (2016) Distribution of rice heat stress in the lower yangtze region in recent 32 years. *Jiangsu Academy of Agricultural Sciences* 44, 97–101.

Tang, S., Zhang, H., Li, L., Liu, X., Chen, L. *et al.* (2018) Exogenous spermidine enhances the photosynthetic and antioxidant capacity of rice under heat stress during early grain-filling period. *Functional Plant Biology* 45, 911.

Tang, Y., Gao, C.C., Gao, Y., Yang, Y., Shi, B. *et al.* (2020) OsNSUN2-mediated 5-methylcytosine mRNA modification enhances rice adaptation to high temperature. *Developmental Cell* 53, 272–286.

Teixeira, E.I., Fischer, G., van Velthuizen, H., Walter, C. and Ewert, F. (2013) Global hot-spots of heat stress on agricultural crops due to climate change. *Agricultural and Forest Meteorology* 170, 206–215.

Terao, T., Miura, S., Yanagihara, T., Hirose, T., Nagata, K. *et al.* (2005) Influence of free-air CO_2 enrichment (FACE) on the eating quality of rice. *Journal of the Science of Food and Agriculture* 85, 1861–1868.

Tian, X.H., Matsui, T., Li, S., Yoshimoto, M., Kobayasi, K. *et al.* (2010) Heat-induced floret sterility of hybrid rice (*Oryza sativa* L.) cultivars under humid and low wind conditions in the field of Jianghan Basin, China. *Plant Production Science* 13, 243–251.

Triboi, E. and Triboi-Blondel, A.M. (2002) Productivity and grain or seed composition: a new approach to an old problem. Invited paper. *European Journal of Agronomy* 16, 163–186.

Uchida, A., Jagendorf, A.T., Hibino, T., Takabe, T. and Takabe, T. (2002) Effects of hydrogen peroxide and nitric oxide on both salt and heat stress tolerance in rice. *Plant Science* 163, 515–523.

Uprety, D.C., Sen, S. and Dwivedi, N. (2010) Rising atmospheric carbon dioxide on grain quality in crop plants. *Physiology and Molecular Biology of Plants* 16, 215–227.

Usui, Y., Sakai, H., Tokida, T., Nakamura, H., Nakagawa, H. *et al.* (2014) Heat-tolerant rice cultivars retain grain appearance quality under free-air CO_2 enrichment. *Rice* 7, 6.

Wada, H., Masumoto-Kubo, C., Gholipour, Y., Nonami, H., Tanaka, F. *et al.* (2014) Rice chalky ring formation caused by temporal reduction in starch biosynthesis during osmotic adjustment under foehn-induced dry wind. *PLoS One* 9, e110374.

Wada, H., Hatakeyama, Y., Onda, Y., Nonami, H., Nakashima, T. *et al.* (2019) Multiple strategies for heat adaptation to prevent chalkiness in the rice endosperm. *Journal of Experimental Botany* 70, 1299–1311.

Wang, B., Zhong, Z., Wang, X., Han, X., Yu, D. *et al.* (2020) Knockout of the OsNAC006 transcription factor causes drought and heat sensitivity in rice. *International Journal of Molecular Sciences* 21, 2288.

Wang, C., Caragea, D., Kodadinne Narayana, N., Hein, N.T., Bheemanahalli, R. *et al.* (2022a) Deep learning based high-throughput phenotyping of chalkiness in rice exposed to high night temperature. *Plant Methods* 18, 9.

Wang, D., Qin, B., Li, X., Tang, D., Zhang, Y.E. *et al.* (2016) Nucleolar DEAD-box RNA helicase TOGR1 regulates thermotolerant growth as a pre-rRNA chaperone in rice. *PLoS Genetics* 12, e1005844.

Wang, F., Tang, Z., Wang, Y., Fu, J., Yang, W. *et al.* (2022b) *Leaf Mutant 7* encoding heat shock protein *OsHSP40* regulates leaf size in rice. *International Journal of Molecular Sciences* 23, 4446.

Wang, H., Lu, S., Guan, X., Jiang, Y., Wang, B. *et al.* (2022c) *Dehydration-responsive element binding protein 1C, 1E, and 1G promote stress tolerance to chilling, heat, drought, and salt in rice. Frontiers in Plant Science* 13, 851731.

Wang, Y., Frei, M., Song, Q. and Yang, L. (2011) The impact of atmospheric CO_2 concentration enrichment on rice quality – a research review. *Acta Ecologica Sinica* 31, 277–282.

Weerakoon, W.M.W., Maruyama, A. and Ohba, K. (2008) Impact of humidity on temperature-induced grain sterility in rice (*Oryza sativa* L.). *Journal of Agronomy and Crop Science* 194, 135–140.

Wei, D., Ye, G., Pan, J., Xiang, J., Huang, J. and Nie, L. (2012) QTL mapping for nitrogen-use efficiency and nitrogen-deficiency tolerance traits in rice. *Plant and Soil* 359, 281–295.

Wei, H., Liu, J., Wang, Y., Huang, N., Zhang, X. *et al.* (2013) A dominant major locus in chromosome 9 of rice (*Oryza sativa* L.) confers tolerance to 48°C high temperature at seedling stage. *Journal of Heredity* 104, 287–294.

Wu, C., Cui, K., Li, Q., Li, L., Wang, W. *et al.* (2021) Estimating the yield stability of heat-tolerant rice genotypes under various heat conditions across reproductive stages: a 5-year case study. *Scientific Reports* 11, 1–11.

Wu, X, Shiroto, Y., Kishitani, S., Ito, Y. and Toriyama, K. (2009) Enhanced heat and drought tolerance in transgenic rice seedlings overexpressing OsWRKY11 under the control of HSP101 promoter. *Plant Cell Reports* 28, 21–30.

Wu, X., Cai, X., Zhang, B., Wu, S., Wang, R. *et al.* (2022) ERECTA regulates seed size independently of its intracellular domain via MAPK-DA1-UBP15 signaling. *Plant Cell* 34, 3773–3789.

Xu, J., Henry, A. and Sreenivasulu, N. (2020a) Rice yield formation under high day and night temperatures – a prerequisite to ensure future food security. *Plant, Cell and Environment* 43, 1595–1608.

Xu, Y., Zhang, L., Ou, S., Wang, R., Wang, Y. *et al.* (2020b) Natural variations of SLG1 confer high-temperature tolerance in indica rice. *Nature Communications* 11, 5441.

Yamakawa, H. and Hakata, M. (2010) Atlas of rice grain filling-related metabolism under high temperature: joint analysis of metabolome and transcriptome demonstrated inhibition of starch accumulation and induction of amino acid accumulation. *Plant and Cell Physiology* 51, 795–809.

Yang, J. and Zhang, J. (2006) Grain filling of cereals under soil drying. *New Phytologist* 169, 223–236.

Yang, J. and Zhang, J. (2010) Grain-filling problem in 'super' rice. *Journal of Experimental Botany* 61, 1–5.

Ye, C., Argayoso, M.A., Redoña, E.D., Sierra, S.N., Laza, M.A. *et al.* (2012) Mapping QTL for heat tolerance at flowering stage in rice using SNP markers. *Plant Breeding* 131, 33–41.

Ye, C., Tenorio, F.A., Redoña, E.D., Morales–Cortezano, P.S., Cabrega, G.A. *et al.* (2015) Fine-mapping and validating qHTSF4.1 to increase spikelet fertility under heat stress at flowering in rice. *Theoretical and Applied Genetics* 128, 1507–1517.

Ye, C., Ishimaru, T., Lambio, L., Li, L., Long, Y. *et al.* (2022) Marker-assisted pyramiding of QTLs for heat tolerance and escape upgrades heat resilience in rice (*Oryza sativa* L.). *Theoretical and Applied Genetics* 135, 1345–1354.

Yoshida, S. (1981) *Fundamentals of Rice Crop Science*. International Rice Research Institute, Los Banos, California.

Yoshimoto, M., Fukuoka, M., Hasegawa, T., Utsumi, M., Ishigooka, Y. *et al.* (2011) Integrated micrometeorology model for panicle and canopy temperature (IM^{2}PACT) for rice heat stress studies under climate change. *Journal of Agricultural Meteorology* 67, 233–247.

Zaidi, S.H.R., Zakari, S.A., Zhao, Q., Khan, A.R., Shah, J.M. *et al.* (2019) Anthocyanin accumulation in black kernel mutant rice and its contribution to ROS detoxification in response to high temperature at the filling stage. *Antioxidants* 8, 510.

Zeng, Y.W., Shen, S.Q., Wang, L.X., Liu, J.F., Pu, X.Y. *et al.* (2005) Correlation of plant morphological and grain quality traits with mineral element contents in Yunnan rice. *Rice Science* 12, 101–106.

Zhang, C.X., Feng, B.H., Chen, T.T., Zhang, X.F., Tao, L.X. *et al.* (2017) Sugars, antioxidant enzymes and IAA mediate salicylic acid to prevent rice spikelet degeneration caused by heat stress. *Plant Growth Regulation* 83, 313–323.

Zhang, C., Li, G., Chen, T., Feng, B., Fu, W. *et al.* (2018) Heat stress induces spikelet sterility in rice at anthesis through inhibition of pollen tube elongation interfering with auxin homeostasis in pollinated pistils. *Rice* 11, 14.

Zhang, G., Chen, L., Zhang, S., Zheng, H. and Liu, G. (2009) Effects of high temperature stress on microscopic and ultrastructural characteristics of mesophyll cells in flag leaves of rice. *Rice Science* 16, 65–71.

Zhang, H., Xu, H., Jiang, Y., Zhang, H., Wang, S. *et al.* (2021) Genetic control and high temperature effects on starch biosynthesis and grain quality in rice. *Frontiers in Plant Science* 12, 757997.

Zhang, T., Yang, L. and Jang, K.F. (2008) QTL mapping for teat tolerance of the tassel period of rice. *Molecular Plant Breeding* 6, 867–873.

Zhao, L., Lei, J., Huang, Y., Zhu, S., Chen, H. *et al.* (2016) Mapping quantitative trait loci for heat tolerance at anthesis in rice using chromosomal segment substitution lines. *Breeding Science* 66, 358–366.

Zhao, Q., Zhou, L., Liu, J., Du, X., Asad, M.-A.-U. *et al.* (2018) Relationship of ROS accumulation and superoxide dismutase isozymes in developing anther with floret fertility of rice under heat stress. *Plant Physiology and Biochemistry* 122, 90–101.

Zhao, X. and Fitzgerald, M. (2013) Climate change: Implications for the yield of edible rice. *PLoS One* 8, e66218.

Zhou, H, Wang, X., Huo, C., Wang, H., An, Z. *et al.* (2019) A quantitative proteomics study of early heat-regulated proteins by two-dimensional difference gel electrophoresis identified OsUBP21 as a negative regulator of heat stress responses in rice. *Proteomics* 19, e1900153.

Zhou, H., Wang, Y., Zhang, Y., Xiao, Y., Liu, X. *et al.* (2022) Comparative analysis of heat-tolerant and heat-susceptible rice highlights the role of *OsNCED1* gene in heat stress tolerance. *Plants* 11, 1062.

Zhu, S., Huang, R., Wai, H.P., Xiong, H., Shen, X. *et al.* (2017) Mapping quantitative trait loci for heat tolerance at the booting stage using chromosomal segment substitution lines in rice. *Physiology and Molecular Biology of Plants* 23, 817–825.

Zhu, Z. and Zhang, H. (2020) It takes NSUN2 to beat the heat in rice. *Developmental Cell* 53, 253–254.

10 Reactive Oxygen Species: Their Generation and Signalling during Abiotic Stress in Plants

Bharati Swain[1], Lipun Sahoo[1], Vivekanand Tiwari[2], Chanchal Kumar[3], Naveen Chandra Joshi[4], Prashant Kumar Singh[5], Shweta[6], Alka Singh[7] and Deepanker Yadav[1]*

[1]*Department of Botany, Guru Ghasidas Vishwavidyalaya, Bilaspur, Chhattisgarh, India; [2]Institute of Plant Sciences, Agricultural Research Organization, Volcani, Israel; [3]Department of Forensic Science, Guru Ghasidas Vishwavidyalaya, Bilaspur, Chhattisgarh, India; [4]Amity Institute of Microbial Technology, Amity University, Noida, Uttar Pradesh, India; [5]Department of Biotechnology, Mizoram Central University, Mizoram, India; [6]Dwarika Prasad Girls Inter College, Prayagraj, Uttar Pradesh, India; [7]Department of Chemistry, Feroze Gandhi College, Raebareli, Uttar Pradesh, India*

Abstract

Plant respond to their environment through a very orchestrated signalling pathway, starting from reception and ending in effect. There are several components that mediate the signalling. Based on their timing, the signal can be primary or secondary. The signalling molecules could be a hormone, an ion, any biomolecule and various reactive oxygen species (ROS). ROS can be categorized as superoxide, peroxide, etc. and can be a by-product of various biochemical reactions mediated by electron transport, like photosynthesis and respiration. The ROS generated in these reactions at a certain level are detoxified by the redox regulatory mechanism in the cytosol and different organelles. This regulatory system is comprised of several enzymes such as hydrogen peroxidase, ascorbate peroxidase (APX), superoxide dismutase (SOD) and some non-enzymatic components. The plant shows ROS-mediated signalling in different abiotic stress responses. Certain proteins which are involved in redox regulation can sense these molecules and initiate various downstream responses like activation of different ion channels, causing a change in the membrane potential. These changes further lead to the activation of different Ca^{2+}-dependent pathways. ROS acts as a double-edged sword; at one point, its production during metabolic and other reactions is required for the signalling, but at another point, its increase beyond a certain level causes damage to different cellular components. So, the ROS cycle balances its production and scavenging through different feedback signalling processes. Different abiotic stresses finally lead to the generation of ROS. Understanding the different sites of ROS generation and ROS-mediated signalling during abiotic stress is important.

Keywords: Reactive oxygen species (ROS), abiotic stress, antioxidants, signalling, oxidative burst and enzymes

*Corresponding author: deepankerbhu@gmail.com

© CAB International 2025. *Soil Health and Nutrition Management*
(eds N.C. Joshi, T. Leustek and P.K. Singh)
DOI: 10.1079/9781800624597.0010

10.1   Introduction

Plants face several changes in their surroundings during their life cycles and they need to continually respond to these changes to maintain growth and development. They have evolved different strategies to survive different conditions. The adaptation to a certain environment can be done through changes at the genetic level. The other strategy is acclimation to short-term changes in the environment. For a plant to survive different stages of its life and successfully reproduce, it needs the help of several environmental cues. These signals are perceived by different parts of the plants and their cells. Environmental cues are perceived through various signalling molecule receptors, which are essential to the signalling cascade (Nair *et al.*, 2019). Conditions, such as high levels of light, create a disbalance due to the overwhelming of photosystems unable to convert light into chemical energy. Similarly, during drought conditions, water deficit in the plant leads to increased ionic toxicity of the cells (Ahuja *et al.*, 2010; Hussain *et al.*, 2018). Further, the NaCl stress causes increased sodium and chloride concentrations in the cell, denaturing many proteins and enzymes (Peleg *et al.*, 2011). Heat causes inhibition of enzyme activity, decoupling of many processes and accumulation of toxic metabolites. Heavy metals competes with essential minerals and can bind with different macromolecules, disrupting their normal functioning. Ultimately, all these abiotic stress conditions lead to the inhibition of photosynthesis due to damage to different components involved in light reactions and carbon fixation (Rossel *et al.*, 2007).

The reactive oxygen species (ROS) generated during stress is the reason behind the damage to different cellular components. These highly reactive oxygen species react and oxidize a wide range of cellular components, including DNA, RNA, proteins and lipids. The ROS have a dual activity; they can have both beneficial and detrimental effects. The positive effect of ROS accumulation is that it activates the signal transduction pathway, which induces the acclimation process. Receptors, such as histidine kinase, receptor-like kinases, G-protein-coupled receptors, receptors for various ROS and other stress-related metabolites and signalling molecules, enable plants to perceive and respond to these

stress conditions (Devireddy *et al.*, 2021). The signalling molecules can be an ion, or an inorganic or organic compound. Plant hormones are master regulators of any physiological and biochemical processes; they are sitting at the top of the signalling cascade. The receptors' perception of these signalling molecules initiates different signalling cascades in the plant. In the second line, there are some inorganic elements like Ca^{2+} and ROS (e.g. superoxide, peroxide, hydroxyl radical) and a few organic molecules like cyclic nucleotide (cyclic adenosine monophosphate and cyclic guanosine monophosphate), inositol triphosphate (IP) and diacylglycerol, etc., which play an important role in signal transduction (Dodd *et al.*, 2010; Czarnocka and Karpiński, 2018; Nair *et al.*, 2019).

The ROS is generally a by-product of many biochemical reactions. Initially, it was considered as a toxic by-product of aerobic metabolism, but it is currently recognized as an essential signalling molecule in the abiotic stress response (Mittler, 2017). NADPH oxidase is a crucial protein associated with ROS-mediated signalling. It is also called a respiratory burst oxidase homologue (RBOH) (Suzuki *et al.*, 2011). Many ROS-detoxifying components of plant cells and their organelles regulate the level of ROS. ROS originates at different sites of the cell where redox reactions take place. It is very important to scavenge the ROS occasionally to avoid the damage caused to the different cell components (Apel and Hirt, 2004; Hasanuzzaman *et al.*, 2020a; Sachdev *et al.*, 2023). At the same time, these molecules also help signal and report the state of any cell organelle. Chloroplasts, peroxisome and mitochondria are the main sources of ROS formation because of their participation in photorespiration (Singh *et al.*, 2019). During abiotic stress, ROS waves cause rapid systemic signalling. The ROS wave is created through a self-propagating wave of ROS production. It begins at the stress site and spreads in the plant. RBOH on the cell surface helps to produce ROS throughout the path (Suzuki *et al.*, 2013). This self-propagating ROS wave helps systemic acquired acclimation to various abiotic stresses like salinity, cold, drought, heat and high light (Mittler *et al.*, 2011, 2022). Due to the crucial role of ROS in signalling during different abiotic stress responses in plants, this chapter will discuss different sites of ROS production inside and outside the cell and the mechanism of ROS transport in the plant. Their negative effect on cellular components is also discussed as is ROS-mediated signalling and its components.

10.2   Different Sites of ROS Generation

The generation of ROS is crucial in plant cells because it is closely related to fundamental metabolic activities. ROS generation and ROS scavenging are interrelated, and cell survival is dependent on this relationship (Choudhary *et al.*, 2020). Subcellular ROS metabolism and its conversion from one form to another, which differs for various ROS types, cellular compartments and even cell types, are necessary for the increased production of ROS under stress situations (Hasanuzzaman *et al.*, 2020b). ROS are generated under both normal and extreme stress conditions at different places in the chloroplasts, peroxisomes, mitochondria, plasma membranes, cell wall and endoplasmic reticulum (Das and Roychoudhury, 2014). ROS are produced in both light and dark conditions. Chloroplasts and peroxisomes serve as crucial locations for the generation of ROS under light, whereas mitochondria are the major producers under dark condition (Foyer and Noctor, 2003; Choudhary *et al.*, 2017).

10.2.1   Chloroplasts

Chloroplasts are the centre of the fixation of CO_2 in reduced hydrocarbons and in generating different precursors of various metabolic pathways. Due to its involvement in transforming light energy to chemical energy, achieved by sequential redox reactions and electron transport, under normal conditions, this flow generates the H^+ ions in the thylakoid. The H^+ gradient is utilized for transport across the thylakoid membrane. However, in adverse conditions like high light, cold and low CO_2, the inhibition of the CO_2 sequestration machinery's excess excitation energy cannot be converted into chemical forms. Different ROS are produced as a result of this imbalance. In normal light conditions, chloroplasts generate ($^1O_2$) species continuously. However, excess light conditions can enhance the synthesis of singlet oxygen species in the chloroplast (Fryer, 2002; Mullineaux *et al.*, 2006; Triantaphylidès and Havaux, 2009). Different stress conditions can lead to increased generation of different ROS species in the chloroplast (Fig. 10.1; Table 10.1). Different proteins can sense this ROS and send the signal to the cell of the redox condition of the chloroplast. Another ROS generated in the chloroplast is the superoxide $O_2^{\cdot-}$. This can be generated due to the incomplete oxidation of O^{-2} after the H_2O photolysis. The excited electron from the photosystem I (PSI), instead of transferring to $NADP^+$, is received by ferredoxin to reduce O_2 to create $O_2^{\cdot-}$. In the chloroplast, the enzyme superoxide dismutase converts the $O_2^{\cdot-}$ to H_2O_2. Finally, the peroxidase present in chloroplast reduces the H_2O_2 in H_2O (Mullineaux *et al.*, 2006; Chang *et al.*, 2009).

10.2.2   Peroxisome

The involvement of peroxisome in photorespiration makes it another site of ROS generation inside the cell (Fig. 10.1; Table 10.1). The generation of glycolate is due to Rubisco's oxidase activity in the chloroplast and its transfer into the peroxisome. The further recycling of glycolate to glycerate through the glyoxalate form by glycolate oxidase leads to the generation of H_2O_2 in the peroxisome (Wituszynska and Karpinski, 2013). The ROS produced in the peroxisome plays a role in seed germination, pollen germination and stomatal movement (Slesak *et al.*, 2012). ROS generated in peroxisomes also help the response to stress conditions like heat, low CO_2 and pathogens. ROS possess different enzymes to minimize the increased ROS levels due to stress conditions (Corpas *et al.*, 2017).

10.2.3   Mitochondria

Some amount of ROS is generated in the mitochondria under normal conditions but its production may increase under different stress conditions (Fig. 10.1; Table 10.1). This results in cell death. This happens during abiotic stresses like heat, drought, hypoxia, heavy metal and biotic stress (Vanlerberghe, 2013; Petrov *et al.*, 2015; Gupta and Igamberdiev, 2016). Mitochondria of

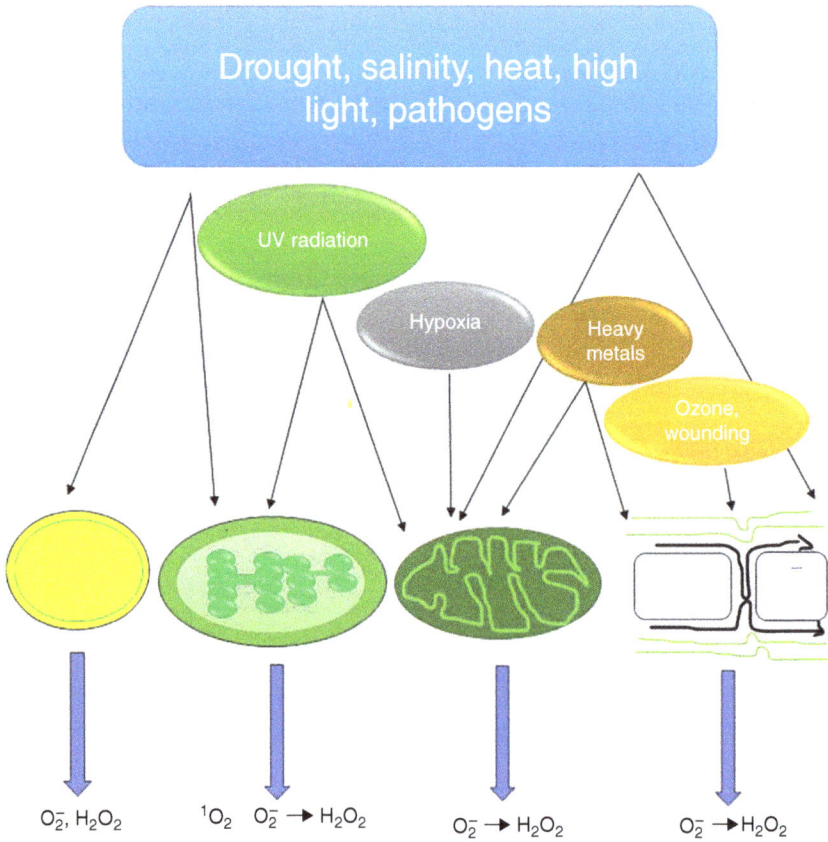

Fig. 10.1. Reactive oxygen species production in response to various environmental conditions in chloroplasts, peroxisomes, mitochondria and apoplasts (Czarnocka and Karpiński, 2018).

plants have alternative oxidase (AOX), which can reduce the O_2 to H_2O without the involvement of complex III, cytochrome c and complex IV. It provides the alternative electron flow method that bypasses the regular electron transport pathway. Thus, AOX1 helps reduce the electron transport chain load (Vanlerberghe, 2013).

10.2.4 Plasma membranes and cell walls

The plasma membrane also works as a site of ROS. Plants have NADPH oxidase in the plasma membrane (Fig. 10.2; Table 10.1). Its apoplastic domain generates superoxide in the apoplast. Its N-terminal regulatory domain is directed towards the cytoplasm (Daudi *et al.*, 2012;

Jakubowska *et al.*, 2015). ROS generated in the apoplast are important in regulating various physiological processes. They also influence the expression of several genes and act as a biological signal (Foreman *et al.*, 2003). Some enzymes in the plant cell wall also generate ROS. These enzymes include class III peroxidases, amine oxidases and quinone reductase (O'Brien *et al.*, 2012).

10.2.5 Other probable origins of ROS

Reactive oxygen species can be generated in the endoplasmic reticulum (ER) because it offers an oxidizing environment to produce disulfide bonds during the folding process of proteins. It has been found that the inner ER membrane-associated

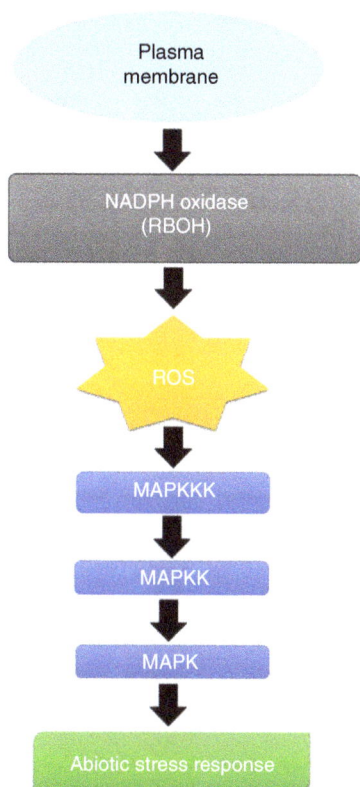

Fig. 10.2. Diagrammatic illustration of reactive oxygen species (ROS)-mediated signalling by activating the mitogen-activated protein kinase (MAPK) cascade under abiotic stress conditions. ROS is a messenger that the respiratory burst oxidase homologue (RBOH) generates in response to an abiotic stressor. ROS molecules cause the activation of MAPKs. The plant responds to abiotic stress by regulating MAPK signalling (Jalmi and Sinha, 2015).

enzyme ER oxidoreductase 1 (ERO1) catalyses the generation of disulfide bonds in secretory proteins and the subsequent production of H_2O_2 (Ozgur et al., 2014; Czarnocka and Karpiński, 2018). Isolated nuclei can also produce H_2O_2 when exposed to calcium (Ashtamker et al., 2007). The action of NADPH oxidase in plants and animals is known to be associated with the plasma membrane. The enzyme generates superoxide either outside the plasma membrane or inside the endosomes by accepting electrons from NADPH at the cytosolic region of the membrane and donating them to molecular oxygen at the opposite side (Leshem et al., 2007).

10.3 Biomolecules Targeted by ROS

Overproduction of ROS causes oxidative bursts and damages biomolecules in unfavourable environmental conditions (Sachdev et al., 2021). ROS rapidly deactivates enzymes, damages important plant cellular organelles and breaks down membranes, ultimately resulting in cell death. ROS also trigger the degradation of pigments, nucleic acids, proteins and lipids (Akeel and Jaleel, 2023). The concentration of a particular biomolecule, the target biomolecule's location relative to the site of ROS generation and the rate constant for the reaction between the target biomolecules are some factors that determine the extent of damage to the biomolecules (Davies, 2005).

10.3.1 Proteins

Proteins regulate plant physiological characteristics, crucial in regulating abiotic stress tolerance (Kosová et al., 2018). Proteins are more vulnerable to oxidation than other biological molecules because of their high rate constants for the reaction and their abundance in living systems (Sachdev et al., 2021). The relatively low oxidation rate causes limited harm to the protein backbone targeted by non-radical antioxidants (Alché, 2019). The site-specific alteration of amino acids such as Lys, Arg, Thr, Pro and Trp and an enhanced susceptibility to proteolytic degradation occur when the ROS concentration exceeds its threshold value (Møller et al., 2007). Protein oxidation is caused by several major ROS, including hydroxyl ($OH^{\cdot}$), alkoxyl ($RO^{\cdot}$), peroxyl ($RO_2^{\cdot}$), hydroperoxyl ($HO_2^{\cdot}$), superoxide ($O_2^{\cdot-}$) and non-radical species like hydrogen peroxide (H_2O_2), singlet oxygen ($^1O_2$), ozone (O_3) and hypochlorous acid ($HOCl$) (Dean et al., 1997). The oxidation of amino acid side chains, specifically those containing thiol groups and sulfur (S) (e.g. oxidation of Met and Cys residue by $OH^{\cdot}$ and $^1O_2$), and the degradation of the peptide backbone, which results in nitrosylation, carbonylation, the formation of disulfide bonds

Table 10.1. Generation of reactive oxygen species (ROS) in different organelles and their involvement in plant stress responses.

ROS	Source of ROS	Stress response	Reference
Superoxide ($O_2^{\cdot-}$)	Mitochondria, chloroplast, peroxisome, plasma membrane, cell wall	Drought, salinity, high light intensity	Tiwari *et al.*, 2017; Choudhary *et al.*, 2020
Hydrogen peroxide (H_2O_2)	Mitochondria, chloroplast, peroxisome, plasma membrane, cell wall	Drought, heat stress, UV radiation, salt	Tiwari *et al.*, 2017; Choudhary *et al.*, 2020
Hydroxyl radical (OH$^{\cdot}$)	Mitochondria, chloroplast, plasma membrane	Salinity, drought	Thomas *et al.*, 2009; Richards *et al.*, 2015; Hernansanz-Agustín and Enríquez, 2021
Alkoxyl (RO$^{\cdot}$)	Plasma membrane	Drought, salinity	Bhattacharjee, 2005; Gill and Tuteja, 2010; Impa *et al.*, 2012
Hydroperoxyl ($HO_2^{\cdot}$)	Chloroplast	Drought, salinity	Gill and Tuteja, 2010; Pospíšil, 2016; Rohman *et al.*, 2019
Singlet oxygen ($^1O_2$)	Plasma membrane, chloroplast, mitochondria, cell wall	Light stress, salinity, drought, heavy metal	Bhattacharjee, 2005; Das and Roychoudhury, 2014; Mittler *et al.*, 2022

and glutathionylation, which modifies protein activity, are involved in direct oxidation by ROS (Das and Roychoudhury, 2014).

10.3.2 Lipid membranes

Lipids make up a significant portion of the plasma membrane, which envelops the cell and helps in its capacity to respond to its variable environment. However, in stressful situations, lipid peroxidation becomes so harmful that it is often considered the sole means of evaluating lipid breakdown when the level of ROS exceeds the threshold value (Das and Roychoudhury, 2014). The ROS, also known as free oxygen radicals, are generally the leading cause of plant lipid peroxidation. The process of lipid peroxidation involves the chemical destruction of polyunsaturated fatty acids (PUFAs) in lipids by oxygen and free radicals, leading to the production of lipoperoxides (Anjum *et al.*, 2015). In lipid peroxidation, the 1,4-pentadiene structure of PUFAs, which is the primary target of the ROS attack on lipids, is either in the free form or esterified to glycerol or cholesterol (Browne and Armstrong, 2002). ROS like $^1O_2$ and OH$^{\cdot}$ are prone to damage PUFAs like linoleic and linolenic acid (Das and Roychoudhury,

2014). The three steps of lipid peroxidation are initiation, propagation and termination. During the initiation step, ROS are produced by the decline of O_2. The ROS produced trigger a series of reactions that produce lipid radicals, and malondialdehyde corresponds to the second stage, propagation. The last step is termination where lipid dimers are ultimately formed from lipid radicals (El-Beltagi and Mohamed, 2013; Das and Roychoudhury, 2014; Anjum *et al.*, 2015; Sachdev *et al.*, 2021).

10.3.3 DNA

Reactive oxygen species are the primary factor causing DNA damage (Imlay and Linn, 1988). ROS have the capability to oxidatively damage DNA present in mitochondria, nuclei and chloroplasts. Chloroplastic and mitochondrial DNA suffer the most from ROS attacks because they lack protective histones and are closely linked to the ROS-generating machinery; in contrast, nuclear DNA in plants is strongly shielded by histones and associated proteins (Halliwell, 2006; Das and Roychoudhury, 2014). ROS also causes the formation of the deoxyribose radical by removing an H atom from the deoxyribose sugar backbone and further breaking the DNA strand (Evans et al., 2004). The OH· radical can interact with purines, pyrimidine bases and even deoxyribose sugar, so it has been found to cause the greatest damage to DNA (Sharma et al., 2012; Soares et al., 2019). $^1O_2$ can react only with the guanine (Sachdev et al., 2021). $^1O_2$ can modify nucleic acids by selectively reacting with deoxyguanosine (Sharma et al., 2012). ROS-induced damage to nucleic acids can lead to variations in signal transduction pathways, synthesis of protein and membrane integrity by interfering with DNA replication and transcription. This can affect cell homeostasis and impair metabolic efficiency (Gill and Tuteja, 2010; Soares et al., 2019).

10.4 ROS Signalling in Plants

Chemically, ROS play a damaging role in different cellular components, but evidence shows its role in signalling (Foyer and Noctor, 2005). Together with Ca^{2+}, ROS play a crucial role in the signalling pathway (Table 10.1). These two amplify each other's signalling and play an essential role in information processing. Singlet oxygen produced in the chloroplast contributes to producing numerous secondary messengers. These secondary messengers elicit various stress-responsive pathways (Asada, 2006). The compounds include the oxidation product of β-carotene derivatives like β-cyclocitral, which is a reactive electrophile that can induce changes in gene expression (Laloi and Havaux, 2015). Different proteins mediate the $^1O_2$-dependent retrograde signalling (Keun et al., 2007). Moreover, transcription factors involved in redox-mediated retrograde signalling have also been identified (Foyer et al., 2014).

The ROS produced by plasma membrane-bound NADPH oxidase (RBOHs) works as a signal during abiotic stress (Laloi et al., 2004). The RBOHs work through a positive feedback mechanism. ROS produced outside the plasma membrane activate the Ca^{2+} channel, causing the influx of Ca^{2+} in the cytosol and an increase in cytosolic Ca^{2+}, which subsequently activates the Ca^{2+}-dependent protein kinase (CDPK). Ultimately, CDPK activates the RBOH to produce more ROS (Mittler et al., 2011; Suzuki et al., 2013; Gilroy et al., 2016). Plants deploy different NADPH

oxidases based on the different stresses they experience. Ten isoforms of RBOH genes have been observed in plants that produce distinct ROS species and, as a result, respond differently to various environmental stimuli (Monshausen et al., 2009; Müller et al., 2009; Pogány et al., 2009; Jalmi and Sinha, 2015). RBOH plays a crucial role in ROS generation in response to pathogens (Simon-Plas et al., 2002). RBOH can be associated with receptor-like kinase, and its phosphorylation is induced during the pathogen response (Kadota et al., 2014). These receptor-like kinases are required for full ROS production during the pathogen response. RBOH has been reported to interact with other regulatory elements (Chen et al., 2017).

The movement of ROS produced in the apoplast to the cytosol is another important factor that regulates ROS signalling. Different plasma membrane aquaporins have been reported to transport the ROS in the cytosol during the pathogen response (Carroll et al., 1998; Hara-Chikuma et al., 2015; Tian et al., 2016). The generation of ROS in adjacent cells can be initiated by the H_2O_2 generated in the apoplast under stress. This can begin a systemic signal called the ROS wave (Mittler et al., 2011; Gilroy et al., 2014). Studies have also indicated that ROS waves are necessary for systemic acquired acclimation (Karpinski et al., 1999) for various abiotic stresses (Zandalinas et al., 2020; Devireddy et al., 2021). The increased cytosolic Ca^{2+} activates RBOH, which can increase extracellular ROS, activating ROS-sensitive non-selective cation channels (NSCC) and annexins, causing the increase in Ca^{2+} present in the cytosol. The NSCC and RBOH jointly promote Ca^{2+} influx by a loop-based activation mechanism. Under stress conditions, annexins were also reported to mediate the rise in cytosolic Ca^{2+}, which is thought to transcriptionally activate salt overly sensitive 1 (SOS1) (Yadav et al., 2018). This annexin mediates a rise in Ca^{2+} levels in the cytosol. It is further sensed by CDPK, which can control the downstream components of Ca^{2+}-dependent abiotic stress signalling and expression of different stress-responsive genes (Yadav et al., 2018).

Plants may use mitochondrial-derived ROS as a signalling molecule (Vanlerberghe et al., 2009), and AOX can modulate their generation during stress. As NADPH oxidase-induced oxidative bursts help in the defence against pathogens, intracellular sources of ROS also play an import- ant role in orchestrating defence responses to pathogens (Vanlerberghe *et al.*, 2002; Ashtamk- er *et al.*, 2007; Van Doorn *et al.*, 2011). One study indicates that the O_2-based signalling pathway may be regulated by AOX (Cvetkovska and Vanlerberghe, 2012, 2013). ROS is involved in signalling and interacts with cellular targets.

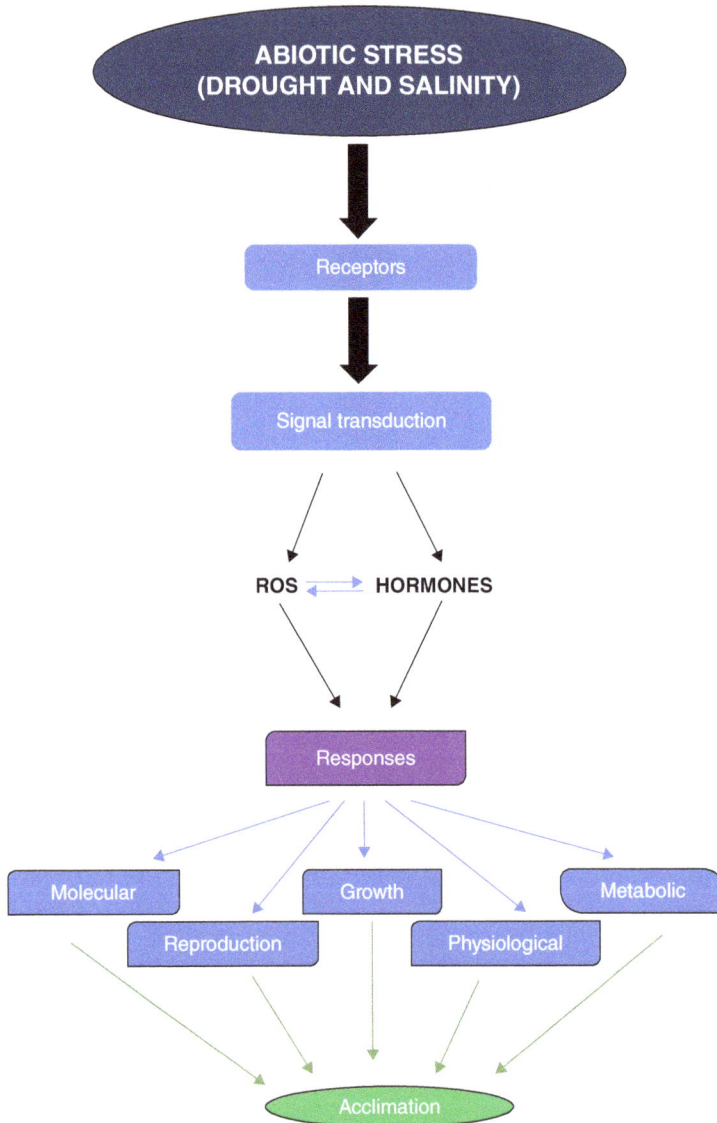

Fig. 10.3. Plants acclimatize to various abiotic stress conditions through the interaction between reactive oxygen species (ROS) and hormone signalling. Several receptors perceive abiotic stress, which can cause alterations in the availability of various hormones and ROS. Changes in hormone levels and ROS promote the activation of signal transduction pathways, which further cause various stress responses in plants. Acclimation is the outcome of distinct molecular, reproduction, growth, physiological and metabolic responses (Devireddy *et al.*, 2021).

Different receptors like STIG1, cysteine-rich receptor-like kinase and leucine-rich receptors were reported for ROS sensing in plants. STIG1 was reported for triggering cell death in stress (Wrzaczek et al., 2009). A membrane-localized sensor of hyperosmotic stress is called OSCA1, a calcium channel whose dysfunction can cause less Ca^{2+} accumulation (Li et al., 2015). Oxidative signal-inducible 1, a serine/threonine protein kinase, is also an important element of ROS signalling; it plays a significant role in ROS signalling through mitogen-activated protein kinase (MAPK) cascade activation (Rentel et al., 2004; Tripathy and Oelmüller, 2012).

The signal transduction occurs through changes in the redox status of the different signalling factors, such as Ca^{2+}, calmodulin (Ca^{2+}-binding protein) and G protein. Also, activating the CDPKs and calcineurin B-like protein (CBL)–CBL interacting protein kinases (CIPKs) in situations of stress generated a signature of H_2O_2 and Ca^{2+} (Raja et al., 2017). MAPK activation is also shown during various environmental stresses (Jalmi and Sinha, 2015). However, the specific activation mechanism of ROS is unclear. Generally, its activation causes phosphorylation and activation of several downstream factors like kinases, phosphatases, different transcription factors and cytoskeleton-associated proteins (Cristina et al., 2010). Due to different stress factors, MAPK kinase kinase (MAPKKK) or MEK kinase 1 (MEKK1) activation can occur in Arabidopsis (Pitzschke et al., 2009). The ROS generated due to these eliciting factors triggers the MAPK cascade (Fig. 10.2). There are different patterns in the case of abiotic and biotic stress; the MEKK1 activates the MKK2-MPK4/6 module in abiotic stress, while it activates the MKK4/5-MPK3/6-VIP1/ACS6 module during biotic stress (Asai et al., 2002).

ROS and hormone interactions are essential during the early and late phases of a plant's abiotic stress response (Fig. 10.3). Initially, the stress is sensed through different sensing mechanisms, which include the alterations in ROS and hormone levels. These alterations may work as a signature for different stimuli, further transducing into the downstream level, and regulate the plant stress response. Further, these metabolic, physiological and molecular changes lead to homeostasis maintenance at these levels and help plants to acclimate and establish a new steady state during stress (Devireddy et al., 2021). ROS can also mediate several cross-talk events among various hormones (Czarnocka and Karpiński, 2018). This helps in different responses to stress conditions and further results in better stress tolerance of plants (Zhou et al., 2014).

Conclusion

The generation of ROS is a normal phenomenon during different metabolic reactions in the cell. It plays a pivotal function in the plant's responses to different environmental cues throughout its life. In this chapter, the various sites of ROS origin are discussed. Furthermore, their interaction with different intracellular and extracellular components has also been covered. Different findings show that their interactions with plant hormones leads to various downstream responses. These ROS-mediated responses help the plant to adjust its metabolism and acclimatize to the changes so they are important for the development and growth of plants under normal and stressful conditions. Despite a lot of information on their possible role and participation in activating different signalling cascades, there are still many gaps in our information. In future research, we need to fill these gaps for a complete understanding of ROS-mediated signalling mechanisms.

References

Ahuja, I., de Vos, R.C.H., Bones, A.M. and Hall, R.D. (2010) Plant molecular stress responses face climate change. *Trends in Plant Science* 15, 664–674.

Akeel, A. and Jaleel, H. (2023) Biomolecules targeted by reactive oxygen species. In: Faizan, M., Hayat, S. and Ahmed, S.M. (eds) *Reactive Oxygen Species: Prospects in Plant Metabolism*. Springer, pp. 43–51.

Alché, J. de D. (2019) A concise appraisal of lipid oxidation and lipoxidation in higher plants. *Redox Biology* 23, 101136.

Anjum, N.A., Sofo, A., Scopa, A., Roychoudhury, A., Gill, S.S. *et al.* (2015) Lipids and proteins – major targets of oxidative modifications in abiotic stressed plants. *Environmental Science and Pollution Research* 22, 4099–4121.

Apel, K. and Hirt, H. (2004) Reactive oxygen species: metabolism, oxidative stress, and signal transduction. *Annual Review of Plant Biology* 55, 373–399.

Asada, K. (2006) Production and scavenging of reactive oxygen species in chloroplasts and their functions. *Plant Physiology* 141, 391–396.

Asai, T., Tena, G., Plotnikova, J., Willmann, M.R., Chiu, W.-L. *et al.* (2002) MAP kinase signalling cascade in *Arabidopsis* innate immunity. *Nature* 415, 977–983.

Ashtamker, C., Kiss, V., Sagi, M., Davydov, O. and Fluhr, R. (2007) Diverse subcellular locations of cryptogein-induced reactive oxygen species production in tobacco Bright Yellow-2 cells. *Plant Physiology* 143, 1817–1826.

Bhattacharjee, S. (2005) Reactive oxygen species and oxidative burst: roles in stress, senescence and signal transduction in plants. *Current Science* 89, 1113–1121.

Browne, R.W. and Armstrong, D. (2002) Simultaneous determination of polyunsaturated fatty acids and corresponding monohydroperoxy and monohydroxy peroxidation products by HPLC. *Methods in Molecular Biology* 186, 13–20.

Carroll, A.D., Moyen, C., Van Kesteren, P., Tooke, F., Battey, N.H. and Brownlee, C. (1998) Ca^{2+}, annexins, and GTP modulate exocytosis from maize root cap protoplasts. *Plant Cell* 10, 1267–1276.

Chang, C.C., Slesak, I., Jordá, L., Sotnikov, A., Melzer, M. *et al.* (2009) *Arabidopsis* chloroplastic glutathione peroxidases play a role in cross talk between photooxidative stress and immune responses. *Plant Physiology* 150, 670–683.

Chen, D., Cao, Y., Li, H., Kim, D., Ahsan, N. *et al.* (2017) Extracellular ATP elicits DORN1-mediated RBOHD phosphorylation to regulate stomatal aperture. *Nature Communications* 8, 2265.

Choudhary, A., Kumar, A. and Kaur, N. (2020) ROS and oxidative burst: roots in plant development. *Plant Diversity* 42, 33–43.

Choudhary, K.K., Chaudhary, N., Agrawal, S.B. and Agrawal, M. (2017) Reactive oxygen species: generation, damage, and quenching in plants during stress. In: Singh, V.P., Singh, S., Tripathi, D.K., Prasad, S.M. and Chauhan, D.K. (eds) *Reactive Oxygen Species in Plants: Boon or Bane – Revisiting the Role of ROS*. Wiley, pp. 98–115.

Corpas, F.J., Barroso, J.B., Palma, J.M. and Rodriguez-Ruiz, M. (2017) Plant peroxisomes: a nitro-oxidative cocktail. *Redox Biology* 11, 535–542.

Cristina, M., Petersen, M. and Mundy, J. (2010) Mitogen-activated protein kinase signaling in plants. *Annual Review of Plant Biology* 61, 621–649.

Cvetkovska, M. and Vanlerberghe, G.C. (2012) Coordination of a mitochondrial superoxide burst during the hypersensitive response to bacterial pathogen in *nicotiana tabacum*. *Plant, Cell and Environment* 35, 1121–1136. x

Cvetkovska, M. and Vanlerberghe, G.C. (2013) Alternative oxidase impacts the plant response to biotic stress by influencing the mitochondrial generation of reactive oxygen species. *Plant, Cell and Environment* 36, 721–732.

Cvetkovska, M., Alber, N.A. and Vanlerberghe, G.C. (2013) The signaling role of a mitochondrial superoxide burst during stress. *Plant Signaling and Behavior* 8, e22749.

Czarnocka, W. and Karpiński, S. (2018) Friend or foe? Reactive oxygen species production, scavenging and signaling in plant response to environmental stresses. *Free Radical Biology and Medicine* 122, 4–20.

Das, K. and Roychoudhury, A. (2014) Reactive oxygen species (ROS) and response of antioxidants as ROS-scavengers during environmental stress in plants. *Frontiers in Environmental Science* 2, 1–13.

Daudi, A., Cheng, Z., O'Brien, J.A., Mammarella, N., Khan, S. *et al.* (2012) The apoplastic oxidative burst peroxidase in *Arabidopsis* is a major component of pattern-triggered immunity. *Plant Cell* 24, 275–287.

Davies, M.J. (2005) The oxidative environment and protein damage. *Biochimica et Biophysica Acta* 1703, 93–109.

Dean, R.T., Fu, S., Stocker, R. and Davies, M.J. (1997) Biochemistry and pathology of radical-mediated protein oxidation. *Biochemical Journal* 324, 1–18.

Devireddy, A.R., Zandalinas, S.I., Fichman, Y. and Mittler, R. (2021) Integration of reactive oxygen species and hormone signaling during abiotic stress. *Plant Journal* 105, 459–476.

Dodd, A.N., Kudla, J. and Sanders, D. (2010) The language of calcium signaling. *Annual Review of Plant Biology* 61, 593–620.

El-Beltagi, H.S. and Mohamed, H.I. (2013) Reactive oxygen species, lipid peroxidation and antioxidative defense mechanism. *Notulae Botanicae Horti Agrobotanici Cluj-Napoca* 41, 44.

Evans, M.D., Dizdaroglu, M. and Cooke, M.S. (2004) Oxidative DNA damage and disease: induction, repair and significance. *Mutation Research* 567, 1–61.

Foreman, J., Demidchik, V., Bothwell, J.H., Mylona, P., Miedema, H. *et al.* (2003) Reactive oxygen species produced by NADPH oxidase regulate plant cell growth. *Nature* 422, 442–446.

Foyer, C.H. and Noctor, G. (2003) Redox sensing and signalling associated with reactive oxygen in chloroplasts, peroxisomes and mitochondria. *Physiologia Plantarum* 119, 355–364.

Foyer, C.H. and Noctor, G. (2005) Oxidant and antioxidant signalling in plants: a re-evaluation of the concept of oxidative stress in a physiological context. *Plant, Cell and Environment* 28, 1056–1071.

Foyer, C.H., Karpinska, B. and Krupinska, K. (2014) The functions of WHIRLY1 and REDOX-RESPONSIVE TRANSCRIPTION FACTOR 1 in cross tolerance responses in plants: a hypothesis. *Philosophical Transactions of the Royal Society of London Series B, Biological Sciences* 369, 20130226.

Fryer, M.J. (2002) Imaging of photo-oxidative stress responses in leaves. *Journal of Experimental Botany* 53, 1249–1254.

Gill, S.S. and Tuteja, N. (2010) Reactive oxygen species and antioxidant machinery in abiotic stress tolerance in crop plants. *Plant Physiology and Biochemistry* 48, 909–930.

Gilroy, S., Białasek, M., Suzuki, N., Górecka, M., Devireddy, A.R. *et al.* (2016) ROS, calcium, and electric signals: key mediators of rapid systemic signaling in plants. *Plant Physiology* 171, 1606–1615.

Gilroy, S., Suzuki, N., Miller, G., Choi, W.-G., Toyota, M. *et al.* (2014) A tidal wave of signals: calcium and ROS at the forefront of rapid systemic signaling. *Trends in Plant Science* 19, 623–630.

Gupta, K.J. and Igamberdiev, A.U. (2016) Reactive nitrogen species in mitochondria and their implications in plant energy status and hypoxic stress tolerance. *Frontiers in Plant Science* 7, 369.

Halliwell, B. (2006) Reactive species and antioxidants. Redox biology is a fundamental theme of aerobic life. *Plant Physiology* 141, 312–322.

Hara-Chikuma, M., Satooka, H., Watanabe, S., Honda, T., Miyachi, Y. *et al.* (2015) Aquaporin-3-mediated hydrogen peroxide transport is required for NF-κB signalling in keratinocytes and development of psoriasis. *Nature Communications* 6, 7454.

Hasanuzzaman, M., Bhuyan, M.H.M.B. and Parvin, K. (2020a) Regulation of ROS metabolism in plants under environmental stress: a review of recent experimental evidence. *International Journal of Molecular Sciences* 21, 1–44.

Hasanuzzaman, M., Bhuyan, M.H.M.B., Zulfiqar, F., Raza, A., Mohsin, S.M. *et al.* (2020b) Reactive oxygen species and antioxidant defense in plants under abiotic stress: revisiting the crucial role of a universal defense regulator. *Antioxidants* 9, 681.

Hernansanz-Agustín, P. and Enríquez, J.A. (2021) Generation of reactive oxygen species by mitochondria. *Antioxidants* 10, 1–18.

Hussain, H.A., Hussain, S., Khaliq, A., Ashraf, U., Anjum, S.A. *et al.* (2018) Chilling and drought stresses in crop plants: implications, cross talk, and potential management opportunities. *Frontiers in Plant Science* 9, 393.

Imlay, J. and Linn, S. (1988) Damage and oxygen radical. *Science* 240, 1302.

Impa, S.M., Nadaradjan, S. and Jagadish, S.V.K. (2012) Drought stress induced reactive oxygen species and anti-oxidants in plants. In: Ahmad, P. and Prasad, M.N.V. (eds) *Abiotic Stress Responses in Plants*. Springer, pp. 131–147.

Jakubowska, D., Janicka-Russak, M., Kabała, K., Migocka, M. and Reda, M. (2015) Modification of plasma membrane NADPH oxidase activity in cucumber seedling roots in response to cadmium stress. *Plant Science* 234, 50–59.

Jalmi, S.K. and Sinha, A.K. (2015) ROS mediated MAPK signaling in abiotic and biotic stress – striking similarities and differences. *Frontiers in Plant Science* 6, 769.

Kadota, Y., Sklenar, J., Derbyshire, P., Stransfeld, L., Asai, S. *et al.* (2014) Direct regulation of the NADPH oxidase RBOHD by the PRR-associated kinase BIK1 during plant immunity. *Molecular Cell* 54, 43–55.

Karpinski, S., Reynolds, H., Karpinska, B., Wingsle, G., Creissen, G. *et al.* (1999) Systemic signaling and acclimation in response to excess excitation energy in *Arabidopsis*. *Science* 284, 654–657.

Keun, P.L., Kim, C., Landgraf, F. and Apel, K. (2007) EXECUTER1- and EXECUTER2-dependent transfer of stress-related signals from the plastid to the nucleus of *Arabidopsis thaliana*. *Proceedings of the National Academy of Sciences USA* 104, 10270–10275.

Kosová, K., Vítámvás, P., Urban, M.O., Prášil, I.T. and Renaut, J. (2018) Plant abiotic stress proteomics: the major factors determining alterations in cellular proteome. *Frontiers in Plant Science* 9, 122.

Laloi, C. and Havaux, M. (2015) Key players of singlet oxygen-induced cell death in plants. *Frontiers in Plant Science* 6, 39.

Laloi, C., Apel, K. and Danon, A. (2004) Reactive oxygen signalling: the latest news. *Current Opinion in Plant Biology* 7, 323–328.

Leshem, Y., Seri, L. and Levine, A. (2007) Induction of phosphatidylinositol 3-kinase-mediated endocytosis by salt stress leads to intracellular production of reactive oxygen species and salt tolerance. *Plant Journal* 51, 185–197.

Li, Y., Yuan, F., Wen, Z., Li, Y., Wang, F. *et al.* (2015) Genome-wide survey and expression analysis of the OSCA gene family in rice. *BMC Plant Biology* 15, 261.

Mittler, R. (2017) ROS are good. *Trends in Plant Science* 22, 11–19.

Mittler, R., Vanderauwera, S., Suzuki, N., Miller, G.A.D., Tognetti, V.B. *et al.* (2011) ROS signaling: the new wave? *Trends in Plant Science* 16, 300–309.

Mittler, R., Zandalinas, S.I., Fichman, Y. and Van Breusegem, F. (2022) Reactive oxygen species signalling in plant stress responses. *Nature Reviews Molecular Cell Biology* 23, 663–679.

Møller, I.M., Jensen, P.E. and Hansson, A. (2007) Oxidative modifications to cellular components in plants. *Annual Review of Plant Biology* 58, 459–481.

Monshausen, G.B., Bibikova, T.N., Weisenseel, M.H. and Gilroy, S. (2009) Ca^{2+} regulates reactive oxygen species production and pH during mechanosensing in *Arabidopsis* roots. *Plant Cell* 21, 2341–2356.

Müller, K., Carstens, A.C., Linkies, A., Torres, M.A. and Leubner-Metzger, G. (2009) The NADPH-oxidase AtrbohB plays a role in *Arabidopsis* seed after-ripening. *New Phytologist* 184, 885–897.

Mullineaux, P.M., Karpinski, S. and Baker, N.R. (2006) Spatial dependence for hydrogen peroxide-directed signaling in light-stressed plants. *Plant Physiology* 141, 346–350.

Nair, A., Chauhan, P., Saha, B. and Kubatzky, K.F. (2019) Conceptual evolution of cell signaling. *International Journal of Molecular Sciences* 20, 3292.

O'Brien, J.A., Daudi, A., Butt, V.S. and Paul Bolwell, G. (2012) Reactive oxygen species and their role in plant defence and cell wall metabolism. *Planta* 236, 765–779.

Ozgur, R., Turkan, I., Uzilday, B. and Sekmen, A.H. (2014) Endoplasmic reticulum stress triggers ROS signalling, changes the redox state, and regulates the antioxidant defence of *Arabidopsis thaliana*. *Journal of Experimental Botany* 65, 1377–1390.

Peleg, Z., Apse, M.P. and Blumwald, E. (2011) Engineering salinity and water-stress tolerance in crop plants: getting closer to the field. *Advances in Botanical Research* 57, 405–443.

Petrov, V., Hille, J., Mueller-Roeber, B. and Gechev, T.S. (2015) ROS-mediated abiotic stress-induced programmed cell death in plants. *Frontiers in Plant Science* 6, 69.

Pitzschke, A., Djamei, A., Bitton, F. and Hirt, H. (2009) A major role of the MEKK1-MKK1/2-MPK4 pathway in ROS signalling. *Molecular Plant* 2, 120–137.

Pogány, M., von Rad, U., Grun, S., Dongó, A., Pintye, A. *et al.* (2009) Dual roles of reactive oxygen species and NADPH oxidase RBOHD in an *Arabidopsis-Alternaria* pathosystem. *Plant Physiology* 151, 1459–1475.

Pospíšil, P. (2016) Production of reactive oxygen species by photosystem II as a response to light and temperature stress. *Frontiers in Plant Science* 7, 1950.

Raja, V., Majeed, U., Kang, H., Andrabi, K.I. and John, R. (2017) Abiotic stress: interplay between ROS, hormones and MAPKs. *Environmental and Experimental Botany* 137, 142–157.

Rentel, M.C., Lecourieux, D., Ouaked, F., Usher, S.L., Petersen, L. *et al.* (2004) OXI1 kinase is necessary for oxidative burst-mediated signalling in *Arabidopsis*. *Nature* 427, 858–861.

Richards, S.L., Wilkins, K.A., Swarbreck, S.M., Anderson, A.A., Habib, N. *et al.* (2015) The hydroxyl radical in plants: from seed to seed. *Journal of Experimental Botany* 66, 37–46.

Rohman, M.M., Islam, M.R., Naznin, T., Omy, S.H., Begum, S. *et al.* (2019) Maize production under salinity and drought conditions: oxidative stress regulation by antioxidant defense and glyoxalase systems. In: Hasanuzzaman, M., Hakeem, K.R., Nahar, K., Alharby, H.F. (eds) *Plant Abiotic Stress Tolerance: Agronomic, Molecular and Biotechnological Approaches*. Springer, pp. 1–34.

Rossel, J.B., Wilson, P.B., Hussain, D., Woo, N.S., Gordon, M.J. *et al.* (2007) Systemic and intracellular responses to photooxidative stress in *Arabidopsis*. *Plant Cell* 19, 4091–4110.

Sachdev, S., Ansari, S.A. and Ansari, M.I. (2023) ROS generation in plant cells orchestrated by stress. In: Sachdev, S., Ansari, S.A. and Ansari, M.I. *Reactive Oxygen Species in Plants: The Right Balance*. Springer, pp. 23–43.

Sachdev, S., Ansari, S.A., Ansari, M.I., Fujita, M. and Hasanuzzaman, M. (2021) Abiotic stress and reactive oxygen species: generation, signaling, and defense mechanisms. *Antioxidants* 10, 1–37.

Sharma, P., Jha, A.B., Dubey, R.S. and Pessarakli, M. (2012) Reactive oxygen species, oxidative damage, and antioxidative defense mechanism in plants under stressful conditions. *Journal of Botany* 2012, 217037.

Simon-Plas, F., Elmayan, T. and Blein, J.P. (2002) The plasma membrane oxidase NtrbohD is responsible for AOS production in elicited tobacco cells. *Plant Journal* 31, 137–147.

Singh, A., Kumar, A., Yadav, S. and Singh, I.K. (2019) Reactive oxygen species-mediated signaling during abiotic stress. *Plant Gene* 18, 100173.

Slesak, I., Slesak, H. and Kruk, J. (2012) Oxygen and hydrogen peroxide in the early evolution of life on earth: in silico comparative analysis of biochemical pathways. *Astrobiology* 12, 775–784.

Soares, C., Carvalho, M.E., Azevedo, R.A. and Fidalgo, F. (2019) Plants facing oxidative challenges – a little help from the antioxidant networks. *Environmental and Experimental Botany* 161, 4–25.

Suzuki, N., Miller, G., Morales, J., Shulaev, V., Torres, M.A. and Mittler, R. (2011) Respiratory burst oxidases: the engines of ROS signaling. *Current Opinion in Plant Biology* 14, 691–699.

Suzuki, N., Miller, G., Salazar, C., Mondal, H.A., Shulaev, E. *et al.* (2013) Temporal-spatial interaction between reactive oxygen species and abscisic acid regulates rapid systemic acclimation in plants. *Plant Cell* 25, 3553–3569.

Thomas, C., Mackey, M.M., Diaz, A.A. and Cox, D.P. (2009) Hydroxyl radical is produced via the fenton reaction in submitochondrial particles under oxidative stress: implications for diseases associated with iron accumulation. *Redox Report* 14, 102–108.

Tian, S., Wang, X., Li, P., Wang, H., Ji, H. *et al.* (2016) Plant aquaporin AtPIP1;4 links apoplastic H_2O_2 induction to disease immunity pathways. *Plant Physiology* 7, 1635–1650.

Tiwari, S., Tiwari, S., Singh, M., Singh, A. and Prasad, S.M. *et al.* (2017) Generation mechanisms of reactive oxygen species in the plant cell: an overview. In: Singh, V.P., Singh, S., Tripathi, D.K., Prasad, S.M. and Chauhan, D.K. (eds) *Reactive Oxygen Species in Plants: Boon or Bane – Revisiting the Role of ROS.* Wiley, pp. 1–22.

Triantaphylidès, C. and Havaux, M. (2009) Singlet oxygen in plants: production, detoxification and signaling. *Trends in Plant Science* 14, 219–228.

Tripathy, B.C. and Oelmüller, R. (2012) Reactive oxygen species generation and signaling in plants. *Plant Signaling and Behavior* 7, 1621–1633.

Van Doorn, W.G., Beers, E.P., Dangl, J.L., Franklin-Tong, V.E., Gallois, P. *et al.* (2011) Morphological classification of plant cell deaths. *Cell Death and Differentiation* 18, 1241–1246.

Vanlerberghe, G.C. (2013) Alternative oxidase: a mitochondrial respiratory pathway to maintain metabolic and signaling homeostasis during abiotic and biotic stress in plants. *International Journal of Molecular Sciences* 14, 6805–6847.

Vanlerberghe, G.C., Robson, C.A. and Yip, J.Y.H. (2002) Induction of mitochondrial alternative oxidase in response to a cell signal pathway down-regulating the cytochrome pathway prevents programmed cell death. *Plant Physiology* 129, 1829–1842.

Vanlerberghe, G.C., Cvetkovska, M. and Wang, J. (2009) Is the maintenance of homeostatic mitochondrial signaling during stress a physiological role for alternative oxidase? *Physiologia Plantarum* 137, 392–406.

Wituszynska, W. and Karpinski, S. (2013) Programmed cell death as a response to high light, UV and drought stress in plants. In: Vahdati, K. and Leslie, C. (eds) *Abiotic Stress – Plant Responses and Applications in Agriculture.* Intech Open, pp. 207–246.

Wrzaczek, M., Brosche, M., Kollist, H. and Kangasjärvi, J. (2009) *Arabidopsis* GRI is involved in the regulation of cell death induced by extracellular ROS. *Proceedings of the National Academy of Sciences USA* 106, 5412–5417.

Yadav, D., Boyidi, P., Ahmed, I. and Kirti, P.B. (2018) Plant annexins and their involvement in stress responses. *Environmental and Experimental Botany* 155, 293–306.

Zandalinas, S.I., Fichman, Y., Devireddy, A.R., Sengupta, S., Azad, R.K. *et al.* (2020) Systemic signaling during abiotic stress combination in plants. *Proceedings of the National Academy of Sciences USA* 117, 13810–13820.

Zhou, J., Wang, J., Li, X., Xia, X.J., Zhou, Y.H. and Shi, K. (2014) H_2O_2 mediates the crosstalk of brassinosteroid and abscisic acid in tomato responses to heat and oxidative stresses. *Journal of Experimental Botany* 65, 4371–4383.

11 Plant Beneficiaries: Root-Dwelling Bacteria

Amanda Nongthombam, Nikena Khwairakpam, Maheshree Maibam, Yurembam Rojiv Singh and Debananda S. Ningthoujam*
Microbial Biotechnology Research Laboratory, Department of Biochemistry, Manipur University, Canchipur, Manipur, India

Abstract

Plants are biotic systems that interact with diverse microorganisms. The microorganisms that reside in and around plants constitute the plant microbiome. The microbiome of plants mainly includes bacteria, fungi and archaea. Plant–microbe interactions may vary from parasitism, competition and commensalism to mutualism. Mutualistic plant–microbe interactions are the most widely studied. The phyllosphere (above ground), endosphere (in plant tissues) and rhizosphere (the zone around the root system) are the three distinct domains identified in the plant microbiome. Bacteria associated with plants influence plant health and development. Plants and bacteria have coevolved complex systems to accommodate mutualistic exchanges. Such interactions benefit plants through abiotic and biotic stress mitigation, growth promotion, nutrient acquisition, etc. Root-associated bacteria may have promising applications in agriculture and pharmaceuticals. Root-dwelling bacteria may inhabit two zones: the rhizosphere and the root endosphere. Rhizospheric bacteria reside in the former, and root endophytic bacteria reside in the latter. Root exudates and mucilage impact the diversity of rhizospheric bacteria, whereas root endophytic bacteria are affected by selective entry mechanisms and biochemical processes. Some rhizobacteria remain tightly anchored to the root surface or the rhizoplane. Since root-associated bacteria are vital for plants, root microbiome studies are an active research area. A detailed understanding of plant–rhizobacteria interactions could lead to novel applications in sustainable agriculture and soil bioremediation.

Keywords: Root microbiome, root-associated bacteria, rhizospheric bacteria, endophytic bacteria, rhizoplane bacteria

11.1 Introduction

Interactions between plants and microbes are essential for the growth and survival of plants in their natural environment. These interactions are critical for the ecology of plants and have a significant impact on ecosystems. It is fascinating to realize that, just as humans have a human microbiome, plants also have their own microbiomes. Bacteria, fungi and archaea are all part of this unique microbiome. The investigation of plant microbiomes can provide a better understanding of how plants thrive and interact with their environment. Plant–microbe interactions vary from parasitism, competition and commensalism to mutualism. The most prevalent is

*Corresponding author: debananda.ningthoujam@gmail.com

© CAB International 2025. *Soil Health and Nutrition Management*
(eds N.C. Joshi, T. Leustek and P.K. Singh)
DOI: 10.1079/9781800624597.0011

commensalism or mutualism, in which one or both species benefit (Wu *et al.*, 2009). Microorganisms that are beneficial and pathogenic, especially bacteria and fungi, often invade plants (Wille *et al.*, 2019). The beneficial interactions mainly include nutrient acquisition, growth stimulation through phytohormones, suppression of pathogens and amelioration of stresses, while harmful interactions damage the host via necrotrophy and various diseases (e.g. blights, scabs, wilts, cankers and soft rots) (Cooper and Gardener, 2006; Dolatabadian, 2020). Microbial communities in the phyllosphere, endosphere, rhizosphere and rhizoplane comprise the plant microbiome. The plant microbiome directly or indirectly impacts the plant root system and, consequently, the entire plant (Goel *et al.*, 2017). Plants and bacteria have coevolved complex systems to accommodate mutualistic exchanges that influence plant health and development. Plant-beneficial bacteria are usually known as plant growth-promoting bacteria.

Root-dwelling bacteria reside in various root zones, such as the rhizosphere, rhizoplane and endosphere. The rhizosphere is the soil area around plant roots that supports a robust bacterial community (Wheatley and Poole, 2018). Plant mucilage and root exudates nurture the rhizospheric microbial community (Kent and Triplett, 2002). Plants recruit beneficial bacteria that may help them combat fungal diseases through the secretion of root exudates discharged into rhizospheric soil (Hussain *et al.*, 2018). Root exudates are diverse substances, including sugars, amino acids, nucleotides, organic acids, flavonoids, mucilage, antimicrobial compounds, enzymes, water and H^+ ions (Ho *et al.*, 2017). A plant species may select a specific group of rhizospheric bacteria (RB) (Ciccazzo *et al.*, 2014). Different bacteria that are firmly connected to the root surface live in the rhizosphere (Goel *et al.*, 2017). The rhizosphere and rhizoplane are so close that both zones are frequently referred to as a continuum (Johri *et al.*, 2003). The rhizoplane is a compartment that promotes nutrient exchange between the soil and plants (Ding *et al.*, 2019). By serving as a selective barrier, the rhizoplane allows only a small percentage of rhizospheric microorganisms to adhere to it, enabling a subset of microbes to constitute the endosphere (Edwards *et al.*, 2015). Endophytic bacteria may enter plant roots after they have established in the rhizosphere and rhizoplane (Hallmann, 2001). Rhizodeposition seems to stimulate an initial community shift in the rhizosphere (Bulgarelli *et al.*, 2013). Rhizobacteria penetrate plant roots and constitute root endophytic bacteria (REB) via a horizontal transmission channel (Ding *et al.*, 2019). However, REB are affected by selective entry mechanisms and biochemical processes. They actively enter plant roots by releasing cell wall-degrading enzymes. REB may also passively enter the root through wounds or by colonizing the root hairs (Hardoim *et al.*, 2015). They possibly enter plant roots through the attachment of exopolysaccharides (EPS), lipopolysaccharides (LPS), structural components and quorum-sensing compounds (Kumar *et al.*, 2020).

Root-dwelling bacteria (RDB) benefits plants through diverse mechanisms, such as biological nitrogen fixation, siderophore and phytohormone production and phosphorus solubilization. They may also elicit induced systemic resistance or produce antibiotics to combat phytopathogens, nematodes or insects (Pineda *et al.*, 2013; Hussain *et al.*, 2018). RDB can also enhance plant tolerance to various abiotic stresses, such as drought, flooding, extreme temperature, salinity and nutrient deficiency. Hence, these beneficial bacteria can be used as biofertilizers.

Agricultural production is the backbone of our society and plays a vital role in sustaining our population. However, improving yield, quality, processing and storage to feed the burgeoning global population is a constant challenge (de Andrade *et al.*, 2023). Agriculture requires the effective use of mineral fertilizers to boost crop productivity, improve soil conditions and increase fertility (Uzakbaevna, 2022). However, synthetic agrochemicals can adversely affect the environment, causing severe water, soil and air pollution and a decrease in soil fertility (Kumar *et al.*, 2022). We must consider alternative methods to fertilize our crops and control their pests. To lessen or eliminate the use of synthetic agrochemicals, it is crucial to explore the application of beneficial bacteria as biofertilizers in agriculture (Babalola, 2010). This approach has the potential to improve agricultural productivity and promote a sustainable and eco-friendly agricultural system. A detailed understanding of plant–microbe interactions could lead to new methods for sustainable agriculture and bioremediation on a global scale.

11.2 Root-Dwelling Bacteria

The plant microbiome is a collective term for microbes that inhabit the phyllosphere, endosphere, rhizosphere and rhizoplane. These microorganisms mainly include bacteria, fungi and archaea, and they have a ripple effect on the root and, subsequently, the entire plant, either directly or indirectly (Goel *et al.*, 2017). RDB reside in rhizocompartments such as the rhizosphere, rhizoplane and root endosphere. The rhizosphere and rhizoplane are are generally referred to as a continuum (Johri *et al.*, 2003). The colonization of bacteria in root zones is triggered by specific compounds present in root exudates (Huang *et al.*, 2014). Root exudates are photosynthates secreted by roots to the rhizosphere, and this process is known as rhizodeposition (Jones *et al.*, 2009). The composition of root exudates includes small molecules such as sugars, amino acids, nucleotides, organic acids, flavonoids, enzymes, antimicrobial compounds, water and H+ ions. They play crucial roles in modulating the root microbial community (Grayston and Campbell, 1996).

RDB can be classified as: (i) rhizospheric bacteria (RB); (ii) rhizoplane bacteria (RPB); or (iii) root endophytic bacteria (REB). *Actinobacteria, Alcaligens, Arthrobacter, Azospirillum, Azotobacter, Bacillota,* Enterobacter, *Klebsiella, Nocardia, Ochrobactrum, Protobacteria, Pseudomonas, Rhodococcus, Serratia* (Saharan and Nehra, 2011; Goel *et al.*, 2017), etc. are the predominant genera of RDB.

11.2.1 Root exudates

Root exudates (REs) are substances that plants release into the rhizosphere. REs greatly influence the development of soil microbes and their functions in the rhizosphere. They function as both attractants and repellents in the rhizosphere. They can also change soil chemical and physical properties (More *et al.*, 2020). Different plant species release unique root exudates, resulting in varying rhizosphere biomes (Rovira, 1969). The quantity of REs a plant produces depends on many factors, such as plant species, age and external stresses (Madhukar *et al.*, 2018). Root exudates can be categorized into organic acids, amino acids, sugars and antimicrobial compounds, as shown in Table 11.1 (Ho

Table 11.1. Types of root exudates.

Types of root exudates	Remarks	Reference
Organic acids	Facilitate root microbiome colonization and biofilm formation; can act as chelating agents Examples: acetic, citric, glutaric, lactic, malic, oxalic, pyruvic, succinic, etc.	Dakora and Phillips, 2002; Yuan *et al.*, 2015
Amino acids	Modulate growth of microorganisms, germination of spores and remodelling of microbial cell walls Tryptophan and methionine are important precursors for indole acetic acid and ethylene Examples: alanine, arginine, aspartic acid, cysteine, glutamine, glycine, lysine, serine, etc.	Arshad and Frankenberger, 1991; Idrees *et al.*, 2020; Vives-Peris *et al.*, 2020
Sugars	Essential carbon source for root-dwelling bacteria Examples: fructose, galactose, glucose, maltose, raffinose, ribose, sucrose, etc.	Vives-Peris *et al.*, 2020; Lopes *et al.*, 2022
Antimicrobial compounds	Phytoanticipins: constitutive antimicrobial compounds that exist in plants even before confronting biotic stress, such as pathogen infections Examples: rhizathalene A, saponin avenin A-1 and saponin α-tomatine Phytoalexins: synthesized and accumulated in plants after infection by microorganisms Examples: terpenoids, phenolics, hydroxy coumarins, scopoletins and camalexins	González-Lamothe *et al.*, 2009

et al. 2017). Abiotic factors such as temperature, soil moisture and pH and biotic factors such as soil microbial secondary metabolism influence root exudate production (Table 11.2).

11.2.2 Rhizospheric bacteria

The rhizosphere is the area of soil directly impacted by the roots of plants (Choudhary *et al.*, 2018). The area of soil beyond the rhizosphere is referred to as the bulk soil (Gobat *et al.*, 2004). The bacterial population in the rhizosphere is 10–1000 times greater than that in bulk soil (Choudhary *et al.*, 2018). The bacteria inhabiting the rhizosphere are called rhizospheric bacteria (Glick, 2012; Ouf *et al.*, 2023). RB positive for plant growth-promoting traits are termed plant growth-promoting rhizobacteria (PGPR). They stimulate the growth of plants by employing various direct and indirect mechanisms: they can enhance the mobilization of nutrients, the production of phytohormones, the development of shoots and roots and defence against various phytopathogens. Furthermore, enzymes

produced by PGPR can detoxify heavy metals and help plants tolerate several abiotic stresses such as drought and salinity (de Andrade *et al.*, 2023). RB predominantly belong to *Proteobacteria* (mainly *Enterobacteriaceae, Rhizobiaceae, Bradyrhizobiaceae, Pseudomonaceae* and *Xanthomonadaceae*), *Acidobacteria* (*Acidobacteriaceae*), *Actinobacteria* (*Actinobaceriaceae, Streptomycetaceae*) and *Bacillota* (*Bacillaceae*) (Ding *et al.*, 2019). The major PGPR species or genera associated with major crops are listed in Table 11.3.

11.2.3 Rhizoplane bacteria

The rhizoplane is the interface between the soil and the root system. The root epidermis and outer cortex are included in the rhizoplane and the area is clogged with soil particles and other microbes. The rhizoplane can be considered the zone that drives nutrient flow and transformation in the rhizosphere (Raynaud *et al.*, 2017). Various bacterial colonization processes are proposed based on the type of bacteria and how they colonize the rhizoplane (Hardoim *et al.*,

Table 11.2. Environmental factors affecting root exudates.

	Remarks	Reference
Abiotic factors		
Temperature	Exudation per unit root mass increases by approximately 2 µg C g^{-1} h^{-1} for every 1°C increase in temperature	Pramanik *et al.*, 2000; Qiao *et al.*, 2014; Leuschner *et al.*, 2022
	Stimulates secretion of benzoic, 4-hydroxy-benzoic, phthalic and palmitic acids from the roots of *Cucumis sativus* at 25–30°C	
	Affects the composition of fine root exudates and the flow of carbon	
Soil moisture	Influences both quantitative and qualitative exudation of organic compounds from plant roots	Hale and Moore, 1980; Marschner, 1995; Leuschner *et al.*, 2022
	Low to moderate drought enhances exudation	
	Drought stimulates the production of root exudates by increasing the soil mechanical pressure on plant roots	
Soil pH	*Lupinus angustifolius* exudes more at alkaline pH when treated with struvite	Pellet *et al.*, 1995; Robles-Aguilar *et al.*, 2019
	Heavy metal stress alters root exudates	
Biotic factors		
Microorganisms	Affect root exudation by influencing root cell permeability and root metabolism	Ho *et al.*, 2017
	Stimulate phenolic compound exudation to improve plant iron uptake in soil with low iron availability	

Table 11.3. Species or genera of plant growth-promoting rhizobacteria (PGPR) associated with major crop plants.

Host plant	Species or genera of PGPR	Reference
Rice	*Bacillus, Micromonospora, Pseudomonas fluorescens, Streptomyces, Rhizobium* spp.	Ding *et al.*, 2019; Nur Mawaddah *et al.*, 2023
Wheat	*Bacillus ellenbachensis, B. megaterium, B. mycoides, B. pumilus, B. simplex, B. subtilis, B. vulgalus*	Truffaut and Vladykov, 1930; Clark, 1940; Hafeez *et al.*, 2006
Maize	*Pseudomonas fluorescens, P. fluorescens* biotype G, *P. putida* biotype A	Shaharoona *et al.*, 2006
Oats and barley	*Bacillus megaterium*	Clark, 1940
Pea	*Azotobacter, Bacillus, Micrococcus, Pseudomonas*	Nazir *et al.*, 2020
Chickpea	*Azotobacter, Azospirillum, Pseudomonas*	Kumar *et al.*, 2019
Soybean	*Bacillus subtilis, B. pumilus, B. sphaericus, Streptomyces manipurensis* strain USC003, *S. panaciradicis* strain 1 MR-8, *S. polychromogenes* strain GXSS21, *S. recifensis* strain ST100, *Streptomyces* sp. strain CAH7	Wahyudi *et al.*, 2011, 2019
Groundnut	*Bacillus halotolerans, B. safensis, B. subtilis*	Sarwar *et al.*, 2020
Tomato	*Bacillus licheniformis, Pseudomonas putida* subgroup B strain 1	García *et al.*, 2004; Gravel *et al.*, 2007
Chilli (*Capsicum annum* L.)	*Bacillus* sp., *Pseudomonas* sp.	Veerapagu *et al.*, 2018
Thale cress (*Arabidopsis thaliana*)	*Pseudomonas protegens*	Li *et al.*, 2021

2008; Compant *et al.*, 2010). Adherence to the rhizoplane includes the establishment of large microcolonies, interrupted biofilms or biofilms (Villacieros *et al.*, 2003; Reinhold-Hurek *et al.*, 2015). However, several bacterial species, including *Agrobacterium, Azospirillum, Pseudomonas, Rhizobium* and *Salmonella*, exhibit a biphasic root attachment mechanism consisting of two phases: (i) the primary attachment phase; and (ii) the secondary attachment phase. The first stage of bacterial attachment involves reversible, weak and non-specific binding to the root surface. This phase is known as the 'settlement', 'adsorption' or 'reversible' phase. The second phase involves a change to a more specific binding mode aided by extracellular fibril synthesis, leading to the accumulation and aggregation of bacteria. This phase is also known as an 'anchoring', 'residence' or 'irreversible' phase (Matthysse, 2014; Wheatley and Poole, 2018).

RPB originate from the surrounding soil (Bulgarelli *et al.*, 2012). Members of the *Alphaproteobacteria* (*Bradyrhizobiaceae, Sphingomonadaceae*), *Betaproteobacteria* (*Comamonadaceae, Methylophilaceae, Oxalobacteraceae*), *Bacteroidota*

(*Sphingobacteriaceae*) and *Gammaproteobacteria* (*Legionellaceae*) are predominant residents of the rhizoplane (Dibbern *et al.*, 2014). RPB can influence plant growth via phosphorus solubilization, cell-wall-degrading enzyme production and phytoremediation (Johri *et al.*, 1999; Giongo *et al.*, 2010). They may also be biocontrol agents against phytopathogens (Deora *et al.*, 2005) and may enhance plant tolerance to abiotic stresses (e.g. extreme pH, salt or temperature). Some examples of the RPB associated with various host plants are listed in Table 11.4.

Many techniques have been employed to explore RPB: light microscopy, scanning electron microscopy (SEM), fluorescent *in situ* hybridization (FISH), double labelling of oligonucleotide probes FISH (DOPE-FISH), catalysed reporter deposition FISH (CARD-FISH), polymerase chain reaction–restriction fragment length polymorphism (PCR-RFLP) and the combination of metagenomics and metatranscriptomics (Muraoka *et al.*, 2000; Eller and Frenzel, 2001; Ikenaga *et al.*, 2002; Compant *et al.*, 2013; Ofek–Lalzar *et al.*, 2014; Schmidt and Eickhorst, 2014).

Table 11.4. Examples of rhizoplane bacteria associated with different plants.

Host plant	Rhizoplane bacteria	Reference
Rice	*Acinetobacter* sp. BR-25, *Acinetobacter* sp. BR-12, *Azospirillum* spp., *Bacillus velezensis*, *Enterobacter* sp. BR-26, *Nitrobacter*, *Nitrosospira* spp., *Klebsiella* sp. BR-15, *Microbacterium* sp. BRS-1, *Pseudomonas* sp. BRS-2, *Pseudomonas fluorescens*	Islam *et al.*, 2007; Ding *et al.*, 2019
Wheat	*Bacillus cereus*, *B. safensis*, *B. stratosphericus*, *B. marisflavi*, *B. subtilis*, *Brevundimonas diminuta*, *Exiguobacterium aurantiacum*, *Pseudomonas geniculata*, *Planomicrobium chinense*, *Rhizobium pusense*, *Serratia nematodiphila*, *Stenotrophomonas rhizophila*	Siddiqa *et al.*, 2016
Maize	*Arthrobacter*, *Azotobacter*, *Bacillus*, *Chryseobacterium lactis*, *C. joostei*, *C. indologenes*, *C. viscerum*, *Listeria*, *Sporolactobacillus*, *Micrococcus*, *Pseudomonas*	Cavaglieri *et al.*, 2009; Kämpfer *et al.*, 2015
Barley	*Nitrobacter* sp. PJNI, *Nitrosolobus* sp. PJAI	Satoh *et al.*, 2003
Soybean	*Rhodanobacter caeni*, *R. panaciterrae*, *R. soli*, *R. spathiphylli*, *R. terrae*	Madhaiyan *et al.*, 2014
Glutinous yam (Japanese yam)	*Chitinophaga polysaccharae* sp. nov.	Han *et al.*, 2014
Sorghum	*Brevundimonas vesicularis*, *Pseudomonas alcaligenes*	Abd El-Ghany *et al.*, 2015
Tree peony	*Agromyces allii*, *Agrobacterium tumefaciens*, *Microbacteria insulae*, *M. neoaurum*, *Sphingobium estrogenivorans*, *Sphingopyxis witflariensis*, *Variovorax koreensis*, *Xanthomonas campestris* pv.	Han *et al.*, 2011
Kangaroo grass	*Bacillus cereus* (KFP9-F), *B. megaterium* (NAS7-L), *B. simplex* (KBS1F-3)	Hassen and Labuschagne, 2010

11.2.4 Root endophytic bacteria

Bacteria or fungi residing within plant tissue without causing adverse effects on the host are called endophytes (Gaiero *et al.*, 2013). REBs originate from the soil. The REB is influenced by the root environment, shape and exudates (Pfeiffer *et al.*, 2017; Compant *et al.*, 2021). The colonization of host tissue by REB is complicated and involves several events. This process typically initiates when specific components of root exudates interact with root bacterial communities (de Weert *et al.*, 2002). Endophytes may be transmitted vertically (from mother plant to seed) or horizontally (from plant to plant or soil to plant). Plants acquire endophytes through the horizontal transmission of soil microbes, which can occur passively or actively (Ding *et al.*, 2019; Hallmann, 2001).

Bacterial colonization usually involves the attachment of a specific strain of REB to the root surface in response to root exudates (Begonia and Kremer, 1994). In this process, structural components of bacteria (e.g. flagella, fimbriae, pili) and secretory products such as LPS, EPS or cell surface polysaccharides (CSP) may be involved (Sauer and Camper, 2001). After adhering to the host surface, the endophytes actively or passively penetrate the root tissue. They enter root tissues passively through wounds in areas of the root tips and lateral roots (Hardoim *et al.*, 2015). Endophytic bacteria may also actively enter plant roots via LPS, EPS, quorum-sensing molecules and other structural components (Duijff *et al.*, 1997; Dörr *et al.*, 1998; Kumar *et al.*, 2020). REB can also penetrate the junctions of rhizodermal cells via enzymes that degrade cell walls, such as xylanases, endoglucanases, cellobiohydrolases and cellulases (Liu *et al.*, 2017; Compant *et al.*, 2021).

The root bacterial microbiome has a greater density of microorganisms than other plant parts (Amend *et al.*, 2019). The densities of bacterial cells in various plant parts are as follows: 10^5–10^7 cells/g in roots, 10^3–10^4 cells/g in stems and leaves and 10^2–10^3 cells/g in flowers, fruits

Table 11.5. Some examples of root endophytic bacteria from various plants.

Host plant	Root endophytic bacteria	Reference
Rice	*Azorhizobium, Azospirillum, Bacillus japonicum, Bradyrhizobium elkanii, Herbaspirillum, Sphingomonas* sp.	Mano and Morisaki, 2008; Hardoim *et al.*, 2015
Wheat	*Agrobacterium, Bacillus, Bradyrhizobium, Polaromonas, Pseudomonas, Serratia, Sphingomonas, Stenotrophomonas, Variovorax*	Robinson *et al.*, 2016; Goel *et al.*, 2017
Maize	*Actinobacteria, Arthrobacter globiformis, Bacillus megaterium, Klebsiella pneumoniae, Microbacterium testaceum*	Chelius and Triplett, 2000; Zinniel *et al.*, 2002; Goel *et al.*, 2017
Pea	*Bacteroides, Enterobacter* sp. MSP10, *Ochrobactrum* sp. MSP9, *Rhizobium, Pseudomonas*	Tariq *et al.*, 2014; Lv *et al.*, 2021
Soyabean	*Bacillus cereus* strain AKAD A1-1, *Microbacterium, Pseudomonas otitidis* strain AKAD A1-2, *Pseudomonas* sp. strain AKAD A1-16, *Psychrobacillus*	Yu *et al.*, 2018; Dubey *et al.*, 2021
Tomato	*Arthrobacter globiformis, Bacillus cereus, B. megaterium, B. pumilus, Paenibacillus polymyxa, Pseudomonas pseudoalcaligenes, P. fluorescens, P. fluorescens* Pf0-1, *Serratia marcescens, Sphingomonas yanoikuyae*	Oku *et al.*, 2012; Hang *et al.*, 2013; Goel *et al.*, 2017
Sweet potato	*Bacillus cereus, Achromobacter xylosoxidans*	Dawwam *et al.*, 2013
Olive	*Pseudomonas* sp.	Mercado-Blanco and Prieto, 2012
Grapevine	*Paraburkholderia phytofirmans* strain PsJN	Compant *et al.*, 2008

and seeds (Compant *et al.*, 2010). The REB phyla predominantly belong to *Acidobacteria, Actinobacteria, Bacillota, Bacteroidota, Chloroflexota, Gemmatimonadota, Planctomycetota, Proteobacteria* and *Verrucomicrobiota* (Samad *et al.*, 2017). Diverse genera and REB species are found in various host plants (Table 11.5).

REB promote plant growth by synthesizing phytohormones, solubilizing phosphate, secreting siderophores and enhancing the abiotic and biotic stress tolerance of the host (Gaiero *et al.*, 2013; Kandel *et al.*, 2017). Additionally, some bacterial endophytes can perform biological nitrogen fixation, facilitating the conversion of dinitrogen gas (N_2) into usable forms of nitrogen accessible to the host (Bhattacharjee *et al.*, 2008; Santi *et al.*, 2013). Endophytes may also indirectly promote plant growth through antagonistic activity against fungal or bacterial pathogens.

Many techniques are used to study REB: transmission electron microscopy (TEM), SEM, FISH and triphenyl tetrazolium chloride vital staining (de Souza *et al.*, 2004; Vendramin *et al.*, 2010; Compant *et al.*, 2011; Thomas, 2011; Goel *et al.*, 2017). Green fluorescent protein tagging and β-β-glucuronidase labelling may be used to trace endophytic bacteria (Robertson-Albertyn *et al.*, 2017). Moreover, next-generation sequencing (NGS), a genomics-based technique, is also an effective tool for monitoring microbial populations in the host (Trujillo *et al.*, 2015) and can be used to rapidly analyse the diversity and composition of microbial communities. Similarly, suppression subtractive hybridization (SSH), a new technique for distinguishing closely related endophytic bacterial species, has also been used in studying REB (Galbraith *et al.*, 2004; Monteiro *et al.*, 2012). Differential expression analysis of endophytes has been performed using SOLiD-SAGE techniques and shotgun metagenomics microarray analysis (Dinkins *et al.*, 2010; Ambrose and Belanger, 2012; Sessitsch *et al.*, 2012).

11.3 Mechanisms of Plant Growth Promotion by Beneficial Root-Dwelling Bacteria

Root-dwelling bacteria are known to enhance soil health and stimulate plant growth. Plant growth-promoting RDB (PGPRDB) impact the growth of plants through both direct and indirect mechanisms (Prashar *et al.*, 2013). Direct mechanisms include phytohormone and siderophore production and nutrient acquisition (biological nitrogen fixation; phosphorus, potassium and sulfur solubilization). PGPRDB may also act indirectly as a biocontrol agent, suppressing phytopathogens via antimicrobial agents (bacteriocin-like peptides, antibiotics) and lytic enzymes (chitinases, β-1,3-glucanases) (Znój et al., 2021). They also induce plant tolerance to stresses such as drought, floods, extreme temperature, salinity and nutrient deficiency (Ho et al., 2017). A flowchart of the mode of action by PGPRDB is shown in Fig. 11.1.

11.3.1 Direct mechanisms

Biological nitrogen fixation

Nitrogen is a crucial macronutrient for plant growth. Biological nitrogen fixation (BNF) is the process by which microorganisms convert atmospheric nitrogen into ammonia that plants can utilize. This process depends on nitrogenase and leghaemoglobin (Hayat et al., 2010; Choudhary et al., 2018). The transfer of electrons during the conversion of N_2 to NH_3 is influenced by dinitrogenase reductase. BNF can be classified as symbiotic (SNF) or non-symbiotic nitrogen fixation (NSNF) (Kim and Rees, 1994; Chandra and Singh, 2016).

SYMBIOTIC N_2 FIXATION. Symbiotic N_2-fixing bacteria (SNFB), including *Rhizobiaceae* members, have mutually beneficial relationships with leguminous plants (Ahemad and Kibret, 2014). These bacteria invade root hairs, stimulate root nodule formation and establish intimate associations with the host. Bacteria convert free N_2 into ammonia within nodules. The bacteria respond chemotactically to flavonoid molecules released by the host. These compounds induce nodulation (*nod*) gene expression in rhizobia, which produces lipochito-oligosaccharides (LCOs),

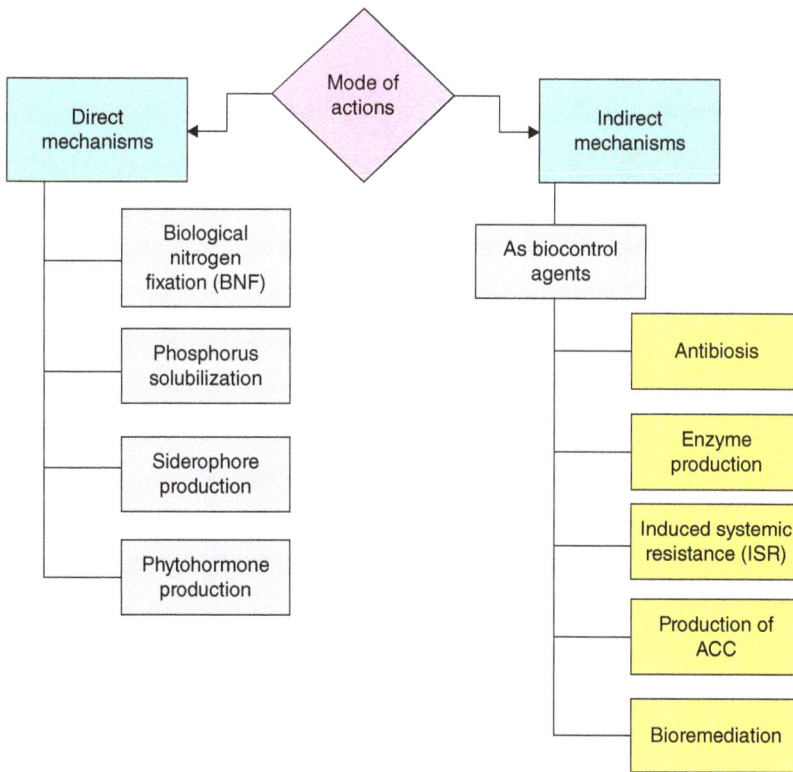

Fig. 11.1. Mode of actions of root-dwelling bacteria. ACC, 1-aminocyclopropane-1-carboxylate.

triggering cell division and the formation of nod-ules (Dakora, 1995; Lhuissier *et al.*, 2001; Mati-ru and Dakora, 2004). SNF has the potential to reduce or eliminate the need for synthetic fertil-izers. Non-legumes may also receive N_2 fixed by rhizobia via the transfer of assimilable nitrogen to intercrops or successive crops rotated with legumes (Snapp *et al.*, 1998; Shah *et al.*, 2003; Hayat, 2005; Hayat *et al.*, 2008a, 2008b). Major SNFB include members of *Allorhizobium*, *Azorhizobium*, *Bradyrhizobium*, *Mesorhizobium*, *Methylobacterium*, *Rhizobium* and *Sinorhizobium* (Sahgal and Johri, 2003).

NON-SYMBIOTIC N_2 FIXATION. Non-symbiotic N_2 fix-ation is performed by free-living bacteria and endophytes (Franche *et al.*, 2009; Bhattacha-ryya and Jha, 2012). Non-symbiotic N_2-fixing bacteria (NSNFB) include members of *Achromo-bacter*, *Arthrobacter*, *Azomonas*, *Azotobacter*, *Bacillus*, *Clostridium*, *Corynebacterium*, *Entero-bacter*, *Gluconacetobacter diazotrophicus*, *Klebsiella*, *Pseudomonas*, *Rhodopseudomonas*, *Rhodospirillum* and *Xanthobacter* (Vessey, 2003; Barriuso *et al.*, 2008).

Phosphorus solubilization

In addition to nitrogen, phosphorus is a critical macronutrient for plants. It regulates physio-logical activities such as respiration, energy transfer, signal transduction, photosynthesis and macromolecule biosynthesis (Khan *et al.*, 2010). Phosphorus sources are abundant in soils in both organic and inorganic forms (Fernández *et al.*, 2007; Ahemad, 2015). However, the bulk of soil phosphorus is insoluble (Bhattacharyya and Jha, 2012).

Therefore, an important plant growth-promoting trait is the ability to convert insoluble phosphates into soluble forms that the host can utilize (Igual *et al.*, 2001; Rodríguez *et al.*, 2006). Phosphate-solubilizing bacteria (PSB) can solu-bilize hydroxyl apatite, rock phosphate, dicalcium phosphate and tricalcium phosphate (Rodríguez and Fraga, 1999; Rodríguez *et al.*, 2006; Singh *et al.*, 2015; Saha *et al.*, 2016; Sureshbabu *et al.*, 2016; Yadav and Sidhu, 2016). PSB may release low-molecular-weight organic acids to solubilize phosphorus. Additionally, PSB may secrete certain agents that dissolve minerals, such as protons, hydroxyl radicals, siderophores and

CO_2. Some PSB may also release extracellular en-zymes to solubilize organic phosphates (Choud-hary *et al.*, 2018). PSB predominantly belong to the genera *Achromobacter*, *Aerobacter*, *Bacillus*, *Burkholderia*, *Erwinia*, *Flavobacterium*, *Micrococ-cus*, *Neptune* and *Rhizobium* (Hayat *et al.*, 2010).

Soil organic phosphorus solubilization is crucial for phosphorus cycling, as 4–90% of total soil phosphorus is organic phosphorus (Khan *et al.*, 2009). Phosphatases, phytases, phosphonatases and other enzymes secreted by PSB facilitate the mineralization of soil phos-phorus from diverse sources, such as phytin or inositol phosphates, nucleic acids and phospho-lipids (Saravanan *et al.*, 2016; Choudhary *et al.*, 2018). PSB include the genera *Bacillus*, *Citrobac-ter*, *Enterobacter*, *Proteus*, *Pseudomonas*, *Rhizo-bium* and *Serratia*, which solubilize organic phos-phates via phosphatases (Abd-Alla, 1994; Thaller *et al.*, 1995; Skraly and Cameron, 1998).

Siderophore production

In addition to being a cofactor for many different enzymatic processes, iron is an important micro-nutrient for the growth of plants. Furthermore, it plays a remarkable role in several physiological processes, for example N_2 fixation, respiration and photosynthesis (Sharma and Johri, 2003). Iron can potentially form oxyhydroxides and in-soluble hydroxides but is primarily found as Fe^{3+} ions in aerobic environments. As a result, neither plants nor microbes can access iron (Rajkumar *et al.*, 2010). Siderophores are low-molecular-weight (~400–1500 Da) compounds with high affinity for Fe^{3+} ions (stability constants ranging between $K = 10^{30}$ and $K = 10^{52}$), membrane re-ceptors and other micronutrients such as mo-lybdenum, manganese, cobalt and nickel. On the bacterial membrane, the siderophore–iron (Fe^{3+}) complex is reduced to Fe^{2+}, which is then re-leased from siderophores into the cell through a gating mechanism (Ahmed and Holmström, 2014).

When the availability of iron is low, microbial siderophores provide iron to plants, enhancing their growth. However, the precise underlying mechanism remains largely unknown. There are two possible mechanisms: (i) highly oxidized microbial siderophores can be reduced to do-nate Fe^{2+} to the plant; in this mechanism, it is speculated that the microbial Fe^{3+}–siderophore complexes may be transported to the apoplast of the root where Fe^{2+} is trapped, leading to high concentrations of iron in the root; (ii) exchange of iron chelated by microbial siderophores with phytosiderophores. This mechanism depends on the concentrations and stability constants of both phytosiderophores and microbial siderophores, as well as the redox conditions and pH of the root environment (Ahmed and Holmström, 2014).

Typically, siderophores are categorized based on the ligands utilized to chelate iron. The main groups of siderophores include carboxylates (e.g. citric acid derivatives), catecholates (phenolates) and hydroxamates. *Agrobacterium tumefaciens, Azotobacter vinelandii, Burkholderia cepacian, Paracoccus denitrificans, Pseudomonas aeruginosa, Pseudomonas putida* and *Rhizobium meliloti* strain DM4 are well-known siderophore producers (Choudhary *et al.*, 2018).

Phytohormone production

Phytohormones are plant-produced organic compounds that affect physiological activities in incredibly low amounts. In addition to plants, several RDB may also secrete phytohormones and phytohormone-like substances (Mandal and Kotasthane, 2014; Porcel *et al.*, 2014). Major phytohormones that support the growth of plants and help plants tolerate stress include indole acetic acid (IAA), gibberellin (GA), cytokinin (CK) and ethylene (Skirycz and Inzé, 2010; Fahad *et al.*, 2015; Chandra and Singh, 2016).

IAA promotes cell elongation, differentiation, extension and division. It is generated by tryptophan transamination and decarboxylation in immature leaves, stems and seeds (Das *et al.*, 2013; Kundan *et al.*, 2015; Kaur *et al.*, 2016). The rhizobacterial IAA affects host biological processes by altering the pool of plant auxins. IAA from bacteria stimulates the growth of plant roots to efficiently access soil nutrients (Glick, 2012). Consequently, the rhizobacterial IAA is acknowledged as a major mediator in the interactions between plants and microbes (Spaepen and Vanderleyden, 2011; Glick, 2014). Bacterial producers of IAA, including members of *Agrobacterium, Bradyrhizobium, Enterobacter, Klebsiella, Pseudomonas* and *Rhizobium* spp., may also synthesize indole-3-acetic aldehyde and indole-3-pyruvic acid (Shilev, 2013).

GA influences the development of flowers and fruits, the growth of leaves and stems, the elongation of shoots, the induction of flowers and the germination and emergence of seeds (Bottini *et al.*, 2004; Spaepen and Vanderleyden, 2011). By inhibiting root elongation and branching and encouraging the formation of root hairs, CK stimulates the division of plant cells and regulates the development of roots (Werner *et al.*, 2003; Riefler *et al.*, 2006). CK-producing bacteria include several species, such as *Azobacter* sp., *Bacillus subtilis, Pseudomonas fluorescens* and *Rhizobium leguminosarum* (Noel *et al.*, 1996; García de Salamone *et al.*, 2001; Aloni *et al.*, 2006; Sokolova *et al.*, 2011; Liu *et al.*, 2013). In plants, ethylene initiates root development, promotes fruit ripening, stimulates seed germination, reduces wilting, activates the synthesis of other phytohormones, inhibits root elongation and promotes leaf abscission. However, under stressful conditions such as drought, heavy metal stress and phytopathogen attack, 'stress ethylene' is produced by plants, causing shrinkage of the root biomass and thereby impairing the plant. 1-Aminocyclopropane-1-carboxylate (ACC) deaminase produced by PGPRDB can alleviate this 'stress ethylene'. Members of *Achromobacter, Acinetobacter, Agrobacterium, Azospirillum, Bacillus, Burkholderia, Enterobacter, Pseudomonas, Rhizobium* and *Serratia*, among others, are prolific producers of ACC deaminase (Glick, 2012; Das *et al.*, 2013; Choudhary *et al.*, 2018).

11.3.2 Indirect mechanisms

Beneficial RDB indirectly promote the growth of plants by serving as biocontrol agents against phytopathogens. This involves the synthesis of lytic enzymes (e.g. chitinases, cellulases, 1,3-glucanases, proteases and lipases) and the production of siderophores, ACC deaminase and antibiotics. PGPRDB may also promote plant growth by inducing host systemic resistance, enabling it to protect against herbivores and necrotrophic pathogens (Pieterse *et al.*, 2001; Glick, 2012; Kundan *et al.*, 2015; Shrivastava *et al.*, 2016; Velázquez *et al.*, 2016).

Antibiosis

The most effective mechanism to control/suppress phytopathogens is the synthesis of antibiotic compounds by beneficial RDB (Shilev, 2013). The antibiotics that suppress root diseases caused by phytopathogens include amphisin, butyrolactones, cyclic lipopeptides, diacetyl phloroglucinol (DAPG), hydrogen cyanide (HCN), kanosamine, oligomycin A, phenazines, phloroglucinols, pyoluteorinphenazine-1-carboxylic acid (PCA), pyrrolnitrin, tensin, viscosinamide, xanthobaccin and zwittermycin A (Nielsen and Sørensen, 2003; Beneduzi *et al.*, 2012). Pyrrolnitrin produced by the *Pseudomonas fluorescens* BL915 strain can control *Rhizoctonia solani* infection in cotton plants (Hill *et al.*, 1994). *P. aureofaciens* produces PCA with biocontrol activity against *Sclerotinia homeocarpa*, infecting creeping bentgrass (Choudhary *et al.*, 2018).

Some studies also reported siderophores with antimicrobial activities preventing the growth of fungi and other phytopathogens (Martínez-Viveros *et al.*, 2010). For example, *P. chlororaphis* YL-1 was reported to produce the siderophore pyoverdine, which has antimicrobial activity against *Bacillus megaterium*, *Erwinia amylovora*, *Xanthomonas oryzae* pv. *oryzae* and *X. oryzae* pv. *oryzicola* (Liu *et al.*, 2021). Pyoverdine from *P. syringae* was shown to have an antagonistic effect on *Caenorhabditis elegans* (Timofeeva *et al.*, 2022). Siderophores produced by the *Bacillus* genus are found to be most effective against the fungi *Bipolarismaydis*, *Colletotrichum graminicola*, *Cercospora zea-maydis* and *Fusarium verticillioides* (Szilagyi-Zecchin *et al.*, 2014).

Enzyme production

Beneficial RDB can produce hydrolytic enzymes that hinder the growth of phytopathogens and serve as an effective biocontrol agents. These enzymes, which are predominantly effective against fungal pathogens, include catalase, cellulases, chitinases, glucanases, proteases and urease (Bowman and Free, 2006). As major components of the cell wall of fungi include chitin and β-glucan, bacteria producing β-glucanases and chitinases can, therefore, inhibit the growth of fungi. *P. fluorescens* LPK2 and *Sinorhizobium fredii* KCC5 produce chitinases and β-glucanases, thus inhibiting *F. udum*, which

causes fusarium wilt in pigeon peas (Kumar *et al.*, 2010).

Induced system resistance

The interaction of plant roots with RB can elicit host resistance against pathogenic microbes, a process known as induced systemic resistance (ISR) (Lugtenberg and Kamilova 2009). PGPRDB mediates ISR by lignifying the cell wall, forming structural barriers (such as the apposition of the cell wall and the deposition of newly formed callose) and accumulating phenolic compounds at the site where the hyphae of the pathogen may penetrate. This quick defence response at fungal entry sites slows the infection process. This provides the host with enough time to develop other defence mechanisms that limit pathogen growth to the root tissue's outermost layers (Conrath *et al.*, 2006; Pastor *et al.*, 2013). ISR triggered by beneficial rhizobacteria provides broad-range resistance against various phytopathogens. PGPRDB that elicit ISR include *Bacillus*, *Pseudomonas*, *Serratia*, *Trichoderma*, etc. (Choudhary *et al.*, 2018).

ACC deaminase production

Beneficial RDB facilitate plant growth by producing ACC deaminase, which converts plant ACC to ammonia and α-ketobutyrate (Arshad and Frankenberger, 1998; Saleem *et al.*, 2007). PGPRDB utilize this enzyme to reduce the level of ethylene stress, thus helping the host maintain a robust root system and to overcome various stresses (Glick, 2014; Souza *et al.*, 2015; Choudhary *et al.*, 2018). Various crops benefit from ACC deaminase-producing rhizobacteria by facilitating plant growth, rhizobial nodulation, mycorrhizal colonization and nitrogen, phosphorus and potassium uptake. Bacterial species/genera such as *Alcaligenes*, *Bacillus pumilus*, *Enterobacter cloacae*, *Methylobacterium fujisawaense*, *Ochrobactrum*, *Pseudomonas*, *Rhizobium* and *Variovorax paradoxus* (Ahmad *et al.*, 2013; Noumavo *et al.*, 2016) are well-known ACC deaminase producers.

Bioremediation

It is beneficial for crop plants to rely on microbes to extract heavy metals from contaminated rhizospheres via remediation (Glick, 2010;

Cabello-Conejo *et al.*, 2014). Several reports of RDB show that they can withstand heavy metals, breakdown soil contaminants and convert them into less harmful substances. *Streptomyces* AR16 was reported to have extremely high zinc resistance. *Serratia* BR780, *Streptomyces* AR36 and *Flavobacterium* PR01 tolerated zinc, cadmium and lead (Kuffner *et al.*, 2008). *Kocuria* sp. CRB15 can withstand heavy metals, for example zinc, cadmium, copper, nickel and lead. This microbe is a game changer, offering a real solution to environmental contamination (Hansda *et al.*, 2017). Table 11.6 lists the major groups of RDB that directly promote plant growth, and Table 11.7 shows some beneficial RDB with biocontrol activity against phytopathogens.

11.4 Conclusion

The ever-burgeoning human population generates a never-ending demand for more food and fodder. Hence, enhancing the productivity of major crops such as rice, wheat and maize is warranted. Traditionally, this has been achieved through massive inputs of synthetic fertilizers and pesticides. However, this may harm the soil ecosystem, animal and human health and other non-target organisms. A better alternative is the use of beneficial RDB, which can help crop production without causing adverse environmental impacts. PGPRDB can actively enhance plant growth by mobilizing nutrients through biological N_2 fixation, phosphorus solubilization and siderophore production and, indirectly, by inhibiting the growth of phytopathogens through cell wall-degrading enzymes. PGPRDB may also elicit ISR or produce antibiotics to antagonize various phytopathogens and stimulate plant growth. However, the current understanding of plant–microbe interactions is still limited. Advancements in metagenomics, amplicon sequencing, metabolomics and the use of pathogen–host interation-based databases have improved our understanding of plant–bacterial interactions.

PGPRDB deployment is an alternative method for the efficient and sensible use of

Table 11.6. Major groups of root-dwelling bacteria causing direct growth promotion.

Mode of action	Crop plant	Bacteria	Reference
Biological nitrogen fixation	Alfalfa	*Sinorhizobium meliloti, Rhizobium meliloti*	Choudhary *et al.*, 2018
	Peas	*Rhizobium leguminosarum* biovar *viceae*	
	Lotus	*Mesorhizobium loti*	
	Rice	*Aulosira fertilissima*	
Phosphorus solubilization and siderophore production	Peanut	*Bacillus velesensis*	Chen *et al.*, 2019
	Groundnut	*Chryseobacterium indologenes, Enterobacter cloacae, E. ludwigii, Klebsiella pneumoniae, Pseudomonas aeruginosa*	Dhole *et al.*, 2016
	Millet	*Bacillus amyloliquefaciens, B. cereus, B. subtilis*	Kushwaha *et al.*, 2020
	Hopbush	*Bacillus idriensis, Pinus taiwanensis, Prunus geniculate, Streptomyces alboniger*	Afzal *et al.*, 2017
Phytohormone production			
IAA	Soybean	*Bacillus, Microbacterium, Lysinibacillus, Psychrobacillus*	Yu, J. *et al.*, 2018
	Lavender	*Bacillus thuringiensis*	Armada *et al.*, 2014
	Mustard	*Pseudomonas aeruginosa, P. putida*	Ahemad and Khan, 2012
GA	Maize	*Azospirillum lipoferum*	Cohen *et al.*, 2009
	Tomato	*Sphingomonas* sp.	Khan *et al.*, 2014
ACC deaminase	Wheat	*Bacillus thuringiensis* AZP2	Timmusk *et al.*, 2014
	Mung bean	*Pseudomonas putida*	Mayak *et al.*, 1999

ACC, 1-aminocyclopropane-1-carboxylate; GA, gibberellin; IAA, indole acetic acid.

Table 11.7. Beneficial root-dwelling bacteria with biocontrol activity against phytopathogens.

Mode of action	Biocontrol agent	Plant pathogen	Crop plant	Reference
Antagonism	*Bacillus cereus* strain AKAD A1-1, *Pseudomonas* sp. strain AKAD A1-16, *P. otitidis* strain AKAD A1-2	*Alternaria alternata, Fusarium oxysporum, Macrophomina phaseolina*	Soybean	Dubey *et al.,* 2021
	Bacillus licheniformis Bl17, *Paenibacillus peoriae* Pa86, *Pseudomonas brassicacearum* Psb101, *P. brassicacearum* Ps169, *P. putida* Ps52	*Rhizotonia solani*	Potato	Bahmani *et al.,* 2021
Antibiotic production	*Pseudomonas* sp.	*Gaeumannomyces graminis, Fusarium oxysporum*	Tomato	Chin-A-Woeng *et al.,* 2003
	Pseudomonas fluorescens BL915	*Rhizotonia solani*	Cotton	Hill *et al.,* 1994
Antibiotic and siderophore production, induction of systemic resistance	*Bacillus cereus, Brevibacterium laterosporus, Serratia marcescens, Pseudomonas fluorescens*	*Pythium ultimum*	Sorghum	Idris *et al.,* 2008
Siderophore and IAA production, antagonism	*Bacillus* sp.	*Fusarium oxysporum* f. sp. *cubense*	Banana	Anusuya and Manimekalai, 2016
Siderophore production	*Azospirillum brasilense*	*Colletotrichum acutatum*	Strawberry	Tortora *et al.,* 2011

IAA, indole acetic acid.

agricultural resources with minimal negative environmental effects. In addition, these bacteria offer a diverse range of opportunities for the development of sustainable agriculture across the globe (Choudhary *et al.,* 2016). Therefore, PGPRDB represent the epitome of an ecological farming system that safeguards the environment and enhances crop productivity. Although beneficial root-associated bacteria have shown potential for improving the stability and productivity of agroecosystems, further research and technological development are urgently warranted to fully explore the promise of PGPRDB. This will lead humanity closer to achieving long-term sustainable agricultural and environmental systems without compromising the health of the planet and its inhabitants.

References

Abd-Alla, M.H. (1994) Solubilization of rock phosphates by *Rhizobium* and *Bradyrhizobium. Folia Microbiologica* 39, 53–56.

Abd El-Ghany T.M.A., Masrahi, Y.S., Alawlaqi, M.M. and Al Abboud, M.A. (2015) Rhizosphere and rhizoplane bacteria isolated from subtropical region of Jazan in Saudi Arabia. *Journal of Biological Chemistry and Research* 32, 934–944.

Afzal, I., Iqrar, I., Shinwari, Z.K. and Yasmin, A. (2017) Plant growth-promoting potential of endophytic bacteria isolated from roots of wild *Dodonaea viscosa* L. *Plant Growth Regulation* 81, 399–408.

Ahemad, M. (2015) Phosphate-solubilizing bacteria-assisted phytoremediation of metalliferous soils: a review. *3 Biotech* 5, 111–121.

Ahemad, M. and Khan, M.S. (2012) Effect of fungicides on plant growth promoting activities of phosphate solubilizing Pseudomonas putida isolated from mustard (Brassica compestris) rhizosphere. *Chemosphere* 86, 945–950.

Ahemad, M. and Kibret, M. (2014) Mechanisms and applications of plant growth promoting rhizobacteria: current perspective. *Journal of King Saud University – Science* 26, 1–20.

Ahmad, M., Zahir, Z.A., Khalid, M., Nazli, F. and Arshad, M. (2013) Efficacy of Rhizobium and Pseudomonas strains to improve physiology, ionic balance and quality of mung bean under salt-affected conditions on farmer's fields. *Plant Physiology and Biochemistry* 63, 170–176.

Ahmed, E. and Holmström, S.J.M. (2014) Siderophores in environmental research: roles and applications. *Microbial Biotechnology* 7, 196–208.

Aloni, R., Aloni, E., Langhans, M. and Ullrich, C.I. (2006) Role of cytokinin and auxin in shaping root architecture: regulating vascular differentiation, lateral root initiation, root apical dominance and root gravitropism. *Annals of Botany* 97, 883–893.

Ambrose, K.V. and Belanger, F.C. (2012) SOLiD-SAGE of endophyte-infected red fescue reveals numerous effects on host transcriptome and an abundance of highly expressed fungal secreted proteins. *PLoS One* 7, e53214.

Amend, A.S., Cobian, G.M., Laruson, A.J., Remple, K., Tucker, S.J. *et al.* (2019) Phytobiomes are compositionally nested from the ground up. *PeerJ* 7, e6609.

Anusuya, V. and Manimekalai, G. (2016) Potential of bacillus isolates as biocontrol agent against fusarium wilt of banana. *World Journal of Pharmaceutical Sciences* 5, 1679–1694.

Armada, E., Roldán, A. and Azcon, R. (2014) Differential activity of autochthonous bacteria in controlling drought stress in native Lavandula and Salvia plants species under drought conditions in natural arid soil. *Microbial Ecology* 67, 410–420.

Arshad, M. and Frankenberger, W.T. (1991) Microbial production of plant hormones. *Plant and Soil* 133, 1–8.

Arshad, M. and Frankenberger, W.T. (1998) Plant growth regulating substances in the rhizosphere: microbial production and functions. *Advances in Agronomy* 62, 46–151.

Babalola, O.O. (2010) Beneficial bacteria of agricultural importance. *Biotechnology Letters* 32, 1559–1570.

Bahmani, K., Hasanzadeh, N., Harighi, B. and Marefat, A. (2021) Isolation and identification of endophytic bacteria from potato tissues and their effects as biological control agents against bacterial wilt. *Physiological and Molecular Plant Pathology* 116, 101692.

Barriuso, J., Ramos Solano, B., Lucas, J.A., Lobo, A.P., García-Villaraco, A. *et al.* (2008) Ecology, genetic diversity and screening strategies of plant growth promoting rhizobacteria (PGPR). In: Ahmad, I., Pichtel, J. and Hayat, S. (eds) *Plant–Bacteria Interactions*. Wiley-VCH, Weinheim, Germany, pp. 1–17.

Begonia, M.F.T. and Kremer, R.J. (1994) Chemotaxis of deleterious rhizobacteria to velvetleaf (*Abutilon theophrasti* Medik.) seeds and seedlings. *FEMS Microbiology Ecology* 15, 227–236.

Beneduzi, A., Ambrosini, A. and Passaglia, L.M.P. (2012) Plant growth-promoting rhizobacteria (PGPR): their potential as antagonists and biocontrol agents. *Genetics and Molecular Biology* 35, 1044–1051.

Bhattacharjee, R.B., Singh, A. and Mukhopadhyay, S.N. (2008) Use of nitrogen-fixing bacteria as biofertiliser for non-legumes: prospects and challenges. *Applied Microbiology and Biotechnology* 80, 199–209.

Bhattacharyya, P.N. and Jha, D.K. (2012) Plant growth-promoting rhizobacteria (PGPR): emergence in agriculture. *World Journal of Microbiology and Biotechnology* 28, 1327–1350.

Bottini, R., Cassán, F. and Piccoli, P. (2004) Gibberellin production by bacteria and its involvement in plant growth promotion and yield increase. *Applied Microbiology and Biotechnology* 65, 497–503.

Bowman, S.M. and Free, S.J. (2006) The structure and synthesis of the fungal cell wall. *BioEssays* 28, 799–808.

Bulgarelli, D., Rott, M., Schlaeppi, K., Ver Loren van Themaat, E., Ahmadinejad, N. *et al.* (2012) Revealing structure and assembly cues for *Arabidopsis* root-inhabiting bacterial microbiota. *Nature* 488, 91–95.

Bulgarelli, D., Schlaeppi, K., Spaepen, S., Ver Loren van Themaat, E. and Schulze-Lefert, P. (2013) Structure and functions of the bacterial microbiota of plants. *Annual Review of Plant Biology* 64, 807–838.

Cabello-Conejo, M.I., Becerra-Castro, C., Prieto-Fernández, A., Monterroso, C., Saavedra-Ferro, A. *et al.* (2014) Rhizobacterial inoculants can improve nickel phytoextraction by the hyperaccumulator *Alyssum pintodasilvae*. *Plant and Soil* 379, 35–50.

Cavaglieri, L., Orlando, J. and Etcheverry, M. (2009) Rhizosphere microbial community structure at different maize plant growth stages and root locations. *Microbiological Research* 164, 391–399.

Chandra, P. and Singh, E. (2016) Applications and mechanisms of plant growth-stimulating rhizobacteria. In: Choudhary, D., Varma, A. and Tuteja, N. (eds) *Plant–Microbe Interaction: An Approach to Sustainable Agriculture*. Springer, Singapore, pp. 37–62.

Chelius, M.K. and Triplett, E.W. (2000) Immunolocalization of dinitrogenase reductase produced by Klebsiella pneumoniae in association with Zea mays L. *Applied and Environmental Microbiology* 66, 783–787.

Chen, L., Shi, H., Heng, J., Wang, D. and Bian, K. (2019) Antimicrobial, plant growth-promoting and genomic properties of the peanut endophyte *Bacillus velezensis* LDO2. *Microbiological Research* 218, 41–48.

Chin-A-Woeng, T.F.C., Bloemberg, G.V. and Lugtenberg, B.J.J. (2003) Phenazines and their role in biocontrol by pseudomonas bacteria. *New Phytologist* 157, 503–523.

Choudhary, M., Ghasal, P.C., Kumar, S., Yadav, R.P., Singh, S. *et al.* (2016) Conservation agriculture and climate change: an overview. In: Bisht, J.K., Meena, V.S., Mishra, P.K. and Pattanayak, A. (eds) *Conservation Agriculture: An Approach to Combat Climate Change in Indian Himalaya*. Springer, Singapore, pp. 1–37.

Choudhary, M., Ghasal, P.C., Yadav, R.P., Meena, V.S., Mondal, T. *et al.* (2018) Towards plant-beneficiary rhizobacteria and agricultural sustainability. In: Meena, V.S. (ed.) *Role of Rhizospheric Microbes in Soil*. Springer, Singapore, pp. 1–46.

Ciccazzo, S., Esposito, A., Rolli, E., Zerbe, S., Daffonchio, D. *et al.* (2014) Different pioneer plant species select specific rhizosphere bacterial communities in a high mountain environment. *SpringerPlus* 3, 1–10.

Clark, F.E. (1940) Notes on types of bacteria associated with plant roots. *Transactions of the Kansas Academy of Science* 43, 75.

Cohen, A.C., Travaglia, C.N., Bottini, R. and Piccoli, P.N. (2009) Participation of abscisic acid and gibberellins produced by endophytic *Azospirillum* in the alleviation of drought effects in maize. *Botany* 87, 455–462.

Compant, S., Kaplan, H., Sessitsch, A., Nowak, J., Ait Barka, E. and Clément, C. (2008) Endophytic colonization of Vitis vinifera L. by Burkholderia phytofirmans strain PsJN: from the rhizosphere to inflorescence tissues. *FEMS Microbiology Ecology* 63, 84–93.

Compant, S., Clément, C. and Sessitsch, A. (2010) Plant growth-promoting bacteria in the rhizo- and endosphere of plants: their role, colonization, mechanisms involved and prospects for utilization. *Soil Biology and Biochemistry* 42, 669–678.

Compant, S., Mitter, B., Colli-Mull, J.G., Gangl, H. and Sessitsch, A. (2011) Endophytes of grapevine flowers, berries, and seeds: identification of cultivable bacteria, comparison with other plant parts, and visualization of niches of colonization. *Microbial Ecology* 62, 188–197.

Compant, S., Muzammil, S., Lebrihi, A. and Mathieu, F. (2013) Visualization of grapevine root colonization by the Saharan soil isolate *Saccharothrix algeriensis* NRRL B-24137 using DOPE-FISH microscopy. *Plant and Soil* 370, 583–591.

Compant, S., Cambon, M.C., Vacher, C., Mitter, B., Samad, A. *et al.* (2021) The plant endosphere world – bacterial life within plants. *Environmental Microbiology* 23, 1812–1829.

Conrath, U., Beckers, G.J., Flors, V., García-Agustín, P., Jakab, G. *et al.* (2006) Priming: getting ready for battle. *Molecular Plant–Microbe Interactions* 19, 1062–1071.

Cooper, J. and Gardener, M. (2006) Bacterial Plant Pathogens and Symptomology. WSU County Extension, SJI. Available at: https://s3.wp.wsu.edu/uploads/sites/2054/2014/04/BacterialPlantPathogens_001.pdf (last accessed November 2024).

Dakora, F.D. (1995) Plant flavonoids: biological molecules for useful exploitation. *Functional Plant Biology* 22, 87.

Dakora, F.D. and Phillips, D.A. (2002) Root exudates as mediators of mineral acquisition in low-nutrient environments. In: Adu-Gyamfi, J.J. (ed.) *Food Security in Nutrient-Stressed Environments: Exploiting Plants' Genetic Capabilities*. Springer, Dordrecht, the Netherlands, pp. 201–213.

Das, A.J., Kumar, M. and Kumar, R. (2013) Plant growth promoting rhizobacteria (PGPR): an alternative of chemical fertilizer for sustainable, environmentally friendly agriculture. *Research Journal of Agricultural Sciences* 4, 21–23.

Dawwam, G.E., Elbeltagy, A., Emara, H.M., Abbas, I.H. and Hassan, M.M. (2013) Beneficial effect of plant growth promoting bacteria isolated from the roots of potato plant. *Annals of Agricultural Sciences* 58, 195–201.

De Andrade, L.A., Santos, C.H.B., Frezarin, E.T., Sales, L.R. and Rigobelo, E.C. (2023) Plant growth-promoting rhizobacteria for sustainable agricultural production. *Microorganisms* 11, 1088.

De Souza, A.O., Pamphile, J.A., Rocha, C.D.M.D. and Azevedo, J.L. (2004) Plant-microbe interactions between maize (*Zea mays* L.) and endophytic microorganisms observed by scanning electron microscopy. *Acta Scientiarum Biological Sciences* 26, 357–359.

De Weert, S., Vermeiren, H., Mulders, I.H.M., Kuiper, I., Hendrickx, N. *et al.* (2002) Flagella-driven chemotaxis towards exudate components is an important trait for tomato root colonization by *Pseudomonas fluorescens*. *Molecular Plant–Microbe Interactions* 15, 1173–1180.

Deora, A., Hashidoko, Y., Islam, M.T. and Tahara, S. (2005) Antagonistic rhizoplane bacteria induce diverse morphological alterations in Peronosporomycete hyphae during *in vitro* interaction. *European Journal of Plant Pathology* 112, 311–322.

Dhole, A., Shelat, H., Vyas, R., Jhala, Y. and Bhange, M. (2016) Endophytic occupation of legume root nodules by *nif*H-positive non-rhizobial bacteria, and their efficacy in the groundnut (*Arachis hypogaea*). *Annals of Microbiology* 66, 1397–1407.

Dibbern, D., Schmalwasser, A., Lueders, T. and Totsche, K.U. (2014) Selective transport of plant root-associated bacterial populations in agricultural soils upon snowmelt. *Soil Biology and Biochemistry* 69, 187–196.

Ding, L.-J., Cui, H.-L., Nie, S.-A., Long, X.-E., Duan, G.-L. *et al.* (2019) Microbiomes inhabiting rice roots and rhizosphere. *FEMS Microbiology Ecology* 95, fiz040.

Dinkins, R.D., Barnes, A. and Waters, W. (2010) Microarray analysis of endophyte-infected and endophyte-free tall fescue. *Journal of Plant Physiology* 167, 1197–1203.

Dolatabadian, A. (2020) Plant–microbe interaction. *Biology* 10, 15.

Dörr, J., Hurek, T. and Reinhold-Hurek, B. (1998) Type IV pili are involved in plant–microbe and fungus–microbe interactions. *Molecular Microbiology* 30, 7–17.

Dubey, A., Saiyam, D., Kumar, A., Hashem, A., Abd Allah, E.F. *et al.* (2021) Bacterial root endophytes: characterization of their competence and plant growth promotion in soybean (*Glycine max* L. Merr.) under drought stress. *International Journal of Environmental Research and Public Health* 18, 931.

Duijff, B.J., Gianinazzi-Pearson, V. and Lemanceau, P. (1997) Involvement of the outer membrane lipopolysaccharides in the endophytic colonization of tomato roots by biocontrol *Pseudomonas fluorescens* strain WCS417r. *New Phytologist* 135, 325–334.

Edwards, J., Johnson, C., Santos-Medellín, C., Lurie, E., Podishetty, N.K. *et al.* (2015) Structure, variation, and assembly of the root-associated microbiomes of rice. *Proceedings of the National Academy of Sciences USA* 112, E911–920.

Eller, G. and Frenzel, P. (2001) Changes in activity and community structure of methane-oxidizing bacteria over the growth period of rice. *Applied and Environmental Microbiology* 67, 2395–2403.

Fahad, S., Hussain, S., Bano, A., Saud, S., Hassan, S. *et al.* (2015) Potential role of phytohormones and plant growth-promoting rhizobacteria in abiotic stresses: consequences for changing environment. *Environmental Science and Pollution Research* 22, 4907–4921.

Fernández, L.A., Zalba, P., Gómez, M.A. and Sagardoy, M.A. (2007) Phosphate-solubilization activity of bacterial strains in soil and their effect on soybean growth under greenhouse conditions. *Biology and Fertility of Soils* 43, 805–809.

Franche, C., Lindström, K. and Elmerich, C. (2009) Nitrogen-fixing bacteria associated with leguminous and non-leguminous plants. *Plant and Soil* 321, 35–59.

Gaiero, J.R., McCall, C.A., Thompson, K.A., Day, N.J., Best, A.S. *et al.* (2013) Inside the root microbiome: bacterial root endophytes and plant growth promotion. *American Journal of Botany* 100, 1738–1750.

Galbraith, E.A., Antonopoulos, D.A. and White, B.A. (2004) Suppressive subtractive hybridization as a tool for identifying genetic diversity in an environmental metagenome: the rumen as a model. *Environmental Microbiology* 6, 928–937.

García, J.A.L., Probanza, A., Ramos, B., Palomino, M.R. and Gutiérrez Mañero, F.J. (2004) Effect of inoculation of *Bacillus licheniformis* on tomato and pepper. *Agronomie* 24, 169–176.

García de Salamone, I.E., Hynes, R.K. and Nelson, L.M. (2001) Cytokinin production by plant growth promoting rhizobacteria and selected mutants. *Canadian Journal of Microbiology* 47, 404–411.

Giongo, A., Beneduzi, A., Ambrosini, A., Vargas, L.K., Stroschein, M.R. *et al.* (2010) Isolation and characterization of two plant growth-promoting bacteria from the rhizoplane of a legume (*Lupinus albescens*) in sandy soil. *Revista Brasileira de Ciência Do Solo* 34, 361–369.

Glick, B.R. (2010) Using soil bacteria to facilitate phytoremediation. *Biotechnology Advances* 28, 367–374.

Glick, B.R. (2012) Plant growth-promoting bacteria: mechanisms and applications. *Scientifica* 2012, 963401.

Glick, B.R. (2014) Bacteria with ACC deaminase can promote plant growth and help to feed the world. *Microbiological Research* 169, 30–39.

Gobat, J.M., Aragno, M. and Matthey, W. (2004) *The Living Soil: Fundamentals of Soil Science and Soil Biology*. Science Publishers.

Goel, R., Kumar, V., Suyal, D.K., Dash, B., Kumar, P. and Soni, R. (2017) Root-associated bacteria: rhizoplane and endosphere. In: Singh, D.P., Singh, H.B. and Prabha, R. (eds) *Plant-Microbe Interactions in Agro-Ecological Perspectives. Vol. 1: Fundamental Mechanisms, Methods and Functions*. Springer, pp. 161–176.

González-Lamothe, R., Mitchell, G., Gattuso, M., Diarra, M.S., Malouin, F. *et al.* (2009) Plant antimicrobial agents and their effects on plant and human pathogens. *International Journal of Molecular Sciences* 10, 3400–3419.

Gravel, V., Antoun, H. and Tweddell, R.J. (2007) Growth stimulation and fruit yield improvement of greenhouse tomato plants by inoculation with *Pseudomonas putida* or *Trichoderma atroviride*: possible role of indole acetic acid (IAA). *Soil Biology and Biochemistry* 39, 1968–1977.

Grayston, S.J. and Campbell, C.D. (1996) Functional biodiversity of microbial communities in the rhizospheres of hybrid larch (*Larix eurolepis*) and Sitka spruce (*Picea sitchensis*). *Tree Physiology* 16, 1031–1038.

Hafeez, F.Y., Yasmin, S., Ariani, D., ur-Rahman, M., Zafar, Y. *et al.* (2006) Plant growth-promoting bacteria as biofertilizer. *Agronomy for Sustainable Development* 26, 143–150.

Hale, M.G. and Moore, L.D. (1980) Factors affecting root exudation Ii: 1970–1978. *Advances in Agronomy* 31, 93–124.

Hallmann, J. (2001) Endophytic bacteria. In: Jeger, M.J. and Spence, N.J. (eds) *Biotic Interactions in Plant–Pathogen Associations*. CABI, Wallingford, UK, pp. 87–119.

Han, J., Song, Y., Liu, Z. and Hu, Y. (2011) Culturable bacterial community analysis in the root domains of two varieties of tree peony (*Paeonia ostii*). *FEMS Microbiology Letters* 322, 15–24.

Han, S.I., Lee, H.J. and Whang, K.S. (2014) *Chitinophaga polysaccharea* sp. nov., an exopolysaccharide-producing bacterium isolated from the rhizoplane of *Dioscorea japonica*. *International Journal of Systematic and Evolutionary Microbiology* 64, 55–59.

Hang, F., Yanchang, L. and Qiongguang, L. (2013) Endophytic bacterial communities in tomato plants with differential resistance to *Ralstonia solanacearum*. *African Journal of Microbiology Research* 7, 1311–1318.

Hansda, A., Kumar, V. and Anshumali, (2017) Influence of Cu fractions on soil microbial activities and risk assessment along Cu contamination gradient. *CATENA* 151, 26–33.

Hardoim, P.R., van Overbeek, L.S. and Elsas, J.D. van (2008) Properties of bacterial endophytes and their proposed role in plant growth. *Trends in Microbiology* 16, 463–471.

Hardoim, P.R., van Overbeek, L.S., Berg, G., Pirttilä, A.M., Compant, S. *et al.* (2015) The hidden world within plants: ecological and evolutionary considerations for defining functioning of microbial endophytes. *Microbiology and Molecular Biology Reviews* 79, 293–320.

Hassen, A.I. and Labuschagne, N. (2010) Root colonization and growth enhancement in wheat and tomato by rhizobacteria isolated from the rhizoplane of grasses. *World Journal of Microbiology and Biotechnology* 26, 1837–1846.

Hayat, R. (2005) Sustainable legume cereal cropping system through management of biological nitrogen fixation in Pothwar. PhD thesis, PMAS Arid Agriculture University, Rawalpindi, Pakistan.

Hayat, R., Ali, S., Ijaz, S.S., Chatha, T.H. and Siddique, M.T. (2008a) Estimation of N_2-fixation of mung bean and mash bean through xylem uriede technique under rainfed conditions. *Pakistan Journal of Botany* 40, 723–734.

Hayat, R., Ali, S., Siddique, M.T. and Chatha, T.H. (2008b) Biological nitrogen fixation of summer legumes and their residual effects on subsequent rainfed wheat yield. *Pakistan Journal of Botany* 40, 711–722.

Hayat, R., Ali, S., Amara, U., Khalid, R. and Ahmed, I. (2010) Soil beneficial bacteria and their role in plant growth promotion: a review. *Annals of Microbiology* 60, 579–598.

Hill, D.S., Stein, J.I., Torkewitz, N.R., Morse, A.M., Howell, C.R. *et al.* (1994) Cloning of genes involved in the synthesis of pyrrolnitrin from Pseudomonas fluorescens and role of pyrrolnitrin synthesis in biological control of plant disease. *Applied and Environmental Microbiology* 60, 78–85.

Ho, Y.N., Mathew, D.C. and Huang, C.C. (2017) Plant–microbe ecology: interactions of plants and symbiotic microbial communities. In: Yousaf, Z. (ed.) *Plant Ecology – Traditional Approaches to Recent Trends*. Intech Open.

Huang, X.-F., Chaparro, J.M., Reardon, K.F., Zhang, R., Shen, Q. *et al.* (2014) Rhizosphere interactions: root exudates, microbes, and microbial communities. *Botany* 92, 267–275.

Hussain, M., Hamid, M.I., Tian, J., Hu, J., Zhang, X. *et al.* (2018) Bacterial community assemblages in the rhizosphere soil, root endosphere and cyst of soybean cyst nematode-suppressive soil challenged with nematodes. *FEMS Microbiology Ecology* 94, fiy142.

Idrees, M., Mohammad, A.R., Karodia, N. and Rahman, A. (2020) Multimodal role of amino acids in micro-bial control and drug development. *Antibiotics* 9, 330.

Idris, H.A., Labuschagne, N. and Korsten, L. (2008) Suppression of Pythium ultimum root rot of sorghum by rhizobacterial isolates from Ethiopia and South Africa. *Biological Control* 45, 72–84.

Igual, J.M., Valverde, A., Cervantes, E. and Velázquez, E. (2001) Phosphate-solubilizing bacteria as inoculants for agriculture: use of updated molecular techniques in their study. *Agronomie* 21, 561–568.

Ikenaga, M., Muraoka, Y., Toyota, K. and Kimura, M. (2002) Community structure of the microbiota associ-ated with nodal roots of rice plants along with the growth stages: estimation by PCR-RFLP analysis. *Biology and Fertility of Soils* 36, 397–404.

Islam, Md.T., Deora, A., Hashidoko, Y., Rahman, A., Ito, T. *et al.* (2007) Isolation and identification of poten-tial phosphate solubilizing bacteria from the rhizoplane of *Oryza sativa* L. cv. BR29 of Bangladesh. *Zeitschrift Für Naturforschung C* 62, 103–110.

Johri, B.N., Sharma, A. and Virdi, J.S. (2003) Rhizobacterial diversity in India and its influence on soil and plant health. *Advances in Biochemical Engineering/Biotechnology* 84, 49–89.

Johri, J.K., Surange, S. and Nautiyal, C.S. (1999) Occurrence of salt, pH, and temperature-tolerant, phosphate-solubilizing bacteria in alkaline soils. *Current Microbiology* 39, 89–93.

Jones, D.L., Nguyen, C. and Finlay, R.D. (2009) Carbon flow in the rhizosphere: carbon trading at the soil–root interface. *Plant and Soil* 321, 5–33.

Kämpfer, P., McInroy, J.A. and Glaeser, S.P. (2015) *Chryseobacterium rhizoplanae* sp. nov. isolated from the rhizoplane environment. *Antonie van Leeuwenhoek* 107, 533–538.

Kandel, S., Joubert, P. and Doty, S. (2017) Bacterial endophyte colonization and distribution within plants. *Microorganisms* 5, 77.

Kaur, H., Kaur, J. and Gera, R. (2016) Plant growth-promoting rhizobacteria: a boon to agriculture. *International Journal of Cell Sceience and Biotechnology* 5, 17–22.

Kent, A.D. and Triplett, E.W. (2002) Microbial communities and their interactions in soil and rhizosphere ecosystems. *Annual Review of Microbiology* 56, 211–236.

Khan, A.A., Jilani, G., Akhtar, M.S., Naqvi, S.S. and Rasheed, M. (2009) Phosphorus solubilizing bacteria: occurrence, mechanisms and their role in crop production. *Journal of Agricultural and Biological Science* 1, 48–58.

Khan, A.L., Waqas, M., Kang, S.M., Al-Harrasi, A., Hussain, J. *et al.* (2014) Bacterial endophyte *Sphingo-monas* sp. LK11 produces gibberellins and IAA and promotes tomato plant growth. *Journal of Microbiology* 52, 689–695.

Khan, M.S., Zaidi, A., Ahemad, M., Oves, M. and Wani, P.A. (2010) Plant growth promotion by phosphate solubilizing fungi – current perspective. *Archives of Agronomy and Soil Science* 56, 73–98.

Kim, J. and Rees, D.C. (1994) Nitrogenase and biological nitrogen fixation. *Biochemistry* 33, 389–397.

Kuffner, M., Puschenreiter, M., Wieshammer, G., Gorfer, M. and Sessitsch, A. (2008) Rhizosphere bacteria affect growth and metal uptake of heavy metal accumulating willows. *Plant and Soil* 304, 35–44.

Kumar, A., Patel, J.S., Meena, V.S. and Ramteke, P.W. (2019) Plant growth-promoting rhizobacteria: strategies to improve abiotic stresses under sustainable agriculture. *Journal of Plant Nutrition* 42, 1402–1415.

Kumar, A., Droby, S., Singh, V.K., Singh, S.K. and White, J.F. (2020) Entry, colonization, and distribution of endophytic microorganisms in plants. In: Kuma, A. and Radhakrishnan, E.E. (eds) *Microbial Endophytes*. Elsevier, pp. 1–33.

Kumar, A., Saha, S., Das, A., Babu, S., Layek, J. *et al.* (2022) Resource management for enhancing nutrient use efficiency in crops and cropping systems of rainfed hill ecosystems of the north-eastern region of India. *Indian Journal of Fertilizers* 18, 1090–1111.

Kumar, H., Bajpai, V.K., Dubey, R.C., Maheshwari, D.K. and Kang, S.C. (2010) Wilt disease management and enhancement of growth and yield of *Cajanus cajan* (L) var. manak by bacterial combinations amended with chemical fertilizer. *Crop Protection* 29, 591–598.

Kundan, R., Pant, G., Jadon, N. and Agrawal, P.K. (2015) Plant growth promoting rhizobacteria: mechanism and current prospective. *Journal of Fertilizers and Pesticides* 06, 9.

Kushwaha, P., Kashyap, P.L., Srivastava, A.K. and Tiwari, R.K. (2020) Plant growth promoting and antifun-gal activity in endophytic bacillus strains from pearl millet (*Pennisetum glaucum*). *Brazilian Journal of Microbiology* 51, 229–241.

Leuschner, C., Tückmantel, T. and Meier, I.C. (2022) Temperature effects on root exudation in mature beech (*Fagus sylvatica L.*) forests along an elevational gradient. *Plant and Soil* 481, 147–163.

Lhuissier, F.G.P., De Ruijter, N.C.A., Sieberer, B.J., Esseling, J.J. and Emons, A.M.C. (2001) Time course of cell biological events evoked in legume root hairs by rhizobium nod factors: state of the art. *Annals of Botany* 87, 289–302.

Li, E., Zhang, H., Jiang, H., Pieterse, C.M.J., Jousset, A. *et al.* (2021) Experimental-evolution-driven identification of *Arabidopsis* rhizosphere competence genes in Pseudomonas protegens. *mBio* 12, e0092721.

Liu, F., Xing, S., Ma, H., Du, Z. and Ma, B. (2013) Cytokinin-producing, plant growth-promoting rhizobacteria that confer resistance to drought stress in *Platycladus orientalis* container seedlings. *Applied Microbiology and Biotechnology* 97, 9155–9164.

Liu, H., Carvalhais, L.C., Crawford, M., Singh, E., Dennis, P.G. *et al.* (2017) Inner plant values: diversity, colonization and benefits from endophytic bacteria. *Frontiers in Microbiology* 8, 2552.

Liu, Y., Dai, C., Zhou, Y., Qiao, J., Tang, B. *et al.* (2021) Pyoverdines are essential for the antibacterial activity of *Pseudomonas chlororaphis* YL-1 under low-iron conditions. *Applied and Environmental Microbiology* 87, e02840-20.

Lopes, L.D., Wang, P., Futrell, S.L. and Schachtman, D.P. (2022) Sugars and jasmonic acid concentration in root exudates affect maize rhizosphere bacterial communities. *Applied and Environmental Microbiology* 88, e0097122.

Lugtenberg, B. and Kamilova, F. (2009) Plant-growth-promoting rhizobacteria. *Annual Review of Microbiology* 63, 541–556.

Lv, X., Wang, Q., Zhang, X., Hao, J., Li, L. *et al.* (2021) Community structure and associated networks of endophytic bacteria in pea roots throughout plant life cycle. *Plant and Soil* 468, 225–238.

Madhaiyan, M., Poonguzhali, S., Saravanan, V.S. and Kwon, S.W. (2014) *Rhodanobacter glycinis* sp. nov., a yellow-pigmented gammaproteobacterium isolated from the rhizoplane of field-grown soybean. *International Journal of Systematic and Evolutionary Microbiology* 64, 2023–2028.

Madhukar, S.M., Raha, P. and Singh, R.K. (2018) Identification of amino acids and sugars in root exudate of mungbean (*Vigna radiata L.*). *Journal of Pharmacognosy and Phytochemistry* 7, 1676–1680.

Mandal, L. and Kotasthane, A.S. (2014) Isolation and assessment of plant growth promoting activity of siderophore producing *Pseudomonas fluorescens* in crops. *International Journal of Agriculture, Environment and Biotechnology* 7, 63.

Mano, H. and Morisaki, H. (2008) Endophytic bacteria in the rice plant. *Microbes and Environments* 23, 109–117.

Marschner, H. (1995) *Mineral Nutrition of Higher Plants*, 2nd edn. Institute of Plant Nutrition University of Hohenheim, Germany.

Martínez-Viveros, O., Jorquera, M.A., Crowley, D.E., Gajardo, G.M.L.M. and Mora, M.L. (2010) Mechanicsms and practical considerations involved in plant growth promotion by rhizobacteria. *Journal of Soil Science and Plant Nutrition* 10, 293–319.

Matiru, V.N. and Dakora, F.D. (2004) Potential use of rhizobial bacteria as promoters of plant growth for increased yield in landraces of African cereal crops. *African Journal of Biotechnology* 3, 1–7.

Matthysse, A.G. (2014) Attachment of agrobacterium to plant surfaces. *Frontiers in Plant Science* 5, 252.

Mayak, S., Tirosh, T. and Glick, B.R. (1999) Effect of wild-type and mutant plant growth-promoting rhizobacteria on the rooting of mung bean cuttings. *Journal of Plant Growth Regulation* 18, 49–53.

Mercado-Blanco, J. and Prieto, P. (2012) Bacterial endophytes and root hairs. *Plant and Soil* 361, 301–306.

Monteiro, R.A., Balsanelli, E., Tuleski, T., Faoro, H., Cruz, L.M. *et al.* (2012) Genomic comparison of the endophyte *Herbaspirillum seropedicae* SmR1 and the phytopathogen *Herbaspirillum rubrisubalbicans* M1 by suppressive subtractive hybridization and partial genome sequencing. *FEMS Microbiology Ecology* 80, 441–451.

More, S.S., Shinde, S.E. and Kasture, M.C. (2020) Root exudates a key factor for soil and plant: an overview. *Pharma Innovation Journal* 8, 449–459.

Muraoka, Y., Hamakawa, E., Toyota, K. and Kimura, M. (2000) Observation of microbial colonization on the surface of rice roots along with their development and degradation. *Soil Science and Plant Nutrition* 46, 491–502.

Nazir, U., Zargar, M., Baba, Z., Mir, S., Mohiddin, F. *et al.* (2020) Isolation and characterization of plant growth promoting rhizobacteria associated with pea rhizosphere in North Himalayan region. *International Journal of Chemical Studies* 8, 1131–1135.

Nielsen, T.H. and Sørensen, J. (2003) Production of cyclic lipopeptides by Pseudomonas fluorescens strains in bulk soil and in the sugar beet rhizosphere. *Applied and Environmental Microbiology* 69, 861–868.

Noel, T.C., Sheng, C., Yost, C.K., Pharis, R.P. and Hynes, M.F. (1996) *Rhizobium leguminosarum* as a plant growth-promoting rhizobacterium: direct growth promotion of canola and lettuce. *Canadian Journal of Microbiology* 42, 279–283.

Noumavo, P.A., Agbodjato, N.A., Baba-Moussa, F., Adjanohoun, A. and Baba-Moussa, L. (2016) Plant growth promoting rhizobacteria: beneficial effects for healthy and sustainable agriculture. *African Journal of Biotechnology* 15, 1452–1463.

Nur Mawaddah, S., Mohd Zafri, A.W. and Sapak, Z. (2023) The potential of Pseudomonas fluorescens as biological control agent against sheath blight disease in rice: a systematic review. *Food Research* 7(Suppl 2), 46–56.

Ofek-Lalzar, M., Sela, N., Goldman-Voronov, M., Green, S.J., Hadar, Y. *et al.* (2014) Niche and host associated functional signatures of the root surface microbiome. *Nature Communications* 5, 49–50.

Oku, S., Komatsu, A., Tajima, T., Nakashimada, Y. and Kato, J. (2012) Identification of chemotaxis sensory proteins for amino acids in Pseudomonas fluorescens Pf0-1 and their involvement in chemotaxis to tomato root exudate and root colonization. *Microbes and Environments* 27, 462–469.

Ouf, S.A., El-Amriti, F., Abu-Elghait, M., Desouky, S. and Mohamed, M.S.M. (2023) Role of plant growth promoting rhizobacteria in healthy and sustainable agriculture. *Egyptian Journal of Botany* 63, 333–359.

Pastor, V., Luna, E., Mauch-Mani, B., Ton, J. and Flors, V. (2013) Primed plants do not forget. *Environmental and Experimental Botany* 94, 46–56.

Pellet, D.M., Grunes, D.L. and Kochian, L.V. (1995) Organic acid exudation as an aluminum-tolerance mechanism in maize (*Zea mays* L.). *Planta* 196, 788–795.

Pfeiffer, S., Mitter, B., Oswald, A., Schloter-Hai, B., Schloter, M. *et al.* (2017) Rhizosphere microbiomes of potato cultivated in the High Andes show stable and dynamic core microbiomes with different responses to plant development. *FEMS Microbiology Ecology* 93, fiw242.

Pieterse, C.M.J., Van Pelt, J.A., Van Wees, S.C.M., Ton, J., Léon-Kloosterziel, K.M. *et al.* (2001) Rhizobacteria-mediated induced systemic resistance: triggering, signalling and expression. *European Journal of Plant Pathology* 107, 51–61.

Pineda, A., Dicke, M., Pieterse, C.M.J. and Pozo, M.J. (2013) Beneficial microbes in a changing environment: are they always helping plants to deal with insects? *Functional Ecology* 27, 574–586.

Porcel, R., Zamarreño, Á.M., García-Mina, J.M. and Aroca, R. (2014) Involvement of plant endogenous ABA in bacillus megaterium PGPR activity in tomato plants. *BMC Plant Biology* 14, 36.

Pramanik, M.H.R., Nagai, M., Asao, T. and Matsui, Y. (2000) Effects of temperature and photoperiod on phytotoxic root exudates of cucumber (*Cucumis sativus*) in hydroponic culture. *Journal of Chemical Ecology* 26, 1953–1967.

Prashar, P., Kapoor, N. and Sachdeva, S. (2013) Biocontrol of plant pathogens using plant growth promoting bacteria. In: *Sustainable Agriculture Reviews* 12, 319–360.

Qiao, M., Xiao, J., Yin, H., Pu, X., Yue, B. *et al.* (2014) Analysis of the phenolic compounds in root exudates produced by a subalpine coniferous species as responses to experimental warming and nitrogen fertilisation. *Chemistry and Ecology* 30, 555–565.

Rajkumar, M., Ae, N., Prasad, M.N.V. and Freitas, H. (2010) Potential of siderophore-producing bacteria for improving heavy metal phytoextraction. *Trends in Biotechnology* 28, 142–149.

Raynaud, X., Eickhorst, T., Nunan, N., Kaiser, C., Woebken, D. *et al.* (2017) Spatial colonization of microbial cells on the rhizoplane. In: *19th EGU General Assembly, EGU2017*. Proceedings from the conference held 23–28 April, 2017 in Vienna, Austria, p. 18887.

Reinhold-Hurek, B., Bünger, W., Burbano, C.S., Sabale, M. and Hurek, T. (2015) Roots shaping their microbiome: global hotspots for microbial activity. *Annual Review of Phytopathology* 53, 403–424.

Riefler, M., Novak, O., Strnad, M. and Schmülling, T. (2006) *Arabidopsis* cytokinin receptor mutants reveal functions in shoot growth, leaf senescence, seed size, germination, root development, and cytokinin metabolism. *Plant Cell* 18, 40–54.

Robertson-Albertyn, S., Alegria Terrazas, R., Balbirnie, K., Blank, M., Janiak, A. *et al.* (2017) Root hair mutations displace the barley rhizosphere microbiota. *Frontiers in Plant Science* 8, 1094.

Robinson, R.J., Fraaije, B.A., Clark, I.M., Jackson, R.W., Hirsch, P.R. *et al.* (2016) Endophytic bacterial community composition in wheat (*Triticum aestivum*) is determined by plant tissue type, developmental stage and soil nutrient availability. *Plant and Soil* 405, 381–396.

Robles-Aguilar, A.A., Pang, J., Postma, J.A., Schrey, S.D., Lambers, H. *et al.* (2019) The effect of pH on morphological and physiological root traits of *Lupinus angustifolius* treated with struvite as a recycled phosphorus source. *Plant and Soil* 434, 65–78.

Rodríguez, H. and Fraga, R. (1999) Phosphate solubilizing bacteria and their role in plant growth promotion. *Biotechnology Advances* 17, 319–339.

Rodríguez, H., Fraga, R., Gonzalez, T. and Bashan, Y. (2006) Genetics of phosphate solubilization and its potential applications for improving plant growth-promoting bacteria. *Plant and Soil* 287, 15–21.

Rovira, A.D. (1969) Plant root exudates. *Botanical Rev* 35, 35–57.

Saha, M., Maurya, B.R., Bahadur, I., Kumar, A. and Meena, V.S. (2016) Can potassium-solubilizing bacteria mitigate the potassium problems in India. In: Meena, V.S., Maurya, B.R., Prakash Verma, J. and Meena, R.S. (eds) *Potassium Solubilizing Microorganisms for Sustainable Agriculture*. Springer, New Delhi, pp. 127–136.

Saharan, B.S. and Nehra, V. (2011) Plant growth promoting rhizobacteria: a critical review. *Life Sciences and Medicine Research* 21, 30.

Sahgal, M. and Johri, B.N. (2003) The changing face of rhizobial systematics. *Current Science* 84, 43–48.

Saleem, M., Arshad, M., Hussain, S. and Bhatti, A.S. (2007) Perspective of plant growth promoting rhizobacteria (PGPR) containing ACC deaminase in stress agriculture. *Journal of Industrial Microbiology and Biotechnology* 34, 635–648.

Samad, A.F.A., Sajad, M., Nazaruddin, N., Fauzi, I.A., Murad, A.M.A. *et al.* (2017) MicroRNA and transcription factor: key players in plant regulatory network. *Frontiers in Plant Science* 8, 565.

Santi, C., Bogusz, D. and Franche, C. (2013) Biological nitrogen fixation in non-legume plants. *Annals of Botany* 111, 743–767.

Saravanan, D., Radhakrishnan, M. and Balagurunathan, R. (2016) Isolation of plant growth promoting substance producing bacteria from Niligiri hills with special reference to phosphatase enzyme. *Journal of Chemical and Pharmaceutical Research* 8, 698–703.

Sarwar, S., Khaliq, A., Yousra, M., Sultan, T., Ahmad, N. *et al.* (2020) Screening of siderophore-producing PGPRs isolated from groundnut (*Arachis hypogaea* L.) rhizosphere and their influence on iron release in soil. *Communications in Soil Science and Plant Analysis* 51, 1680–1692.

Satoh, K., Yanagida, T., Isobe, K., Tomiyama, H., Takahashi, R. *et al.* (2003) Effect of root exudates on growth of newly isolated nitrifying bacteria from barley rhizoplane. *Soil Science and Plant Nutrition* 49, 757–762.

Sauer, K. and Camper, A.K. (2001) Characterization of phenotypic changes in Pseudomonas putida in response to surface-associated growth. *Journal of Bacteriology* 183, 6579–6589.

Schmidt, H. and Eickhorst, T. (2014) Detection and quantification of native microbial populations on soil-grown rice roots by catalyzed reporter deposition-fluorescence in situ hybridization. *FEMS Microbiology Ecology* 87, 390–402.

Sessitsch, A., Hardoim, P., Döring, J., Weilharter, A., Krause, A. *et al.* (2012) Functional characteristics of an endophyte community colonizing rice roots as revealed by metagenomic analysis. *Plant–Microbe Interactions* 25, 28–36.

Shah, Z., Shah, S.H., Peoples, M.B., Schwenke, G.D. and Herridge, D.F. (2003) Crop residue and fertiliser N effects on nitrogen fixation and yields of legume–cereal rotations and soil organic fertility. *Field Crops Research* 83, 1–11.

Shaharoona, B., Arshad, M. and Zahir, Z.A. (2006) Effect of plant growth promoting rhizobacteria containing ACC-deaminase on maize (*Zea mays L.*) growth under axenic conditions and on nodulation in mung bean (*Vigna radiata* L.). *Letters in Applied Microbiology* 42, 155–159.

Sharma, A. and Johri, B.N. (2003) Growth promoting influence of siderophore-producing pseudomonas strains GRP3A and PRS9 in maize (*Zea mays* L.) under iron limiting conditions. *Microbiological Research* 158, 243–248.

Shilev, S. (2013) Soil rhizobacteria regulating the uptake of nutrients and undesirable elements by plants. In: Arora, N.K. (ed.) *Plant Microbe Symbiosis: Fundamentals and Advances*. Springer, New Delhi, pp. 147–167.

Shrivastava, M., Srivastava, P.C. and D'Souza, S.F. (2016) KSM soil diversity and mineral solubilization, in relation to crop production and molecular mechanism. In: Meena, V.S., Maurya, B.R., Prakash Verma, J. and Meena, R.S. (eds) *Potassium Solubilizing Microorganisms for Sustainable Agriculture*. Springer, New Delhi, pp. 221–234.

Siddiqa, A., Rehman, Y. and Hasnain, S. (2016) Rhizoplane microbiota of superior wheat varieties possess enhanced plant growth-promoting abilities. *Frontiers in Biology* 11, 481–487.

Singh, N.P., Singh, R.K., Meena, V.S. and Meena, R.K. (2015) Can we use maize (*Zea mays*) rhizobacteria as plant growth promoter? *Vegetos* 28, 86.

Skirycz, A. and Inzé, D. (2010) More from less: plant growth under limited water. *Current Opinion in Biotechnology* 21, 197–203.

Skraly, F.A. and Cameron, D.C. (1998) Purification and characterization of a *Bacillus licheniformis* phosphatase specific for D-α-glycerophosphate. *Archives of Biochemistry and Biophysics* 349, 27–35.

Snapp, S.S., Aggarwal, V.D. and Chirwa, R.M. (1998) Note on phosphorus and cultivar enhancement of biological nitrogen fixation and productivity of maize/bean intercrops in Malawi. *Field Crops Research* 58, 205–212.

Sokolova, M.G., Akimova, G.P. and Vaishlya, O.B. (2011) Effect of phytohormones synthesized by rhizosphere bacteria on plants. *Applied Biochemistry and Microbiology* 47, 274–278.

Souza, R. de, Ambrosini, A. and Passaglia, L.M.P. (2015) Plant growth-promoting bacteria as inoculants in agricultural soils. *Genetics and Molecular Biology* 38, 401–419.

Spaepen, S. and Vanderleyden, J. (2011) Auxin and plant–microbe interactions. *Cold Spring Harbor Perspectives in Biology* 3, a001438–a001438.

Sureshbabu, K., Amaresan, N. and Kumar, K. (2016) Amazing multiple function properties of plant growth promoting rhizobacteria in the rhizosphere soil. *International Journal of Current Microbiology and Applied Sciences* 5, 661–683.

Szilagyi-Zecchin, V.J., Ikeda, A.C., Hungria, M., Adamoski, D., Kava-Cordeiro, V. et al. (2014) Identification and characterization of endophytic bacteria from corn (*Zea mays* L.) roots with biotechnological potential in agriculture. *AMB Express* 4, 26.

Tariq, M., Hameed, S., Yasmeen, T., Zahid, M. and Zafar, M. (2014) Molecular characterization and identification of plant growth promoting endophytic bacteria isolated from the root nodules of pea (*Pisum sativum* L.). *World Journal of Microbiology and Biotechnology* 30, 719–725.

Thaller, M.C., Berlutti, F., Schippa, S., Iori, P., Passariello, C. et al. (1995) Heterogeneous patterns of acid phosphatases containing low-molecular-mass polypeptides in members of the family enterobacteriaceae. *International Journal of Systematic Bacteriology* 45, 255–261.

Thomas, P. (2011) Intense association of non-culturable endophytic bacteria with antibiotic-cleansed in vitro watermelon and their activation in degenerating cultures. *Plant Cell Reports* 30, 2313–2325.

Timmusk, S., Abd El-Daim, I.A., Copolovici, L., Tanilas, T., Kännaste, A. et al. (2014) Drought-tolerance of wheat improved by rhizosphere bacteria from harsh environments: enhanced biomass production and reduced emissions of stress volatiles. *PLoS One* 9, e96086.

Timofeeva, A.M., Galyamova, M.R. and Sedykh, S.E. (2022) Bacterial siderophores: classification, biosynthesis, perspectives of use in agriculture. *Plants* 11, 3065.

Tortora, M.L., Díaz-Ricci, J.C. and Pedraza, R.O. (2011) Azospirillum brasilense siderophores with antifungal activity against Colletotrichum acutatum. *Archives of Microbiology* 193, 275–286.

Truffaut, G. and Vladykov, V. (1930) The microflora of a wheat rhizosphere. *Comptes Rendus Hebdomadaires des Séances de l'Académie des Sciences* 190, 824–826.

Trujillo, M.E., Riesco, R., Benito, P. and Carro, L. (2015) Endophytic actinobacteria and the interaction of Micromonospora and nitrogen fixing plants. *Frontiers in Microbiology* 6, 1341.

Uzakbaevna, I.A. (2022) The effect of unconventional fertilizers on the growth and development of cotton. *International Journal on Integrated Education* 5, 226–229.

Veerapagu, M., Jeya, K.R., Priya, R. and Vetrikodi, N. (2018) Isolation and screening of plant growth promoting rhizobacteria from rhizosphere of Chilli. *Journal of Pharmacognosy and Phytochemistry* 7, 3444–3448.

Velázquez, E., Silva, L.R., Ramírez-Bahena, M.H. and Peix, A. (2016) Diversity of potassium-solubilizing microorganisms and their interactions with plants. In: Meena, V.S., Maurya, B.R., Prakash Verma, J. and Meena, R.S. (eds) *Potassium Solubilizing Microorganisms for Sustainable Agriculture*. Springer, New Delhi, pp. 99–110.

Vendramin, E., Gastaldo, A., Tondello, A., Baldan, B., Villani, M. et al. (2010) Identification of two fungal endophytes associated with the endangered orchid *Orchis militaris* L. *Journal of Microbiology and Biotechnology* 20, 630–636.

Vessey, J.K. (2003) Plant growth promoting rhizobacteria as biofertilizers. *Plant and Soil* 255, 571–586.

Villacieros, M., Power, B., Sánchez-Contreras, M., Lloret, J., Oruezabal, R.I. et al. (2003) Colonization behaviour of Pseudomonas fluorescens and Sinorhizobium meliloti in the alfalfa (*Medicago sativa*) rhizosphere. *Plant and Soil* 251, 47–54.

Vives-Peris, V., de Ollas, C., Gómez-Cadenas, A. and Pérez-Clemente, R.M. (2020) Root exudates: from plant to rhizosphere and beyond. *Plant Cell Reports* 39, 3–17.

Wahyudi, A.T., Astuti, R.P., Widyawati, A., Meryandini, A. and Nawangsih, A.A. (2011) Characterization of Bacillus sp. strains isolated from rhizosphere of soybean plants for their use as potential plant growth for promoting rhizobacteria. *Journal of Microbiology and Antimicrobials* 3, 34–40.

Wahyudi, A.T., Priyanto, J.A., Afrista, R., Kurniati, D., Astuti, R.I. *et al.* (2019) Plant growth promoting activity of actinomycetes isolated from soybean rhizosphere. *OnLine Journal of Biological Sciences* 19, 1–8.

Werner, T., Motyka, V., Laucou, V., Smets, R., Van Onckelen, H. *et al.* (2003) Cytokinin-deficient transgenic *Arabidopsis* plants show multiple developmental alterations indicating opposite functions of cytokinins in the regulation of shoot and root meristem activity. *Plant Cell* 15, 2532–2550.

Wheatley, R.M. and Poole, P.S. (2018) Mechanisms of bacterial attachment to roots. *FEMS Microbiology Reviews* 42, 448–461.

Wille, L., Messmer, M.M., Studer, B. and Hohmann, P. (2019) Insights to plant–microbe interactions provide opportunities to improve resistance breeding against root diseases in grain legumes. *Plant, Cell and Environment* 42, 20–40.

Wu, C.H., Bernard, S.M., Andersen, G.L. and Chen, W. (2009) Developing microbe–plant interactions for applications in plant-growth promotion and disease control, production of useful compounds, remediation and carbon sequestration. *Microbial Biotechnology* 2, 428–440.

Yadav, B.K. and Sidhu, A.S. (2016) Dynamics of potassium and their bioavailability for plant nutrition. In: Meena, V.S., Maurya, B.R., Prakash Verma, J. and Meena, R.S. (eds) *Potassium Solubilizing Microorganisms for Sustainable Agriculture*. Springer, New Delhi, pp. 187–201.

Yu, J., Yu, Z.H., Fan, G.Q., Wang, G.H. and Liu, X.B. (2018) Isolation and characterization of indole acetic acid producing root endophytic bacteria and their potential for promoting crop growth. *Journal of Agricultural Science and Technology* 18, 1381–1391.

Yuan, J., Zhang, N., Huang, Q., Raza, W., Li, R. *et al.* (2015) Organic acids from root exudates of banana help root colonization of PGPR strain *Bacillus amyloliquefaciens* NJN-6. *Scientific Reports* 5, 13438.

Zinniel, D.K., Lambrecht, P., Harris, N.B., Feng, Z., Kuczmarski, D. *et al.* (2002) Isolation and characterization of endophytic colonizing bacteria from agronomic crops and prairie plants. *Applied and Environmental Microbiology* 68, 2198–2208.

Znój, A., Grzesiak, J., Gawor, J., Gromadka, R. and Chwedorzewska, K.J. (2021) Bacterial communities associated with *Poa annua* roots in central European (Poland) and Antarctic settings (King George Island). *Microorganisms* 9, 811.

12 Soil Health and its Impact on Plant Phenology

Yash Mangla[1] and Charu Khosla Gupta[2]*

[1]*Department of Botany, Kirori Mal College, University of Delhi, Delhi, India;*
[2]*Department of Botany, Acharya Narendra Dev College, University of Delhi, Delhia, India*

Abstract

Phenology studies the changes in the timing of plant life-cycle events, including budburst, flowering, fruiting, dormancy, migration and hibernation. Over the last few decades, the phenology of plants has rapidly changed due to global climate change, resulting in altered ecosystems. Most phenology studies have focused on climatic conditions, but soil health, a significant factor in plant phenology, has been under-represented. Key factors influencing plant phenology include soil moisture, temperature, nutrient availability, microorganisms, forest fires and grazing. Soil moisture is a crucial link between precipitation and phenological patterns. Low soil moisture delays phenological events, with these effects becoming more pronounced in subsequent seasons. Climate change increases soil temperatures, reducing soil moisture and shortening root lifespans, resulting in thinner, deeper roots. Soil nutrient content also critically impacts phenology. Soil degradation, characterized by decreased nutrients, above-ground biomass and richness, delays reproductive phenology in annual species. Microorganisms, such as arbuscular mycorrhizal fungi, significantly influence vegetative development and seedling phenology, as seen in mango trees. Also, phenology is directly affected through foraging and trampling; and indirectly by altering nitrogen and phosphorus concentrations in plants and soils, thereby regulating plant growth. Grazing also accelerates nutrient cycling through animal excreta, promoting plant growth and reproduction. In fire-prone biomes like savannahs, phenology is a functional trait characterizing plant community responses to fire. Although fire's direct influence on phenology is limited, it affects fruiting seasonality. Soil burned by fire leads to fewer individuals participating in each phenological phase and reduces flower and fruit production, as observed in a neotropical savannah community.

Keywords: Soil health, phenoevents, climate change, soil temperature and moisture, soil nutrients, nutrient availability

12.1 Introduction

Plant phenology, also known as the timing of key events, encompasses a range of processes such as plant growth, development and reproduction. While the growth and development stages include seed germination, vegetative growth, leaf shedding and regrowth, the reproductive stages involve events such as flowering initiation, pollination, fruit formation and seed maturation, along with seed dispersal and their germination (Shivanna and Tandon, 2014).

*Corresponding author: charukhoslagupta@andc.du.ac.in

© CAB International 2025. *Soil Health and Nutrition Management*
(eds N.C. Joshi, T. Leustek and P.K. Singh)
DOI: 10.1079/9781800624597.0012

Researchers typically examine these events concerning specific seasons or months and they tend to follow predictable patterns for each plant species. Consequently, comprehensive records can be compiled for individual species and specific types of forests and plant communities. These studies are instrumental in tracking changes in the timing of these events over time (Gray and Ewers, 2021). In recent years, this natural phenomenon has become less predictable due to the effects of global climate change, which encompasses a variety of environmental changes, including rising CO_2 levels, temperature fluctuations, shifts in precipitation patterns, alterations in soil conditions (such as nutrient availability and temperature) and changes in light availability. These factors can directly or indirectly influence plant phenology (Gray and Ewers, 2021).

In the existing literature, the primary focus has been on temperature fluctuations and increased CO_2 levels, which impact the phenology of plants in various ways. These changes result in shifts in the altitudinal distribution of plants and altered interactions between plants and their pollinators, often causing a mismatch in their availability to each other (Ogilvie and Forrest, 2017; Stemkovski et al., 2020). Comparatively fewer studies have been devoted to understanding the influence of soil health on plant phenology. In this context, soil health refers to soil's ongoing ability to function as a vital, living ecosystem that sustains plants, animals

and humans, as defined by the United States Department of Agriculture (USDA, n.d.). This definition by USDA implies that soil serves as a dynamic medium owing to the interplay of biotic and abiotic factors associated with it. Generally, the physical factors, that is, grazing and fire, are recognized for affecting soil properties including cycling and availability of various nutrients. Soil temperature, too, plays a significant role as it represents one of the key abiotic factors that not only influences other biotic factors (such as microbial activity and soil moisture) but also alters the physical characteristics of the soil. Consequently, these factors impact plant growth and phenological events like flowering, fruit set and seed germination. The absence or untimely/limited availability of nutrients at different stages in the life history of plants often leads to changes in plant phenology, resulting in alterations in the growth and reproduction patterns (Gungula et al., 2003; Classen et al., 2015).

Interactions among multiple soil factors further complicate the modelling and prediction of changes in plant phenology related to soil conditions, as highlighted by Piao et al. in 2019. Shifts in phenological events, particularly in crop species, can reduce yields or even lead to crop failure (Ishtiaq et al., 2022). This chapter primarily reviews existing studies that explore various factors, including soil type, temperature, moisture and organisms, the impact of fire, the availability of nutrients, and their subsequent influence on plant phenology (Fig. 12.1; Table 12.1).

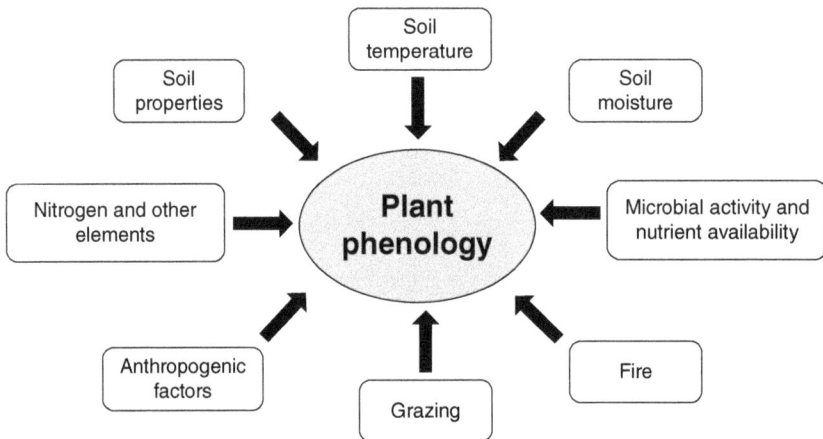

Fig. 12.1. Factors that influence the phenology of plants. Several factors are interrelated and correlated with other factors and have a synergistic impact on plant growth and phenoevents.

Table 12.1. Summary of parameters affecting soil health and their specific influences on plant phenology.

Soil parameter	Effect
Soil temperature	Increase in soil temperature primarily due to changing climate leads to a decreased soil moisture, reduced flower production per plant (e.g. *Oxalis pes-caprae*) and shortened lifespan of many roots. It may also cause delayed bud break with onset of spring (e.g. *Populus tremuloides*). Higher temperature may lead to an early start of growing season in herbaceous species
Soil moisture	Increase in above-ground biomass (e.g. *Stipa grandis*), delayed senescence, early onset of flowering (e.g. *Leymus chinensis*) and delayed autumnal phenology (e.g. *P. tremuloides*) are some of the consequences of a reduced or an altered soil moisture availability. Low soil moisture delays the phenological events with effects of this water stress being more pronounced in subsequent seasons. Drought-stressed plants exhibit earlier blooming and produce fewer flowers, limited nutrient absorption from soil results in reduced nectar and pollen production
Soil nitrogen	Degraded habitats, reduced below-ground nutrients and deteriorated above-ground communities lead to delayed reproduction and phenology. Reduced nutrients in the soil significantly delay the timing of the first flowering and later extend its duration (e.g. *Anemone trullifolia* var. *linearis*, *Caltha scaposa* and *Trollius farreri*). Soil nutrient availability in conjunction with precipitation changes the flowering period (e.g. *Helianthus annuus* and *Solidago*). More nutrient and more water may lead to delayed flowering affecting the overall phenology
Grazing	Grazing changes physical properties of soil such as soil composition and its structure. Grazing and other activities with similar impact (land mowing) result in extended flowering and fruiting phases (e.g. *A. trullifolia* var. *linearis*, *C. scaposa* and *T. farreri*). Grazing promotes plant growth and reproduction by foraging and trampling. Animal excreta enhances the nitrogen and phosphorus content of the soil
Soil microorganisms	Soil microorganisms, arbuscular mycorrhizae and some soil fungi positively influence the vegetative development and phenology in many plant species. The presence of soil microbes like *Glomus* and *Azotobacter* may promote vegetative growth (e.g. in mango seedlings) especially when soil nutrient availability is limited
Fire	Fire, especially in grasslands, influences flower and fruit production and the recruitment of trees in savannahs and thereby changes the timing of plant reproduction at both the landscape and community levels. Fire may also disrupt the reproductive synchronization of species across the landscape. It burns most individuals leading to a reduction in the number of plants and eventually lowering flower and fruit production

Furthermore, changes in plant phenology have significant implications for ecosystem carbon cycles and ecosystem feedback to the climate. However, quantifying these impacts remains challenging (Piao *et al.*, 2019). Additionally, this chapter provides a perspective on global climate change and its effects on soil parameters, leading to changes in plant phenology.

12.2 Soil Properties and their Impact on Plants and their Environment

12.2.1 Soil temperature

Soil serves as a heat reservoir during the day and a heat source at night. Generally, dark-coloured soils tend to absorb more radiant heat than light-coloured soils. Fluctuations in soil temperature, both on a daily or seasonal basis, significantly impact moisture levels and microbial activity in the soil. This, in turn, has implications for plants' availability and acquisition of nutrients, ultimately affecting their growth. The soil type, along with its elevation, has notably influenced the number of flowers produced per plant in *Oxalis pes-caprae* in the Western Cape, South Africa (Haukka *et al.*, 2013). In a study conducted by Falk and his team in 2020, observations suggested that defoliation in the previous season and the availability of soil nutrients directly influenced the timing of bud break in the following spring for different genotypes of *Populus tremuloides* trees planted on the University of

Wisconsin-Madison campus. Both prior-season defoliation and low soil nutrient levels delayed the onset of bud break, consequently impacting the overall phenology of the aspen trees.

Some earlier studies have also reported similar delays in phenology in various host species in response to early defoliation, which eventually results in reduced nutrient availability in the soil (den Herder *et al.*, 2009). This delay in bud break phenology may be attributed to nutrient stress resulting from foliar damage. In a recent study conducted at the Botanical Garden of Jena, researchers described herbaceous plant species displaying more significant variability in phenology compared with trees during spring. Their investigation suggested that the timing of phenological events in herbaceous species is more influenced by their height, with shorter species tending to flower earlier than taller ones, as they are likely to not be overshadowed by larger plants (Horbach *et al.*, 2023). Their analysis revealed that herbaceous vegetation experiences more pronounced variations in microclimatic conditions and a higher degree of variability in environmental factors, including shade, soil properties and competition, than in trees. This heightened variability is believed to account for the greater diversity of phenological stages observed in herbaceous species.

During spring, it was also observed at the Botanical Garden of Jena that soil temperatures exceeded air temperatures (Lembrechts *et al.*, 2022). This discrepancy between air and soil temperatures is thought to contribute to herbaceous species' earlier initiation of the growing season. These herbaceous plants seemed to exhibit a more pronounced response to changes in soil temperature when compared with trees, as highlighted by Lembrechts *et al.* (2022) and Mašková *et al.* (2022). This suggests that the higher soil temperature may play a role in herbaceous species' early start of the growing season, as they appear more sensitive to soil temperature fluctuations than trees.

12.2.2 Soil moisture

Water availability in the soil directly affects plants' uptake of nutrients. Consequently, soil moisture has a direct impact on plant phenology. Recent climate changes, leading to unpredictable rainfall patterns, have resulted in insufficient precipitation in many regions, with more pronounced effects (Dey *et al.*, 2020). Moreover, extreme precipitation events and rising temperatures have introduced seasonal variations in soil moisture levels, subsequently affecting their phenology. Such alterations in the phenology of grasses in the grasslands of western Sydney, Australia, which corresponded to changes in soil moisture due to precipitation, were also observed by Yang *et al.* (2023). Their study revealed that increased precipitation during the early growing season led to higher above-ground biomass in the rhizomatous grass *Leymus chinensis*, and consequently it advanced the flowering date of the species.

Similarly, increased precipitation during the late growing season increased above-ground biomass in the grass *Stipa grandis* by delaying senescence. Likewise, Chauhan *et al.* (2019) also found empirical evidence supporting the influence of soil moisture on flowering time in chickpea and wheat crops. Low soil moisture conditions have also been associated with delayed autumnal phenology in trembling aspen (*Populus tremuloides* Michx.) (Inoue *et al.*, 2020).

In arid and semi-arid regions, soil moisture availability is a critical limiting factor for vegetation growth. During the summer on the Mongolian Plateau plains, soil moisture plays a pivotal role in extending the growing season, thereby bringing about changes in the phenological cycles of plants (Luo *et al.*, 2021). This research underlines the significant impact of soil moisture dynamics and drought on the phenological characteristics of vegetation in the Mongolian Plateau, emphasizing the vital role of soil moisture in shaping vegetation phenology. An increase in soil moisture is directly associated with an extension of the growing season. Another research team embarked on a study to investigate the effects of climate warming and nitrogen enrichment on the spring phenology of vegetation in three distinct types of grasslands in Inner Mongolia, China. Their findings revealed that the response of the start of the growing season (SOS) to warming exhibited variations contingent on soil moisture levels. They further concluded that in semi-arid grasslands, the pivotal factor influencing the SOS response to climate warming was soil moisture availability rather than the soil's nitrogen content (Liu *et al.*, 2022).

It is intriguing to note that moisture in the soil not only impacts plant phenology but also influences the phenological cycles of insects. Ma *et al.* (2017) documented that the onset of favourable soil moisture conditions affected the timing of phenological events and also influenced the success of larval escape in the peach fruit moth, *Carposina sasakii*, which is known as one of the most destructive insect pests in orchards across East Asia. The research group observed that larvae exhibited higher success rates in escaping from their surroundings when soil moisture levels were within a moderate range. They proposed that the time required for larvae to escape decreased consistently as soil moisture levels increased. The number of larvae successfully escaping increased during optimal soil moisture conditions, highlighting that most could only escape successfully when sufficient soil moisture was available. Given the limited number of studies, it can be inferred that the impact of soil moisture availability on phenological events may be species-specific.

12.2.3 Soil nitrogen and other elements

Typically, plants have a substantial demand for nitrogen, and the availability of nitrogen to plants is influenced or facilitated by various factors such as moisture, nutrient recycling and microbial activity. It is believed that nitrogen limitation and excess nitrogen can impact plant growth and, subsequently, the timing of phenological events. In an experimental study conducted in a Tibetan alpine meadow, the phenological responses of three early flowering species (*Anemone trullifolia* var. *linearis, Caltha scaposa* and *Trollius farreri*) were investigated over a span of three consecutive years. The study revealed that adding nitrogen to the soil led to a significant delay in the timing of the first flowering, extending it by approximately 11 days. Additionally, the accumulation of litter in the soil also influenced phenology, particularly concerning the timing of flowering and fruiting (Liu *et al.*, 2017).

Chemical fertilizers, in conjunction with soil moisture, have been reported to modify plants' nutrient uptake and subsequent utilization (Walter, 2018). During extended dry periods, it has been observed that drought-stressed plants tend to exhibit earlier blooming, produce fewer flowers and limit their nutrient absorption from the soil, resulting in reduced nectar and pollen production (Thuma *et al.*, 2023). Conversely, adequate water can increase plant biomass and flower production (Zhang *et al.*, 2020). However, excessive water can lead to prolonged soil saturation, inhibiting plant growth due to reduced soil oxygen, root loss, nitrogen leaching and restricted nutrient uptake by plants (Thuma *et al.*, 2023).

In an experimental study involving wild sunflower (*Helianthus annuus*) and golden rod (*Solidago* spp.), the researchers manipulated water and supplemental nutrient availability throughout the plants' growing season. The study determined that soil nutrients and rainfall could change the plants' flowering times, a key phenological phase ranging from 2 to 18 days. However, the response in terms of flowering was specific to each plant species. While water and soil nutrients influenced the phenology of *Helianthus* and *Solidago*, they did not significantly impact plant growth or the number of flowers produced when the plants were grown under drought conditions. Additionally, a study demonstrated that soil properties significantly affected plant phenology and influenced plant yield, with the cadmium concentration in the soil being a notable factor (Ata-Ul-Karim *et al.*, 2020).

12.2.4 Grazing, soil health and phenology

While it is well established that various biotic and abiotic factors play a significant role in shaping the phenology of plant species, grazing also exerts regulatory influence on phenology, directly and indirectly, by impacting various aspects of vegetation growth conditions. Wang *et al.* (2023) have recently brought attention to the impact of grazing on vegetation phenology by its effects on the physical properties of soil and the alteration of soil composition and structure in the Tibetan Plateau region. Their study focused on three species: *Anemone trullifolia* var. *linearis, Caltha scaposa* and *Trollius farreri*, which displayed extended flowering and fruiting phases when subjected to land mowing practices similar to grazing. It was also noted that alterations in surface litter resulting from grazing activities in alpine grasslands were likely to affect vegetation phenology and lead to vegetation degradation. Specifically, the first flowering and first fruiting times increased by 4.7 and 7.4 days across all three species, respectively (Liu *et al.*, 2017).

12.2.5   Soil microorganisms

Soil microorganisms are pivotal in regulating nutrient conversions and providing vital nutrients to plants. They also promote the peaceful coexistence of neighbouring organisms and contribute to managing plant populations. Alterations in the dynamics between soil microorganisms and plants can have profound consequences on the composition of plant communities and the overall functioning of ecosystems, including plant phenology (Classen *et al.*, 2015).

These interactions between microorganisms and plants in the soil can be categorized as unfavourable when they result in reduced plant growth and positive when they stimulate lush and enhanced growth cycles. Changes in soil temperature and moisture levels indirectly influence plant phenology by influencing the relationships between soil microorganisms and plants. The microbial communities associated with plant roots are anticipated to significantly impact phenology, plant survival and the expression of various functional traits (Wagner *et al.*, 2014; Classen *et al.*, 2015). Plants that form mycorrhizal associations demonstrate exceptional efficiency in nutrient uptake from the soil, rendering them relatively less vulnerable to diseases and more productive, especially in nutrient-poor conditions. These enhancements in nutrient availability can subsequently result in changes in plant phenology.

The effects on the phenology of mango seedlings in conditions of reduced soil nutrient availability and the presence of soil microbes like *Glomus* and *Azotobacter* also provided insights into the relationship between microbes and phenology (Sharma *et al.*, 2014). Specifically, the presence of *Glomus fasciculatum* and *Azotobacter chroococcum* strain Z1 promoted the vegetative growth of mango seedlings when soil nutrient conditions were constrained. Furthermore, Forey *et al.* (2015) examined the role of soil organisms, particularly *Collembola* (microarthropods), as potential drivers of flowering in plants. They suggested that *Collembola* could influence plant phenology by modifying the allocation of resources to reproductive processes, which is itself influenced by shifts in nutrient availability in the soil. Significantly, *Collembola* has been shown to play a beneficial role in nutrient cycling within the soil, thereby further shaping plant phenological patterns.

12.2.6   Fire

While precipitation, photoperiod and temperature are typically recognized as the primary drivers of plant phenology, disturbances in the biochemical composition of the soil, caused naturally and/or by human activities, such as fire, can also significantly impact the phenological patterns of certain seasonal tropical plants. An investigation in the Madagascar mountain area revealed that fire can affect plants' phenology, reducing individuals' participation in various phenological events (Alvarado *et al.*, 2014). The study indicated that the influence of fire on flower and fruit production is particularly notable in areas that experience frequent burning. The researchers also identified surface and ground fires as major disturbances affecting the mortality and recruitment of trees in savannahs, thereby confirming the influence of fire on the timing of plant reproduction at the landscape and community levels. They further clarified that fire decreased the percentage of individuals involved in each phenophase, leading to significantly lower flower and fruit production in areas with frequent fires. As a result, the increased frequency of fires disrupted the reproductive synchronization of species across the landscape.

Similarly, a recent study focusing on fire-prone ecosystems, such as savannahs, identified plant phenology as a functional trait that characterizes the responses of plant communities to fire (Valentin-Silva *et al.*, 2021). The research group noted that while the fire did not alter the average timing of vegetative and reproductive phenophases at the community level, it did induce changes in fruiting seasonality. It was observed that plants responded differently to burning, especially at the end of the dry season when most anthropogenic fires occur, and this response was associated with the specific soil characteristics of the plant's habitat. The researchers stressed the importance of evaluating the effects of fire on plant phenology, considering variations in fire intensity across different seasons and under various fire regimes.

12.3 Climate Change and Soil Health: Repercussions on Plant Phenology

Temperature plays a pivotal role as a driver of global climate change. Rising temperatures have been noted to expedite plant growth and reproduction, resulting in shifts in plant phenology. However, the magnitude of this alteration varies among plant species, developmental stages and ecosystems (Cleland et al., 2006; Nord and Lynch, 2009).

The increased soil temperature induced by global warming substantially challenges crop physiology and production (Wu et al., 2022). Furthermore, soil temperature significantly affects root growth and respiration, which, in turn, impacts the availability of nutrients to plants. With rising temperatures, nutrient mineralization rates are expected to increase, eventually enhancing nutrient availability. Conversely, elevated soil temperature may lead to reduced soil moisture levels or a decrease in organic matter input into the soil, further affecting the availability of resources (Borner et al., 2008). These alterations in resource availability are presumed to result in changes in plant phenology.

A study was conducted on phenological responses in seven different plant species in northern Japan in the context of soil warming (Ishioka et al., 2013). They experimented with a deciduous forest, where the soil was artificially warmed by 5°C higher than in the control plots. The study involved seven understorey species with varying leaf habits, including one evergreen shrub, one semi-evergreen fern, one summer-deciduous shrub and four summer-green herbs. The results showed delayed new leaf emergence in the evergreen shrub, accelerated senescence of overwintering leaves in the semi-evergreen fern, and quicker leaf shedding in the summer-deciduous shrub. In the case of the four summer-green species, only the earliest

leaf-out species exhibited advancement in the initiation of growth. At the same time, the overall duration of the growing season remained unchanged.

Numerous studies have explored the impact of climate change on the reproductive phenology of plants, with the majority of these studies indicating that rising temperatures lead to earlier flowering. However, a study by Suonan et al. (2021) presented a contrasting perspective. Their research demonstrated that warming and habitat degradation in an alpine meadow delayed the onset of reproductive phenology in the annual plant Chenopodium glaucum, native to the Tibetan Plateau. While warming was responsible for the delay in the first fruit-setting day, degradation affected all three reproductive phenophases. Changes in soil nutrients and alterations in plant community structure were identified as factors influencing these shifts in reproductive phenology. The study revealed that degraded habitats, reduced below-ground soil nutrients and a deteriorated above-ground plant community, all components of soil health, collectively contributed to the delayed reproductive phenology of this annual plant.

Root growth dynamics have been the subject of various studies concerning soil temperature. The effect of soil temperature on silver birch seedlings confirmed that while root growth is influenced to some extent by intrinsic factors of the tree, it is predominantly driven by soil temperature (Jouni et al., 2022). In another study, a field-scale soil warming experiment on Cunninghamia lanceolata offered clear evidence that a 4°C increase in temperature significantly reduced the lifespan of fine roots in the plant (Jiang et al., 2022). While the elevated soil temperature directly contributed to the shortened lifespan of fine roots under warming conditions, leading to alterations in the plant's phenological cycle, other factors such as decreased soil moisture, deeper root penetration, smaller root diameter and changes in the seasonal patterns of root emergence also played a role in this transformation.

Wu et al. (2022) explored the impact of elevated soil temperature on the phenology of winter wheat plants at China Agricultural University. The soil temperature was increased by 3.8°C by installing a heating cable at a depth of 20 cm. The study's results revealed that soil warming led to an advancement of the anthesis date of the wheat plants by 1 week, promoting their growth and dry matter transportation before anthesis. However, the research also uncovered adverse effects of soil warming during the post-anthesis phase. It resulted in a reduction in the duration of the post-anthesis period and the leaf area index, net photosynthesis and spectral vegetation indexes. Consequently, dry matter accumulation and grain filling after anthesis were negatively affected, decreasing the 1000-grain weight and harvest index. Additionally, the root weight, length, surface area densities and root-to-shoot ratio during the post-anthesis period all experienced a decrease under the conditions of soil warming. As an outcome of these combined effects, the grain yield underwent a significant decline of 35.2% during the dry year of 2018/19 as a direct result of the impact of soil warming on the winter wheat plants.

12.4 Conclusion and Future Perspectives

Though the current chapter summarizes existing studies and their findings regarding soil health and its effects on plant phenology, it is important to note that these studies have limitations in drawing definitive conclusions. Nevertheless, it is evident that soil is a constantly changing factor influenced by factors such as habitat, time and prevailing environmental conditions like temperature, moisture, etc. Therefore, the consequences of alterations in

soil health may vary depending on the specific habitat or species involved. The essential elements that play a pivotal role in shaping plant phenology, such as soil moisture, temperature, humidity and photoperiod, collectively constitute the soil–plant–atmosphere continuum (SPAC). Interactions among these SPAC components form dynamic feedback loops, resulting in altered plant phenology. For instance, soil moisture levels affect plant water absorption, subsequently impacting leaf expansion and transpiration rates, thereby influencing atmospheric humidity. Changes in atmospheric humidity can, in turn, affect soil moisture dynamics. Similarly, alterations in atmospheric temperature can influence soil temperature, thereby affecting nutrient availability and microbial activity in the soil, ultimately influencing plant growth and phenology. In essence, the SPAC functions as an interconnected system where environmental factors interact to regulate plant phenology, with each component exerting influence on and being influenced by the others.

Acknowledgements

Both authors are thankful to their respective college principals for constant encouragement and motivation. The authors also acknowledge the editors' invitation to contribute to this chapter and express deep gratitude to reviewers for their constructive suggestions.

References

Alvarado, S.T., Buisson, E., Rabarison, H., Rajeriarison, C., Birkinshaw, C. *et al.* (2014) Fire and the reproductive phenology of endangered Madagascar sclerophyllous tapia woodlands. *South African Journal of Botany* 94, 79–87.

Ata-Ul-Karim, S.T., Cang, L., Wang, Y. and Zhou, D. (2020) Effects of soil properties, nitrogen application, plant phenology, and their interactions on plant uptake of cadmium in wheat. *Journal of Hazardous Materials* 384, 121452.

Borner, A.P., Kielland, K. and Walker, M.D. (2008) Effects of simulated climate change on plant phenology and nitrogen mineralization in Alaskan arctic tundra. *Arctic, Antarctic and Alpine Research* 40, 27–38.

Chauhan, Y.S., Ryan, M., Chandra, S. and Sadras, V.O. (2019) Accounting for soil moisture improves prediction of flowering time in chickpea and wheat. *Science Reports* 9, 7510.

Classen, A.T., Sundqvist, M.K., Henning, J.A., Newman, G.S., Moore, J.A.M. *et al.* (2015) Direct and indirect effects of climate change on soil microbial and soil microbial–plant interactions: what lies ahead? *Ecosphere* 6, 130.

Cleland, E.E., Chiariello, N.R., Loarie, S.R., Mooney, H.A. and Field, C.B. (2006) Diverse responses of phenology to global changes in a grassland ecosystem. *Proceedings of the National Academy of Sciences USA* 103, 13740–13744.

Den Herder, M, Bergström, R., Niemelä, P., Danell, K. and Lindgren, M. (2009) Effects of natural winter browsing and simulated summer browsing by moose on growth and shoot biomass of birch and its associated invertebrate fauna. *Annales Zoologici Fennici* 46, 63–74.

Dey, R., Gallant, A.J.E. and Lewis, S.C. (2020) Evidence of a continent-wide shift of episodic rainfall in Australia. *Weather and Climate Extremes* 29, 100274.

Falk, M.A., Donaldson, J.R., Stevens, M.T., Raffa, K.F. and Lindroth, R.L. (2020) Phenological responses to prior-season defoliation and soil-nutrient availability vary among early- and late-flushing aspen (*Populus tremuloides* Michx.) genotypes. *Forest Ecology and Management* 458, 117771.

Forey, E., Coulibaly, S.F.M. and Chauvat, M. (2015) Flowering phenology of a herbaceous species (*Poa annua*) is regulated by soil Collembola. *Soil Biology and Biochemistry* 90, 30–33.

Gray, R.E.J. and Ewers, R.M. (2021) Monitoring forest phenology in a changing world. *Forests* 12, 297.

Gungula, D.T., Kling, J.G. and Togun, A.O (2003) CERES-maize predictions of maize phenology under nitrogen-stressed conditions in Nigeria. *Agronomy Journal* 95, 892–899.

Haukka, A.K., Dreyer, L.L. and Esler, K.J. (2013) Effect of soil type and climatic conditions on the growth and flowering phenology of three Oxalis species in the Western Cape, South Africa. *South African Journal of Botany* 88, 152–163.

Horbach, S., Rauschkolb, R. and Römermann, C. (2023) Flowering and leaf phenology are more variable and stronger associated to functional traits in herbaceous compared to tree species. *Flora* 300, 152218.

Inoue, S., Dang, Q.-L., Man, R. and Tedla, B. (2020) Photo-period, CO_2 and soil moisture interactively affect phenology in trembling aspen: implications to climate change-induced migration. *Environmental and Experimental Botany* 180, 104269.

Ishioka, R., Muller, O., Hiura, T. and Kudo, G. (2013) Responses of leafing phenology and photosynthesis to soil warming in forest-floor plants. *Acta Oecologica* 51, 34–41.

Ishtiaq, M., Maqbool, M., Muzamil, M., Casini, R., Alataway, A. et al. (2022) Impact of climate change on phenology of two heat-resistant wheat varieties and future adaptations. *Plants* 11, 1180.

Jiang, Q., Jia, L., Wang, X., Chen, W., Xiong, D. et al. (2022) Soil warming alters fine root lifespan, phenology, and architecture in a *Cunninghamia lanceolata* plantation. *Agricultural and Forest Meteorology* 327, 109201.

Jouni, K., Timo, D., Tarja, L., Sirpa, P., Raimo, S. and Tapani, R. (2022) Separating the effects of air and soil temperature on silver birch. Part I. Does soil temperature or resource competition determine the timing of root growth? *Tree Physiology* 42, 2480–2501.

Lembrechts, J.J., van den Hoogen, J., Aalto, J., Ashcroft, M.B., De Frenne, P. et al. (2022) Global maps of soil temperature. *Global Change Biology* 28, 3110–3144.

Liu, Y., Miao, R., Chen, A., Miao, Y., Liu, Y. et al. (2017) Effects of nitrogen addition and mowing on reproductive phenology of three early-flowering forb species in a Tibetan alpine meadow. *Ecological Engineering* 99, 119–125.

Liu, Z., Fu, Y.H., Shi, X., Lock, T.R., Kallenbach, R.L. et al. (2022) Soil moisture determines the effects of climate warming on spring phenology in grasslands. *Agricultural and Forest Meteorology* 323, 109039.

Luo, M., Meng, F., Sa, C., Duan, Y., Bao, Y. et al. (2021) Response of vegetation phenology to soil moisture dynamics in the Mongolian Plateau. *CATENA* 206, 105505.

Ma, G., Tian, B.-L., Zhao, F., Wei, G.-S., Hoffmann, A.A. et al. (2017) Soil moisture conditions determine phenology and success of larval escape in the peach fruit moth, *Carposina sasakii* (Lepidoptera, Carposinidae): implications for predicting drought effects on a diapausing insect. *Applied Soil Ecology* 110, 65–72.

Mašková, T., Herben, T., Hošková, K. and Koubek, T. (2022) Shoot senescence in herbaceous perennials of the temperate zone: identifying drivers of senescence pace and shape. *Journal of Ecology* 110, 1296–1311.

Nord, E.A. and Lynch, J.P. (2009) Plant phenology: a critical controller of soil resource acquisition. *Journal of Experimental Botany* 60, 1927–1937.

Ogilvie, J.E. and Forrest, J.R. (2017) Interactions between bee foraging and floral resource phenology shape bee populations and communities. *Current Opinion in Insect Science* 21, 75–82.

Piao, S., Liu, Q., Chen, A., Janssens, I.A., Fu, Y. et al. (2019) Plant phenology and global climate change: current progresses and challenges. *Global Change Biology* 25, 1922–1940.

Sharma, S.D., Kumar, P. and Yadav, S.K. (2014) Glomus–azotobacter association affects phenology of mango seedlings under reduced soil nutrient supply. *Scientia Horticulturae* 173, 86–91.

Shivanna, K.R. and Tandon, R. (2014) *Reproductive Ecology of Flowering Plants: A Manual*. Springer, New Delhi.

Stemkovski, M., Pearse, W.D., Griffin, S.R., Pardee, G.L., Gibbs, J. *et al.* (2020) Bee phenology is predicted by climatic variation and functional traits. *Ecology Letters* 23, 1589–1598.

Suonan, J., Cui, S., Lv, W., Wang, W., Li, B. *et al.* (2021) Degradation rather than warming delays onset of reproductive phenology of annual *Chenopodium glaucum* on the Tibetan Plateau. *Agricultural and Forest Meteorology* 311, 108688.

Thuma, J.A., Duff, C., Pitera, M., Januario, N., Orians, C.M. *et al.* (2023) Nutrient enrichment and rainfall affect plant phenology and floral resource availability for pollinators. *Frontiers in Ecology and Evolution* 11, 1150736.

USDA (n.d.) Soil health. Available at: https://www.nrcs.usda.gov/conservation-basics/natural-resource-concerns/soils/soil-health (last accessed November 2024).

Valentin-Silva, A., Alves, V.N., Tunes, P. and Guimarães, E. (2021) Fire does not change sprouting nor flowering, but affects fruiting phenology in a neotropical savanna community. *Flora* 283, 151901.

Wagner, M.R., Lundberg, D.S., Coleman-Derr, D., Tringe, S.G., Dangl, S.G. *et al.* (2014) Natural soil microbes alter flowering phenology and the intensity of selection on flowering time in a wild Arabidopsis relative. *Ecology Letters* 17, 717–726.

Walter, J. (2018) Effects of changes in soil moisture and precipitation patterns on plant-mediated biotic interactions in terrestrial ecosystems. *Plant Ecology* 219, 1449–1462.

Wang, L., Li, P., Li, T., Zhou, X., Liu, Z. *et al.* (2023) Grazing alters vegetation phenology by regulating regional environmental factors on the Tibetan Plateau. *Agriculture, Ecosystems and Environment* 351, 108479.

Wu, G., Ling, J., Liu, Z.-X., Xu, Y.-P., Chen, X.-M. *et al.* (2022) Soil warming and straw return impacts on winter wheat phenology, photosynthesis, root growth, and grain yield in the North China Plain. *Field Crops Research* 283, 108545.

Yang, J., Medlyn, B.E., Barton, C.V.M., Churchill, A.C., De Kauwe, M.G. *et al.* (2023) Green-up and brown-down: modelling grassland foliage phenology responses to soil moisture availability. *Agricultural and Forest Meteorology* 328, 109252.

Zhang, J., Zuo, X., Zhao, X., Ma, J. and Medina-Roldán, E. (2020) Effects of rainfall manipulation and nitrogen addition on plant biomass allocation in a semiarid sandy grassland. *Scientific Reports* 10, 9026.

13 Soil Health and Sustainable Agriculture: Concept and Practices

Usha Sabharwal[1], Piyush Kant Rai[2] and Kamlesh Choure[2]*

[1]*Department of Life Sciences, Parul Institue of Applied Sciences, Parul University, Waghodia, Vadodara, Gujarat, India;* [2]*Department of Biotechnology, AKS University, Sherganj, Satna, Madhya Pradesh, India*

Abstract

The close connection between soil health and sustainable agriculture is pivotal in modern food production systems. Maintaining good soil is crucial for crop growth, yield improvement and natural resource protection. Many sustainable farming practices, like reducing soil erosion, enhancing water retention and minimizing chemical inputs, contribute significantly to soil health improvement. Soil health refers to the soil's ability to sustain environmental quality, support plant and animal productivity and uphold ecosystem integrity. Sustainable agricultural practices ensure fertile, productive soil for future generations, promoting long-term environmental and food supply sustainability. Additionally, healthy soil aids in climate change mitigation by reducing greenhouse gas emissions and sequestering carbon. Sustainable agriculture aims to balance nutrient inputs and outputs by ensuring optimal nutrient levels for crops while managing soil health through proper tillage, irrigation, residue management, weed control and crop rotation. All these practices can enhance soil quality and fertility and minimize adverse environmental impacts from agriculture. Thus, sustainable farming methods lead to healthier, more productive land and increased agricultural yields. Soil health directly influences crop yields, water retention, nutrient cycling, erosion control, pest resistance and resilience, making it a critical aspect of sustainable agriculture. Adopting sustainable practices can yield long-term cost–benefits by reducing fertilizer and pesticide inputs, while carbon sequestration in soil aids in climate change mitigation. Farmers embracing sustainable methods can increase output, reduce environmental impact and contribute to a more resilient food production system. Understanding this relationship empowers farmers and policy makers to make informed decisions supporting sustainable agriculture and long-term soil health.

Keywords: Soil health, soil erosion, water retention, climate change, environmental sustainability

13.1 Introduction

The condition of the soil is a crucial element of sustainable agriculture as it directly affects food production and the services provided by the ecosystem (Powlson *et al.* 2011). Optimal soil conditions facilitate the growth and productivity of plants and animals while preserving water and air quality, fostering the ecosystem's overall well-being (Tahat *et al.*, 2020). It includes chemical, physical and biological characteristics and operates as a living system. Modifying soil characteristics in agricultural environments can enhance or impair soil functioning (Tahat *et al.*,

*Corresponding author: kamlesh.chaure@gmail.com

© CAB International 2025. *Soil Health and Nutrition Management*
(eds N.C. Joshi, T. Leustek and P.K. Singh)
DOI: 10.1079/9781800624597.0013

2020; Yang *et al.*, 2020). Prolonged utilization of artificial fertilizers can result in soil acidification and the build-up of salts, which can have detrimental effects on soil quality (Krasilnikov *et al.*, 2022).

Additionally, the manufacturing of artificial nitrogen fertilizer is a significant contributor to the release of greenhouse gases, hence exacerbating climate change (Menegat *et al.*, 2022). To ensure soil health, it is important to consider the diversity of organisms living in the soil (Tahat *et al.*, 2020). Intensive agricultural methods, such as annual tillage, can decrease the presence of soil creatures like earthworms and mites, which are important for the ecosystem (Sofo *et al.*, 2020). In contrast, low-intensity farming and conservation approaches help maintain soil health and reduce environmental harm (Cárceles Rodríguez *et al.*, 2022).

The economic ramifications of soil health and conservation are significant. Implementing soil conservation practices at the farm level results in substantial cost savings downstream, including decreased expenses for river dredging and flood control (Panagos *et al.*, 2024). This highlights the significance of prioritising soil health for environmental and economic sustainability (Tahat *et al.*, 2020). The health of the soil is crucial for sustainable agriculture as it directly impacts food production and the stability of ecosystems (Shahane and Shivay, 2021; Telo Da Gama, 2023). The process includes conserving the diversity of soil organisms and adopting methods that promote soil health while limiting adverse effects on the ecosystem (Cárceles Rodríguez *et al.*, 2022) (Fig. 13.1). Investing in soil health is crucial for global agricultural systems, as it improves the environment and generates significant economic rewards (Das *et al.*, 2022).

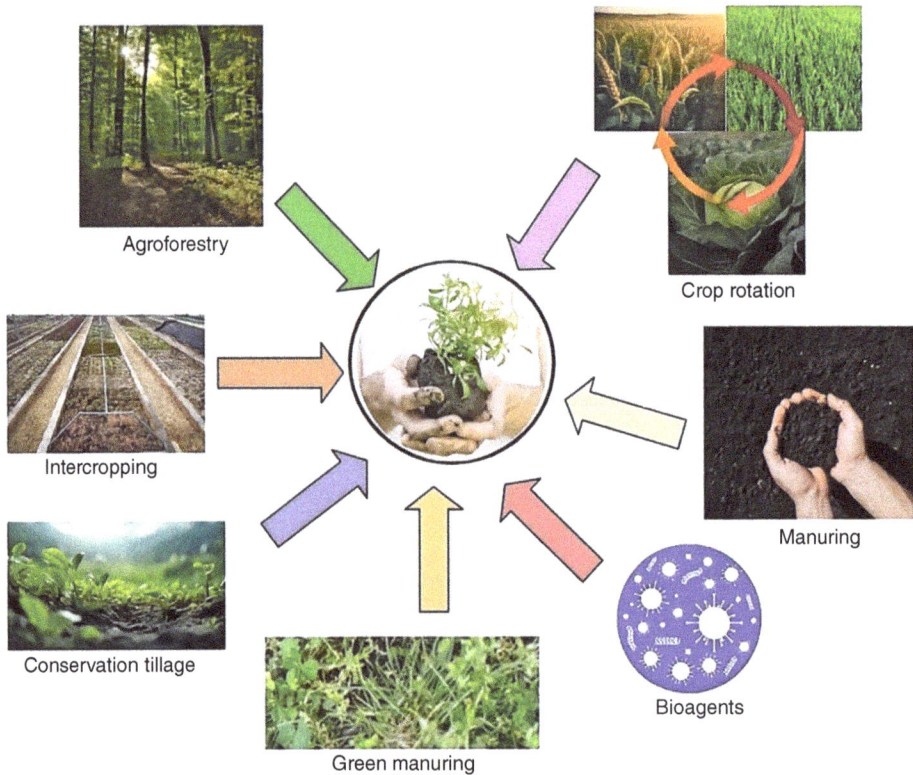

Fig. 13.1. Graphic representation of sustainable agriculture.

13.2 Soil Health Indicators

All soil health indicators are quantifiable characteristics that affect the ability of soil to support crop growth and to perform environmental activities (Shahane and Shivay, 2021). The indicators encompass the soil's many physical, chemical and biological aspects, including nutrient levels, pH, organic matter concentration and microbial activity and biological qualities responsive to land management, natural disturbances and chemical pollutants (Zaghloul *et al.*, 2019; Bhaduri *et al.*, 2022; Zhang *et al.*, 2024a) (Fig. 13.2). By monitoring these indicators, farmers and land managers can gather valuable information to make well-informed decisions to enhance soil health and overall productivity (Karunathilake *et al.*, 2023; Zyngier *et al.*, 2024).

Chemical indicators such as pH levels are measured using a pH meter to determine soil acidity or alkalinity while total organic carbon is analysed in laboratories to quantify the amount of organic carbon in the soil (Alvarez, 2024). Simultaneously, the nutrient levels of soil samples can be analysed to ascertain the presence of essential nutrients like nitrogen, phosphorus and potassium (Liu *et al.*, 2024c). The major physical indicators for soil structure can be evaluated by observing soil aggregates and their stability using techniques such as visual examination or tools like a penetrometer (Mora-Motta

et al., 2024). Usually, soil texture is determined through tactile examination to categorize soil as sandy, loamy or clayey, including biological indicators like the number of earthworms, which indicates soil health and biological activity (Thiruchelve *et al.*, 2024).

The soil microbiome is used to analyse the variety and quantity of microorganisms by using phospholipid fatty acid analysis or metagenomics sequencing (Kandeler, 2024). Also, soil respiration involves measuring the carbon dioxide emission rate from the soil, indicating microbial activity and organic material decomposition. All these techniques provide valuable insights into soil health by assessing multiple components of soil quality. Integrating measurements from chemical, physical and biological markers allows farmers and soil scientists to comprehensively evaluate soil health, track changes over time and make informed decisions to improve soil quality and productivity (Kantola *et al.*, 2023).

Farmers can see variations in soil health over time by consistently monitoring these indicators and adapting their management strategies accordingly (Eze *et al.*, 2021; Mikhailova *et al.*, 2024). Understanding the relationships between these variables can also aid in formulating sustainable soil management practices that foster enduring productivity and environmental sustainability (Mazzocchi, 2020; Liu *et al.*,

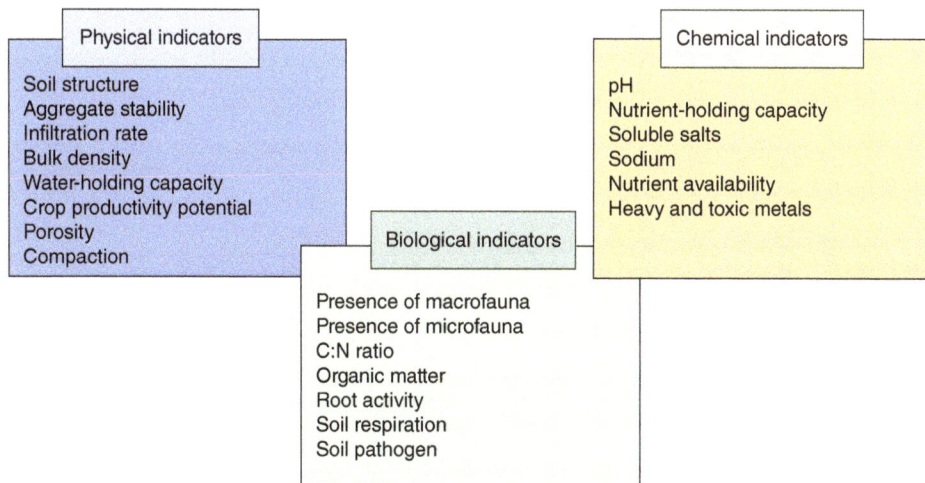

Fig. 13.2. Soil health indicators in agriculture.

2024d). Typical markers of soil health include the overall amount of organic matter in the soil, the amount of organic matter that is actively decomposing, the rate at which the soil releases carbon dioxide through respiration, the ability of soil particles to stick together and resist erosion, the amount of nitrogen that may be released from organic matter and the presence of enzymes in the soil (Zhang *et al.*, 2024c). These indicators offer valuable information about soil biology and its impact on plant growth, as well as the physical and chemical qualities of the soil and the process of carbon sequestration (Kodaparthi *et al.*, 2024; Zheng *et al.*, 2024).

Some commonly used techniques by farmers for evaluating soil health include the visual examination of soil aggregates, root penetration, water infiltration and earthworm abundance (Ball *et al.*, 2017). A penetrometer can also be used which assesses soil compaction, indicative of soil health status, or the use of a test to measure the water infiltration rate using a set volume of water and a timer (Hassan and Beshr, 2024). To assess soil microbial activity farmers can bury a cotton cloth for a specified period, observing the decomposition extent. The soil organic matter measurement is a test undertaken in a specified laboratory to evaluate soil organic matter alongside other chemical and biological factors; this is distinct from routine fertility tests (Bogush and Kourtchev, 2024).

The soil enzyme activity test assesses enzyme activity as an indicator of soil biological activity (Franzluebbers, 2020). Most soil respiration will quantify carbon dioxide emissions, reflecting soil microbial activity and organic material decomposition (Romeijn *et al.*, 2019).

Soil microbiome analysis examines microorganism variety and quantity using techniques like phospholipid fatty acid analysis or metagenomic sequencing, providing valuable soil condition insights (Wydro, 2022). Soil aggregate stability indicates the soil health, organic matter levels and nutrient cycling capacity by assessing soil aggregates and their ability to retain structure under disruption.

13.2.1 Some common soil health indicators used by farmers

Farmers often rely on perceptible soil health indicators, including crop attributes and pigmentation,

with soil colour and texture being the most frequently recognized characteristics (Wydro, 2022). The qualitative assessments are typically rated on a scale from 1 to 5. Collaboration with field personnel and partners aids in conducting qualitative soil health assessments (Dasgupta *et al.*, 2024). Various methods are employed to assess soil physical characteristics, such as the 'hoe test' for digging ease, soil structure for aggregate presence and size, soil compaction for hardpan identification and soil crusting for crust formation observation (Entz *et al.*, 2022). Additionally, water movement and holding capacity are evaluated (Bashir *et al.*, 2021). Vegetation presence and soil colour analysis offer insights into organic matter quantity and condition (Huang *et al.*, 2020).

13.2.2 Soil health indicators differ from soil quality indicators

Soil health indicators and soil quality indicators are closely associated concepts that are frequently used interchangeably. Nevertheless, certain distinctions exist between the two. Soil health indicators are a specific group of soil quality indicators that assess the effectiveness of soil in performing its tasks, as direct measurement of soil function is often impossible (Juhos *et al.*, 2023). Measuring soil quality involves identifying soil attributes that are influenced by how the soil is managed, have an impact on or are related to environmental results, and can be accurately evaluated within specific technological and budgetary limitations (Nungula *et al.*, 2024). Soil health indicators encompass chemical, physical and biological aspects that assess the soil's capacity to carry out essential tasks, including nutrient cycling, water relations, buffering, physical stability and support, habitat provision, biodiversity maintenance and filtration capabilities.

Conversely, soil quality indicators are more comprehensive and include health indicators (Zhang *et al.*, 2024a). All soil quality indicators are measurements that show how well an ecosystem is functioning in a specific soil. They are features of the soil that can change quickly when the soil is disturbed. The soil analysis can be either qualitative or quantitative and offers valuable insights into various aspects of soil health and composition (Dasgupta *et al.*, 2024) as it provides information about the balance between soil solution and exchange sites, the well-being of plants, the nutritional needs of plants and soil animals, the presence of soil contaminants and their accessibility to animals and plants, the hydrological properties of the soil, nutrient availability, erosion levels and the organisms involved in the soil food web responsible for organic matter decomposition and nutrient cycling (Blanco-Velázquez and Anaya-Romero, 2024).

Conservation agriculture commonly employs physical, chemical and biological indicators to assess soil health (Agyei *et al.*, 2024; Hassan *et al.*, 2024). These indicators evaluate the effectiveness of soil function, which is frequently challenging to quantify directly. Soil carbon is an essential indication that encompasses all three categories and has the most universally acknowledged impact on soil quality as it is connected to all soil functions (Poddar *et al.*, 2024). The evaluation of these indicators is based on a rating system ranging from 1 to 5, where 5 represents the highest degree of desirability and 1 represents the lowest (worst) level of soil health. Farmers collaborate with field personnel and partners to conduct qualitative soil health assessments.

The chemical indicators utilized in conservation agriculture encompass pH levels, organic matter content and nutrient availability (Juhos *et al.*, 2023) and are assessed using farmers' observations and scientific measurements (Ibrahim *et al.*, 2024). The pH level is a crucial indicator because it impacts the availability of nutrients and the activity of microorganisms. The organic matter content is a crucial indicator that is directly associated with soil fertility and biological activity (Chaudhry *et al.*, 2024).

13.2.3 Challenges to achieving sustainable soil management

Some challenges to achieving sustainable soil management include the need to adapt to changing environmental conditions caused by climate change, such as severe drought and increased flooding, inappropriate agricultural practices, destruction of forest areas, land use changes contributing to increased carbon in the atmosphere, inadequate capacity, knowledge

and experience, lack of education and awareness, policy and socio-cultural constraints, low adoption of improved technologies, high rates of land degradation, inadequate infrastructure, variable performance of bacterial biostimulants, poor shelf-life of packaged food, limited understanding of the interactions of plants and the environment, soil erosion, loss of soil organic carbon, loss of biological activity and diversity and contamination (Skendžić *et al.*, 2021; Löbmann *et al.*, 2022; Grigorieva *et al.*, 2023).

13.3 Preventing Soil Contamination in Agriculture

To mitigate soil contamination in agriculture, a range of strategies can be utilized including adopting sustainable farming methods and the responsible use of agrochemicals (Fig. 13.3). It is essential to minimize soil pollution by avoiding the incorrect application of fertilizers, insecticides and herbicides (Aqeel *et al.*, 2014). This can be done by making agricultural decisions based on soil tests, creating plant barriers and adopting sustainable approaches such as soil fertility and pest management (Gamage *et al.*, 2023; Fuentes-Peñailillo *et al.*, 2024). Efficient waste management and appropriate rubbish disposal, including hazardous and recyclable materials, is crucial to avoid soil pollution (Barathi *et al.*, 2024). This entails the efficient breakdown of organic waste to decrease the presence of pathogenic microbes and promote the recycling of nutrients (Rastogi *et al.*, 2020; Xiao *et al.*, 2024).

Furthermore, it is crucial to limit the accumulation of waste on agricultural lands and to effectively handle animal dung and liquid waste to protect the integrity of water bodies. The use of eco-friendly fertilizers and pesticides can efficiently reduce soil-borne pathogen diversity (Rastogi *et al.*, 2020). Soil tests can ascertain the necessary inputs, while practices such as sustainable soil management and integrated soil fertility management improve nutrient recycling and decrease dependence on agrochemicals. Preventative measures employing personal protective equipment (PPE) such as gloves, eyewear and masks can protect farmers' health and help prevent soil pollution (Tahat *et al.*, 2020). Efficiently handling animal manure is crucial to prevent soil pollution, taking into account elements such

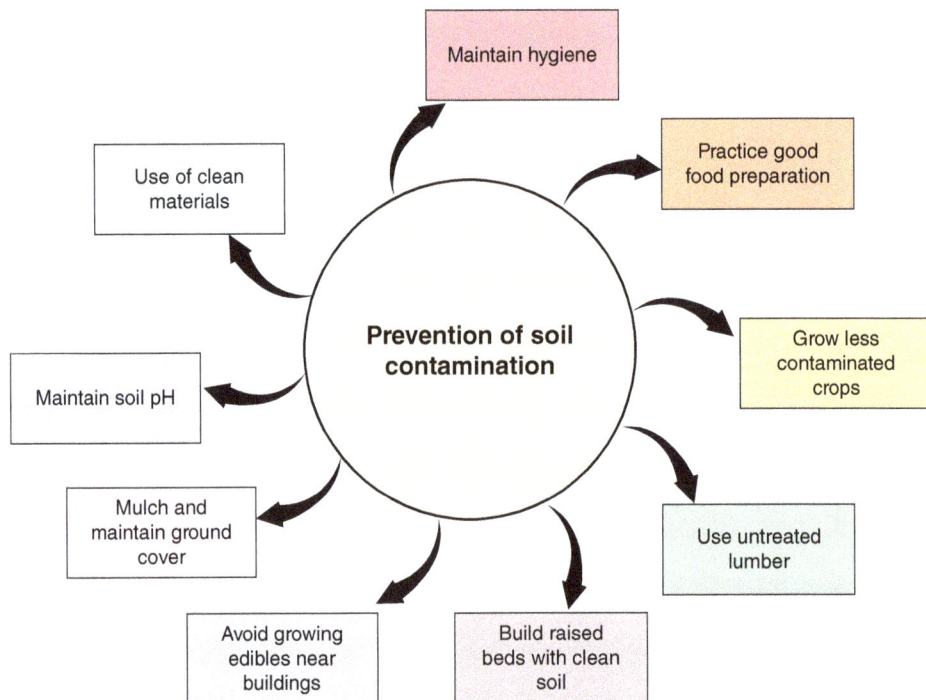

Fig. 13.3. Different ways of preventing soil contamination in agriculture.

as timing, quantity and placement to protect the quality of water bodies (Mehmet Tuğrul, 2020). By limiting machinery usage and adopting sustainable soil management approaches it is possible to reduce soil compaction and contamination. It is also important to reduce pollution caused by neighbouring activities (Kumar *et al.*, 2021; Jorat *et al.*, 2024). Strategies should be enforced to prevent pollution resulting from the disposal of plastic mulch, outdated pesticides and animal feed containing heavy metals (Juhos *et al.*, 2023). The implementation of plant barriers and using supplementary measures will help to defend fields from potential contamination caused by surrounding activity. Crops should not be cultivated in regions that are contaminated (Rashid *et al.*, 2023), for example in fields exposed to harmful substances from mining operations, factories or urban areas (Raffa and Chiampo, 2021). Watershed initiatives cooperating with diverse stakeholders to reduce nutrient pollution in water and air are important (Raffa and Chiampo, 2021). Farmers can

assume leadership by actively engaging in such watershed projects alongside state governments, agricultural associations, environmental organizations, educational institutions, non-profit organizations and community groups (Mello *et al.*, 2021).

13.3.1 Economic challenges to implementing sustainable soil management practices

The economic challenges to implementing sustainable soil management include balancing short-term economic gains with long-term sustainability goals (Muhie, 2022). Adopting sustainable soil management practices may require upfront costs, such as investing in new equipment or changing farming practices, which may not yield immediate economic benefits (Aznar-Sánchez *et al.*, 2020; Serebrennikov *et al.*, 2020). Additionally, there may be a lack of financial incentives or subsidies to support adopting sustainable soil management practices, making it difficult for farmers to justify the investment (Piñeiro et al., 2020). Furthermore, there may be a lack of awareness or understanding of the long-term economic benefits of sustainable soil management practices, such as increased soil health and productivity, reduced input costs and improved resilience to climate change (Liu et al., 2018; Shah and Wu, 2019). This can make it difficult for farmers to make informed decisions about adopting such practices (Mehmet Tuğrul, 2020). Finally, there may be challenges in measuring and quantifying the economic benefits of sustainable soil management practices, making it difficult to demonstrate their value to farmers, policy makers and other stakeholders (Struik and Kuyper, 2017; Hou et al., 2020; Mehmet Tuğrul, 2020). Developing widely accepted, easily applicable methods for recording and assessing soil quality, soil functions and the ecosystem services provided by soils can help to address this challenge (Tahat et al., 2020; Ros et al., 2022).

Addressing these economic challenges requires a multifaceted approach, including education and awareness-raising efforts, financial incentives and subsidies and the development of tools and methods for measuring and quantifying the economic benefits of sustainable soil management practices. An obstacle to implementing these strategies is balancing immediate economic benefits and long-term sustainability objectives (Nicastro et al., 2024; Osman et al., 2024). To effectively tackle these economic challenges, a comprehensive strategy is needed. This strategy should encompass various elements such as educational campaigns, financial incentives and subsidies, and the creation of tools and techniques to accurately assess and quantify the economic advantages of sustainable soil management practices (Cantú et al., 2021; Köninger et al., 2021).

13.3.2 Improvement in soil biodiversity

Each farmer has the opportunity to enhance soil biodiversity through a variety of practices aimed at nurturing and enriching the diverse life within the soil. One fundamental approach minimizes soil disturbance (Köninger et al., 2021). By reducing the frequency and intensity

of tillage, farmers can foster an environment where soil organisms can flourish and sustain their populations effectively (Juhos et al., 2023). Another pivotal strategy is to increase crop diversity (Xue et al., 2024). Planting a range of crops, including cover crops, augments the diversity of soil organisms while bolstering soil health (Fowler et al., 2024). Diversified crop rotations further aid in mitigating pests and diseases specific to particular plant species (Agyei et al., 2024).

Integrating livestock into farming systems is also beneficial. This approach promotes nutrient cycling and organic matter distribution, enhancing soil biodiversity. Furthermore, managing organic matter is crucial (Gamage et al., 2023). Incorporating crop residues, compost or manure into the soil fosters thriving populations of surface-feeding creatures like earthworms (Urra et al., 2019). Limiting the use of chemical fertilizers and pesticides mitigates negative impacts on soil organisms, thereby promoting soil biodiversity (Tudi et al., 2021). Implementing conservation practices is equally essential. Minimum or no-till techniques, maintaining and augmenting perennial plants and judicious use of fertilizers contribute to sustaining and enhancing soil biodiversity (Wydro, 2022). Farmers can actively foster soil biodiversity by diligently implementing these practices, consequently improving their soil's overall health and productivity (Bertola et al., 2021; Löbmann et al., 2022).

13.3.3 Some alternatives to synthetic fertilizers in agriculture

Agricultural practitioners can use substitute fertilizers instead of synthetic fertilizers to ameliorate soil quality and minimize adverse environmental impacts (Ning et al., 2022). The various alternatives mostly include manure from livestock excreta, such as that from cows, pigs and chickens, which can furnish essential nutrients to crops (Gržinić et al., 2023). However, it is crucial to compost the manure properly and apply it in appropriate quantities to prevent excessive nutrient runoff. The compost, derived from organic matter like food waste, yard debris and animal manure, can enhance soil quality by improving its structure, fertility and water retention capacity

(Gržinić *et al.*, 2023). Also, cover crops, such as clover, rye and vetch, can be intercropped with cash crops to augment soil fertility by introducing organic matter and nitrogen (Sayara *et al.*, 2020).

Much green manure incorporates buckwheat, mustard and radish added to the soil to bolster its organic content and nutritional levels through tillage. Bone meal is a naturally occurring substance containing phosphorus and calcium, which can enhance root development and fruit production in crops (Yang *et al.*, 2024b). Rock phosphate, a naturally occurring phosphorus substance, can be utilized as a fertilizer that gradually releases nutrients. Seaweed can also be used as a natural reservoir of minerals and growth hormones (Gu *et al.*, 2023). Worm castings, or vermicompost, are naturally occurring substances containing essential nutrients such as nitrogen, phosphate and potassium (Yang *et al.*, 2024b). These nutrients are advantageous for enhancing soil quality, fertility and structure (Gu *et al.*, 2023). Biochar, a carbon-rich substance produced through the pyrolysis of organic waste, can improve soil quality by enhancing its structure, fertility and water retention capacity (Rehman *et al.*, 2023). These alternatives can provide essential nutrients to crops while mitigating the environmental impacts of synthetic fertilizers (Köninger *et al.*, 2021). Nonetheless, organic fertilizers may require additional labour and oversight compared with artificial fertilizers, and their effectiveness may vary depending on the specific crop and soil conditions (Hou *et al.*, 2020).

13.3.4 Reducing the use of synthetic pesticides and fertilizers

Farmers can reduce their dependence on artificial fertilizers and pesticides by adopting alternative approaches that promote soil health and natural pest management (Muhie, 2022), including the use of organic fertilizers for transitioning from synthetic fertilizers to organic alternatives such as manure and compost (Skendžić *et al.*, 2021). Organic fertilizers supply essential nutrients to the soil and improve soil vitality, leading to long-term enhancements in crop productivity (Barathi *et al.*, 2024).

Employing cover crops can reduce the need for synthetic fertilizers by incorporating organic matter into the soil, improving soil structure and enhancing soil biodiversity (Upadhyay *et al.*, 2024). Additionally, cover crops can suppress weed growth and harmful organisms, reducing reliance on herbicides and insecticides. Implementing crop rotation can decrease farmers' dependence on synthetic fertilizers and pesticides (Jaworski *et al.*, 2023). Crop rotation disrupts pest and disease cycles while improving soil health. Integrated pest management (IPM) is a comprehensive pest control strategy that integrates cultural, biological and chemical methods. By promoting natural pest management techniques such as habitat modification, biological control and cultural practices, IPM can reduce reliance on synthetic pesticides (Karlsson Green *et al.*, 2020).

Precision agriculture, which utilizes advanced technologies to optimize agricultural productivity and minimize input costs through techniques like variable rate application and precision irrigation, can reduce the need for synthetic fertilizers and pesticides (Abiri *et al.*, 2023). Implementing agroecological practices such as permaculture, agroforestry and organic farming can decrease reliance on synthetic inputs by promoting biodiversity, improving soil health and fostering natural pest management (Karunathilake *et al.*, 2023; Dönmez *et al.*, 2024).

Governments can implement legislative measures that incentivize farmers to reduce their reliance on synthetic fertilizers and pesticides through various policy measures (Yang *et al.*, 2024a). These may include subsidies for organic farming, tax credits for sustainable agriculture practices and regulations restricting synthetic inputs. By implementing these strategies, farmers can reduce their dependence on artificial fertilizers and pesticides, improve soil quality and promote sustainable agricultural practices (Jacquet *et al.*, 2022).

13.4 Sustainable Agriculture: Concepts and Practices

Sustainable agriculture endeavours to achieve consistent and uninterrupted production while safeguarding resources for future generations

(Jacquet *et al.*, 2022). This endeavour underscores the significance of enhancing food chain productivity, protecting environmental resources, bolstering individual well-being, promoting economic growth, strengthening ecosystem and community resilience and supporting governmental efforts and regulations (Joseph Sekhar *et al.*, 2024).

The primary objectives of sustainable agriculture revolve around ensuring food and fibre security, maintaining soil fertility, preserving biodiversity, enhancing ecological conditions, preventing pollution, fostering rural economic growth, improving farmers' health, raising environmental awareness and promoting responsibility. The evaluation of sustainable agriculture hinges on three key aspects: environmental (agri-ecological), social (social–territorial) and economic scales (Lakhiar *et al.*, 2024). These pillars emphasize environmentally friendly farming practices, social equity and financial viability (Kodaparthi *et al.*, 2024).

Numerous sustainable agriculture practices contribute to achieving these objectives, including crop rotation, integrated weed management, permaculture, polyculture, crop residue management, biodynamic and organic farming, livestock–crop integration, intercropping, mulching, conservation tillage, biofuel use, agroforestry and urban agriculture (Table 13.1). These approaches aim to reduce dependence on artificial fertilizers and pesticides by advocating for natural pest management measures, enhancing soil health and promoting biodiversity. Sustainable agriculture prioritizes using renewable energy sources, minimizes land use and reduces pollution to promote long-term agricultural sustainability (Kodaparthi *et al.*, 2024).

Sustainable agricultural techniques can positively impact the environment in multiple ways. Decreased chemical inputs through the reduction of use of synthetic fertilizers, insecticides and herbicides (Raffa and Chiampo, 2021) minimizes the discharge of chemicals into water bodies, mitigating the potential harm to aquatic ecosystems. It also helps prevent soil pollution, safeguarding soil health and biodiversity (Bashir *et al.*, 2021; Chaudhry *et al.*, 2024).

Soil conservation involves implementing techniques such as crop rotation, cover cropping and conservation tillage to preserve the integrity

and fertility of the soil (Bashir *et al.*, 2021). This mitigates soil erosion, preventing sediment deposition in water bodies and safeguarding the integrity of aquatic ecosystems (Pandey *et al.*, 2024).

Sustainable practices, including drip irrigation, rainwater collection and precision irrigation, can achieve water conservation in agriculture. These methods effectively maximize water utilization (Kreitzman *et al.*, 2022). By reducing water wastage and minimizing water pollution caused by agricultural runoff, these measures actively contribute to the conservation of freshwater resources and the preservation of aquatic habitats (Hirschfeld and Van Acker, 2021). Biodiversity conservation fosters the cultivation of various crops, creating environments that support the presence of advantageous insects, birds and other forms of wildlife (Franzluebbers and Martin, 2022; Ihenetu *et al.*, 2024). This enhances the variety of life forms on farms, bolstering the ability of ecosystems to withstand and recover from disturbances while safeguarding indigenous species' existence (Mane *et al.*, 2024).

Techniques like agroforestry, rotational grazing and the utilization of cover crops can enhance the process of carbon sequestration in both soils and plants (Pandey *et al.*, 2024). This process aids in reducing climate change by extracting carbon dioxide from the atmosphere and sequestering it in agricultural environments (Kreitzman *et al.*, 2022). Sustainable agriculture practices frequently favour energy-efficient methods, such as organic farming, which depends on natural processes instead of energy-intensive inputs. In addition, implementing practices such as no-till farming can help minimize reliance on fossil fuel machinery, further reducing greenhouse gas emissions (Bhattacharya and Pandey, 2024).

Enhancing pollinator health can be done by implementing sustainable agricultural methods that reduce pesticide usage and encourage a wide range of flowering plants to help maintain and promote the well-being of pollinator populations such as bees and butterflies (Martin *et al.*, 2019). Robust pollinator populations are crucial for the propagation of numerous crops and the preservation of indigenous ecosystems. In general, sustainable agriculture techniques aim to protect the environment by reducing harm to

Table 13.1. Different sustainable approaches used in agriculture practices.

Strategy	Description	Reference
Hydroponics	An agricultural technique where plant roots are immersed in a solution containing abundant nutrients, enabling optimal water utilization and increased crop production	Son *et al.*, 2024
Aeroponics	Developed by NASA, this technique enables plants to grow suspended in mist, reducing water consumption by 98% and eliminating the necessity for pesticides and fertilizers	Mittal and Bhukal, 2024
Aquaponics	A self-contained system that integrates plants and fish, sustainably enabling crop cultivation and optimizing resource use	Chaudhary and Anand, 2024
Precision agriculture	Technology that uses GPS and data analytics to enhance field management, resulting in increased crop yield and reduced wastage	Karunathilake *et al.*, 2023
Artificial intelligence (AI) and machine learning	Improving agricultural efficiency by maximizing the utilization of resources, reducing waste and automating labour-intensive activities	Yeasmin *et al.*, 2024
Drones	Used in precision farming to detect crop concerns early and use resources efficiently, resulting in less environmental impact	Karunathilake *et al.*, 2023
Renewable energy sources	Integrates renewable energy sources, such as solar panels, geothermal energy and biogas, into agriculture to decrease environmental impacts and improve resource utilization	Joseph Sekhar *et al.*, 2024
Soil conditioners, enhancers and activators	These inputs alter soil structure and properties, improve soil minerals and provide a rich food source for soil microorganisms, leading to enhanced overall soil quality, better root development and improved water-holding capacity	Shashirekha *et al.*, 2024
Nanobubble technology	Enhances soil structure by increasing soil flocculation, reducing soil compaction, improving water infiltration and nutrient mobility and improving soil health and crop yields	Li *et al.*, 2024a
Food forests	Unique and rewarding approach to growing food, focusing on preserving the environment's natural resources and ecosystems, producing nutrient-rich food and minimizing the use of harmful chemicals and pesticides	Brower-Toland *et al.*, 2024
Agroforestry	Sustainable practice that can eventually result in increased crop yields due to improved soil health, nutrient cycling and water retention	Villat and Nicholas, 2024
Integrated pest management (IPM)	Sustainable agricultural practice that seeks to control pests while reducing reliance on chemical pesticides, combining biological, chemical, physical and crop-specific management strategies	Muhie, 2022
Sustainable animal husbandry	Acknowledges the significance of livestock as they generate organic fertilizer through their manure; enhances productivity in regions with traditional livestock management practices, for example using rotational grazing	Gržinić *et al.*, 2023

ecosystems, preserving natural resources and improving the ability of agricultural systems to withstand climate change and other environmental difficulties (Katumo *et al.*, 2022).

13.4.1   Soil erosion reduction

Cover crops are crucial in mitigating soil erosion by providing a protective layer over the soil. This layer significantly reduces the likelihood of erosion caused by wind and water, particularly during periods of heavy precipitation and strong winds, such as in early spring and late autumn when typical cash crops are not actively growing (Katumo *et al.*, 2022). Living and dead plant material is essential in limiting the effects of rainfall and wind on soil erosion (He *et al.*, 2018). Moreover, cover crops shield soil aggregates from the erosive effects of rainfall by mitigating soil aggregate disintegration. Cover crops effectively reduce wind speeds at ground level and decrease water velocity in the runoff, offering a comprehensive defence against wind and water erosion (Roofchaee *et al.*, 2024). This multifaceted approach underscores the importance of integrating cover crops into sustainable agricultural practices to preserve soil health and to mitigate the environmental impacts of erosion (Li *et al.*, 2024b).

Cover crops produce more plant material than plants growing independently. This abundance contributes to several erosion-reducing mechanisms (Crack *et al.*, 2024). For example, cover crops release water vapour and enhance water absorption into the soil, thereby reducing the amount of water flowing over the surface and the speed at which it flows (Pamuru *et al.*, 2024). Research indicates that doubling the velocity of runoff water in a stream increases the carrying capacity of water, or the stream's ability to transport soil material and nutrients, by a significant factor.

13.4.2 Some challenges that farmers face when transitioning to sustainable agriculture practices

One of the challenges that farmers face when transitioning to sustainable agriculture practices is a lack of knowledge and awareness. Farmers need to learn new techniques, understand soil health and implement conservation practices (Zhang *et al.*, 2024b).

Financial constraints for transitioning to sustainable agriculture often involve upfront investments in infrastructure, equipment and alternative farming practices, which can be financially challenging for farmers, especially those with limited resources (Kalyani *et al.*, 2024; Mana *et al.*, 2024). Creating market demand

and a suitable infrastructure for sustainably produced agricultural products can be challenging, as establishing markets that value and pay a premium for sustainable products may not be readily available (Gemtou *et al.*, 2024).

Scaling up sustainable practices to meet global food demands can be a significant challenge as these practices often require more labour-intensive methods and may initially yield lower outputs (Gemtou *et al.*, 2024).

Sustainable farming practices must be tailored to local systems, which can vary significantly across climates, growing seasons, pests, diseases and crop varieties (Chen *et al.*, 2024). Measurement tools for sustainable practices are not common, making it challenging to monitor and assess the impact of these practices effectively (Liu *et al.*, 2024a). Farmers may have a limited understanding of soil health and organic matter and the ability to monitor carbon fluxes through a system, which is crucial for verifying carbon offset programmes (Thanekar *et al.*, 2024).

Managing emissions from soil management, livestock, rice paddies and fertilizer application requires a comprehensive understanding of total emissions through the food system, which can be complex and challenging to address effectively (Deshavath *et al.*, 2024). Connectivity with global trade can impact sustainable agriculture practices, as changing standards in one part of the trade can have ripple effects on other countries, leading to economic impacts (Šeremešić *et al.*, 2024). For example, the increase of Mediterranean agricultural commerce, specifically the increase in irrigated crop cultivation, has substantially raised competitiveness for scarce water resources in that area (Llop and Ponce-Alifonso, 2016). The irrigated agriculture sector currently represents a significant proportion of overall water usage. Specifically, it is the largest consumer of freshwater in the Mediterranean region, accounting for 21% of the total area used for cultivating crops (Mancuso *et al.*, 2019). All these irrigated agricultures have depleted around 23.2% of the blue water resources in the Mediterranean basin. From 1981 to 2001, Syria experienced a 124% rise in irrigated land, while Algeria, Jordan and Libya saw increases of 114% and 109%, respectively. Also, in eastern and southern Mediterranean countries, irrigated agriculture accounted for 81% of the total water demand (Mallareddy *et al.*, 2023).

The swift growth of irrigated agriculture, propelled by globalization and agricultural trade, has resulted in a shortage of water that limits other economic endeavours. The presence and dependability of water resources provide a constraint on the economic progress of numerous water-deprived Mediterranean nations. Water scarcity is especially severe in southern and eastern Mediterranean countries, as well as in certain catchments areas in the north such as south-east Spain and the Ebro Depression (Margat and Gun, 2013). The combination of increased irrigated production, tourism and urbanization has led to significant strain on water reservoirs. However, if irrigated agriculture does not adjust to the changing climate conditions in the region, it risks exacerbating the already critical water crisis (de Oliveira *et al.*, 2020).

These are major concerns for the nature of transitioning to sustainable agriculture and the need for support, education, financial incentives and market demand to facilitate a successful shift towards more sustainable farming practices (Sharma *et al.*, 2021).

13.4.3   Soil health benefits

Sustainable agriculture practices can benefit soil health by improving soil structure, increasing water infiltration, enhancing nutrient cycling, reducing erosion and promoting biodiversity (Scarlato *et al.*, 2024). Practices such as no-till or reduced-till farming, cover cropping, crop rotation, composting and pasture management are key ways for farmers and ranchers to improve soil health (Bier *et al.*, 2024). For example, no-till farming can reduce soil erosion, save time and money on inputs and improve the resiliency of working land (Ibrahim *et al.*, 2024). Cover crops can build soil structure, protect water quality, suppress pests and improve a farm's bottom line (Scarlato *et al.*, 2024). Conservation tillage can improve soil structure, reduce erosion and enhance nutrient cycling. Composting can improve soil health by adding organic matter and promoting biological activity (Xue *et al.*, 2024). Pasture management can improve soil health by promoting healthy plant communities, increasing organic matter and reducing erosion. By adopting these sustainable agriculture practices, farmers and ranchers can

improve soil health, enhance productivity and promote long-term sustainability (Pandey *et al.*, 2024).

13.4.4   Overcoming the challenges of transitioning to sustainable agriculture practices

Farmers can overcome the hurdles of transitioning to sustainable agriculture by embracing various strategies and approaches (Kumar *et al.*, 2021; Deshavath *et al.*, 2024). Utilizing research innovation is key to this (Tahat *et al.*, 2020), by tapping into research and innovation, farmers can better understand the challenges and develop scalable solutions to facilitate their shift towards sustainable agriculture (Bertola *et al.*, 2021). Technological advancements such as robotic devices, satellite data and artificial intelligence offer effective means to monitor crop and livestock conditions (Köninger *et al.*, 2021). Moreover, adopting sustainable agriculture innovation can attract significant investments (Johnraja *et al.*, 2024; Kodaparthi *et al.*, 2024).

Implementing regenerative practices and nature-based solutions is another crucial step. These methods can significantly improve soil health, reduce greenhouse gas emissions and promote farm biodiversity (Köninger *et al.*, 2021; Frasconi *et al.*, 2024). Practices like cover cropping, conservation tillage, crop rotation and agroforestry contribute to this effort (Kamyab *et al.*, 2024; Yahyah *et al.*, 2024).

Offering training and education plays a vital role in fostering the adoption of sustainable agriculture techniques (Löbmann *et al.*, 2022). Providing instruction and knowledge to farmers and other stakeholders in the food value chain helps them understand the benefits of sustainable agriculture practices and learn how to implement them effectively (Struik and Kuyper, 2017; Gamage *et al.*, 2023). This includes research on regenerative approaches, soil health management and sustainable food systems (Mello *et al.*, 2021; Yang *et al.*, 2024a).

Enhancing transparency and traceability is essential for farmers and consumers (Gemtou *et al.*, 2024). Transparency and traceability tools, such as labelling and certification schemes, blockchain technology and other means of providing information about food products' origin and environmental impact, help stakeholders make informed decisions (Muhie, 2022; Yahyah *et al.*, 2024). Promoting interdisciplinary collaboration is also crucial. Cooperation among farmers, policy makers, researchers and other stakeholders can accelerate the transition to sustainable agriculture (Gemtou *et al.*, 2024; Li *et al.*, 2024b). This involves collaborations between farmers and researchers, public–private partnerships and initiatives that facilitate the exchange of knowledge, foster innovation and encourage investment in sustainable agriculture (Hathaway, 2016; Jacquet *et al.*, 2022). With the right input, farmers can effectively address the challenges of transitioning to sustainable agriculture practices (Cao and Solangi, 2023) and this contributes to building a more sustainable food system that benefits individuals, the environment and future generations (Tahat *et al.*, 2020; Šeremešić *et al.*, 2024).

13.4.5 Sustainable agriculture strategies

Planting vegetation during off-seasons helps protect soil, mitigate erosion, enhance fertility, restrict weeds and improve water retention (Hemkemeyer *et al.*, 2024). Crop rotation by alternating crops helps to prevent soil depletion, maintain nutrient levels, reduce pest susceptibility and enhance biodiversity. Conservation tillage uses minimal soil disturbance techniques like no-till or reduced tillage to preserve moisture, soil composition, carbon capture and to reduce costs (Entz *et al.*, 2022). Agroforestry by introducing trees and shrubs into agricultural landscapes helps to improve soil fertility, manage water cycles, provide habitats and generate additional crops or revenue (Jaworski *et al.*, 2023). IPM uses various pest control methods, such as biological control and cultural practices, to reduce reliance on chemicals and to protect beneficial insects (Karlsson Green *et al.*, 2020). Precision agriculture uses technology such as GPS and drones to optimize resource utilization, track soil diversity, monitor crop development and administer resources efficiently (Poddar *et al.*, 2024). Organic farming avoids the use of artificial pesticides, fertilizers and genetically modified organisms while prioritizing soil fertility, ecosystem diversity and environmental

sustainability (Park *et al.*, 2024). Rotational grazing through systematically moving animals between pastures allows vegetation regeneration, maintains soil quality, prevents overgrazing, optimizes nutrient cycling and boosts pasture productivity. These practices collectively contribute to resilient, environmentally sustainable and economically prosperous agricultural systems benefiting farmers and ecosystems.

13.4.6 Some common misconceptions about sustainable agriculture practices

False beliefs about agriculture often arise from limited understanding or misinformation regarding its environmental impact and modern farming practices. We can promote a more accurate understanding of sustainable agriculture and its benefits by debunking these myths.

- **Myth 1:** *US agriculture is responsible for a significant amount of greenhouse gas emissions.* In fact, in 2019, US agriculture accounted for approximately 10% of the country's greenhouse gas emissions, making it the least polluting significant industry in the USA (Bhatti *et al.*, 2024).
- **Myth 2:** *Agriculture has a negative impact on the environment.* Scientists continuously improve farming operations to be more environmentally friendly and sustainable. Precision agriculture and low-till and no-till farming are employed to reduce water runoff and enhance soil quality (Shao, 2024).
- **Myth 3:** *Genetic alteration in agriculture is detrimental.* But the fact is that genetic modification is a precise and effective method for selectively breeding plants and animals with desirable characteristics (Gosai *et al.*, 2024). This process enhances food safety by enabling them to resist diseases and pests better (Brunner *et al.*, 2024).
- **Myth 4:** *Regenerative farming approaches decrease farm profit margins.* Although there may be initial expenses, such as adopting cover crops, growers often reduce their expenditure on synthetic fertilizers and observe significant improvements in soil health when implementing regenerative farming practices correctly (Brunner *et al.*, 2024).

- **Myth 5:** *Corporate farms are equivalent to industrial farms.* In fact approximately 98% of farms in the USA are classified as family farms, which contribute nearly 88% of the overall value of agricultural production in the country (Ball *et al.*, 2017).

These misconceptions can be dispelled by educating individuals about advances in sustainable agriculture and the positive impacts of modern farming techniques on the environment and food production.

13.4.7 Transition to sustainable agriculture practices

Farmers looking to transition to sustainable agriculture practices can access various resources and support systems to aid them. Many governments offer programmes and incentives to support farmers adopting sustainable practices. These may include grants, subsidies and technical assistance for implementing conservation practices, transitioning to organic farming or embracing renewable energy technologies. Farmers can enquire about such programmes from local agricultural extension offices or government agricultural departments (Hannam, 2024).

Non-profit organizations focus on promoting sustainable agriculture and providing resources and support to farmers. These organizations may offer training programmes, workshops, demonstration farms and technical assistance on sustainable farming practices (Manjula and Sharma, 2024). Examples include the Sustainable Agriculture Research and Education (SARE) programme in the USA and organizations like the Sustainable Agriculture Network (SAN) and the Soil Association in the UK (Köninger *et al.*, 2021).

Agricultural research institutions and universities often investigate sustainable farming practices and may offer farmers resources, information publications and training opportunities. Farmers can access research findings, attend workshops and collaborate with researchers to implement sustainable farm practices (Veenstra *et al.*, 2024). All certification programmes, such as organic or fair trade certification, can guide and support farmers in adopting sustainable

practices. These programmes often offer technical assistance, training and market access opportunities for certified farmers (Bhattacharyya *et al.*, 2024).

Financial institutions provide loans or financing options specifically tailored for farmers transitioning to sustainable agriculture. These loans may have favourable terms and conditions to support investments in equipment, infrastructure or transitioning to organic production (Veenstra *et al.*, 2024).

Peer networks or farmer groups focused on sustainable agriculture can provide valuable support and information-sharing opportunities. Farmers can learn from each other's experiences, exchange best practices and access collective resources and expertise. Websites, webinars and online courses offer information, tools and guidance on various aspects of sustainable farming, including soil health, water management, crop rotation and agroecology (Damian *et al.*, 2024).

Consumers are increasingly willing to pay a premium for food produced sustainably, providing farmers with opportunities to enter premium markets (Xie *et al.*, 2024). Sustainable techniques like crop rotation and cover cropping improve soil health, leading to long-term increases in crop yields (Siddique *et al.*, 2024). No-till farming and cover cropping help reduce losses from weather-related events like droughts, pests and weeds (Siddique *et al.*, 2024). Sustainable methods like no-till farming and covering cropping sequester carbon in the soil, mitigating climate change effects and potentially generating income through carbon credits (Villat and Nicholas, 2024). Sustainable agriculture helps maintain clean water, wildlife habitats and recreational opportunities, adding market value through ecosystem services. Enhanced agricultural land makes it more valuable, with long-term investments and profits (Jose *et al.*, 2024). Transitioning to sustainable practices may initially incur expenses. Still, research shows that production costs become equal or cheaper within a few years due to improved soil fertility and reduced reliance on purchased inputs. Higher prices and government financial aid can offset extra costs (Jose *et al.*, 2024; Villat and Nicholas, 2024).

Small-scale farms can compete by directly selling products through channels like farm stands, community supported agriculture (CSA) outlets and food hubs, securing retail revenue instead of relying on commodity pricing. Value-added products further enhance earnings. In conclusion, sustainable agriculture offers environmental and economic benefits, providing an intelligent approach to restoring agricultural output and supporting rural communities responsibly (Saparova *et al.*, 2024).

13.4.8 Economic benefits of sustainable agriculture practices for small-scale farmers

Sustainable agriculture practices can yield substantial economic advantages for small-scale producers (Liu *et al.*, 2024b). The advantages include using natural fertilizers and pest management measures, which can reduce costs by minimizing reliance on costly synthetic inputs, lowering production expenses (Hassan *et al.*, 2024). Small-scale farmers can optimize their crop production by implementing crop rotation, cover cropping and IPM strategies. These practices improve soil health, reduce pest and disease issues and ultimately increase crop yields (Löbmann *et al.*, 2022).

Farmers can enter premium markets where consumers are increasingly willing to pay more for food that is produced sustainably, which allows farmers to charge higher prices and attract more customers (Bertola *et al.*, 2021; Chao *et al.*, 2024). Sustainable agriculture strategies, such as agroforestry and conservation tillage, might enhance the ability of small-scale farmers to cope with climate change and mitigate the effects of weather-related losses (Fowler *et al.*, 2024). Preservation of natural resources can be achieved by using techniques such as water conservation and sustainable irrigation. By adopting these practices, small-scale farmers can contribute to protecting natural resources for future generations (Agyei *et al.*, 2024). Sustainable agriculture practices frequently depend on locally sourced resources and have the potential to generate employment opportunities within the local community, thus enhancing the local economy (Agyei *et al.*, 2024). The economic advantages can assist small-scale farmers in improving their quality of life and positively impact a more environmentally friendly food system (Mandal *et al.*, 2024). Nevertheless, adopting sustainable agriculture practices may necessitate

initial capital inputs and acquiring new knowledge and resources. Policy makers, investors and agricultural groups can significantly assist small-scale farmers in adopting sustainable farming methods (Crack *et al.*, 2024).

13.4.9 Economic benefits of sustainable agriculture for large-scale farmers

Effective irrigation technologies, like mulching and drip irrigation, can save water expenses and usage (Toumi *et al.*, 2024). Long-term cost reductions can be achieved by enhancing soil health, lowering erosion and increasing water retention through crop rotation, cover crops and reduced tillage (Kodaparthi *et al.*, 2024). By reducing the need for chemical pesticides, IPM techniques can save money and protect the environment (Kallali *et al.*, 2024). Less reliance on chemical fertilizers for manure and composting are two organic farming techniques that can increase soil fertility and lessen the demand for pricey chemical fertilizers (Villat and Nicholas, 2024). Higher agricultural yields can result from sustainable agriculture techniques, including agroforestry and precision agriculture (Pandey *et al.*, 2024). Increased productivity can also be obtained through sustainable livestock management techniques like rotational grazing, which helps the environment and saves money (Rashid *et al.*, 2023).

13.5 Conclusion

Soil health indicators encompass physical, chemical and biological aspects crucial for farmers to assess soil quality and productivity. These indicators, including nutrient levels, pH, organic matter concentration and microbial activity, inform decisions on sustainable soil management practices. By utilizing qualitative assessments and tailored strategies based on soil health indicators, farmers can improve soil health, enhance crop development and ensure nutrient availability, addressing regional variations and challenges like climate change and inadequate practices.

Sustainable soil management practices, such as cover cropping, crop rotation and IPM, are vital in mitigating soil erosion, promoting soil fertility and reducing environmental impacts. Despite challenges like knowledge gaps and economic constraints, resources such as government programmes and peer networks can support farmers in transitioning to sustainable agriculture, offering economic advantages such as reduced production costs, enhanced productivity and access to high-value markets. Sustainable agriculture fosters responsible farming practices, financial benefits and environmental stewardship for current and future generations.

References

Abiri, R., Rizan, N., Balasundram, S.K., Shahbazi, A.B. and Abdul-Hamid, H. (2023) Application of digital technologies for ensuring agricultural productivity. *Heliyon* 9, e22601.

Agyei, B., Sprunger, C.D., Anderson, E., Curell, C. and Singh, M.P. (2024) Farm-level variability in soil biological health indicators in Michigan is dependent on management and soil properties. *Soil Science Society of America Journal* 88, 326–338.

Alvarez, R. (2024) A quantitative review of the effects of residue removing on soil organic carbon in croplands. *Soil and Tillage Research* 240, 106098.

Aqeel, M., Jamil, M. and Yusoff, I. (2014) Soil contamination, risk assessment and remediation. In: Hernandez Soriano, M.C. (ed.) *Environmental Risk Assessment of Soil Contamination*. IntechOpen.

Aznar-Sánchez, J.A., Velasco-Muñoz, J.F., López-Felices, B. and del Moral-Torres, F. (2020) Barriers and facilitators for adopting sustainable soil management practices in mediterranean olive groves. *Agronomy* 10, 506.

Ball, B.C., Guimarães, R.M.L., Cloy, J.M., Hargreaves, P.R., Shepherd, T.G. *et al.* (2017) Visual soil evaluation: a summary of some applications and potential developments for agriculture. *Soil and Tillage Research* 173, 114–124.

Barathi, S., Sabapathi, N., Kandasamy, S. and Lee, J. (2024) Present status of insecticide impacts and eco-friendly approaches for remediation-a review. *Environmental Research* 240, 117432.

Bashir, O., Ali, T., Baba, Z.A., Rather, G.H., Bangroo, S.A. *et al.* (2021) Soil organic matter and its impact on soil properties and nutrient status. In: Dar, G.H., Bhat, R.A., Mehmood, M.A. and Hakeem, K.R. (eds) *Microbiota and Biofertilizers*, Vol. 2. Springer, pp. 129–159.

Bertola, M., Ferrarini, A. and Visioli, G. (2021) Improvement of soil microbial diversity through sustainable agricultural practices and its evaluation by -omics approaches: a perspective for the environment, food quality and human safety. *Microorganisms* 9, 1400.

Bhaduri, D., Sihi, D., Bhowmik, A., Verma, B.C., Munda, S. *et al.* (2022) A review on effective soil health bio-indicators for ecosystem restoration and sustainability. *Frontiers in Microbiology* 13, 938481.

Bhattacharya, S. and Pandey, M. (2024) Deploying an energy efficient, secure and high-speed sidechain-based TinyML model for soil quality monitoring and management in agriculture. *Expert Systems with Applications* 242, 122735.

Bhattacharyya, P.N., Sandilya, S.P., Sarma, B., Pandey, A.K., Dutta, J. *et al.* (2024) Biochar as soil amendment in climate-smart agriculture: opportunities, future prospects, and challenges. *Journal of Soil Science and Plant Nutrition* 24, 135–158.

Bhatti, U.A., Bhatti, M.A., Tang, H., Syam, M.S., Awwad, E.M. *et al.* (2024) Global production patterns: understanding the relationship between greenhouse gas emissions, agriculture greening and climate variability. *Environmental Research* 245, 118049.

Bier, R.L., Daniels, M., Oviedo-Vargas, D., Peipoch, M., Price, J.R. *et al.* (2024) Agricultural soil microbiomes differentiate in soil profiles with fertility source, tillage, and cover crops. *Agriculture, Ecosystems and Environment* 368, 109002.

Blanco-Velázquez, F.J. and Anaya-Romero, M. (2024) Review of digital solutions for soil contamination management by mining activities. In: Ortega-Calvo, J.J. and Coulon, F. (eds) *Soil Remediation Science and Technology*. Springer, pp. 133–159.

Bogush, A.A. and Kourtchev, I. (2024) Disposable surgical/medical face masks and filtering face pieces: source of microplastics and chemical additives in the environment. *Environmental Pollution* 348, 123792.

Brower-Toland, B., Stevens, J.L., Ralston, L., Kosola, K. and Slewinski, T.L. (2024) A crucial role for technology in sustainable agriculture. *ACS Agricultural Science and Technology* 4, 283–291.

Brunner, M., Zeisler, C., Neu, D., Rotondo, C., Rubbmark, O.R. *et al.* (2024) Trap crops enhance the control efficacy of Metarhizium brunneum against a soil-dwelling pest. *Journal of Pest Science* 97, 1633–1645.

Cantú, A., Aguiñaga, E. and Scheel, C. (2021) Learning from failure and success: the challenges for circular economy implementation in SMEs in an emerging economy. *Sustainability* 13, 1529.

Cao, J. and Solangi, Y.A. (2023) Analyzing and prioritizing the barriers and solutions of sustainable agriculture for promoting sustainable development goals in China. *Sustainability* 15, 8317.

Cárceles Rodríguez, B., Durán-Zuazo, V.H., Soriano Rodríguez, M., García-Tejero, I.F., Gálvez Ruiz, B. *et al.* (2022) Conservation agriculture as a sustainable system for soil health: a review. *Soil Systems* 6, 87.

Chao, J., Li, T., Yin, H. and Wang, Z. (2024) Adoption of multiple sustainable agricultural practices among farmers in Northwest, China. *Cogent Food and Agriculture* 10.

Chaudhary, A. and Anand, S. (2024) Soilless cultivation: a distinct vision for sustainable agriculture. *Artificial Intelligence and Smart Agriculture Advances in Geographical and Environmental Sciences* 2024, 337–368.

Chaudhry, H., Vasava, H.B., Chen, S., Saurette, D., Beri, A. *et al.* (2024) Evaluating the soil quality index using three methods to assess soil fertility. *Sensors* 24, 864.

Chen, F., Zhang, W., Chen, R., Jiang, F., Ma, J. *et al.* (2024) Adapting carbon neutrality: tailoring advanced emission strategies for developing countries. *Applied Energy* 361, 122845.

Crack, L.E., Larkin-Kaiser, K.A., Phillips, A.A. and Edwards, W.B. (2024) Knowledge and awareness assessment of bone loss and fracture risk after spinal cord injury. *Journal of Spinal Cord Medicine* 47, 306–312.

Damian, C.S., Devarajan, Y. and Jayabal, R. (2024) Biodiesel production in India: prospects, challenges, and sustainable directions. *Biotechnology and Bioengineering* 121, 894–902.

Das, B.S., Wani, S.P., Benbi, D.K., Muddu, S. and Bhattacharyya, T. (2022) Soil health and its relationship with food security and human health to meet the sustainable development goals in India. *Soil Security* 8, 100071.

Dasgupta, S., Lavanya, V., Chakraborty, S. and Ray, D.P. (2024) Contemporary use of sensors for soil qualitative and quantitative assessment in the context of climate change. In: Pathak, H., Chatterjee, D., Saha, S. and Das, B. (eds) *Climate Change Impacts on Soil-Plant-Atmosphere Continuum*. Springer, pp. 183–207.

De Oliveira, M.A.T., Santos, J.C. and Lemos, R. (2020) 80,000 years of geophysical stratigraphic record at the Serra da Capivara National Park, in northeastern Brazil: uncovering hidden deposits and landforms at a canyon's floor. *Journal of South American Earth Sciences* 104, 102691.

Deshavath, N.N., Goud, V.V. and Veeranki, V.D. (2024) Commercialization of 2G bioethanol as a transportation fuel for the sustainable energy, environment, and economic growth of India: theoretical and empirical assessment of bioethanol potential from agriculture crop residues. *Biomass Conversion and Biorefinery* 14, 3551–3563.

Dönmez, D., Isak, M.A., İzgü, T. and Şimşek, Ö. (2024) Green horizons: navigating the future of agriculture through sustainable practices. *Sustainability* 16, 3505.

Entz, M.H., Stainsby, A., Riekman, M., Mulaire, T.R., Kirima, J.K. *et al.* (2022) Farmer participatory assessment of soil health from conservation agriculture adoption in three regions of East Africa. *Agronomy for Sustainable Development* 42, 97.

Eze, S., Dougill, A.J., Banwart, S.A., Sallu, S.M., Smith, H.E. *et al.* (2021) Farmers' indicators of soil health in the African highlands. *CATENA* 203, 105336.

Fowler, A., Basso, B., Maureira, F., Millar, N., Ulbrich, R. *et al.* (2024) Spatial patterns of historical crop yields reveal soil health attributes in US Midwest fields. *Scientific Reports* 14, 465.

Franzluebbers, A.J. (2020) Soil mass and volume affect soil-test biological activity estimates. *Soil Science Society of America Journal* 84, 502–511.

Franzluebbers, A.J. and Martin, G. (2022) Farming with forages can reconnect crop and livestock operations to enhance circularity and foster ecosystem services. *Grass and Forage Science* 77, 270–281.

Frasconi, C., Fontanelli, M. and Antichi, D. (2024) Smart strategies and technologies for sustainability and biodiversity in herbaceous and horticultural crops. *Agronomy* 14, 528.

Fuentes-Peñailillo, F., Gutter, K., Vega, R. and Carrasco, G. 2024) Transformative technologies in digital agriculture: leveraging internet of things, remote sensing, and artificial intelligence for smart crop management. *Journal of Sensor and Actuator Networks* 13, 39.

Gamage, A., Gangahagedara, R., Gamage, J., Jayasinghe, N., Kodikara, N. *et al.* (2023) Role of organic farming for achieving sustainability in agriculture. *Farming System* 1, 100005.

Gemtou, M., Kakkavou, K., Anastasiou, E., Fountas, S., Pedersen, S.M. *et al.* (2024) Farmers' transition to climate-smart agriculture: a systematic review of the decision-making factors affecting adoption. *Sustainability* 16, 2828.

Gosai, H.G., Jadeja, F., Sharma, A. and Jain, S. (2024) Occurrence and toxicity of organic microcontaminants in agricultural perspective: an overview. In: Bhadouria, R., Tripathi, S., Singh, P., Singh, R. and Singh, H.P. (eds) *Organic Micropollutants in Aquatic and Terrestrial Environments*. Springer, pp. 107–126.

Grigorieva, E., Livenets, A. and Stelmakh, E. (2023) Adaptation of agriculture to climate change: a scoping review. *Climate* 11, 202.

Gržinić, G., Piotrowicz-Cieślak, A., Klimkowicz-Pawlas, A., Górny, R.L., Ławniczek-Wałczyk, A. *et al.* (2023) Intensive poultry farming: a review of the impact on the environment and human health. *Science of the Total Environment* 858, 160014.

Gu, C., Lv, W., Liao, X., Brooks, M., Li, Y. *et al.* (2023) Green manure amendment increases soil phosphorus bioavailability and peanut absorption of phosphorus in red soil of South China. *Agronomy* 13, 376.

Hannam, I. (2024) Sustainable soil management and soil carbon sequestration. In: Ginzky, H. *et al.* (eds) *International Yearbook of Soil Law and Policy 2022*. Springer, pp. 3–33.

Hassan, M. and Beshr, E. (2024) Predicting soil cone index and assessing suitability for wind and solar farm development in using machine learning techniques. *Scientific Reports* 14, 2924.

Hassan, S., Karaila, G.K., Singh, P., Meenatchi, R., Venkateswaran, A.S. *et al.* (2024) Implications of myco-nanotechnology for sustainable agriculture – applications and future perspectives. *Biocatalysis and Agricultural Biotechnology* 57, 103110.

Hathaway, M.D. (2016) Agroecology and permaculture: addressing key ecological problems by rethinking and redesigning agricultural systems. *Journal of Environmental Studies and Sciences* 6, 239–250.

He, Y., Presley, D.R., Tatarko, J. and Blanco-Canqui, H. (2018) Crop residue harvest impacts wind erodibility and simulated soil loss in the central great plains. *GCB Bioenergy* 10, 213–226.

Hemkemeyer, M., Schwalb, S.A., Berendonk, C., Geisen, S., Heinze, S. *et al.* (2024) Potato yield and quality are linked to cover crop and soil microbiome, respectively. *Biology and Fertility of Soils* 60, 525–545.

Hirschfeld, S. and Van Acker, R. (2021) Review: ecosystem services in permaculture systems. *Agroecology and Sustainable Food Systems* 45, 794–816.

Hou, D., Bolan, N.S., Tsang, D.C.W., Kirkham, M.B. and O'Connor, D. (2020) Sustainable soil use and management: an interdisciplinary and systematic approach. *Science of the Total Environment* 729, 138961.

Huang, W., González, G. and Zou, X. (2020) Earthworm abundance and functional group diversity regulate plant litter decay and soil organic carbon level: a global meta-analysis. *Applied Soil Ecology* 150, 103473.

Ibrahim, H.T.M., Modiba, M.M., Dekemati, I., Gelybó, G., Birkás, M. *et al.* (2024) Status of soil health indicators after 18 years of systematic tillage in a long-term experiment. *Agronomy* 14, 278.

Ihenetu, S.C., Li, G., Mo, Y. and Jacques, K.J. (2024) Impacts of microplastics and urbanization on soil health: an urgent concern for sustainable development. *Green Analytical Chemistry* 8, 100095.

Jacquet, F., Jeuffroy, M.-H., Jouan, J., Le Cadre, E., Litrico, I. *et al.* (2022) Pesticide-free agriculture as a new paradigm for research. *Agronomy for Sustainable Development* 42, 8.

Jaworski, C.C., Thomine, E., Rusch, A., Lavoir, A.-V., Wang, S. *et al.* (2023) Crop diversification to promote arthropod pest management: a review. *Agriculture Communications* 1, 100004.

Johnraja, J.I., Leelipushpam, P.G.J., Shirley, C.P. and Princess, P.J.B. (2024) Impact of cloud computing on the future of smart farming. *Signals and Communication Technology Intelligent Robots and Drones for Precision Agriculture* 2014, 391–420.

Jorat, M.E., Minto, A., Tierney, I. and Gilmour, D. (2024) Future carbon-neutral societies: minimising construction impact on groundwater-dependent wetlands and peatlands. *Sustainability* 16, 7713.

Jose, S., Renuka, N., Ratha, S.K., Kumari, S. and Bux, F. (2024) Bioprospecting of microalgae from agricultural fields and developing consortia for sustainable agriculture. *Algal Research* 78, 103428.

Joseph Sekhar, S., Samuel, M.S., Glivin, G., Le, T. and Mathimani, T. (2024) Production and utilization of green ammonia for decarbonizing the energy sector with a discrete focus on sustainable development goals and environmental impact and technical hurdles. *Fuel* 360, 130626.

Juhos, K., Nugroho, P.A., Jakab, G., Prettl, N., Kotroczó, Z. *et al.* (2023) A comprehensive analysis of soil health indicators in a long-term conservation tillage experiment. *Soil Use and Management* 40, e12942.

Kallali, N.S., Ouijja, A., Goura, K., Laasli, S.-E., Kenfaoui, J. *et al.* (2024) From soil to host: discovering the tripartite interactions between entomopathogenic nematodes, symbiotic bacteria and insect pests and related challenges. *Journal of Natural Pesticide Research* 7, 100065.

Kalyani, Y., Vorster, L., Whetton, R. and Collier, R. (2024) Application scenarios of digital twins for smart crop farming through cloud–fog–edge infrastructure. *Future Internet* 16, 100.

Kamyab, H., SaberiKamarposhti, M., Hashim, H. and Yusuf, M. (2024) Carbon dynamics in agricultural greenhouse gas emissions and removals: a comprehensive review. *Carbon Letters* 34, 265–289.

Kandeler, E. (2024) Physiological and biochemical methods for studying soil biota and their functions. In: Paul, A. (ed.) *Soil Microbiology, Ecology and Biochemistry*. Elsevier, pp. 193–227.

Kantola, I.B., Blanc-Betes, E., Masters, M.D., Chang, E., Marklein, A. *et al.* (2023) Improved net carbon budgets in the US Midwest through direct measured impacts of enhanced weathering. *Global Change Biology* 29, 7012–7028.

Karlsson Green, K., Stenberg, J.A. and Lankinen, Å. (2020) Making sense of integrated pest management (IPM) in the light of evolution. *Evolutionary Applications* 13, 1791–1805.

Karunathilake, E.M.B.M., Le, A.T., Heo, S., Chung, Y.S. and Mansoor, S. (2023) The path to smart farming: innovations and opportunities in precision agriculture. *Agriculture* 13, 1593.

Katumo, D.M., Liang, H., Ochola, A.C., Lv, M., Wang, Q.-F. *et al.* (2022) Pollinator diversity benefits natural and agricultural ecosystems, environmental health, and human welfare. *Plant Diversity* 44, 429–435.

Kodaparthi, A., Kondakindi, V.R., Kehkashaan, L., Belli, M.V., Chowdhury, H.N. *et al.* (2024) Topsoil regeneration and bio-sequestration. In: Aransiola, S.A., Babaniyi, B.R., Aransiola, A.B. and Maddela, N.R. (eds) *Prospects for Soil Regeneration and Its Impact on Environmental Protection*. Springer, pp. 123–157.

Köninger, J., Lugato, E., Panagos, P., Kochupillai, M., Orgiazzi, A. *et al.* (2021) Manure management and soil biodiversity: towards more sustainable food systems in the EU. *Agricultural Systems* 194, 103251.

Krasilnikov, P., Taboada, M.A., and Amanullah (2022) Fertilizer use, soil health and agricultural sustainability. *Agriculture* 12, 462.

Kreitzman, M., Chapman, M., Keeley, K.O. and Chan, K.M.A. (2022) Local knowledge and relational values of midwestern woody perennial polyculture farmers can inform tree-crop policies. *People and Nature* 4, 180–200.

Kumar, R., Verma, A., Shome, A., Sinha, R., Sinha, S. *et al.* (2021) Impacts of plastic pollution on ecosystem services, sustainable development goals, and need to focus on circular economy and policy interventions. *Sustainability* 13, 9963.

Lakhiar, I.A., Yan, H., Zhang, J., Wang, G., Deng, S. *et al.* (2024) Plastic pollution in agriculture as a threat to food security, the ecosystem, and the environment: an overview. *Agronomy* 14, 548.

Li, M., Zhu, G., Liu, Z., Li, L., Wang, S. *et al.* (2024a) Hydrogen fertilization with hydrogen nanobubble water improves yield and quality of cherry tomatoes compared to the conventional fertilizers. *Plants* 13, 443.

Li, W., Liu, Y., Duan, J., Liu, G., Nie, X. *et al.* (2024b) Leguminous cover orchard improves soil quality, nutrient preservation capacity, and aggregate stoichiometric balance: a 22-year homogeneous experimental site. *Agriculture, Ecosystems and Environment* 363, 108876.

Liu, G., Deng, X. and Zhang, F. (2024a) The spatial and source heterogeneity of agricultural emissions highlight necessity of tailored regional mitigation strategies. *Science of the Total Environment* 914, 169917.

Liu, M., Zhong, T. and Lyu, X. (2024b) Spatial spillover effects of "new farmers" on diffusion of sustainable agricultural practices: evidence from China. *Land* 13, 119.

Liu, Q., Sun, W., Zeng, Q., Zhang, H., Wu, C. *et al.* (2024c) Integrated processes for simultaneous nitrogen, phosphorus, and potassium recovery from urine: a review. *Journal of Water Process Engineering* 59, 104975.

Liu, T., Bruins, R. and Heberling, M. (2018) Factors influencing farmers' adoption of best management practices: a review and synthesis. *Sustainability* 10, 432.

Liu, Y., Huang, X. and Liu, Y. (2024d) Detection of long-term land use and ecosystem services dynamics in the loess hilly-gully region based on artificial intelligence and multiple models. *Journal of Cleaner Production* 447, 141560.

Llop, M. and Ponce-Alifonso, X. (2016) Water and agriculture in a mediterranean region: the search for a sustainable water policy strategy. *Water* 8, 66.

Löbmann, M.T., Maring, L., Prokop, G., Brils, J., Bender, J. *et al.* (2022) Systems knowledge for sustainable soil and land management. *Science of the Total Environment* 822, 153389.

Mallareddy, M., Thirumalaikumar, R., Balasubramanian, P., Naseeruddin, R., Nithya, N. *et al.* (2023) Maximizing water use efficiency in rice farming: a comprehensive review of innovative irrigation management technologies. *Water* 15, 1802.

Mana, A.A., Allouhi, A., Hamrani, A., Rehman, S., el Jamaoui, I. *et al.* (2024) Sustainable AI-based production agriculture: exploring AI applications and implications in agricultural practices. *Smart Agricultural Technology* 7, 100416.

Mancuso, G., Lavrnić, S. and Toscano, A. (2019) Reclaimed water to face agricultural water scarcity in the Mediterranean area: an overview using Sustainable Development Goals preliminary data. *Advances in Chemical Pollution, Environmental Management and Protection* 5, 113–143.

Mandal, M., Roy, A., Das, S., Rakwal, R., Agrawal, G.K. *et al.* (2024) Food waste-based bio-fertilizers production by bio-based fermenters and their potential impact on the environment. *Chemosphere* 353, 141539.

Mane, S., Das, N., Singh, G., Cosh, M. and Dong, Y. (2024) Advancements in dielectric soil moisture sensor calibration: a comprehensive review of methods and techniques. *Computers and Electronics in Agriculture* 218, 108686.

Manjula, M. and Sharma, D. (2024) Teaching sustainability in agriculture to students of development. In: Leal Filho, W., Dibbern, T., de Maya, S.R., Alarcón-del-Amo, MdC. and Rives, L.M. (eds) *The Contribution of Universities Towards Education for Sustainable Development*. Springer, pp. 471–487.

Margat, J. and Gun, J.V.D. (2013) *Groundwater Around the World*. CRC Press.

Martin, E.A., Feit, B., Requier, F., Friberg, H. and Jonsson, M. (2019) Assessing the resilience of biodiversity-driven functions in agroecosystems under environmental change. *Advances in Ecological Research* 60, 59–123.

Mazzocchi, F. (2020) A deeper meaning of sustainability: insights from indigenous knowledge. *Anthropocene Review* 7, 77–93.

Mehmet Tuğrul, K. (2020) Soil management in sustainable agriculture. In: Hasanuzzaman, M., Teixeira Filho, M.C.M., Fujita, M. and Nogueira, T.A.R. (eds) *Sustainable Crop Production*. IntechOpen.

Mello, I., Laurent, F., Kassam, A., Marques, G.F., Okawa, C.M.P. *et al.* (2021) Benefits of conservation agriculture in watershed management: participatory governance to improve the quality of no-till systems in the Paraná 3 watershed, Brazil. *Agronomy* 11, 2455.

Menegat, S., Ledo, A. and Tirado, R. (2022) Greenhouse gas emissions from global production and use of nitrogen synthetic fertilisers in agriculture. *Scientific Reports* 12, 14490.

Mikhailova, E.A., Zurqani, H.A., Lin, L., Hao, Z., Post, C.J. *et al.* (2024) Possible integration of soil information into land degradation analysis for the United Nations (UN) land degradation neutrality (LDN) concept: a case study of the contiguous United States of America (USA). *Soil Systems* 8, 27.

Mittal, R. and Bhukal, S. (2024) Sustainable approach for agriculture and environmental remediation using hydroponics and their perspectives. In: Kumar, N. (ed.) *Hydroponics and Environmental Bioremediation*. Springer, pp. 65–90.

Mora-Motta, D., Llanos-Cabrera, M.P., Chavarro-Bermeo, J.P., Ortíz-Morea, F.A. and Silva-Olaya, A.M. (2024) Visual evaluation of soil structure is a reliable method to detect changes in the soil quality of Colombian Amazon pasturelands. *Soil Science Society of America Journal* 88, 527–539.

Muhie, S.H. (2022) Novel approaches and practices to sustainable agriculture. *Journal of Agriculture and Food Research* 10, 100446.

Nicastro, R., Papale, M., Fusco, G.M., Capone, A., Morrone, B. and Carillo, P. (2024) Legal barriers in sustainable agriculture: valorization of agri-food waste and pesticide use reduction. *Sustainability* 16, 8677.

Ning, L., Xu, X., Zhang, Y., Zhao, S., Qiu, S. *et al.* (2022) Effects of chicken manure substitution for mineral nitrogen fertilizer on crop yield and soil fertility in a reduced nitrogen input regime of North-Central China. *Frontiers in Plant Science* 13, 1050179.

Nungula, E.Z., Mugwe, J., Massawe, B.H.J. *et al.* (2024) Morphological, pedological and chemical characterization and classification of soils in Morogoro District, Tanzania. *Agricultural Research* 13, 266–276.

Osman, A.I., Fang, B., Zhang, Y., Liu, Y., Yu, J. *et al.* (2024) Life cycle assessment and techno-economic analysis of sustainable bioenergy production: a review. *Environmental Chemistry Letters* 22, 1115–1154.

Pamuru, S.T., Morash, J., Lea-Cox, J.D., Ristvey, A.G., Davis, A.P. *et al.* (2024) Nutrient transport, shear strength and hydraulic characteristics of topsoils amended with mulch, compost and biosolids. *Science of the Total Environment* 918, 170649.

Panagos, P., Matthews, F., Patault, E., De Michele, C., Quaranta, E. *et al.* (2024) Understanding the cost of soil erosion: an assessment of the sediment removal costs from the reservoirs of the European Union. *Journal of Cleaner Production* 434, 140183.

Pandey, A., Tiwari, P., Manpoong, C. and Jatav, H.S. (2024) Agroforestry for restoring and improving soil health. In: Jatav, H.S., Rajput, V.D., Minkina, T., Van Hullebusch, E.D. and Dutta, A. (eds) *Agroforestry to Combat Global Challenges*. Springer, pp. 147–164.

Park, S., Lee, S., Kim, T., Choi, A., Lee, S. *et al.* (2024) Development strategy of non-GMO organism for increased hemoproteins in Corynebacterium glutamicum: a growth-acceleration-targeted evolution. *Bioprocess and Biosystems Engineering* 47, 549–556.

Piñeiro, V., Arias, J., Dürr, J., Elverdin, P., Ibáñez, A.M. *et al.* (2020) A scoping review on incentives for adoption of sustainable agricultural practices and their outcomes. *Nature Sustainability* 3, 809–820.

Poddar, R., Sen, A., Sarkar, A., Patra, S.K. and Hossain, A. (2024) Climate-smart advanced technological interventions in field crop production under problematic soil for sustainable agricultural development. In: Chakraborty, R., Mathur, P. and Roy, S. (eds) *Food Production, Diversity, and Safety Under Climate Change*. Springer, pp. 199–210.

Powlson, D.S., Gregory, P.J., Whalley, W.R., Quinton, J.N., Hopkins, D.W. *et al.* (2011) Soil management in relation to sustainable agriculture and ecosystem services. *Food Policy* 36, S72–87.

Raffa, C.M. and Chiampo, F. (2021) Bioremediation of agricultural soils polluted with pesticides: a review. *Bioengineering* 8, 92.

Rashid, A., Schutte, B.J., Ulery, A., Deyholos, M.K., Sanogo, S. *et al.* (2023) Heavy metal contamination in agricultural soil: environmental pollutants affecting crop health. *Agronomy* 13, 1521.

Rastogi, M., Nandal, M. and Khosla, B. (2020) Microbes as vital additives for solid waste composting. *Heliyon* 6, e03343.

Rehman, S.U., De Castro, F., Aprile, A., Benedetti, M. and Fanizzi, F.P. (2023) Vermicompost: enhancing plant growth and combating abiotic and biotic stress. *Agronomy* 13, 1134.

Romeijn, P., Comer-Warner, S.A., Ullah, S., Hannah, D.M. and Krause, S. (2019) Streambed organic matter controls on carbon dioxide and methane emissions from streams. *Environmental Science and Technology* 53, 2364–2374.

Roofchaee, A.S., Abrishamkesh, S., Fazeli, M. and Shabanpour, M. (2024) Optimizing biochar application: effects of placement method, particle size, and application rate on soil physical properties and soil loss. *Journal of Soils and Sediments* 24, 1541–1555.

Ros, G.H., Verweij, S.E., Janssen, S.J.C., De Haan, J. and Fujita, Y. (2022) An open soil health assessment framework facilitating sustainable soil management. *Environmental Science and Technology* 56, 17375–17384.

Saparova, G., Khan, G.D. and Joshi, N.P. (2024) Linking farmers to markets: assessing small-scale farmers' preferences for an official phytosanitary regime in the Kyrgyz republic. *Economic Analysis and Policy* 81, 696–708.

Sayara, T., Basheer-Salimia, R., Hawamde, F. and Sánchez, A. (2020) Recycling of organic wastes through composting: process performance and compost application in agriculture. *Agronomy* 10, 1838.

Scarlato, M., Rieppi, M., Alliaume, F., Illarze, G., Bajsa, N. *et al.* (2024) Towards the development of cover crop–reduced tillage systems without herbicides and synthetic fertilizers in onion cultivation: promising but challenges remain. *Soil and Tillage Research* 240, 106061.

Serebrennikov, D., Thorne, F., Kallas, Z. and McCarthy, S.N. (2020) Factors influencing adoption of sustainable farming practices in Europe: a systemic review of empirical literature. *Sustainability* 12, 9719.

Šeremešić, S., Dolijanović, Ž., Tomaš Simin, M., Milašinović Šeremešić, M., Vojnov, B. *et al.* (2024) Articulating organic agriculture and sustainable development goals: Serbia case study. *Sustainability* 16, 1842.

Shah, F. and Wu, W. (2019) Soil and crop management strategies to ensure higher crop productivity within sustainable environments. *Sustainability* 11, 1485.

Shahane, A.A. and Shivay, Y.S. (2021) Soil health and its improvement through novel agronomic and innovative approaches. *Frontiers in Agronomy* 3, 680456.

Shao, H. (2024) Agricultural greenhouse gas emissions, fertilizer consumption, and technological innovation: a comprehensive quantile analysis. *Science of the Total Environment* 926, 171979.

Sharma, G., Shrestha, S., Kunwar, S. and Tseng, T.-M. (2021) Crop diversification for improved weed management: a review. *Agriculture* 11, 461.

Shashirekha, V., Sowmiya, V., Malleswar, R.B. and Seshadr, S. (2024) Conserving soil microbial population and sustainable agricultural practices – polymers in aid of safe delivery, protection, population enhancement, and maintenance. In: Sa, T. (ed.) *Beneficial Microbes for Sustainable Agriculture Under Stress Conditions*. Elsevier, pp. 313–358.

Siddique, K.H.M., Bolan, N., Rehman, A. and Farooq, M. (2024) Enhancing crop productivity for recarbonizing soil. *Soil and Tillage Research* 235, 105863.

Skendžić, S., Zovko, M., Živković, I.P., Lešić, V. and Lemić, D. (2021) The impact of climate change on agricultural insect pests. *Insects* 12, 440.

Sofo, A., Mininni, A.N. and Ricciuti, P. (2020) Soil macrofauna: a key factor for increasing soil fertility and promoting sustainable soil use in fruit orchard agrosystems. *Agronomy* 10, 456.

Son, N., Chen, C.-R. and Syu, C.-H. (2024) Towards artificial intelligence applications in precision and sustainable agriculture. *Agronomy* 14, 239.

Struik, P.C. and Kuyper, T.W. (2017) Sustainable intensification in agriculture: the richer shade of green. A review. *Agronomy for Sustainable Development* 37, 39.

Tahat, M.M., Alananbeh, K.M., Othman, Y.A. and Leskovar, D.I. (2020) Soil health and sustainable agriculture. *Sustainability* 12, 4859.

Telo Da Gama, J. (2023) The role of soils in sustainability, climate change, and ecosystem services: challenges and opportunities. *Ecologies* 4, 552–567.

Thanekar, U., Sacks, G., Ruffini, O., Reeve, B. and Blake, M.R. (2024) Local government stakeholders' perceptions of potential policy actions to influence both climate change and healthy eating in Victoria: a qualitative study. *Health Promotion Journal of Australia* 35, 1158–1173.

Thiruchelve, S.R., Chandran, S., Kumar, V. and Chandramohan, K. (2024) Assessment of land use and land cover dynamics and its impact in direct runoff generation estimation using SCS CN method. *Acta Geophysica* 72, 4415–4430.

Toumi, I., Ghrab, M., Zarrouk, O. and Nagaz, K. (2024) Impact of deficit irrigation strategies using saline water on soil and peach tree yield in an arid region of Tunisia. *Agriculture* 14, 377.

Tudi, M., Daniel Ruan, H., Wang, L., Lyu, J., Sadler, R. *et al.* (2021) Agriculture development. Pesticide application and its impact on the environment. *International Journal of Environmental Research and Public Health* 18, 1112.

Upadhyay, S.K., Singh, G., Rani, N., Rajput, V.D., Seth, C.S. *et al.* (2024) Transforming bio-waste into value-added products mediated microbes for enhancing soil health and crop production: perspective views on circular economy. *Environmental Technology and Innovation* 34, 103573.

Urra, J., Alkorta, I. and Garbisu, C. (2019) Potential benefits and risks for soil health derived from the use of organic amendments in agriculture. *Agronomy* 9, 542.

Veenstra, J., Coquet, Y., Melot, R. and Walter, C. (2024) A European stakeholder survey on soil science skills for sustainable agriculture. *European Journal of Soil Science* 75, e13449.

Villat, J. and Nicholas, K.A. (2024) Quantifying soil carbon sequestration from regenerative agricultural practices in crops and vineyards. *Frontiers in Sustainable Food Systems* 7.

Wydro, U. (2022) Soil microbiome study based on DNA extraction: A review. *Water* 14, 3999.

Xiao, N., Kong, L., Wei, M., Hu, X. and Li, O. (2024) Innovations in food waste management: from resource recovery to sustainable solutions. *Waste Disposal and Sustainable Energy* 6, 401–417.

Xie, Y., Chen, Z., Khan, A. and Ke, S. (2024) Organizational support, market access, and farmers' adoption of agricultural green production technology: evidence from the main kiwifruit production areas in Shaanxi Province. *Environmental Science and Pollution Research* 31, 12144–12160.

Xue, B., Wu, R., Liu, B., An, H., Gao, R. *et al.* (2024) Nutrient supplementation changes chemical composition of soil organic matter density fractions in desert steppe soil in northern China. *Soil and Tillage Research* 241, 106107.

Yahyah, H., Kameri-Mbote, P. and Kibugi, R. (2024) Implications of pesticide use regulation on soil sustainability in Uganda. *Soil Security* 16, 100133.

Yang, T., Siddique, K.H.M. and Liu, K. (2020) Cropping systems in agriculture and their impact on soil health – a review. *Global Ecology and Conservation* 23, e01118.

Yang, X., Dai, X. and Zhang, Y. (2024a) The government subsidy policies for organic agriculture based on evolutionary game theory. *Sustainability* 16, 2246.

Yang, Y., Zhang, J., Chang, X., Chen, L., Liu, Y. *et al.* (2024b) Green manure incorporation enhanced soil labile phosphorus and fruit tree growth. *Frontiers in Plant Science* 15, 1356224.

Yeasmin, S., Dipto, A.R., Zakir, A.B., Shovan, S.D., Suvo, M.A.H. *et al.* (2024) Nanopriming and AI for sustainable agriculture: boosting seed germination and seedling growth with engineered nanomaterials, and smart monitoring through deep learning. *ACS Applied Nano Materials* 7, 8703–8715.

Zaghloul, A., Saber, M. and El-Dewany, C. (2019) Chemical indicators for pollution detection in terrestrial and aquatic ecosystems. *Bulletin of the National Research Centre* 43, 156.

Zhang, J., Dyck, M., Quideau, S.A. and Norris, C.E. (2024a) Assessment of soil health and identification of key soil health indicators for five long-term crop rotations with varying fertility management. *Geoderma* 443, 116836.

Zhang, K., Li, X., Chang, B., Song, S. and Liu, C. (2024b) Use of coal gasification coarse slag to amend the soil water, nutrient and salt of waste dump in the northwest arid region: feasibility and potential. *Environment, Development and Sustainability* 2024.

Zhang, S., Gong, W., Wan, X., Li, J., Li, Z. *et al.* (2024c) Influence of organic matter input and temperature change on soil aggregate-associated respiration and microbial carbon use efficiency in alpine agricultural soils. *Soil Ecology Letters* 6, 230220.

Zheng, X., Wei, L., Lv, W., Zhang, H., Zhang, Y. *et al.* (2024) Long-term bioorganic and organic fertilization improved soil quality and multifunctionality under continuous cropping in watermelon. *Agriculture, Ecosystems and Environment* 359, 108721.

Zyngier, R.L., Archibald, C.L., Bryan, B.A. *et al.* (2024) Knowledge co-production for identifying indicators and prioritising solutions for food and land system sustainability in Australia. *Sustainability Science* 19, 1897–1919.

14 Determinants of Soil Health and its Role in Environmental Sustainability

Maneesh Kumar[1]*, Priyanka Kumari[2] and Arti Kumari[3]
[1]*Department of Biotechnology, Magadh University, Bodh Gaya, Bihar, India;*
[2]*Department of Biotechnology, Gautam Buddha University, Greater Noida, Uttar Pradesh, India;* [3]*Department of Biotechnology, Patna Women's College, Patna, Bihar, India*

Abstract

Soil health is a critical determinant of environmental sustainability, significantly influencing agricultural productivity, ecosystem resilience and global nutrient cycling. This summary addresses the key determinants of soil health and its central role in promoting sustainable land management practices. The determinants of soil health, which include factors such as soil structure, texture, pH, organic matter content, microbial activity, water-holding capacity and erosion control, collectively determine the soil's ability to support plant growth, nutrient cycling and water dynamics. These factors dynamically interact with each other, influence each other and shape the soil's overall health. The role of these factors in environmental sustainability cannot be overstated. Healthy soils support diverse and robust plant ecosystems, promote biodiversity and sequester atmospheric carbon dioxide. Optimal nutrient cycling promoted by microbial activity maintains soil fertility, reduces the need for chemical fertilizers and mitigates nutrient pollution in water bodies. Soil structure and erosion control prevent soil degradation, protect valuable topsoil and maintain water quality by reducing runoff and sedimentation. Recognition of the interaction between soil health and environmental sustainability underscores the need for responsible land management. Conservation practices such as no-till, cover cropping and adding organic matter improve soil structure, organic matter content and microbial diversity. These soil health practices contribute to achieving ecological balance, sustainable agriculture and resilient ecosystems.

Keywords: Soil health, sustainability, biodiversity, soil determinant

14.1 Introduction

Soil health is a critical aspect of environmental sustainability. It directly affects the productivity, resilience and overall functioning of ecosystems, which determine the ability of soil to provide essential services to people and the environment (Herrick and Wander, 1997; Lal, 2020). It is an essential requirement for sustainable agricultural production and contributes to global food security. The determinants of soil health are a complex interplay of various physical, chemical and biological factors that influence soil fertility, structure and nutrient cycling (Osman *et al.*, 2022). Healthy soils are the foundation for agricultural production, soil restoration, water quality, carbon sequestration, energy supply, wildlife habitat and healthy plant growth. Poorly managed

Corresponding author: kumar.maneesh11@gmail.com

DOI: 10.1079/9781800624597.0014

soils can result in a number of environmental and economic losses (Arriaga *et al.*, 2017). Protecting and improving soil health is essential to the productive functioning of our environment and the sustainability of food production, water supply and air supply. Healthy soils can reduce the need for fertilizers and chemical inputs. Soil management techniques that help promote soil health include crop rotation, cover cropping, minimum tillage, grass irrigation, compost application and other conservation practices (Herrick and Wander, 1997; Veum *et al.*, 2015). These techniques help protect the soil, improve nutrient cycling and minimize runoff and erosion. A popular strategy for improving soil health is integrating crop rotations and grazing, which can return important nutrients and organic matter to the soil. In addition, soil mapping and testing knowledge are key to making informed soil management decisions. Finally, promoting conservation tillage, reduced chemical use and prudent fertilization can help protect soil health and prevent land degradation (Bhattacharyya *et al.*, 2015).

Soil health is an important factor in environmental sustainability because it directly affects the availability of essential ecosystem services and natural resources. Healthy soil is important in mitigating climate change because it stores carbon, controls water runoff and serves as a buffer against extreme weather events such as floods and drought (Teague and Kreuter, 2020). Without healthy soils, landscapes are more vulnerable to environmental damage and degradation, which can lead to various environmental and economic problems. In addition to providing essential services and natural resources, soil health is essential to growing crops and providing food for human consumption. Therefore, soil health must be maintained to ensure the long-term sustainability of the environment (Visser *et al.*, 2019).

14.2   Sustainable Land Management and Soil Quality

Creating sustainable land management systems requires considering how people benefit from them, how efficiently they use resources and how well they maintain the environment in good condition for people and most other species. The task of designing agricultural management systems that can meet the needs of both food and fibre production and ecological conservation is challenging. The concept of soil quality as the primary link between land management strategies and the main goals of sustainable agriculture is central to this field. In short, the basic indicator of environmentally sound land use is assessing soil quality or health and the trend of change over time. While it is common knowledge that healthy soil increases crop yields, fewer people know that a lack of care can negatively impact air and water quality (Liniger *et al.*, 2017; Doran *et al.*, 2019). Intensive land management practices have led to an imbalance in the soil's carbon, nitrogen and water cycles, threatening the quality of surface and subsurface water supplies in many regions worldwide. The rate at which nitrogen enters terrestrial ecosystems has nearly doubled in the last 30 years due to human intervention in the nitrogen cycle, leading to a massive increase in nitrogen release to the atmosphere and subsequent transport via rivers, estuaries and coastal seas. Tillage, crop rotation, the use of pesticides and fertilizers and other land management practices affect the purity of water. Changes in the ability of soil to produce or consume important atmospheric gases such as carbon dioxide, nitrous oxide and methane are another way these management practices can affect air quality. Understanding the effects of land management on soil functioning is important because of the current threat of global climate change and ozone depletion due to increased levels of atmospheric greenhouse gases and changed hydrological cycles (Rathore, 2018; Seleiman and Hafez, 2021).

14.3   Basic Determinants of Soil Health

The basic components that affect soil health are important indicators of soil ecosystem health and function. Several factors influence the ability of soils to support plant growth, maintain biodiversity and provide ecosystem services. It is fair to consider soil health as part of ecosystem health. A thriving ecosystem has consistent nutrient cycling, energy flows, stability and tolerance to disturbance or stress (Büenemann *et al.*, 2018; Pires, 2023). Management and land use decisions affect soil health, affecting production, the environment and plant and animal health. Good soil health management considers all functions, not just crop productivity. Soil microorganisms are probably the most important determinants of soil health because of their close relationship with plants and animals (Jangir *et al.*, 2019).

Contrary to the early 20th century assumption that most microorganisms inhibit plant and animal growth and cause disease, plants grown in biologically rich soils can provide nutrient-rich, healthful food. Increased crop yields and more efficient animal growth due to reduced or no synthetic pesticide use have driven biotechnology advances. However, little is known about how biotechnology affects soil health and ecological services (Chagnon *et al.*, 2015; Bünemann *et al.*, 2018).

14.3.1   Soil structure: a crucial determinant of soil health

Healthy and productive terrestrial ecosystems depend on soil structure, that is, the organization of soil particles into aggregates. Plant development, water flow, nutrient availability and ecosystem functionality depend on a well-structured soil. Precipitation has to move through the soil to reach plant roots and replenish groundwater (Hartmann and Six, 2023). A well-structured soil contributes to drainage and prevents root rot and other plant stresses. It directly affects root growth: a loose, well-distributed soil allows roots to penetrate and find nutrients and water, which promotes plant health and growth. This characteristic significantly affects the habitat of soil organisms. Soil aggregates and air spaces support a diverse microbial population. Nutrient cycling and decomposition of organic matter by these microbes influence soil fertility (Van Antwerpen *et al.*, 2022). Compaction from intensive agriculture and development can disrupt soil structure. Compacted soil limits water, root and microbiological activity. Soil and ecosystem health depends on maintaining or recovering soil structure. Minimizing tillage, cover cropping and adding organic matter improves soil structure. These methods increase stable aggregation, aeration, water movement and microbial diversity (Lynch *et al.*, 2021).

14.3.2   Soil texture: a defining determinant of soil health

Soil texture, the ratio of sand, silt and clay particles, determines soil health and influences agricultural productivity and ecosystem dynamics. It has a great influence on the water-holding capacity of the soil, the availability of nutrients and plant growth. The ratio of sand, silt and clay particles affects water retention and drainage. Sandy soils with larger particles drain quickly but retain little water. Clay soils with smaller particles retain water well but clump and drain poorly. A balanced particle size combination gives clay soils good water-holding capacity and drainage which is critical for nutrient availability. Clay particles can store potassium and calcium due to their large surface area and strong negative charge (Williams *et al.*, 2020). On the other hand, sandy soils contain fewer negatively charged sites, so nutrients are easily washed out. Sandy soils help to compensate for nutrient deficiencies and surpluses and have a major impact on root penetration. Root growth in loose, well-structured soils helps plants conserve water and nutrients. In contrast, poor soil texture hinders root development, which affects growth and crop yields. Soil texture affects ecosystem dynamics beyond agriculture, for example it influences the population and diversity of soil microbes by shaping their habitat. These microbes depend on nutrient cycling, organic waste decomposition and disease control. Effective land management requires an understanding of soil composition. Organic matter improves the water-holding capacity of sandy soils and the drainage of clay soils. Nutrient management solutions tailored to soil conditions optimize crop nutrition. Soil texture determines soil health, which affects water availability, nutrient cycling and plant growth. Understanding and regulating soil health enables sustainable agriculture, strong ecosystems and a more resilient environment (Yu *et al.*, 2017; Lehmann *et al.*, 2020).

14.3.3 Soil pH: a crucial gauge of soil health

Soil pH, a measure of soil acidity or alkalinity, has a major impact on soil health and plant growth. It determines nutrient availability, microbial activity and overall balance of the ecosystem. A balanced pH is critical for plants to uptake optimally; excessively acidic or alkaline conditions can impair nutrient solubility and inhibit plant growth. Soil pH also affects microbial communities and influences their nutrient cycling and disease control functions. Different plants thrive in specific pH ranges; maintaining an appropriate pH ensures robust growth and vigour. Regular monitoring and pH management through lime application or acidification practices are essential to cultivating an environment conducive to healthy vegetation and sustainable land use (Neina, 2019).

14.3.4 Organic matter: an essential cornerstone of soil health

Soil health, healthy ecosystems and productive agriculture depend on organic matter. It contains plant and animal wastes in various stages of decomposition and provides nutrients, energy and habitat for many soil organisms. The formation of aggregates promotes soil structure, water infiltration, aeration and rooting. It also encourages water retention, prevents drying out and keeps plant roots moist (Jangir *et al.*, 2019). Organic matter stores nutrients and releases them through degradation and mineralization. A diverse assemblage of microbes contributes to nutrient cycling, disease control and soil health maintenance. It promotes plant symbiotic soil fungi and bacteria. These bacteria improve nutrient uptake, plant resistance and defence against pathogens. Intensive agriculture and deforestation reduce organic matter and degrade soil structure, nutrient content and microbial diversity. Restoring soil health requires covering crops, composting and reducing tillage. It improves nutrient availability, water retention and microbial diversity, making it essential to soil health. To ensure healthy ecosystems and resilient agricultural systems, sustainable land management practices must prioritize organic matter accumulation and conservation (Lal and Stewart, 2013).

14.3.5 Nutrient content: fueling soil health and ecosystem vitality

Soil health, crop development, ecosystem dynamics and sustainable land management depend on nutrient levels. Healthy plants and ecosystems require nitrogen, phosphorus and potassium. Plant growth and development require balanced nutrients. Adequate nutrients promote photosynthesis, root growth and plant health. Imbalances and deficiencies can limit growth, reduce production and increase susceptibility to diseases and pests. This affects plant health, ecosystem dynamics and resilience (Yu *et al.*, 2017). Plants, bacteria and other living organisms engage in nutrient cycling. A healthy nutrient cycle ensures soil fertility and ecological stability by providing necessary nutrients. Nutrient runoff into streams leads to eutrophication and dangerous algal blooms. Precise fertilization and soil testing are necessary to reduce these impacts and increase crop yields. Nutrients in soil are tied to organic material and microbial activity. When organic material decomposes, nutrients are released. The nutrient cycle relies on microbes breaking down organic material and providing plant nutrients. Balanced nutrients and competent nutrient management are critical for resilient landscapes, productive agriculture and natural resource conservation (Saleem *et al.*, 2023).

14.3.6 Microbial activity: the hidden engine of soil health

Microbial activity controls ecosystems, nutrient cycling and plant growth in soil. Soil life and function depend on microorganisms' diversity, interactions and metabolism. In nutrient cycling, they decompose organic matter and simple compounds that plants can take up. This mechanism enriches the soil for plant growth and ecological productivity. Microbes suppress disease by producing antimicrobial molecules and displacing bacteria. Balanced microbial populations promote plant resilience and reduce chemical use (Sokol *et al.*, 2022). Microorganisms form extracellular compounds that bind soil particles into aggregates and improve soil structure (Table 14.1). The aggregates provide water, root penetration and aeration to

Table 14.1. Microbes are involved in boosting the soil fertility.

Microbes	Role and examples	Reference
Bacteria	Various bacterial species contribute to soil fertility by fixing atmospheric nitrogen and converting it into forms that plants can use. Examples include *Rhizobium*, *Azotobacter* and *Clostridium*	Itelima *et al.*, 2018
Fungi	Mycorrhizal fungi form symbiotic relationships with plant roots, enhancing nutrient uptake, particularly phosphorus. *Trichoderma* and *Penicillium* species also aid in organic matter decomposition	El-Maraghy *et al.*, 2021
Actinomycetes	These soil bacteria are vital in breaking down complex organic matter, releasing nutrients in forms accessible to plants. *Streptomyces* is a well-known example	Mitra *et al.*, 2022
Nitrogen-fixing bacteria	Free-living nitrogen-fixing bacteria like *Azotobacter* and *Cyanobacteria* enrich the soil with nitrogen, a vital nutrient for plant growth	Thomas and Singh, 2019
Phosphate-solubilizing bacteria	These bacteria release bound phosphorus from soil particles, increasing availability for plants. *Bacillus* and *Pseudomonas* species are examples	Elhaissoufi *et al.*, 2022
Nitrifying bacteria	Nitrifiers convert ammonium into nitrate, an essential form of nitrogen for plants. *Nitrosomonas* and *Nitrobacter* are common nitrifying bacteria	Frincu *et al.*, 2016
Cellulose-degrading microbes	These microbes decompose plant residues, releasing nutrients and organic matter back into the soil. *Cellulomonas*, *Bacillus* and *Trichoderma* species are involved	Arora *et al.*, 2013; Grzyb *et al.*, 2020
Symbiotic nitrogen-fixing bacteria	Leguminous plants partner with bacteria like *Rhizobium* or *Bradyrhizobium* to form root nodules that fix atmospheric nitrogen, enriching the soil	Sharma *et al.*, 2020
Saprophytic fungi	These fungi decompose dead plant material, returning nutrients to the soil. White rot fungi like *Pleurotus* and *Agaricus* are examples	Singh and Vyas, 2021
Arbuscular mycorrhizal fungi	They form symbiotic relationships with plant roots, aiding in nutrient uptake, especially phosphorus, in exchange for sugars from plants	Li and Cai, 2021
Denitrifying bacteria	These bacteria convert nitrates into nitrogen gas, preventing excessive nitrate accumulation that could lead to pollution. *Pseudomonas* and *Paracoccus* species are involved	Rana *et al.*, 2019
Acid-producing bacteria	Some bacteria produce organic acids that facilitate the release of bound nutrients from soil particles	Sindhu *et al.*, 2022

plants. Land use, agriculture and pollution are changing microbial communities and affecting nutrient cycling and ecology. Synthetic fertilizers and pesticides harm microbial diversity and function. Soil-covering crops, less tillage and organic matter improve microbial diversity and maintain ecology. These practices promote good bacteria. Unnoticed, microbial activity maintains soil health and ecosystem function. Recognition of their importance supports the use of sustainable soil management that maintains and enhances microbial diversity, creat-

ing resilient and productive ecosystems that are productive (Dubey *et al.*, 2019).

14.3.7 Water-holding capacity: a crucial sustainer of soil health

The water-holding capacity of soil is an indicator of its health, as it determines its ability to store and release water to plants. This has major implications for agriculture, ecology and water management. The best soils can store sufficient water for plant growth and drain excess water. This property is critical in drought-prone or high-precipitation areas because it irrigates plants without flooding them (Neina, 2019). Soil type and structure affect water retention. A balanced mix of sand, silt and clay particles retains the most water and provides water to plants during dry seasons. Water retention minimizes soil erosion. Water-retaining soil anchors plants and prevents fertile topsoil from washing away during heavy rains, protecting resources and aquatic habitats. Water retention is critical for agricultural irrigation and optimizes water use to avoid over- or under-irrigation, which leads to yield loss, nutrient leaching and environmental damage. Compaction and the loss of organic matter reduce the soil's water-holding capacity, agricultural yield and climate sensitivity. Soil cover and reduced tillage improve soil health and water-holding capacity. These practices improve soil structure, organic matter and water retention (Gavrilescu, 2021).

14.3.8 Compaction: soil health's silent adversary

Compaction is a critical factor in soil health and significantly impacts soil structure, plant growth and overall ecosystem functionality. This phenomenon occurs when soil particles are compacted, reducing pore spaces and altering soil properties. Compacted soil impedes root development and restricts plant access to water, nutrients and oxygen. This impairs plant growth, resulting in lower yields and reduced agricultural productivity. Water movement is impeded in compacted soils, resulting in poor drainage and increased surface runoff. This can lead to erosion, nutrient loss, reduced water availability for plants and aquatic ecosystems and increases the risk of flooding. Microbial activity, a cornerstone of soil health, is suppressed in compacted soils due to limited oxygen availability. This affects nutrient cycling, organic matter decomposition and overall soil fertility (Hartmann and Six, 2023).

Heavy machinery, foot traffic and inadequate tillage compact the soil. Architecture, agriculture and urbanization compact and change soil dynamics. Reducing soil compaction

lowers the impact of these factors. Organic matter, low tillage and ground cover improve soil structure and reduce compaction (Wendel et al., 2022).

14.3.9 Erosion and runoff: guardians of soil health

Erosion and runoff are critical determinants of soil health, exerting a profound influence on soil structure, nutrient cycling and ecosystem resilience. These processes, driven by natural factors and human activities, play a pivotal role in shaping the health and productivity of landscapes. Erosion, the removal of topsoil by wind or water, depletes fertile layers essential for plant growth. It exposes subsoil, which often lacks the nutrients and organic matter crucial for sustaining healthy vegetation. Runoff occurs when excess water flows over the soil surface instead of infiltrating the ground. This can carry away valuable nutrients, sediments and pollutants, impacting water quality and aquatic ecosystems downstream. Both erosion and runoff disrupt soil structure, diminishing its ability to retain water nutrients and support plant roots. This compromises soil health, reducing agricultural yields and decreasing ecosystem resilience. Human activities like deforestation, agriculture and urbanization exacerbate erosion and runoff by disturbing natural vegetation cover and altering land surfaces. Climate change intensifies its effects by increasing the frequency and intensity of rainfall events (Zinngrebe et al., 2020).

14.4 Soil Biodiversity and Soil Health

Biodiversity and soil health are closely linked. Biodiversity helps various species and habitats function as a system, and this helps maintain soil health. The diversity of life in the soil is an example of the intricate web of biological activity that results from the interaction of plants and small creatures. This activity helps retain and infiltrate water, prevents soil erosion, nourishes plants, controls pests and diseases and recycles soil organic matter (Dubey et al., 2019; Guerra et al., 2021). Thus, soil biodiversity determines good soil, which is essential for long-term agricultural productivity. However, it has been noted that loss of soil biodiversity results from intensive agricultural activity. Degradation of the natural environment and associated global warming are blamed for this. The decomposition of organic matter by various living organisms produces soil organic matter, which is a critical component of healthy soil structure and good soil quality (Fan *et al.*, 2021; Leal Filho *et al.*, 2023). When the abundance of plants and animals varies, and there is a great diversity of species in a region, the turnover rate of organic matter increases and soil quality improves. In addition, species with deep root systems can improve soil stability and the movement of nutrients and water in the deeper soil layers (Lynch and Wojciechowski, 2015; Sharma *et al.*, 2016). The diversity of organisms found in or on the soil is also important. A healthy and diverse population of microorganisms can prevent soil-borne pathogens, improve soil structure and water-holding capacity and promote nutrient cycling. This leads to improved soil health and a resilient, more productive environment (Paz-Ferreiro and Fu, 2016; Crowther *et al.*, 2019).

14.4.1   Vital link for sustainable landscapes

One of the most important ecosystems that depend on biodiversity for its vitality is soil. Often overlooked as mere soil, soil is a complex and dynamic living system teeming with a multitude of organisms. The interplay between biodiversity and soil health is critical to maintaining productive landscapes, supporting agriculture, promoting ecosystem resilience and protecting the overall well-being of our planet (Teague and Kreuter, 2020; Janzen *et al.*, 2021). Biodiversity plays a multi-faceted role in maintaining and improving soil health. First, it contributes to nutrient cycling. Plant species have different nutrient needs, and various soil microorganisms help the breakdown of organic matter, releasing important nutrients into the soil. This efficient nutrient cycling maintains soil fertility and gives plants the resources to thrive. Second, biodiversity supports the decomposition of organic matter (Mallick and Chakraborty, 2018). Microorganisms in the soil, including bacteria and fungi,

help the breakdown of dead plant material, animal waste and other organic compounds, facilitating the release of nutrients and improving soil structure. This process improves the water-holding capacity, aeration and overall stability of the soil, making it more resistant to erosion and degradation (Berlinches de Gea *et al.*, 2023). Here, soil biodiversity is a natural defence against pests and diseases. Beneficial organisms such as predatory insects, nematodes and certain microorganisms help regulate populations of harmful pests and pathogens, reducing the need for chemical pesticides. This ecological balance promotes sustainable agriculture while minimizing negative impacts on human health and the environment (Singh *et al.*, 2019).

14.4.2   Link to ecosystem resilience

Soil biodiversity contributes to ecosystem resilience, enabling it to withstand environmental stresses and disturbances. In diverse ecosystems, different species have unique functional properties and different tolerances to environmental conditions. This diversity ensures that the ecosystem can continue to function and adapt in facing challenges such as drought, extreme temperatures or pollution. An ecosystem can recover from disturbances such as a forest fire or flood (Bauhus *et al.*, 2017). The greater the biodiversity, the more resources are available to the ecosystem for recovery. Biodiversity provides greater species diversity, which helps maintain stable populations and abundances and allows for a wide range of possible interactions and ecological processes. Species may adapt differently to environmental changes due to different specialization patterns within their habitat. This could allow the ecosystem to support a variety of niches and species and act as a buffer against environmental stresses (Pouteau *et al.*, 2015). For example, if an ecosystem is significantly damaged, species with similar specializations may be similarly affected, resulting in greater ecosystem degradation. On the other hand, greater biodiversity provides for more specialized species, so species with different specializations may be affected differently. Over time, greater diversity can help an ecosystem's structure and function recover (Moreno-Mateos *et al.*, 2012).

14.4.3 Conservation and restoration efforts

Global conservation and restoration initiatives are gaining momentum due to the critical importance of biodiversity to soil health. Several practices, including crop rotation diversification, cover crops, agroforestry, reduced tillage and organic amendments, can help increase soil biodiversity (Pouteau *et al.*, 2015). These practices provide habitat and food sources for various organisms and promote their presence and activity in the soil. In addition, good soil promotes the development of mutually beneficial associations between species, creating a robust and effective nutrient cycling and waste management system. To maintain the functionality and health of soils into the future, conservation and restoration efforts must focus on promoting biodiversity and providing habitat, nutrients and other resources (Dubey *et al.*, 2019).

14.5 Sustainable Soil Management and Biodiversity Conservation

Sustainable soil management and biodiversity conservation are fundamental to achieving long-term agricultural productivity, environmental sustainability and global food security. Soil provides the foundation for crop growth and important ecosystem services, while biodiversity is critical in maintaining ecosystem health and resilience. Biodiversity is essential for the health of soil and human populations (McLennon *et al.*, 2021). It helps provide food, medicine, fuel, shelter and ecosystem services. Biodiversity also contributes to resilience, which is the ability to withstand and repair damage from disasters. We must understand the importance of sustainable soil management practices for biodiversity conservation, the reciprocal relationship between soil health and biodiversity, and potential strategies for integrating soil management and biodiversity conservation in agricultural and land management systems. One way to manage soil for sustainable and biodiversity conservation is to minimize the use of chemical pesticides and fertilizers and instead use natural methods such as crop rotation, companion planting, soil amendments and cover crops. These methods

help promote microbial diversity, soil structure and soil fertility (Jangir *et al.*, 2019). To enhance ecological sustainability, native plants and/or permaculture crops could be planted that require fewer inputs and can support beneficial insect species. This helps promote biodiversity and allows the land to stay in its natural state. Gradual restoration of soil organic matter is enabled by agricultural practices that reduce soil disturbance (such as not ploughing) and return plant residues to the soil (Van Looy *et al.*, 2019). The population of beneficial organisms in the soil generally increases when tillage is reduced. However, the amount of beneficial soil microbes tends to decrease when the soil is compacted, vegetation cover is low or there is not enough plant litter covering the soil surface (Wang *et al.*, 2020).

14.5.1 Importance of sustainable soil management

Sustainable soil management practices are critical to maintaining soil health, optimizing agricultural productivity and reducing environmental impacts. By using sustainable soil management practices such as organic farming, cover cropping and reduced tillage, farmers can improve soil fertility, structure and water-holding capacity (Pouteau *et al.*, 2015). By improving nutrient cycling, reducing erosion and increasing carbon sequestration, sustainable soil management helps slow global warming. It helps sustain beneficial soil organisms that increase nutrient availability and maintain healthy ecosystems. Sustainable soil management practices also reduce the need for artificial inputs, keep water clean and increase crop yields, all of which contribute to the long-term viability of agricultural systems (Lobry de Bruyn *et al.*, 2017).

14.5.2 Reciprocal relationship between soil health and biodiversity

Soil health and biodiversity are closely linked and mutually beneficial. Healthy soils provide habitat for various soil organisms, including bacteria, fungi, earthworms and arthropods, which contribute to nutrient cycling, decomposition of organic matter and formation of soil structures. As microbial and plant life diversifies, the soil becomes more resilient to climatic changes and can better resist disease and pest infestation. A healthy and biodiverse soil system can help improve soil fertility, increase water infiltration and retention and reduce sediment and chemical runoff (Dubey *et al.*, 2019). This improves nutrient availability, soil fertility and overall ecosystem functioning. The presence of diverse plant species in agroecosystems, in turn, promotes greater aboveground biodiversity, including insects, birds and mammals. Plant diversity contributes to pollination, pest control and overall ecosystem resilience (Fig. 14.1). All this biodiversity also helps improve soil productivity, as a diversity of organisms and plant varieties provide a variety of essential nutrients for plant growth. By understanding and promoting the link between soil health and biodiversity, farmers can ensure that their soils are productive, resilient and healthy well into the future (Pouteau *et al.*, 2015).

14.6 Future of Sustainable Soil Management and Biodiversity Conservation

Today's agricultural scientists face a daunting challenge in the face of an ever-growing world population, shrinking availability of arable land, changing weather patterns, pervasive pollution and other forms of environmental degradation. Thanks to biotechnology, alternative ways to improve the natural environment and agricultural systems will become available. Current agricultural production systems can use less fertilizers and pesticides, improving soil, air and water quality. Biotechnology as a strategic tool can help breed new strains of crops with higher yields and better resistance to environmental stresses (Tonin *et al.*, 2018).

Understanding the determinants of soil health and their implications for environmental sustainability is promising. In the face of climate

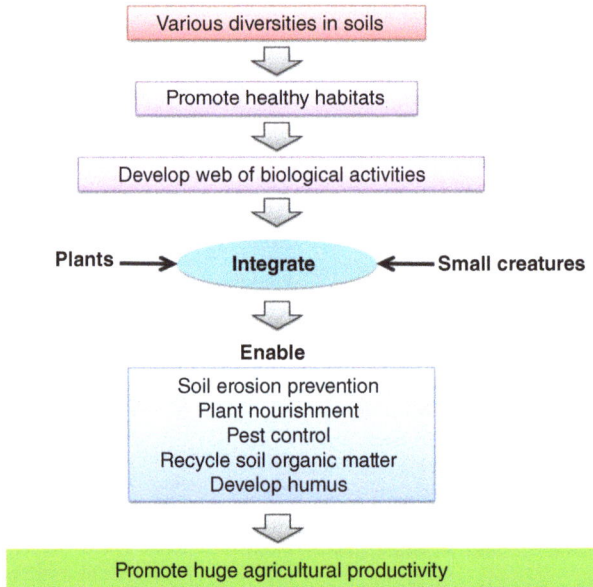

Fig. 14.1. Soil diversity and role in agricultural products. To make agriculture more productive, you need to build a living web of biological processes that includes plants and tiny animals such as microbes, insects and earthworms. This method creates a symbiotic environment where beneficial organisms help transport nutrients, keep the soil healthy and keep pests away. Farmers can reduce their reliance on chemicals, improve soil health and increase food yields in a way that does not harm the environment by promoting biodiversity and natural interactions. This all-round system not only makes farming more resilient but also ensures that it is productive and beneficial to the environment in the long term.

change, population growth and resource depletion, soil health and environmental well-being must be considered. Technologies such as precision agriculture and soil monitoring systems can detect and manage soil health in real-time. Genomic methods can reveal complex microbial interactions and improve our understanding of decomposition of nutrient cycling and organic matter (Loreau *et al.*, 2021). Regenerative agriculture and agroecological methods can restore degraded soils, sequester carbon and reduce erosion. Educating people about soil health can also promote sustainable land management. Integrated efforts by policy makers, researchers and stakeholders can improve soil health and environmental sustainability. These prospects enable robust and harmonious coexistence with our planet's resources.

14.7 Conclusion

Sustainable soil management and biodiversity conservation are intertwined with agricultural sustainability and environmental stewardship. Farmers can improve soil health, increase agricultural productivity and minimize environmental impacts by using sustainable soil management practices. Environmental sustainability depends on soil health, which affects productivity, resilience and ecosystem functioning.

Sustainable agricultural production and global food security depend on it. Physical, chemical and biological factors influence soil fertility, structure and nutrient cycling. Poor soil management imposes environmental and economic costs. Crop rotation, cover cropping, minimum tillage, grass irrigation, compost application and other soil management practices protect soil, promote nutrient cycling and reduce runoff and erosion. Crop rotations and grazing can also improve soil health. Mitigating climate change requires healthy soils to store carbon, control water runoff and protect against harsh weather. Without healthy soils, landscapes are more vulnerable to environmental damage and degradation, leading to environmental and economic problems. Maintaining soil health for long-term environmental sustainability is critical to growing crops and feeding people. Sustainable land management strategies consider how people benefit, how efficiently they use resources and how well they protect the environment. Soil quality is the essential link between land management and the goals of sustainable agriculture. Intensive land management has disrupted soil carbon, nitrogen and water cycles and threatens surface and subsurface water resources. With global climate change, ozone depletion from greenhouse gases and altered hydrological cycles, it is critical to understand how land management affects soil function.

References

Arora, N.K., Tewari, S. and Singh, R. (2013) Multifaceted plant-associated microbes and their mechanisms diminish the concept of direct and indirect PGPRs. In: Arora, N.K. (ed.) *Plant Microbe Symbiosis: Fundamentals and Advances*. Springer, New Delhi, pp. 411–449.

Arriaga, F.J., Guzman, J. and Lowery, B. (2017) Conventional agricultural production systems and soil functions. In: Al-Kaisi, M.M. and Lowery, B. (eds) *Soil Health and Intensification of Agroecosytems*. Academic Press, pp. 109–125.

Bauhus, J., Forrester, D.I., Gardiner, B., Jactel, H., Vallejo, R. *et al.* (2017) Ecological stability of mixed-species forests. In: Pretzsch, H., Forrester, D.I. and Bauhus, J. (eds) *Mixed-Species Forests: Ecology and Management*. Springer, pp. 337–382.

Berlinches de Gea, A., Hautier, Y. and Geisen, S. (2023) Interactive effects of global change drivers as determinants of the link between soil biodiversity and ecosystem functioning. *Global Change Biology* 29, 296–307.

Bhattacharyya, R., Ghosh, B., Mishra, P., Mandal, B., Rao, C. *et al.* (2015) Soil degradation in India: challenges and potential solutions. *Sustainability* 7, 3528–3570.

Bünemann, E.K., Bongiorno, G., Bai, Z., Creamer, R.E., De Deyn, G. *et al.* (2018) Soil quality – a critical review. *Soil Biology and Biochemistry* 120, 105–125.

Chagnon, M., Kreutzweiser, D., Mitchell, E.A.D., Morrissey, C.A., Noome, D.A. *et al.* (2015) Risks of large-scale use of systemic insecticides to ecosystem functioning and services. *Environmental Science and Pollution Research International* 22, 119–134.

Crowther, T.W., van den Hoogen, J., Wan, J., Mayes, M.A., Keiser, A.D. *et al.* (2019) The global soil community and its influence on biogeochemistry. *Science* 365, eaav0550.

Doran, J.W., Jones, A.J., Arshad, M.A. and Gilley, J.E. (2019) Determinants of soil quality and health. In: Ratta, R. and Lal, R. (eds) *Soil Quality and Soil Erosion.* CRC Press, pp. 17–36.

Dubey, A., Malla, M.A., Khan, F., Chowdhary, K., Yadav, S. *et al.* (2019) Soil microbiome: a key player for conservation of soil health under changing climate. *Biodiversity and Conservation* 28, 2405–2429.

El-Maraghy, S.S., Tohamy, A.T. and Hussein, K.A. (2021) Plant protection properties of the plant growth promoting fungi (PGPF): mechanisms and potentiality. *Current Research in Environmental and Applied Mycology* 11, 391–415.

Elhaissoufi, W., Ghoulam, C., Barakat, A., Zeroual, Y. and Bargaz, A. (2022) Phosphate bacterial solubilization: a key rhizosphere driving force enabling higher P use efficiency and crop productivity. *Journal of Advanced Research* 38, 13–28.

Fan, K., Delgado-Baquerizo, M., Guo, X., Wang, D., Zhu, Y.G. *et al.* (2021) Biodiversity of key-stone phylotypes determines crop production in a 4-decade fertilization experiment. *ISME Journal* 15, 550–561.

Frincu, M., Dumitrache, C., Cîmpeanu, P.C. and L.P.C., M. (2016) Study regarding nitrification in experimental aquaponic system. *Journal of Young Scientist* 4, 27–32.

Gavrilescu, M. (2021) Water, soil, and plants interactions in a threatened environment. *Water* 13, 2746.

Grzyb, A., Wolna-Maruwka, A. and Niewiadomska, A. (2020) Environmental factors affecting the mineralization of crop residues. *Agronomy* 10, 12.

Guerra, C.A., Bardgett, R.D., Caon, L., Crowther, T.W., Delgado-Baquerizo, M. *et al.* (2021) Tracking, targeting, and conserving soil biodiversity. *Science* 371, 239–241.

Hartmann, M. and Six, J. (2023) Soil structure and microbiome functions in agroecosystems. *Nature Reviews Earth and Environment* 4, 4–18.

Herrick, J.E. and Wander, M.M. (1997) Relationships between soil organic carbon and soil quality in cropped and rangeland soils: the importance of distribution, composition, and soil biological activity. In: Lal, R., Kimble, J.M., Follett, R.F. and Stewart, B.A. (eds) *Soil Processes and the Carbon Cycle.* CRC Press, pp. 405–425.

Itelima, J.U., Bang, W.J., Onyimba, I.A., Sila, M.D. and Egbere, O.J. (2018) Bio-fertilizers as key player in enhancing soil fertility and crop productivity: a review. *Direct Research Journal of Agriculture and Food Science* 6, 73–83.

Jangir, C.K., Kumar, S. and Meena, R.S. (2019) Significance of soil organic matter to soil quality and evaluation of sustainability. In: Meena, R.S. (ed.) *Sustainable Agriculture.* Scientific Publisher, Jodhpur, pp. 357–381.

Janzen, H.H., Janzen, D.W. and Gregorich, E.G. (2021) The 'soil health' metaphor: illuminating or illusory? *Soil Biology and Biochemistry* 159, 108167.

Lal, R. (2020) Soil quality and sustainability. In: Lal, R., Blum, W.E.H., Valentin, C. and Stewart, B.A. (eds) *Methods for Assessment of Soil Degradation.* CRC Press, pp. 17–30.

Lal, R. and Stewart, B.A. (eds) (2013) *Principles of Sustainable Soil Management in Agroecosystems.* CRC Press.

Leal Filho, W., Nagy, G.J., Setti, A.F.F., Sharifi, A., Donkor, F.K. *et al.* (2023) Handling the impacts of climate change on soil biodiversity. *Science of the Total Environment* 869, 161671.

Lehmann, J., Bossio, D.A., Kögel-Knabner, I. and Rillig, M.C. (2020) The concept and future prospects of soil health. *Nature Reviews. Earth and Environment* 1, 544–553.

Li, M. and Cai, L. (2021) Biochar and arbuscular mycorrhizal fungi play different roles in enabling maize to uptake phosphorus. *Sustainability* 13, 3244.

Liniger, H., Mekdaschi, R., Moll, P. and Zander, U. (2017) *Making Sense of Research for Sustainable Land Management.* Centre for Development and Environment (CDE), University of Bern and Helmholtz-Centre for Environmental Research GmbH–UFZ, Germany.

Lobry de Bruyn, L., Jenkins, A. and Samson-Liebig, S. (2017) Lessons learnt: sharing soil knowledge to improve land management and sustainable soil use. *Soil Science Society of America Journal* 81, 427–438.

Loreau, M., Barbier, M., Filotas, E., Gravel, D., Isbell, F. *et al.* (2021) Biodiversity as insurance: from concept to measurement and application. *Biological Reviews of the Cambridge Philosophical Society* 96, 2333–2354.

Lynch, J.P. and Wojciechowski, T. (2015) Opportunities and challenges in the subsoil: pathways to deeper rooted crops. *Journal of Experimental Botany* 66, 2199–2210.

Lynch, J.P., Strock, C.F., Schneider, H.M., Sidhu, J.S., Ajmera, I. *et al.* (2021) Root anatomy and soil resource capture. *Plant and Soil* 466, 21–63.

Mallick, P.H. and Chakraborty, S.K. (2018) Forest, wetland and biodiversity: revealing multi-faceted ecological services from ecorestoration of a degraded tropical landscape. *Ecohydrology and Hydrobiology* 18, 278–296.

McLennon, E., Dari, B., Jha, G., Sihi, D. and Kankarla, V. (2021) Regenerative agriculture and integrative permaculture for sustainable and technology driven global food production and security. *Agronomy Journal* 113, 4541–4559.

Mitra, D., Mondal, R., Khoshru, B., Senapati, A., Radha, T.K. *et al.* (2022) Actinobacteria-enhanced plant growth, nutrient acquisition, and crop protection: advances in soil, plant, and microbial multifactorial interactions. *Pedosphere* 32, 149–170.

Moreno-Mateos, D., Power, M.E., Comín, F.A. and Yockteng, R. (2012) Structural and functional loss in restored wetland ecosystems. *PLoS Biology* 10, e1001247.

Neina, D. (2019) The role of soil ph in plant nutrition and soil remediation. *Applied and Environmental Soil Science* 2019, 1–9.

Osman, A.I., Fawzy, S., Farghali, M., El-Azazy, M., Elgarahy, A.M. *et al.* (2022) Biochar for agronomy, animal farming, anaerobic digestion, composting, water treatment, soil remediation, construction, energy storage, and carbon sequestration: a review. *Environmental Chemistry Letters* 20, 2385–2485.

Paz-Ferreiro, J. and Fu, S. (2016) Biological indices for soil quality evaluation: perspectives and limitations. *Land Degradation and Development* 27, 14–25.

Pires, D., Orlando, V., Collett, R.L., Moreira, D., Costa, S.R. *et al.* (2023) Linking nematode communities and soil health under climate change. *Sustainability* 15, 11747.

Pouteau, R., Hulme, P.E. and Duncan, R.P. (2015) Widespread native and alien plant species occupy different habitats. *Ecography* 38, 462–471.

Rana, A., Pandey, R.K. and Ramakrishnan, B. (2019) Enzymology of the nitrogen cycle and bioremediation of toxic nitrogenous compounds. In: Bhatt, P. (ed.) *Smart Bioremediation Technologies*. Academic Press, pp. 45–61.

Rathore, R. (2018) Investigating the impact of soil tillage and crop rotation on the bacterial microbiome associated with winter oilseed rape under Irish agronomic conditions. Doctoral dissertation, Institute of Technology, Carlow, Ireland.

Saleem, S., Mushtaq, N.U., Rasool, A., Shah, W.H., Tahir, I. *et al.* (2023) Plant nutrition and soil fertility: physiological and molecular avenues for crop improvement. In: Aftab, T. and Rehman, K. (eds) *Sustainable Plant Nutrition*. Academic Press, pp. 23–49.

Seleiman, M.F. and Hafez, E.M. (2021) Optimizing inputs management for sustainable agricultural development. In: Awaad, H., Abu-hashim, M., Negm, A. (eds) *Mitigating Environmental Stresses for Agricultural Sustainability in Egypt*. Springer, pp. 487–507.

Sharma, N., Bohra, B., Pragya, N., Ciannella, R., Dobie, P. *et al.* (2016) Bioenergy from agroforestry can lead to improved food security, climate change, soil quality, and rural development. *Food and Energy Security* 5, 165–183.

Sharma, V., Bhattacharyya, S., Kumar, R., Kumar, A., Ibañez, F. *et al.* (2020) Molecular basis of root nodule symbiosis between Bradyrhizobium and 'crack-entry' legume groundnut (Arachis hypogaea L.). *Plants* 9, 276.

Sindhu, S.S., Sehrawat, A. and Glick, B.R. (2022) The involvement of organic acids in soil fertility, plant health and environment sustainability. *Archives of Microbiology* 204, 720.

Singh, A., Bhardwaj, R. and Singh, I.K. (2019) Biocontrol agents: potential of biopesticides for integrated pest management. In: Giri, B., Prasad, R., Wu, Q.-S. and Varma, A. (eds) *Biofertilizers for Sustainable Agriculture and Environment*. Springer, pp. 413–433.

Singh, C. and Vyas, D. (2021) Biodegradation by fungi for humans and plants nutrition. In: Mendes, K.F., de Sousa, R.N. and Mielke, K.C. (eds) *Biodegradation Technology of Organic and Inorganic Pollutants*. IntechOpen.

Sokol, N.W., Slessarev, E., Marschmann, G.L., Nicolas, A., Blazewicz, S.J. *et al.* (2022) Life and death in the soil microbiome: how ecological processes influence biogeochemistry. *Nature Reviews. Microbiology* 20, 415–430.

Teague, R. and Kreuter, U. (2020) Managing grazing to restore soil health, ecosystem function, and ecosystem services. *Frontiers in Sustainable Food Systems* 4.

Thomas, L. and Singh, I. (2019) Microbial biofertilizers: types and applications. In: Giri, B., Prasad, R., Wu, Q.-S. and Varma, A. (eds) *Biofertilizers for Sustainable Agriculture and Environment*. Springer, pp. 1–19.

Tonin, A.M., Pozo, J., Monroy, S., Basaguren, A., Pérez, J. *et al.* (2018) Interactions between large and small detritivores influence how biodiversity impacts litter decomposition. *Journal of Animal Ecology* 87, 1465–1474.

Van Antwerpen, R., van Heerden, P.D.R., Keeping, M.G., Titshall, L.W., Jumman, A. *et al.* (2022) A review of field management practices impacting root health in sugarcane. *Advances in Agronomy* 173, 79–162.

Van Looy, K., Tonkin, J.D., Floury, M., Leigh, C., Soininen, J. *et al.* (2019) The three Rs of river ecosystem resilience: resources, recruitment, and refugia. *River Research and Applications* 35, 107–120.

Veum, K.S., Kremer, R.J., Sudduth, K.A., Kitchen, N.R., Lerch, R.N. *et al.* (2015) Conservation effects on soil quality indicators in the Missouri Salt River Basin. *Journal of Soil and Water Conservation* 70, 232–246.

Visser, S., Keesstra, S., Maas, G., de Cleen, M. and Molenaar, C. (2019) Soil as a basis to create enabling conditions for transitions towards sustainable land management as a key to achieve the SDGs by 2030. *Sustainability* 11, 6792.

Wang, H., Wang, S., Yu, Q., Zhang, Y., Wang, R. *et al.* (2020) No tillage increases soil organic carbon storage and decreases carbon dioxide emission in the crop residue-returned farming system. *Journal of Environmental Management* 261, 110261.

Wendel, A.S., Bauke, S.L., Amelung, W. and Knief, C. (2022) Root-rhizosphere-soil interactions in biopores. *Plant and Soil* 475, 253–277.

Williams, H., Colombi, T. and Keller, T. (2020) The influence of soil management on soil health: an on-farm study in southern Sweden. *Geoderma* 360, 114010.

Yu, O.-Y., Harper, M., Hoepfl, M. and Domermuth, D. (2017) Characterization of biochar and its effects on the water holding capacity of loamy sand soil: comparison of hemlock biochar and switchblade grass biochar characteristics. *Environmental Progress and Sustainable Energy* 36, 1474–1479.

Zinngrebe, Y., Borasino, E., Chiputwa, B., Dobie, P., Garcia, E. *et al.* (2020) Agroforestry governance for operationalising the landscape approach: connecting conservation and farming actors. *Sustainability Science* 15, 1417–1434.

15 Role of Non-coding RNAs in Plant Nutrition and Growth

Sakshi Arora*, Priyasha Das and Sakshi Singh

Amity Institute of Microbial Technology, Amity University, Noida, Uttar Pradesh, India

Abstract

Small RNAs are generally regulatory RNAs that influence the expression of genes. siRNAs and miRNAs are the two most common small RNAs that regulate gene expression; however, miRNAs contribute more during plant growth and development compared with siRNAs. miRNAs are 22–24 nucleotides long and influence all aspects of plant growth and development, as well as response to biotic and abiotic stresses and tolerance mechanisms. For optimum plant growth and development, adequate uptake of minerals from soil and their transport within plants is essential. Numerous studies have identified miRNAs under different nutrient starvation conditions expressed differentially. These miRNAs modulate the expression of various genes and transcriptional factors involved in maintaining nutrient homeostasis, thus permitting plants to adapt and thrive in nutrition-deprived soil. To develop plant varieties with high tolerance against specific nutrient deficiencies, it is essential to understand the role and mechanism of miRNAs during nutrient homeostasis and plant growth. Thus, in this chapter, we discuss the role of various miRNAs under different nutrient stresses and during the maintenance of meristem activity.

Keywords: Non-coding RNA, miRNAs, nutrient deficiency, shoot apical meristem formation, gene expression regulation

15.1 Introduction

The coding sequence in the eukaryotic genome is only 2%, whereas 90% of the genome undergoes transcription (Pauli *et al.*, 2011; Yu *et al.*, 2019). The remaining non-translated RNAs are identified as non-coding RNA (ncRNAs). They are evolutionary diverse and originate from intergenic regions, pseudogenes, repetitive sequences and transposons (Ariel *et al.*, 2015). Numerous scientific investigations have confirmed the extensive involvement of ncRNAs in

regulating gene expression at the transcriptional, post-transcriptional and translational levels. Therefore, they have become crucial bioactive molecules in shaping genomic and phenotypic diversities (Eddy, 2001; Cech and Steitz, 2014; Peschansky and Wahlestedt, 2014). The origin of ncRNAs is possibly attributed to the duplication of genome and transposable elements, random hairpin structures, pseudogenes, RNA viruses, DNA repeats and regulatory regions of genes (Allen *et al.*, 2004; Voinnet, 2009; Waititu *et al.*, 2020). Depending on the origin, biogenesis, size

*Corresponding author: arora.17sakshi@gmail.com

© CAB International 2025. *Soil Health and Nutrition Management*
(eds N.C. Joshi, T. Leustek and P.K. Singh)
DOI: 10.1079/9781800624597.0015

and function, ncRNAs can be categorized into several classes. Size-wise, they can be divided into two groups – small non-coding RNA (<200 nucleotides) and long non-coding RNA (>200 nucleotides). Functionally, ncRNA can be classified into housekeeping and regulatory categories. The housekeeping ncRNAs, including rRNA, tRNA, snRNA and snoRNA, are generally involved in cellular and ribosomal functions. In contrast, regulatory ncRNAs, including miRNAs, siRNAs, piRNAs and lncRNAs, participate in the manipulation of gene expression (Fig. 15.1).

15.2 Small RNAs: Origin, Biogenesis and Functions

Small non-coding RNAs, with a length typically less than 200 nucleotides (usually ranging from 20 to 30 nucleotides), are long double-stranded RNAs that regulates gene expression through chromatin modification, post-transcription gene silencing or translation inhibition (Zhan and Meyers, 2023) (Fig. 15.2). Various types of small RNAs are identified, including but not limited to micro RNAs (miRNAs), small interfering RNAs (siRNAs), piwi RNAs (piRNAs), small temporal RNAs (stRNAs), tiny non-coding RNAs (tncRNAs) and small modular RNAs (smRNAs). miRNAs and siRNAs have been extensively characterized in plants, whereas piRNAs have been only affirmed in animal germ cells (Kim *et al.*, 2009; Guleria *et al.*, 2011). Therefore, this chapter will focus on the biogenesis and functions of miRNAs and siRNAs.

miRNAs constitute a category of non-coding RNAs, measuring 22–24 nucleotides in length, that flaunt evolutionary conservation. They are crucial in moderating gene expression in response to external and internal stimuli. Generally, miRNAs exert their impact at the post-transcriptional level, predominantly by impeding transcription through mRNA cleavage. Alternatively, they may hinder translation or, in rare instances, induce chromatin modification (Baumberger and Baulcombe, 2005; Chen, 2005). The biogenesis of miRNAs commences with transcribing a primary miRNA (pri-miRNA) from the genic sequence by DNA polymerase II. Subsequently, the Dicer-like 1 (DCL1) enzyme cleaves the pri-miRNA, resulting in the formation of a stem-loop structure, referred to as the precursor miRNA (pre-miRNA) (Chen, 2005; Jung *et al.*, 2009). Second cleavage by DCL1 results in the production of a miRNA: the

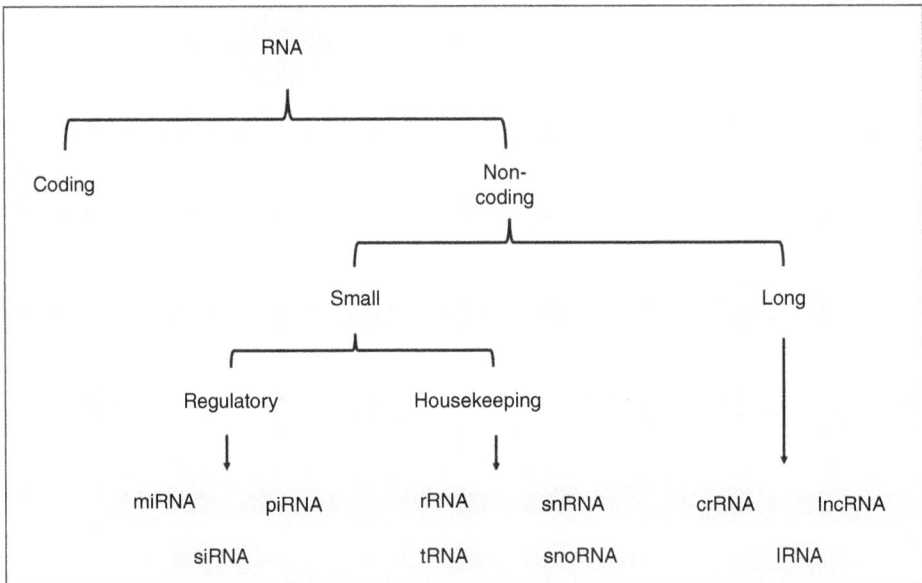

Fig. 15.1. Categorization of RNAs depending on function and size.

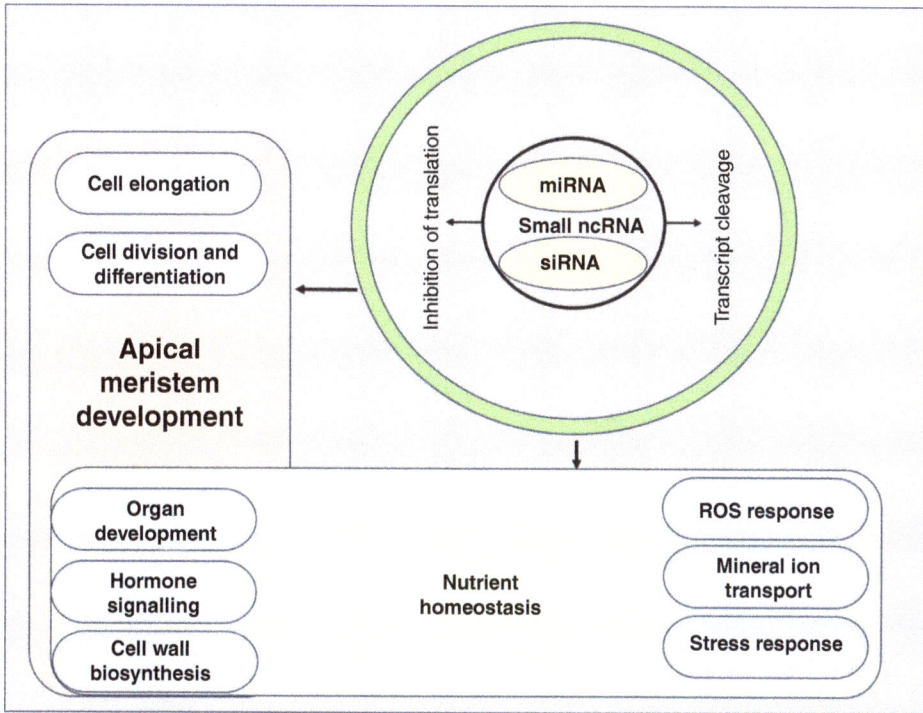

Fig. 15.2. Representation of various cellular processes regulated by small RNAs during maintenance of nutrient homeostasis and meristem development.

miRNA duplex, which is transported from the nucleus to the cytoplasm by the HASTY protein (Bartel, 2004; Jung *et al.*, 2009).

Helicase-mediated processing transforms the RNA duplex into a single-stranded mature miRNA molecule. This mature miRNA interacts with the AGO1 (ARGONAUTE) protein, leading to the formation of a ribonucleoprotein complex recognized as the RNA-induced silencing complex (RISC) (Bartel, 2004; Kurihara and Watanabe, 2004). The RISC guides the mature miRNA to its target gene, where the miRNA either cleaves the mRNA of the target gene or inhibits its translation (Bartel, 2004). Notably, miRNAs often target mRNAs encoding transcription factors.

Similarly, siRNAs, measuring 20–22 nucleotides in length, function as small RNAs that govern post-transcriptional gene expression. siRNAs and miRNAs share similar mechanisms of action but differ significantly in origin and biogenesis. The initiation of siRNA biogenesis is triggered in response to double-stranded RNA

(dsRNA) derived from viral RNA, DNA repeats or transposable elements. A Dicer enzyme cleaves the dsRNA into siRNA, yielding small dsRNA molecules with characteristic overhangs (Bernstein *et al.*, 2001; Kamthan *et al.*, 2015). The small dsRNA molecules interact with the siRNA-induced silencing complex (siRISC), constituting AGO and other effector proteins. The sense strand of the double-stranded siRNA is degraded, exposing the antisense strand to bind with complementary mRNA (Kamthan *et al.*, 2015). The AGO protein of the RISC then degrades the target mRNA or inhibits the translation of mRNA that showed perfect complementation to the siRNA. siRNAs are further categorized into three groups, namely tasiRNAs (trans-acting siRNAs), hc-siRNA (heterochromatic siRNAs) and nat-siRNAs (natural antisense siRNAs), depending on their origin and biogenesis. ta-siRNAs are endogenous siRNAs that are similar to miRNAs in terms of biogenesis and regulation mechanisms. They originate from the miRNA cleavage of TAS target genes and subsequently target other related genes (Yoshikawa, 2013).

In the context of several plant developmental processes, including cell division, differentiation, root initiation, flowering, flower development, vascular development, leaf development and seed development, small RNAs play an integral role (Bartel, 2004; Chen, 2005) (Fig. 15.2). Additionally, small RNAs respond to environmental fluctuations during biotic and abiotic stress conditions (Sunkar *et al.*, 2012; Shriram *et al.*, 2016).

15.3 Role of Small RNAs in Nutritional Uptake and Accumulation

Nutrient uptake and homeostasis are essential for optimum plant growth and development (Paul *et al.*, 2015). Plants require at least 17 elements, categorized into macronutrients and micronutrients, based on the quantity requirements. Macronutrients encompass nitrogen (N), potassium (K), calcium (Ca), phosphorus (P), magnesium (Mg), sulfur (S), oxygen (O), carbon (C) and hydrogen (H). In contrast, elements such as boron (B), iron (Fe), chlorine (Cl), manganese (Mn), copper (Cu), zinc (Zn), molybdenum (Mo) and nickel (Ni) are needed in trace amounts, and hence are referred to as micronutrients. The roots of plants predominantly acquire most of these nutrients from the soil, and subsequent transport occurs (Islam *et al.*, 2022). Nutrient stress arises from the unavailability of mineral ions for absorption by plant roots, thereby hindering growth and development. Plants have evolved various mechanisms to counteract the adverse effects of nutrient starvation and to maintain homeostasis. Nutrient starvation triggers signalling cascades, activating a complex regulatory mechanism involving transcription factors, small RNAs and genes. The modified expression of these components enables plants to detect nutrient stress and adapt to maintain growth and development by altering various structural, biochemical and physiochemical mechanisms (Paul *et al.*, 2015; Islam *et al.*, 2022) (Fig. 15.2).

Small RNAs, particularly miRNAs, play a significant role in plant growth and development and in response to biotic and abiotic stresses.

Nutrient deficiency, considered abiotic stress, elicits an adaptation response involving various miRNAs that act as signalling molecules or regulate the assimilation and transport of nutrients from the soil. This chapter discusses the significant role of various miRNAs in various responses to nutrient starvation.

15.3.1 Nitrogen

Nitrogen is an indispensable element for plant growth and development, serving as a key component of nucleic acids, protein, chlorophyll and phytohormones. N is assimilated by plants as nitrate (NO_3^-), ammonium (NH_4^+) or urea from the soil. Leguminous plants form a symbiotic association with microorganisms, facilitating N fixation by converting environmental N_2 into NH_4 within nodules which is subsequently taken up by plants (Andrews *et al.*, 2013). Numerous miRNAs have been identified as pivotal contributors to the response to N starvation and the formation of nodules in leguminous plants. Genome-wide screening of miRNAs under N starvation in *Arabidopsis*, rice and maize revealed differential expression patterns of several miRNAs, including miR156, miR160, miR167, miR169, miR171, miR172, miR319, miR395, miR399, miR826, miR827, miR829, miR839, miR846, miR850, miR857 and miR863 (Xu *et al.*, 2011; Liang *et al.*, 2012; Nischal *et al.*, 2012). In barley, the availability of excess N for growth exhibited significant upregulation of miR408, miR164, miR396, miR399, miR827 and miR528 and downregulation of miR393 (Grabowska *et al.*, 2020).

Overexpression of miR393 in rice resulted in the down-expression of *TIR1* and *AFBs* in the axillary bud, implicating its involvement in N-mediated tillering by suppressing auxin signalling (Li *et al.*, 2016). In N-deficient conditions, *ARF8* in *Arabidopsis*, a target of miR167, was overexpressed in pericycle and root cap cells, contributing to lateral root development and enhanced N assimilation (Gifford *et al.*, 2008). Remarkably, it was observed that under N stress, the expression of miR399 and miR827, along with their target genes, underwent significant alterations. Both the miRNAs are associated with phosphate starvation stress response, suggesting a cross-talk between N and phosphate stress response (Paul *et al.*, 2015). Under N-deficient conditions, the knockout of miR396e/f in rice increased grain size, yield and biomass. Similarly, overexpression of miR396 target *GRFs* resulted in enhanced grain size, yield and panicle branching, underscoring the significant role of miR396 in N assimilation (Zhang *et al.*, 2019). miR164 along with its target *NAC1* is also reported to influence N uptake. In apple, *NAC1* regulates expression of *NRT2.4*, a nitrate transporter. Moreover, it was observed that plant tolerance against N starvation increased in a NAC1 silenced line (Wang *et al.*, 2024). In sugarcane roots subjected to low N treatment, miR156 was significantly upregulated. Moreover, the overexpression of sugarcane miR156 in *Arabidopsis* resulted in increased root length and surface area. Additionally, miR156-overexpressed plants exhibited enhanced N assimilation, attributed to the increased expression of *NRT1.1*, a nitrate transporter gene, as well as N assimilation genes *NR1*, *NIR1* and *GS* and enzymes NR and GS (Gao *et al.*, 2022). miR169 has been implicated

in the regulation of N uptake and mobilization. Its role involves the degradation of the NFYA transcription factor, which is necessary for the expression of *NRT1.1* and *NRT2.1*. In *Arabidopsis* and rice, miR169 was downregulated under N deficiency, whereas overexpression of miR169 resulted in low N accumulation, leading to leaf yellowing (Zhao *et al.*, 2011). Moreover, in rice, overexpression of miR169 reduces the expression of NF-YA5 transcription factor and consequently the expression of *NRT1*, *GS1/2* (*Glutamate Synthetase 1/2*) and *GOGAT1/2* (*Glutamine Oxoglutarate Aminotransferase 1/2*) (Seo *et al.*, 2023) (Fig. 15.3, Table 15.1).

miR826 was observed to be significantly increased under N deficiency, leading to the downregulation of its target gene *AOP2* (*Alkenyl Hydroxalkyl Producing 2*). The overexpression of miR826 in *Arabidopsis* conferred tolerance against N starvation accompanied by increased biomass, root development, chlorophyll content and decreased glucosinolate and anthocyanin contents (He *et al.*, 2014). miR3979, a rice-specific miRNA, was reported

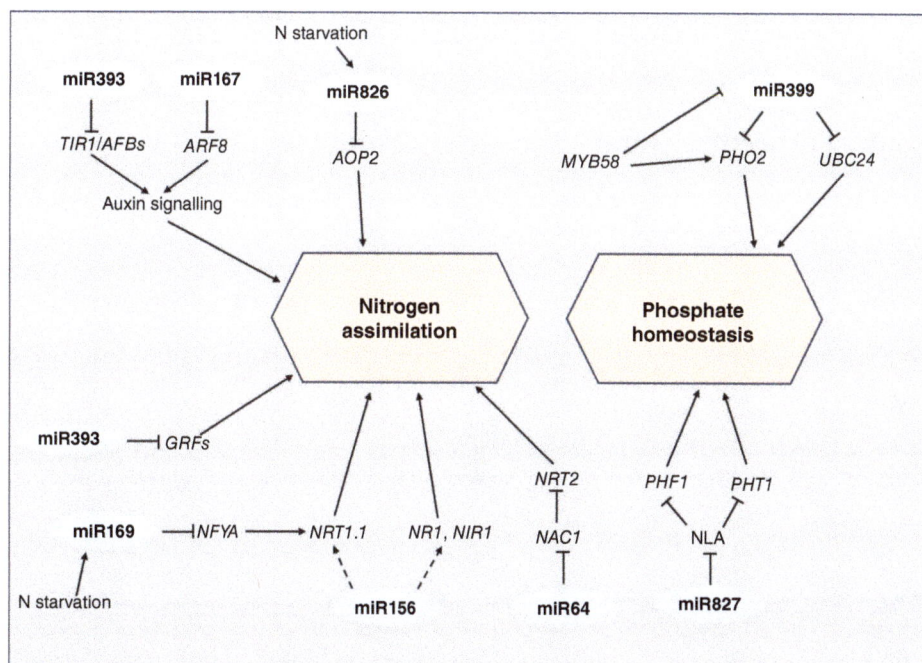

Fig. 15.3. Representation of certain miRNAs that influence nitrogen and phosphate assimilation and transport.

Table 15.1. Some important miRNAs that play an important role in different nutrient assimilation, homeostasis and apical meristem development.

Small RNA	Genes altered	Cellular/molecular function affected	Reference
Nitrogen			
miR393	*TIR1, AFBs*	Auxin signalling	Li *et al.*, 2016
miR167	*ARF8, ARF6*	Auxin signalling and lateral root development	Gifford *et al.*, 2008
miR396	*GRFs*	Enhanced panicle branching and yield	Zhang *et al.*, 2019
miR156	*NRT1.1, NR1, NIR1, GS*	Nitrogen transport and assimilation	Gao *et al.*, 2022
miR169	*NRT1.1, NRT2.1*	Nitrogen transport and assimilation	Zhao *et al.*, 2011
miR826	*AOP2*	Glucosinolate biosynthesis	He *et al.*, 2014
miR3979	*AnPRT*	Tryptophan synthesis	Jeong *et al.*, 2011
miR408	*LAC3, LAC13, PLC, cupredoxin, SOD1A*	Photosynthesis, nitrogen assimilation	Xu *et al.*, 2011; Liang *et al.*, 2012
Phosphate			
miR399	*PHO2, UBC24*	Phosphate transport, protein degradation	Gao *et al.*, 2010
miR827	*NLA, SPX-MFS1, SPX-MFS2* (in rice only)	Membrane transportation	Wang *et al.*, 2012; Lin *et al.*, 2013
Potassium			
miR408	*LAC3, LAC13, PLC, cupredoxin, SOD1A*	Potassium uptake, photosynthesis, ROS scavenging	Zhao *et al.*, 2020
miR168	*AGO1A*	Potassium uptake, root and leaf development	Liu *et al.*, 2020
miR160	*ARF*	Root development, auxin signalling	Zhao *et al.*, 2020; Li *et al.*, 2021
Sulfur			
miR395	*APS1, APS3, APS4, SULTR;1*	Sulfate assimilation and transport	Jones-Rhoades and Bartel, 2004; Kawashima *et al.*, 2011
miR408	*GSTU25*	Sulfate assimilation	Kumar *et al.*, 2023
Iron			
miR164	*IRT1, FRO2*	Iron transport, ferric reductase activity, root development	Du *et al.*, 2022
miR408	*LAC17, B-GLU23, LAC3, LAC12, LAC13, PLC*	Cell wall development and lignification	Gao *et al.*, 2022
Copper			
miR408	*PLC, LAC, UCL*	Regulation of copper homeostasis	Perea-García *et al.*, 2021
miR398	*CSD1/2*	Response to copper starvation	Sun *et al.*, 2020; Perea-García *et al.*, 2021
miR156	*SPL3*	Response to copper starvation	Perea-García *et al.*, 2021
Boron			
miR397	*LAC4, LAC17*	Cell wall biosynthesis	Huang *et al.*, 2016
miR319	*MYB*	Modulating root architecture	Huang *et al.*, 2016
miR171	*SCARECROW-like*	Modulating root architecture	Huang *et al.*, 2016

Continued

Table 15.1. Continued.

Small RNA	Genes altered	Cellular/molecular function affected	Reference
miR396	ATPase	Modulating root architecture	Huang et al., 2016
miR172	SPL	Hormone signalling (ethylene and jasmonic acid)	Kayihan et al., 2019
Root growth and development			
miR160	ARF10, ARF16, ARF17	Auxin signalling, root growth, cell division and elongation	Wang et al., 2005
miR164	NAM/CUC1/CUC2	Lateral root and branching	Guo et al., 2005
miR167	ARF6/ARF8	Adventitious root development, root production	Gutierrez et al., 2009; Arora et al., 2020
miR393	TIR1, AFB1/2/3	Lateral root development, primary root length elongation	Jiang et al., 2022
miR390	AR2/3/4	Lateral root development	Hobecker et al., 2017
miR165/166	PHB, PHV	Maintain cytokinin homeostasis and cell differentiation	Bertolotti et al., 2021
miR396	GRF	Regulator of SCN, stem cell conversion to TACs	Rodriguez et al., 2015
miR156	SPL	RAM development	Barrera-Rojas et al., 2019
Shoot apical meristems			
miR165/166	PHB, PHV, REV	Maintenance of SAM	Byrne, 2006
miR164	CUC, CUC2	Regulate leaf serration, determine organ boundary, shoot meristem establishment	Nikovics et al., 2006; Wang et al., 2021
miR171	HAM, CLV1, CLC3, WOX4	SAM development, shoot branching, seedling growth	Han and Zhou, 2022
miR394	LCR	SAM development and maintenance	Knauer et al., 2013

RAM, root apical meristem; ROS, reactive oxygen species; SAM, shoot apical meristem; SCN, stem cell niche; TACs, transit amplifying cells.

to have a significant role in N absorption by targeting the *AnPRT* (*Anthranilate Phosphosribosyl Transferase*) gene involved in tryptophan synthesis (Jeong et al., 2011). In *Arabidopsis* and *Zea mays*, the expression of miR408 was repressed, whereas the respective target genes – *LAC* (*LAC3* and *LAC13*), *PLC*, *cupredoxin* and *SOD1A* – exhibited enhanced expression under N starvation (Table 15.1) (Xu et al., 2011; Liang et al., 2012).

miRNAs are also reported to play a significant role in N fixation. In one of the earliest studies, various miRNAs, including miR172, miR166, miR396, miR159, miR393, miR168 and miR169, were differentially expressed during nodule formation and the interaction of soybean with *Bradyrhizobium japonicum* (Subramanian et al., 2008).

15.3.2 Phosphorus

Phosphate is another indispensable macromolecule essential for plant growth, serving as a structural component in nucleic acids, membranes and ATP, an energy-providing molecule. Plants acquire phosphate in its inorganic form (Pi), and various enzymes released by plants either solubilize it or liberate it from organic substances (Paul et al., 2015). Numerous studies conducted across diverse plant species, including *Arabidopsis*, rice, wheat, barley, maize, soybean and tomato, have highlighted the significant role of miRNAs in phosphate (Pi) homeostasis (Paul et al., 2015).

miR399 has also been identified as a key regulator of Pi homeostasis directly by targeting PHO2, a phosphate transporter and UBC24, a protein degradation enzyme. Under Pi starvation, miR399 was induced in the shoot, where it repressed the expression of *PHO2*, thus impeding phosphate mobilization and resulting in its accumulation in the shoot (Pant *et al.*, 2008). Tomato plants exhibiting overexpression of miR399 demonstrated increased phosphate accumulation in both the roots and shoots, accompanied by enhanced proton transport from roots and excretion from shoots (Gao *et al.*, 2010). Similarly, in *Z. mays*, miR399 was also reported to influence phosphate homeostasis (Wang *et al.*, 2023). It was also revealed that, in rice, overexpression of *MYB58* resulted in increased expression of *PHO2* and decreased the expression of miR399 (Baek *et al.*, 2024).

miR827, a regulator targeting *NLA* (*Nitrogen Limitation Adaptation*), is also reported to play a significant role in maintaining P homeostasis. The miR827–NLA module exerted a substantial influence on the expression of *PHF1* (*PHD Finger protein 1*) and *PHT1* (*Phosphate Transporter 1*) genes, crucial components in phosphate transport (Lin *et al.*, 2013) (Fig. 15.3). *NLA* induced the ubiquitination of PHT1, a membrane-bound protein that aids in Pi acquisition. Consequently, miR827 expression undergoes upregulation during P-starvation conditions, while the *nla* mutant results in extreme P deposition under N starvation (Lin *et al.*, 2013). In rice, miR827 has been reported to target *SPX-MFS1* and *SPX-MFS2*, both belonging to the subgroup of the SPX domain (named after proteins SYG1/PHO81/XPR1) and the MFS (Major Facility Superfamily) protein, encompassing various secondary membrane transporters (Wang *et al.* 2012). The miR827-SPX-MFS domain is also involved in Pi homeostasis, and expression of miR827 showed upregulation under P starvation (Table 15.1) (Wang *et al.*, 2012).

Additionally, miR778, miR828 and miR2111 exhibited increased expression under P deficiency in *Arabidopsis* (Pant *et al.*, 2009). Notably, the expression of miR2111 and miR827 exhibited opposing patterns under N- and P-starvation conditions (Liang *et al.*, 2012). In soybean roots, P starvation led to the upregulation of miR159 and the downregulation of miR319, miR396, miR398 and miR1507 (Zeng *et al.*, 2010). Furthermore, miR408 displayed species-dependent differential expression under P deficiency: it displayed upregulation in

Arabidopsis and *Glycine max* but was repressed in *Triticum aestivum* (Gao *et al.*, 2022).

15.3.3 Potassium

Potassium is essential in various signalling pathways and metabolic activities and is crucial for plant growth, development, reproduction and response to abiotic and biotic stress. Cellular K uptake occurs in the form of K^+ through multiple ion transporters. Genome-wide identification of miRNAs under K deficiency identified 69, 36 and 65 differential miRNAs in wheat, sugarcane and barley, respectively (Zhao *et al.*, 2020; Ye *et al.*, 2021; Zhang *et al.*, 2022).

In tobacco, miR408 overexpression lines exhibited increased K uptake, biomass, photosynthesis and reactive oxygen species (ROS) scavenging (Zhao *et al.*, 2020). Meanwhile, overexpression of miR168 in tomatoes led to enhanced root and leaf development and increased K content under K-starvation stress compared with controls (Liu *et al.*, 2020). Similar results were observed in lines overexpressing *AGO1A*, a target of miR168 (Liu *et al.*, 2020). In *Arabidopsis*, overexpression of miR160 reduced tolerance against K starvation and inhibited root development. The expression of miR393 was upregulated in peanuts, whereas miR408 was repressed in wheat under K-deficiency conditions (Li *et al.*, 2021; Zhao *et al.*, 2020).

15.3.4 Sulfur

Sulfur serves as a constituent of several amino acids and plays a crucial role in protein production. Additionally, it is essential to activate several enzymes, including chlorophyll, which is vital for photosynthesis. Furthermore, it is a component of several secondary metabolites, making it indispensable for plant growth and development and responding to abiotic and biotic metabolism and signalling (Narayan *et al.*, 2023).

S is uptaken and transported in the form of SO_4^{2-} from the soil, and its deficiency has a negative impact on yield and quality. Mild S deficiency does not impact yield but significantly impacts quality as it substantially decreases protein production (Narayan *et al.*, 2023). Sulfate deficiency resulted in altered expression of enzymes involved in sulfate assimilation, including ATP sulfurylase (ATPS), adenosine 5-phosphosulfate reductase (APR) and several sulfate transporters, including low-affinity sulfate transporter SULTR2;1. In *Arabidopsis*, S deficiency resulted in increased expression of miR395, which is reported to target *APS1*, *APS3*, *APS4* and *SULTR2;1* (Jones-Rhoades and Bartel, 2004; Kawashima *et al.*, 2009, 2011).

However, it was observed that under S deficiency, both miR395 and *SULTR2;1* showed elevated expression in *Arabidopsis* roots. This contradiction was resolved at the cellular level, revealing that miR395 was expressed more in xylem parenchyma cells but exhibited minimal target gene expression. Conversely, *SULTR2;1* exhibited high expression in xylem parenchyma cells and minimal expression of miR395 (Kawashima *et al.*, 2009). Furthermore, it was noted that miR395 induction under S deficiency was regulated by SLIM1 (Sulphur Limitation 1) transcription factor (Kawashima *et al.*, 2009).

miR408 is also reported to play a significant role during S assimilation by targeting *GSTU25* (*glutathione S-transferase*). In *Arabidopsis*, miR408 overexpression resulted in S-sensitive plants, whereas miR408 knockout resulted in S-tolerant plants (Kumar *et al.*, 2023). The study conducted by Huang *et al.* (2010) reported that S deficiency in *Brassica napus* resulted in differential expression of miR156, miR160, miR164, miR167, miR168 and miR394.

15.3.5 Other minerals

In addition to macronutrients, various microelements are essential for achieving optimum plant yield and quality, including Mn, Fe, Zn, Cu, B and Mg.

In plants, Mn acts as an enzyme cofactor and catalytic element in metalloenzyme clusters, including oxygen-evolving complex (OEC) in photosystem II. Consequently, Mn is significant in diverse cellular processes within plants, including photosynthesis, respiration, hormonal signalling, ROS response and adaptation to biotic stress (Alejandro *et al.*, 2020). Mn deficiency impairs plant growth, reduces photosynthesis and renders plants more susceptible to biotic and abiotic stress (Alejandro *et al.*, 2020).

A miRNA screening conducted under Mn toxicity in *Phaseolus vulgaris* revealed the differential expression of 37 miRNAs including miR156, miR160, miR167, miR172, miR390, miR394, miR1508, miR1515, miR1532 and miR1510 (Valdés-López *et al.*, 2010).

Fe is another indispensable element that plays a significant role in various metabolic processes, including respiration, photosynthesis and DNA synthesis. It also serves as a prosthetic group for various enzymes. Several differential miRNAs have been identified under Fe-deficiency conditions in citrus plants, *Arabidopsis* and common beans, including miR167, miR164, miR156, miR397, miR398 and miR408, utilizing sequencing, northern blot analysis and quantitative expression analysis (Valdés-López *et al.*, 2010; Jin *et al.*, 2021; Bakirbas *et al.*, 2023). In *Arabidopsis*, under Fe starvation, miR164 expression was significantly repressed. However, the miR164 mutant, under Fe-starvation conditions, exhibited overexpression of *IRT1* (*Iron-regulated Transporter 1*) and *FRO2* (*Ferric Reduction Oxidase 2*) and increased ferric reductase activity and displayed augmented primary root length and lateral root numbers, suggesting the crucial role of miR164 in maintaining Fe homeostasis (Du *et al.*, 2022). The *Arabidopsis* plants exhibiting miR408 overexpression displayed Fe sensitivity, upregulation of lignification-related genes, including *LAC17* and *B-GLU23*, and the downregulation of *LAC3*, *LAC12*, *LAC13* and *PLC* (Gao *et al.*, 2022). In wheat, miR1130 was reported to target ferroportin 1 (FPN1) and to contribute to iron homeostasis (Table 15.1) (Sharma *et al.*, 2024).

Cu serves as a crucial cofactor in various enzymes, including plastocyanin, SOD (superoxide dismutase), laccase and cytochrome C oxidase, and its deficiency or excess can significantly adversely impact plant growth and yield. Several miRNAs, including miR397, miR408 and miR857, have been reported to have crucial roles in maintaining Cu homeostasis by regulating the expression of various proteins and genes involved in this process (Abdel-Ghany and Pilon, 2008). miR408 targets Cu-containing proteins of the phytocyanin family, including *cupredoxin*, *PLC*, *LAC* and *UCL* and is suggested to have a significant contribution to the regulation of Cu homeostasis. In *Arabidopsis*, the expression of miR408 was upregulated, whereas the target genes were downregulated under Cu-deficiency conditions (Perea-García *et al.*, 2021). In *Arabidopsis*, miR398 and miR156 expression was upregulated under severe Cu starvation. Simultaneously, the expression of their respective target genes – *CSD1/2* (*Cu/Zn Superoxide Dismutase*) and *SPL3* – exhibited downexpression (Sun *et al.*, 2020; Perea-García *et al.*, 2021).

miR398 and *CSD* also exhibited distinct expression patterns under Zn deficiency. Zn is another micronutrient that influences plant growth by serving as a cofactor for various enzymes and participating in ion transport and metabolic and physiological processes (Hamzah Saleem *et al.*, 2022). In *Sorghum bicolor*, Zn deficiency led to the upregulation of several miRNAs, including miR166, miR171, miR172, miR319, miR398, miR408 and miR399 (Li *et al.*, 2013). miRNA transcriptomic analysis study of *Brassica juncea* roots and rice seedlings exposed to Zn starvation revealed differential expression of 101 and 68 miRNAs, respectively, compared with the control, including miR399, miR845, miR171, miR397 and miR398 (Shi *et al.*, 2013; Zhang *et al.*, 2019).

In plants, B participates in various processes, including lipid, protein and nucleic acid metabolism, cell division and cell wall development and maintenance. In *Citrus sinensis*, 91 and 81 miRNAs were upregulated and downregulated under B-deficiency conditions, respectively (Lu *et al.*, 2015). miR159, miR393, miR782, miR3946, miR7539 and miR160 were upregulated and elucidated an adaptation response by targeting NAC and auxin signalling-associated transcription factors. Conversely, miR408, miR5037, miR164, miR3446, miR3946, miR6260, miR5929 and miR6214 exhibited downregulation and activated various stress and ROS responses (Lu *et al.*, 2015).

In a miRNA profiling study of *Citrus sinensis* and *C. grandis* treated with B, it was uncovered that miR397 plays a crucial role in developing B tolerance by influencing secondary cell wall biosynthesis, targeting the *LAC4* and *LAC17* genes (Huang *et al.*, 2016). Further, it was revealed that in the root of citrus, miR319, miR171 and miR396 that target *MYB*, *SCARECROW-like* and an *ATPase* gene, respectively, were also involved in developing B tolerance by modulating root architecture (Huang *et al.*, 2019). In *Arabidopsis*, under mild B toxicity,

miRNAs associated with jasmonic acid and ethylene signalling, including miR172 and miR319, exhibited increased expression, shedding light on the intricate regulatory network involved in the plant's response to B stress (Kayihan *et al.*, 2019).

Mg serves as a crucial cofactor for over 300 enzymes, encompassing kinases, phosphatases, ATPases, DNA polymerases and ribulose-1,5-bisphosphate carboxylase, and allosterically modulates various processes, including respiration, photosynthesis and metabolism. The deficiency of Mg in citrus crops leads to a pronounced loss in yield and quality. miRNA transcriptome screening of leaves and roots of *C. sinesis* exposed to Mg deficiency identified 146 and 170 differentially expressed miRNA, respectively (Ma *et al.*, 2016; Liang *et al.*, 2017). In these investigations, it was uncovered that miR164, miR3946, miR5158, miR7812 and miR5742 were downregulated under Mg starvation in leaves and resulted in upregulation of nutrient-stress responsive genes (Liang *et al.*, 2017). The differential expression of miR395, miR1077, miR3946, miR5158, miR1044, miR1077, miR1160, miR6190, miR5029 and miR3437 influenced cell transport in both leaves and roots while miR158, miR5261 and miR6485 maintained root growth (Ma *et al.*, 2016; Liang *et al.*, 2017).

15.4 Role of Small RNAs in Plant Growth

Plant growth is a multifaceted process contingent upon various internal factors, encompassing genetics and phytohormonal signalling, and external influencers, including light, temperature, nutritional availability and biotic and abiotic stress. Throughout plants' growth phases, a rapid progression of cell division, differentiation and elongation occurs, constituting an intricately regulated process. The region of the plant where cells rapidly divide is called meristem and is involved in plant primary growth. The apical meristem, situated at the tips of the plant, is responsible for the growth of both the shoot and root (Aichinger *et al.*, 2012). Small RNAs (sRNAs) are key in orchestrating plant growth by finely governing cellular and molecular processes such as cell division and elongation, hormonal signalling and the activation of genes associated with development. This discussion delves into the intricate details of the role these sRNAs play in plant growth by regulating the meristematic region's development.

15.4.1 Shoot apical meristem

A small RNA sequencing study of the woody plant *Populus tomentosa* has unveiled the participation of 193 functionally known miRNAs. These miRNAs exert a profound influence on the growth and development of the shoot apical meristem (SAM) primarily through the regulation of phytohormone signalling genes, developmental genes and genes associated with cellular processes (Cui *et al.*, 2019). SAM's development, maintenance and conversion to flower meristem are regulated mainly by the *STM–WUS–CLV* (*Shoot Meristemless–Wushel–Clavata*) pathway (Gaillochet and Lohmann, 2015). The expression of these three crucial genes is intricately regulated by a multitude of other genes and small RNAs, particularly miRNAs. Within the organizing centre (OC), the regulation of *WUS* expression is orchestrated through the miR394–*LCR* (*Leaf Curling Responsiveness*) module. miR394 is produced in the L1 layer and diffuses to the OC, suppressing *LCR* activity (in the L3 layer). *LCR*, in turn, acts as a suppressor of *WUS* activity within the OC (Knauer *et al.*, 2013).

HD-ZIP III transcription factors, including *PHB* (*Phabulosa*), *PHV* (*Phavoluta*), *REV* (*Revoluta*) and *CNA* (*Corona*), play a crucial role in the maintenance of the SAM (Byrne, 2006). These HD-ZIP III transcription factors are regulated by miR165/166 (Yadav *et al.*, 2021). Moreover, *AGO1* and *AGO10* antagonistically regulate the expression of miR165/166, where *AGO1* promotes miR165/166 activity, whereas *AGO10* diminishes the miRNA activity (Zhu *et al.*, 2011). Noteworthy is the existence of a positive feedback loop between the HD-ZIP III transcription factor REV and *AGO10*, with REV promoting the expression of *AGO10* (Brandt *et al.*, 2013). Interestingly, in the abaxial side of leaves, expression of miR166/165 is downregulated by miR390-*AGO7-Tas3* tasiRNA (Cheng *et al.*, 2021).

When one looks at miR164, it targets a total of six genes, including *CUC1* and *CUC2* (*Cup-shaped Cotyledon1* and *2*), which belong to the NAC transcriptional factor family. *CUC1* and *CUC2* play integral roles in the establishment of shoot meristem, determining organ boundaries and regulating leaf serration (Nikovics *et al.*, 2006; Wang *et al.*, 2021). Notably, it was reported that miR171, which targets *HAM* (*Hairy Meristem*), plays a significant role in the development of SAM in plants, including *Arabidopsis*, tomato, rice and barley (Han and Zhou, 2022). Overexpression of miR171 in *Arabidopsis*, rice and tomato resulted in altered expression of meristem regulators, including *CLV1*, *CLV3* and *WOX4*. Consequently, these plants exhibited defects in SAM growth and development, along with reduced shoot branching and seedling growth (Han and Zhou, 2022).

15.4.2 Root growth and development

The longitudinal growth of roots is intricately reliant on maintaining a delicate equilibrium between rapid cell division and the subsequent processes of elongation and differentiation occurring within the stem cells present in the root apical meristem (RAM). Within the RAM, these actively dividing stem cells are situated in the stem cell niche (SCN), positioned at the distal end of the root. These self-renewing cells undergo asymmetrical divisions, giving rise to transit amplifying cells (TACs). The TACs, in turn, undergo symmetrical divisions for a predetermined number of cycles before eventually ceasing division, subsequently elongating and differentiating into specialized cell types (Benfey and Scheres, 2000; Ivanov and Dubrovsky, 2013). The orchestration of root growth and development involves regulating various associated transcription factors and genes by several sRNAs. These sRNAs exert their influence by negatively modulating the expression of key molecular components. Notably, these sRNAs can traverse cell boundaries through plasmodesmata, facilitating their movement across cells (Skopelitis *et al.*, 2017).

The phytohormone auxin serves as a promoter of root growth and development, with miRNAs associated with auxin signalling, including miR160, miR164, miR167, miR390 and miR393, playing pivotal roles in these processes. In *Arabidopsis*, miR160 overexpression and double loss-of-function mutation resulted in reduced root growth, malformed root cap formation and uncontrolled cell division and elongation (Wang *et al.*, 2005). miR160 targets three auxin-responsive factors, including *ARF10, ARF16* and *ARF17* (Mallory *et al.*, 2005; Wang *et al.*, 2005).

Furthermore, other auxin signalling-associated miRNAs, including miR164 and miR167, which target *NAM/CUC1/CUC2* and *ARF6/ARF8*, respectively, also play a significant role in regulating root growth (Guo *et al.*, 2005; Gutierrez *et al.*, 2009; Arora *et al.*, 2019). miR167 influences adventitious root development, and the diminution of miR167 activity leads to increased root production (Gutierrez *et al.*, 2009; Arora *et al.*, 2020). On the other hand, miR164 regulates lateral root and branching (Guo *et al.*, 2005; Gutierrez *et al.*, 2009).

The auxin receptors facilitate auxin transport amongst plant cells, primarily *TIR1* (*Transport Inhibitor Response I*) and *AFB1/2/3* (*Auxin signalling F-box 1/2/3*). The expression of these receptors is regulated by miR393 (Jiang *et al.*, 2022). Increased miR393 expression increased rice's primary root length and lateral root development (Bian *et al.*, 2012). Additionally, in *Arabidopsis* and *Medicago truncatula*, miR390 was reported to play an important role in lateral root development by regulating the expression of *AR2/3/4* via tasiRNAs (Marin *et al.*, 2010; Hobecker *et al.*, 2017).

The transcriptional factors PHB and PHV, targets of miR165/166, play a crucial role in root growth and development by regulating cell differentiation and maintaining cytokinin homeostasis (Bertolotti *et al.*, 2021). In *Arabidopsis*, miR165/166-insensitive PHB and PHV mutants exhibited an elevated rate of cell differentiation, resulting in reduced root length and smaller RAM size (Dello Ioio *et al.*, 2012). Furthermore, in *Arabidopsis*, miR396 has been identified as a regulator of the SCN and stem cell

conversion to TACs by regulating the expression of transcription factor growth regulating factor (GRF) (Rodriguez *et al.*, 2015). Amongst the variants of miR156, which regulates the expression of the SPL gene family, miR156a and miR156c are reported to participate in the regulation of RAM. In *Arabidopsis*, the expression of miR156-resistant *SPL* gene or diminution of miR156 resulted in increased RAM size, leading to longer roots (Barrera-Rojas *et al.*, 2019).

Thus, these findings highlight the importance of small RNAs, especially miRNAs, in developing the apical meristem, nutrient uptake and homeostasis, influencing proper growth and development.

15.5   Conclusion

Small RNAs significantly impact plant growth and development as they regulate gene expression at post-transcription or translational levels. miRNAs are small RNA molecules that influence all aspects of plant growth and development and response to biotic and abiotic stress, including nutrient starvation. Numerous high-throughput sequencing studies have identified various miRNAs that have a crucial role in the adaptation of plants during nutrient-deficient conditions by maintaining nutrient homeostasis. Identifying miRNAs in independent studies under several nutrient deficiency conditions has revealed that specific miRNAs are expressed under every nutrient stress and regulate general functions such as hormone signalling, cell division, cell elongation and ROS scavengers. However, some miRNAs are differentially expressed under a specific element deficiency and regulate the protein expression required for specific element assimilation and transport. To ensure food security and quality, it is crucial to understand the involvement of miRNAs during nutritional stress responses and to employ them to create crop varieties exhibiting higher yield and quality in nutritionally deprived soil.

References

Abdel-Ghany, S.E. and Pilon, M. (2008) MicroRNA-mediated systemic down-regulation of copper protein expression in response to low copper availability in *Arabidopsis*. *Journal of Biological Chemistry* 283, 15932–15945.

Aichinger, E., Kornet, N., Friedrich, T. and Laux, T. (2012) Plant stem cell niches. *Annual Review of Plant Biology* 63, 615–636.

Alejandro, S., Höller, S., Meier, B. and Peiter, E. (2020) Manganese in plants: from acquisition to subcellular allocation. *Frontiers in Plant Science* 11, 300.

Allen, E., Xie, Z., Gustafson, A.M., Sung, G.-H., Spatafora, J.W. *et al.* (2004) Evolution of microRNA genes by inverted duplication of target gene sequences in *Arabidopsis thaliana*. *Nature Genetics* 36, 1282–1290.

Andrews, M., Raven, J.A. and Lea, P.J. (2013) Do plants need nitrate? The mechanisms by which nitrogen form affects plants. *Annals of Applied Biology* 163, 174–199.

Ariel, F., Romero-Barrios, N., Jégu, T., Benhamed, M. and Crespi, M. (2015) Battles and hijacks: noncoding transcription in plants. *Trends in Plant Science* 20, 362–371.

Arora, S., Pandey, D.K. and Chaudhary, B. (2019) Target-mimicry based diminution of miRNA167 reinforced flowering-time phenotypes in tobacco *via* spatial-transcriptional biases of flowering-associated miRNAs. *Gene* 682, 67–80.

Arora, S., Singh, A.K. and Chaudhary, B. (2020) Target-mimicry based miRNA167-diminution ameliorates cotton somatic embryogenesis *via* transcriptional biases of auxin signaling associated miRNAs and genes. *Plant Cell, Tissue and Organ Culture* 141, 511–531.

Baek, D., Hong, S., Kim, H.J., Moon, S., Jung, K.H. *et al.* (2024) OsMYB58 negatively regulates plant growth and development by regulating phosphate homeostasis. *International Journal of Molecular Sciences* 25, 2209.

Bakirbas, A., Castro-Rodriguez, R. and Walker, E.L. (2023) The small RNA component of *Arabidopsis thaliana* phloem sap and its response to iron deficiency. *Plants* 12, 2782.

Barrera-Rojas, C.H., Rocha, G.H.B., Polverari, L., Pinheiro Brito, D.A., Batista, D.S. *et al.* (2019) MiR156-targeted SPL10 controls *Arabidopsis* root meristem activity and root-derived de novo shoot regeneration via cytokinin responses. *Journal of Experimental Botany* 71, 934–950.

Bartel, D.P. (2004) MicroRNAs: genomics, biogenesis, mechanism, and function. *Cell* 116, 281–297.

Baumberger, N. and Baulcombe, D.C. (2005) *Arabidopsis* ARGONAUTE1 is an RNA slicer that selectively recruits microRNAs and short interfering RNAs. *Proceedings of the National Academy of Sciences USA* 102, 11928–11933.

Benfey, P.N. and Scheres, B. (2000) Root development. *Current Biology* 10, R813–5.

Bernstein, E., Caudy, A.A., Hammond, S.M. and Hannon, G.J. (2001) Role for a bidentate ribonuclease in the initiation step of RNA interference. *Nature* 409, 363–366.

Bertolotti, G., Scintu, D. and Dello Ioio, R. (2021) A small cog in a large wheel: crucial role of miRNAs in root apical meristem patterning. *Journal of Experimental Botany* 72, 6755–6767.

Bian, H., Xie, Y., Guo, F., Han, N., Ma, S. *et al.* (2012) Distinctive expression patterns and roles of the miRNA393/TIR1 homolog module in regulating flag leaf inclination and primary and crown root growth in rice (*Oryza sativa*). *New Phytologist* 196, 149–161.

Brandt, R., Xie, Y., Musielak, T., Graeff, M. and Stierhof, Y.-D. (2013) Control of stem cell homeostasis via interlocking microRNA and microProtein feedback loops. *Mechanisms of Development* 130, 25–33.

Byrne, M.E. (2006) Shoot meristem function and leaf polarity: the role of class III HD-ZIP genes. *PLoS Genetics* 2, e89.

Cech, T.R. and Steitz, J.A. (2014) The noncoding RNA revolution – trashing old rules to forge new ones. *Cell* 157, 77–94.

Chen, X. (2005) MicroRNA biogenesis and function in plants. *FEBS Letters* 579, 5923–5931.

Cheng, Y.-J., Shang, G.-D., Xu, Z.-G., Yu, S., Wu, L.-Y. *et al.* (2021) Cell division in the shoot apical meristem is a trigger for miR156 decline and vegetative phase transition in *Arabidopsis*. *Proceedings of the National Academy of Sciences USA* 118, e2115667118.

Cui, J., Lu, W., Lu, Z., Ren, S. and Zhao, B. (2019) Identification and analysis of microRNAs in the SAM and leaves of *Populus tomentosa*. *Forests* 10, 130.

Dello Ioio, R., Galinha, C., Fletcher, A.G., Grigg, S.P., Molnar, A. *et al.* (2012) A PHABULOSA/cytokinin feedback loop controls root growth in *Arabidopsis*. *Current Biology* 22, 1699–1704.

Du, Q., Lv, W., Guo, Y., Yang, J., Wang, S. *et al.* (2022) MIR164b represses iron uptake by regulating the *NAC domain transcription factor5-Nuclear Factor Y, Subunit A8* module in *Arabidopsis*. *Plant Physiology* 189, 1095–1109.

Eddy, S.R. (2001) Non-coding RNA genes and the modern RNA world. *Nature Reviews Genetics* 2, 919–929.

Gaillochet, C. and Lohmann, J.U. (2015) The never-ending story: from pluripotency to plant developmental plasticity. *Development* 142, 2237–2249.

Gao, N., Su, Y., Min, J., Shen, W. and Shi, W. (2010) Transgenic tomato overexpressing ath-miR399d has enhanced phosphorus accumulation through increased acid phosphatase and proton secretion as well as phosphate transporters. *Plant and Soil* 334, 123–136.

Gao, S., Yang, Y., Yang, Y., Zhang, X., Su, Y. *et al.* (2022) Identification of low-nitrogen-related miRNAs and their target genes in sugarcane and the role of *miR156* in nitrogen assimilation. *International Journal of Molecular Sciences* 23, 13187.

Gifford, M.L., Dean, A., Gutierrez, R.A., Coruzzi, G.M. and Birnbaum, K.D. (2008) Cell-specific nitrogen responses mediate developmental plasticity. *Proceedings of the National Academy of Sciences USA* 105, 803–808.

Grabowska, A., Smoczynska, A., Bielewicz, D., Pacak, A., Jarmolowski, A. *et al.* (2020) Barley microRNAs as metabolic sensors for soil nitrogen availability. *Plant Science* 299, 110608.

Guleria, P., Mahajan, M., Bhardwaj, J. and Yadav, S.K. (2011) Plant small RNAs: biogenesis, mode of action and their roles in abiotic stresses. *Genomics, Proteomics and Bioinformatics* 9, 183–199.

Guo, H.-S., Xie, Q., Fei, J.-F. and Chua, N.-H. (2005) MicroRNA directs mRNA cleavage of the transcription factor NAC1 to downregulate auxin signals for *Arabidopsis* lateral root development. *Plant Cell* 17, 1376–1386.

Gutierrez, L., Bussell, J.D., Păcurar, D.I., Schwambach, J. and Păcurar, M. (2009) Phenotypic plasticity of adventitious rooting in *Arabidopsis* is controlled by complex regulation of AUXIN RESPONSE FACTOR transcripts and microRNA abundance. *Plant Cell* 21, 3119–3132.

Hamzah Saleem, M., Usman, K., Rizwan, M., Al Jabri, H. and Alsafran, M. (2022) Functions and strategies for enhancing zinc availability in plants for sustainable agriculture. *Frontiers in Plant Science* 13, 1033092.

Han, H. and Zhou, Y. (2022) Function and regulation of microRNA171 in plant stem cell homeostasis and developmental programing. *International Journal of Molecular Sciences* 23, 2544.

He, H., Liang, G., Li, Y., Wang, F. and Yu, D. (2014) Two young microRNAs originating from target duplication mediate nitrogen starvation adaptation via regulation of glucosinolate synthesis in *Arabidopsis thaliana*. *Plant Physiology* 164, 853–865.

Hobecker, K.V., Reynoso, M.A., Bustos-Sanmamed, P., Wen, J., Mysore, K.S. *et al.* (2017) The microRNA390/TAS3 pathway mediates symbiotic nodulation and lateral root growth. *Plant Physiology* 174, 2469–2486.

Huang, J.-H., Qi, Y.-P., Wen, S.-X., Guo, P., Chen, X.-M. *et al.* (2016) Illumina microRNA profiles reveal the involvement of miR397a in *Citrus* adaptation to long-term boron toxicity via modulating secondary cell-wall biosynthesis. *Scientific Reports* 6, 22900.

Huang, J.-H., Lin, X.-J., Zhang, L.-Y., Wang, X.-D., Fan, G.-C. *et al.* (2019) MicroRNA sequencing revealed citrus adaptation to long-term boron toxicity through modulation of root development by miR319 and miR171. *International Journal of Molecular Sciences* 20, 1422.

Huang, S.Q., Xiang, A.L., Che, L.L., Chen, S. and Li, H. (2010) A set of miRNAs from *Brassica napus* in response to sulphate deficiency and cadmium stress. *Plant Biotechnology Journal* 8, 887–899.

Islam, W., Tauqeer, A., Waheed, A. and Zeng, F. (2022) MicroRNA mediated plant responses to nutrient stress. *International Journal of Molecular Sciences* 23, 2562.

Ivanov, V.B. and Dubrovsky, J.G. (2013) Longitudinal zonation pattern in plant roots: conflicts and solutions. *Trends in Plant Science* 18, 237–243.

Jeong, D.-H., Park, S., Zhai, J., Gurazada, S.G.R., De Paoli, E. *et al.* (2011) Massive analysis of rice small RNAs: mechanistic implications of regulated microRNAs and variants for differential target RNA cleavage. *Plant Cell* 23, 4185–4207.

Jiang, J., Zhu, H., Li, N., Batley, J. and Wang, Y. (2022) The miR393-target module regulates plant development and responses to biotic and abiotic stresses. *International Journal of Molecular Sciences* 23, 9477.

Jin, L.-F., Yarra, R., Yin, X.-X., Liu, Y.-Z. and Cao, H.-X. (2021) Identification and function prediction of iron-deficiency-responsive microRNAs in citrus leaves. *3 Biotech* 11, 121.

Jones-Rhoades, M.W. and Bartel, D.P. (2004) Computational identification of plant microRNAs and their targets, including a stress-induced miRNA. *Molecular Cell* 14, 787–799.

Jung, J.-H., Seo, P.J. and Park, C.-M. (2009) MicroRNA biogenesis and function in higher plants. *Plant Biotechnology Reports* 3, 111–126.

Kamthan, A., Chaudhuri, A., Kamthan, M. and Datta, A. (2015) Small RNAs in plants: recent development and application for crop improvement. *Frontiers in Plant Science* 6, 208.

Kawashima, C.G., Yoshimoto, N., Maruyama-Nakashita, A., Tsuchiya, Y.N., Saito, K. *et al.* (2009) Sulphur starvation induces the expression of microRNA-395 and one of its target genes but in different cell types. *Plant Journal* 57, 313–321.

Kawashima, C.G., Matthewman, C.A., Huang, S., Lee, B., Yoshimoto, N. *et al.* (2011) Interplay of SLIM1 and miR395 in the regulation of sulfate assimilation in *Arabidopsis*. *Plant Journal* 66, 863–876.

Kayihan, D.S., Kayihan, C. and Özden Çiftçi, Y. (2019) Moderate level of toxic boron causes differential regulation of microRNAs related to jasmonate and ethylene metabolisms in *Arabidopsis thaliana*. *Turkish Journal of Botany* 43, 167–172.

Kim, V.N., Han, J. and Siomi, M.C. (2009) Biogenesis of small RNAs in animals. *Nature Reviews Molecular Cell Biology* 10, 126–139.

Knauer, S., Holt, A.L., Rubio-Somoza, I., Tucker, E.J. and Hinze, A. (2013) A protodermal miR394 signal defines a region of stem cell competence in the *Arabidopsis* shoot meristem. *Developmental Cell* 24, 125–132.

Kumar, R.S., Sinha, H., Datta, T., Asif, M.H. and Trivedi, P.K. (2023) MicroRNA408 and its encoded peptide regulate sulfur assimilation and arsenic stress response in *Arabidopsis*. *Plant Physiology* 192, 837–856.

Kurihara, Y. and Watanabe, Y. (2004) Arabidopsis micro-RNA biogenesis through Dicer-like 1 protein functions. *Proceedings of the National Academy of Sciences USA* 101, 12753–12758.

Li, L., Li, Q., Davis, K.E., Patterson, C., Oo, S. *et al.* (2021) Response of root growth and development to nitrogen and potassium deficiency as well as microRNA-mediated mechanism in peanut (*Arachis hypogaea* L.). *Frontiers in Plant Science* 12, 695234.

Li, X., Xia, K., Liang, Z., Chen, K., Gao, C. *et al.* (2016) MicroRNA393 is involved in nitrogen-promoted rice tillering through regulation of auxin signal transduction in axillary buds. *Scientific Reports* 6, 32158.

Li, Y., Zhang, Y., Shi, D., Liu, X., Qin, J. *et al.* (2013) Spatial-temporal analysis of zinc homeostasis reveals the response mechanisms to acute zinc deficiency in *Sorghum bicolor*. *New Phytologist* 200, 1102–1115.

Liang, G., He, H. and Yu, D. (2012) Identification of nitrogen starvation-responsive microRNAs in *Arabidopsis thaliana*. *PLoS One* 7, e48951.

Liang, W.-W., Huang, J.-H., Li, C.-P., Yang, L.-T., Ye, X. *et al.* (2017) MicroRNA-mediated responses to long-term magnesium-deficiency in *Citrus sinensis* roots revealed by Illumina sequencing. *BMC Genomics* 18, 657.

Lin, W.-Y., Huang, T.-K. and Chiou, T.-J. (2013) Nitrogen limitation adaptation, a target of microRNA827, mediates degradation of plasma membrane-localized phosphate transporters to maintain phosphate homeostasis in *Arabidopsis*. *Plant Cell* 25, 4061–4074.

Liu, X., Tan, C., Cheng, X., Zhao, X., Li, T. *et al.* (2020) miR168 targets Argonaute1A mediated miRNAs regulation pathways in response to potassium deficiency stress in tomato. *BMC Plant Biology* 20, 477.

Lu, Y.-B., Qi, Y.-P., Yang, L.-T., Guo, P., Li, Y. *et al.* (2015) Boron-deficiency-responsive microRNAs and their targets in *Citrus sinensis* leaves. *BMC Plant Biology* 15, 271.

Ma, C.-L., Qi, Y.-P., Liang, W.-W., Yang, L.-T., Lu, Y.-B. *et al.* (2016) MicroRNA regulatory mechanisms on *Citrus sinensis* leaves to magnesium-deficiency. *Frontiers in Plant Science* 7, 201.

Mallory, A.C., Bartel, D.P. and Bartel, B. (2005) MicroRNA-directed regulation of *Arabidopsis* AUXIN RESPONSE FACTOR17 is essential for proper development and modulates expression of early auxin response genes. *Plant Cell* 17, 1360–1375.

Marin, E., Jouannet, V., Herz, A., Lokerse, A.S., Weijers, D. *et al.* (2010) miR390, *Arabidopsis* TAS3 tasiRNAs, and their AUXIN RESPONSE FACTOR targets define an autoregulatory network quantitatively regulating lateral root growth. *Plant Cell* 22, 1104–1117.

Narayan, O.P., Kumar, P., Yadav, B., Dua, M. and Johri, A.K. (2023) Sulfur nutrition and its role in plant growth and development. *Plant Signaling and Behavior* 18, 2030082.

Nikovics, K., Blein, T., Peaucelle, A., Ishida, T., Morin, H. *et al.* (2006) The balance between the MIR164A and CUC2 genes controls leaf margin serration in *Arabidopsis*. *Plant Cell* 18, 2929–2945.

Nischal, L., Mohsin, M., Khan, I., Kardam, H. and Wadhwa, A. (2012) Identification and comparative analysis of microRNAs associated with low-N tolerance in rice genotypes. *PLoS One* 7, e50261.

Pant, B.D., Buhtz, A., Kehr, J. and Scheible, W.-R. (2008) MicroRNA399 is a long-distance signal for the regulation of plant phosphate homeostasis. *Plant Journal* 53, 731–738.

Pant, B.D., Musialak-Lange, M., Nuc, P., May, P., Buhtz, A. *et al.* (2009) Identification of nutrient-responsive *Arabidopsis* and rapeseed microRNAs by comprehensive real-time polymerase chain reaction profiling and small RNA sequencing. *Plant Physiology* 150, 1541–1555.

Paul, S., Datta, S.K. and Datta, K. (2015) miRNA regulation of nutrient homeostasis in plants. *Frontiers in Plant Science* 6, 232.

Pauli, A., Rinn, J.L. and Schier, A.F. (2011) Non-coding RNAs as regulators of embryogenesis. *Nature Reviews Genetics* 12, 136–149.

Perea-García, A., Andrés-Bordería, A., Huijser, P. and Peñarrubia, L. (2021) The copper-microRNA pathway is integrated with developmental and environmental stress responses in *Arabidopsis thaliana*. *International Journal of Molecular Sciences* 22, 9547.

Peschansky, V.J. and Wahlestedt, C. (2014) Non-coding RNAs as direct and indirect modulators of epigenetic regulation. *Epigenetics* 9, 3–12.

Rodriguez, R.E., Ercoli, M.F., Debernardi, J.M., Breakfield, N.W., Mecchia, M.A. *et al.* (2015) MicroRNA miR396 regulates the switch between stem cells and transit-amplifying cells in *Arabidopsis* roots. *Plant Cell* 27, 3354–3366.

Seo, J.S., Kim, S.H., Shim, J.S., Um, T., Oh, N. *et al.* (2023) The rice NUCLEAR FACTOR-YA5 and microRNA169a module promotes nitrogen utilization during nitrogen deficiency. *Plant Physiology* 194, 491–510.

Sharma, S., Joon, R., Singh, A.P., Tyagi, D., Kaur, G. *et al.* (2024) MicroRNA Tae-miR1130p targets wheat ferroportin1 (*TaFPN1*) in the absence of iron-responsive element/iron-regulatory protein1 module. *bioRxiv* doi: 10.1101/2024.05.06.592689.

Shi, D., Zhang, Y., Ma, J., Li, Y. and Xu, J. (2013) Identification of zinc deficiency-responsive microRNAs in *Brassica juncea* roots by small RNA sequencing. *Journal of Integrative Agriculture* 12, 2036–2044.

Shriram, V., Kumar, V., Devarumath, R.M., Khare, T.S. and Wani, S.H. (2016) MicroRNAs as potential targets for abiotic stress tolerance in plants. *Frontiers in Plant Science* 7, 817.

Skopelitis, D.S., Benkovics, A.H., Husbands, A.Y. and Timmermans, M.C.P. (2017) Boundary formation through a direct threshold-based readout of mobile small RNA gradients. *Developmental Cell* 43, 265–273.

Subramanian, S., Fu, Y., Sunkar, R., Barbazuk, W.B., Zhu, J.-K. *et al.* (2008) Novel and nodulation-regulated microRNAs in soybean roots. *BMC Genomics* 9, 160.

Sun, Z., Shu, L., Zhang, W. and Wang, Z. (2020) Cca-mir398 increases copper sulfate stress sensitivity via the regulation of *CSD* mRNA transcription levels in transgenic *Arabidopsis thaliana*. *PeerJ* 8, e9105.

Sunkar, R., Li, Y.-F. and Jagadeeswaran, G. (2012) Functions of microRNAs in plant stress responses. *Trends in Plant Science* 17, 196–203.

Valdés-López, O., Yang, S.S., Aparicio-Fabre, R., Graham, P.H., Reyes, J.L. *et al.* (2010) MicroRNA expression profile in common bean (*Phaseolus vulgaris*) under nutrient deficiency stresses and manganese toxicity. *New Phytologist* 187, 805–818.

Voinnet, O. (2009) Origin, biogenesis, and activity of plant microRNAs. *Cell* 136, 669–687.

Waititu, J.K., Zhang, C., Liu, J. and Wang, H. (2020) Plant non-coding RNAs: origin, biogenesis, mode of action and their roles in abiotic stress. *International Journal of Molecular Sciences* 21, 8401.

Wang, C., Huang, W., Ying, Y., Li, S., Secco, D. *et al.* (2012) Functional characterization of the rice SPX-MFS family reveals a key role of OsSPX-MFS1 in controlling phosphate homeostasis in leaves. *New Phytologist* 196, 139–148.

Wang, J., Bao, J., Zhou, B., Li, M., Li, X. *et al.* (2021) The osa-miR164 target OsCUC1 functions redundantly with OsCUC3 in controlling rice meristem/organ boundary specification. *New Phytologist* 229, 1566–1581.

Wang, J.-W., Wang, L.-J., Mao, Y.-B., Cai, W.-J., Xue, H.-W. *et al.* (2005) Control of root cap formation by microRNA-targeted auxin response factors in *Arabidopsis*. *Plant Cell* 17, 2204–2216.

Wang, X., Yuan, D., Liu, Y., Liang, Y., He, J. *et al.* (2023) INDETERMINATE1 autonomously regulates phosphate homeostasis upstream of the miR399-Zmpho2 signaling module in maize. *Plant Cell* 35, 2208–2231.

Wang, X., Zhou, Y., Chai, X., Foster, T.M., Deng, C.H. *et al.* (2024) miR164-MhNAC1 regulates apple root nitrogen uptake under low nitrogen stress. *New Phytologist* 242, 1218–1237.

Xu, Z., Zhong, S., Li, X., Li, W., Rothstein, S.J. *et al.* (2011) Genome-wide identification of microRNAs in response to low nitrate availability in maize leaves and roots. *PLoS One* 6, e28009.

Yadav, A., Kumar, S., Verma, R., Lata, C., Sanyal, I. *et al.* (2021) microRNA 166: an evolutionarily conserved stress biomarker in land plants targeting HD-ZIP family. *Physiology and Molecular Biology of Plants* 27, 2471–2485.

Ye, Z., Zeng, J., Long, L., Ye, L. and Zhang, G. (2021) Identification of microRNAs in response to low potassium stress in the shoots of Tibetan wild barley and cultivated. *Current Plant Biology* 25, 100193.

Yoshikawa, M. (2013) Biogenesis of trans-acting siRNAs, endogenous secondary siRNAs in plants. *Genes and Genetic Systems* 88, 77–84.

Yu, Y., Zhang, Y., Chen, X. and Chen, Y. (2019) Plant noncoding RNAs: hidden players in development and stress responses. *Annual Review of Cell and Developmental Biology* 35, 407–431.

Zeng, H.Q., Zhu, Y.Y., Huang, S.Q. and Yang, Z.M. (2010) Analysis of phosphorus-deficient responsive miRNAs and cis-elements from soybean (*Glycine max* L.). *Journal of Plant Physiology* 167, 1289–1297.

Zhan, J. and Meyers, B.C. (2023) Plant small RNAs: their biogenesis, regulatory roles, and functions. *Annual Review of Plant Biology* 74, 21–51.

Zhang, J., Zhou, Z., Bai, J., Tao, X., Wang, L. *et al.* (2019) Disruption of *MIR396e* and *MIR396f* improves rice yield under nitrogen-deficient conditions. *National Science Review* 7, 102–112.

Zhang, N., Feng, X., Zeng, Q., Lin, H., Wu, Z. *et al.* (2021) Integrated analysis of miRNAs associated with sugarcane responses to low-potassium stress. *Frontiers in Plant Science* 12, 750805.

Zhao, M., Ding, H., Zhu, J.-K., Zhang, F. and Li, W.-X. (2011) Involvement of miR169 in the nitrogen-starvation responses in *Arabidopsis*. *New Phytologist* 190, 906–915.

Zhao, Y., Xu, K., Liu, G., Li, S., Zhao, S. *et al.* (2020) Global identification and characterization of miRNA family members responsive to potassium deprivation in wheat (*Triticum aestivum* L.). *Scientific Reports* 10, 15812.

Zhu, H., Hu, F., Wang, R., Zhou, X., Sze, S.-H. *et al.* (2011) *Arabidopsis* Argonaute10 specifically sequesters miR166/165 to regulate shoot apical meristem development. *Cell* 145, 242–256.

16 Current Appraisal of Microbial Agents Against Phytopathogens for Sustainable Agriculture

Nikena Khwairakpam, Yurembam Rojiv Singh, Amanda Nongthombam, Shantirani Thokchom and Debananda S. Ningthoujam*

Microbial Biotechnology Research Laboratory, Department of Biochemistry, Manipur University, Canchipur, Manipur, India

Abstract

Conventional agriculture is currently far from sustainable, considering the vagaries of crop yield under worsening climate change conditions and different abiotic and biotic stresses. Moreover, intensive and long-term use of traditional agricultural practices causes the depletion of natural resources. Numerous studies indicate that sustainable agriculture is more suitable for managing a growing population's needs without compromising the environment's integrity. Biotic stress caused by phytopathogens, mainly fungal pathogens, adversely affects plant performance by causing plant diseases. This necessitates heavy agrochemical inputs, which negatively impact the ecosystem. Microbes are critical agricultural players and have shown promise as biocontrol agents and plant growth promoters. Plant-associated microorganisms are essential in warding off phytopathogens naturally, thus promoting plant growth and enhancing stress tolerance while sustaining the ecosystem's health. Currently, the most available biocontrol agents are strains of *Pseudomonas*, *Bacillus*, *Streptomyces* and *Trichoderma*. These microbes deploy different modes of action to defend plants, such as antibiotic production, hyperparasitism, niche competition and induced systemic resistance (ISR). Though the benefits of microbes as biocontrol agents against phytopathogens are widely reported, their commercial use is still limited. Several companies are emerging to provide bio-based formulations for crop protection. Companies such as Isagro, AgraQuest, Agrovet and others are introducing microbial products into the commercial market. This chapter overviews microbial agents' modes of action against phytopathogens and highlights major commercially available microbial biobactericides and biofungicides.

Keywords: Antimicrobial compounds, biocontrol agent, microbial biopesticide, sustainable agriculture, phytopathogen

16.1 Introduction

A burgeoning global population and worsening climatic change in the 21st century gravely threaten global food security. Conventional agriculture (CA) uses synthetic fertilizers and pesticides to maximize yield. Prolonged application of synthetic chemicals leads to depletion of soil fertility due to salinization, acidification, nutrient imbalance, metal contamination and loss of soil microbial biodiversity (Ortiz and Sansinenea, 2022). Water bodies also get contaminated due

*Corresponding author: debananda.ningthoujam@gmail.com

© CAB International 2025. *Soil Health and Nutrition Management*
(eds N.C. Joshi, T. Leustek and P.K. Singh)
DOI: 10.1079/9781800624597.0016

to runoff from chemically treated areas, leading to eutrophication and loss of aquatic life (El Gayar, 2021). Using such water sources for drinking causes diseases such as 'blue baby syndrome' caused by nitrate-contaminated water (Knobeloch *et al.*, 2000). The build-up of hazardous residues in soil, water, plants and animals arises from the persistent use of synthetic agrochemicals (Aktar *et al.*, 2009; Ayilara *et al.*, 2023).

The continued consumption of chemicals by farmers challenges the sustainability of the ecosystem and the maintenance of good-quality crop production. Neglecting the harmful effects of CA is not affordable in the long run. Sustainable agriculture (SA) is a newer and more holistic approach, a better alternative to CA. It is a system that integrates the preservation of environmental quality with food production in an economically profitable and more eco-friendly way (Fenibo *et al.*, 2021). SA minimizes synthetic fertilizer and pesticide use, emphasizes biodiversity conservation, utilizes natural interactions to ensure crop health and performance and wards off pests and pathogens (Gomiero *et al.*, 2011; Nicholls and Altieri, 2013). SA offers significant advantages over CA in meeting agricultural demands while cost-effectively

maintaining the environment (Tal, 2018; Fenibo *et al.*, 2022) (Fig. 16.1).

Advancements in science have also provided sustainable alternatives to chemical fertilizers and pesticides, encouraging the practice of SA. Microbes are well recognized by researchers as important agricultural participants. Several bacteria and fungi, both natural and recombinant, have been investigated for their biocontrol and plant growth-promoting potentials and are commercially accessible as biocontrol agents and biofertilizers, respectively. The food system involves several parties, including agricultural producers, distributors who store large amounts of produce and consumers who bear the direct effects from the produce. For successful SA practices, all parties must opt for methods promoting environmental conservation and human well-being. Pesticides constitute the backbone of the agri-food sector, and their use is viewed by many as a barrier to achieving sustainability; the primary concerns arise when looking at the adverse effects on human health and the environment. Hence, alternatives to classic pesticides constitute a significant area of research. Enzymes, siderophores and antimicrobial substances are all examples of microbial products

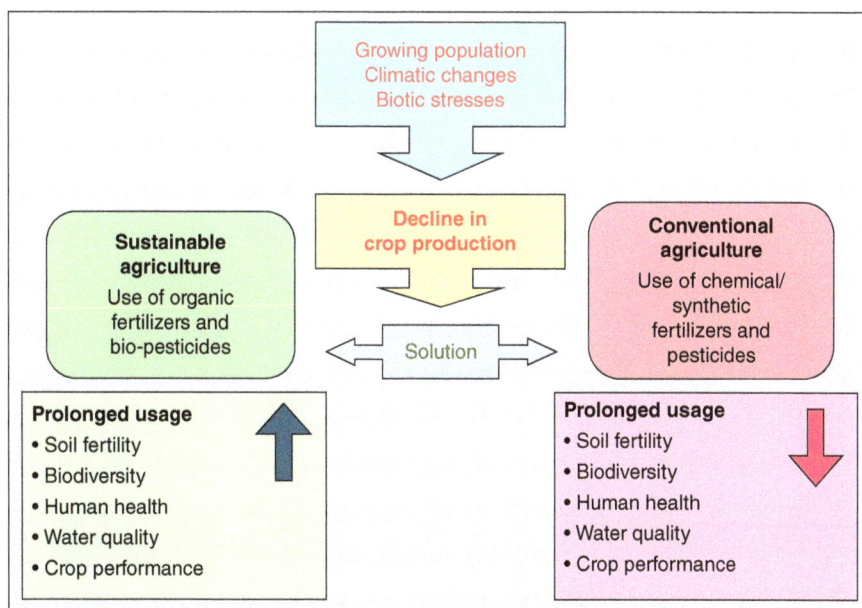

Fig. 16.1. Diagram highlighting sustainable agriculture's significance over conventional agriculture.

that can activate plant defence systems as biopesticides without the common issues of traditional pesticide usage. As a result, microbial agents are suitable alternatives for harmful chemical pesticides.

16.2   Pesticides

Crop health degradation hugely impacts agricultural production, causing pandemics and epidemics culminating in food crises such as the Irish and Bengal famines (Teja *et al.*, 2023). Phytopathogen-induced biotic stressors severely reduce crop output, resulting in a 21–30% reduction in agricultural productivity (El-Saadony *et al.*, 2022). Plant pathogens predominantly include fungi, bacteria, nematodes, protozoa and viruses. Farmers rely on pesticides for disease control, limiting harvestable food loss and preventing postharvest losses by protecting stored grains and crops from infestations and spoilage (Spurgeon *et al.*, 2020; Bajsa *et al.*, 2023). Pesticides are crucial in modern agriculture for ensuring productivity by boosting grain production, regulating crop growth and controlling plant pests and diseases (Aktar *et al.*, 2009; Blair *et al.*, 2015). Pesticides may be classified as fungicides, bactericides, nematicides, herbicides, etc. (Aktar *et al.*, 2009).

16.2.1   Synthetic pesticides

Synthetic pesticides (SPs) can have positive and negative consequences in the agricultural and environmental sectors (Lopes-Ferreira *et al.*, 2022). SPs exhibit quick inhibition of the target pest, making it an efficient way to control the pest population. They have a more extended residual activity, providing greater persistent control under field conditions. While they are typically lethal to their intended target, unintended consequences are also reported (de Gomes *et al.*, 2020; Goçalves and Delabona, 2022). The contamination of SPs in food and drinking water can lead to environmental risks and short-term (e.g. eye and skin irritation, headache, nausea) and long-term (e.g. cancer, asthma, diabetes) health issues. Understanding pesticide risk is

challenging due to various factors such as exposure duration, toxicity and environmental characteristics of the area (Damalas and Eleftherohorinos, 2011; Pathak *et al.*, 2022). Therefore, more severe regulations for SP trade are now in place (Handford *et al.*, 2015).

16.2.2   Biopesticides

Where conventional systems depend on synthetic chemicals to combat disease and pests, sustainable systems use organic and biological products as preventive strategies. The toxicity of traditional pesticides and their decreased efficacy due to the emergence of resistant phytopathogens and the destruction of natural enemies of phytopathogens have encouraged research into safer and more ecologically friendly alternatives such as biopesticides (Vero *et al.*, 2023). Biological products are considered better for the environment and human health, making them a better choice for SA (Paret *et al.*, 2015). Biopesticides derived from plants and microbes are increasingly promoted by researchers (Blair *et al.*, 2015). Biopesticides are mainly of three types: microbial biopesticides (MBs), biochemical biopesticides (BBs) and plant-incorporated protectants (PIPs) (Samada and Tambunan, 2020; Liu *et al.*, 2021; Ram *et al.*, 2021). They represent around 5% of the global pesticide market, with MBs being the predominant type (Fenibo *et al.*, 2021; Pathma *et al.*, 2021). With a 14–15% compound annual growth rate, the usage of biopesticides is estimated to outpace chemical pesticides soon (Marrone, 2014; Chakraborty *et al.*, 2023) (Fig. 16.2).

16.2.3   Microbial biopesticides

The role of MBs in SA is an active area of research. MBs are based on bacteria, fungi, viruses and other organisms (nematodes and protozoa). Several research groups and firms have discovered, developed and commercialized multiple MBs over the last few decades (Marrone, 2014; Ruiu, 2018). In the global and domestic commercial market, bacterial MBs occupy 60% and 29%, fungal MBs 27% and 66%, viral MBs 10% and 4% and other MBs 3% and 1% market share,

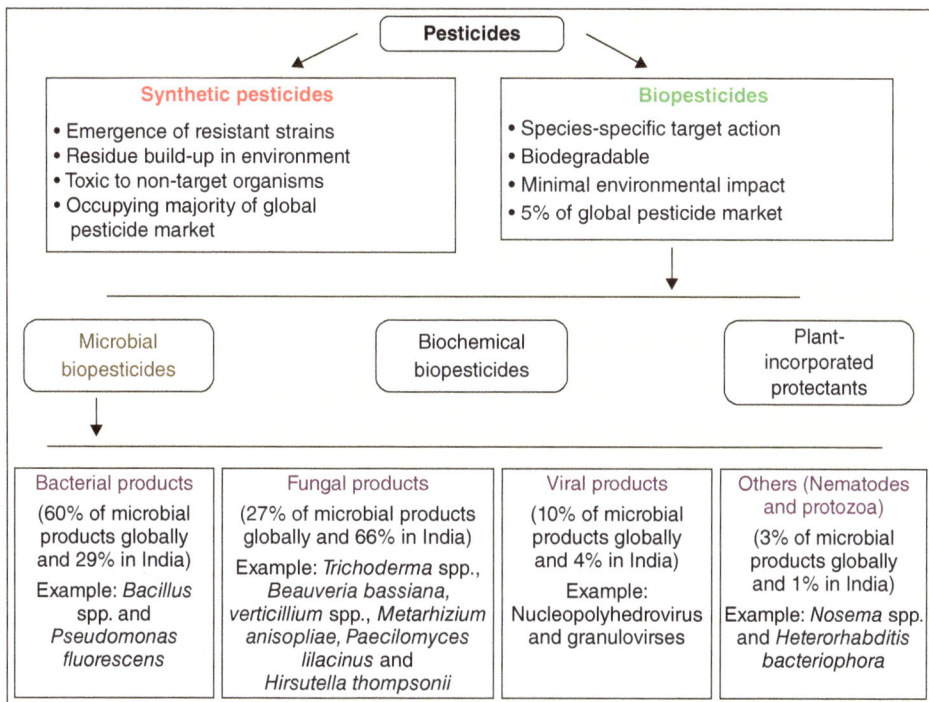

Fig. 16.2. Types of pesticides and microbial biopesticide classification.

respectively (Ashishie and Ashishie, 2018; Chakraborty *et al.*, 2023). The use of MBs to control plant diseases is gaining increasing attention. Compared with traditional chemical pesticides, they develop pest-specific toxins and inhibit pathogenic microbes through antagonism or other non-toxic mechanisms of action (Seenivasagan and Babalola, 2021). The microbial antagonists used in such biopesticides can be produced on an industrial scale by large-scale fermentation. Since they are typically self-sustaining for a short period, the frequency of application is as per requirements (Glare and O'Callaghan, 2019; Vero *et al.*, 2023). MBs are species-specific and non-pathogenic to non-target microorganisms. They can be natural or genetically modified strains. Commonly applied MBs are non-toxic and effective in small quantities. MBs pose a lower risk of resistance development, are user-friendly and have minimal environmental impact (Kumari *et al.*, 2014; Thakur *et al.*, 2020). Increasing research towards the wider use of MBs in SA is urgently warranted.

16.3 Biocontrol Activities of Microbes

Plants have evolved complex defence systems to combat phytopathogens. Successful pathogens, on the other hand, have evolved strategies to circumvent host defences (Sun *et al.*, 2023). Research into novel plant disease management strategies, particularly biological control, is an active area of research relevant to SA (Barratt *et al.*, 2018). Deployment of MBs in agriculture is an effective and eco-friendly method of suppressing phytopathogens. Microbial products classified as biocontrol agents (BCAs) are natural enemies of phytopathogens and antagonists of plant pathogens or other organisms mediating disease control. The traditional biocontrol strategies of MBs involve using microorganisms as active ingredients formulated and released for the biocontrol of phytopathogens, primarily bacterial and fungal pathogens. A BCA's mode of action includes antagonizing pathogens, causing niche competition and boosting plant health.

Therefore, a microbial biocontrol agent (MBCA) is defined similarly to probiotics as microorganisms that confer health benefits to the host plant when administered adequately (Hill *et al.*, 2014; Legein *et al.*, 2020). MBCAs help control phytopathogens via different modes of action, which may be grouped into direct and indirect mechanisms (Heimpel and Mills, 2017; Hashemi *et al.*, 2022). The direct mechanisms include antibiosis and hyperparasitism, while the indirect modes include competition and induced systemic resistance (Köhl *et al.*, 2019; Stenberg *et al.*, 2021) (Fig. 16.3).

16.3.1   Direct biocontrol activity

Direct modes involve interactions between the BCA and the pathogen. The two most common modes of direct biocontrol are:

1. Antibiosis. This refers to an antagonistic interaction between two organisms. The MBCA directly suppresses the pathogen by producing antibiotics. Antibiotics are produced and released in the environment in small quantities as antimicrobial chemicals, toxins and volatile compounds to inhibit or kill plant pathogens (Upmanyu and Malviya, 2020). About 60% of all reported bioactive molecules are naturally derived from microorganisms (Pham *et al.*, 2019). The major contributors of bioactive molecules are actinobacteria, endophytic fungi and other bacteria (Hutchings *et al.*, 2019). The number of antitbiotics is still increasing as new ones are discovered or an old strain is found in a new niche (Bérdy, 2005).

2. Hyperparasitism. This occurs when the MBCA becomes a facultative hyperparasite, that is, a parasite whose host is a phytopathogen. Hyperparasites directly infect and destroy the phytopathogen through enzymes degrading the cell walls of pathogens, especially fungi. Examples of such fungal lytic enzymes include proteases, cellulases and chitinases (Someya *et al.*, 2007).

16.3.2   Indirect biocontrol activity

The indirect mode of action involves an indirect interaction between the MBCA and the host, leading to improved host fitness and resistance

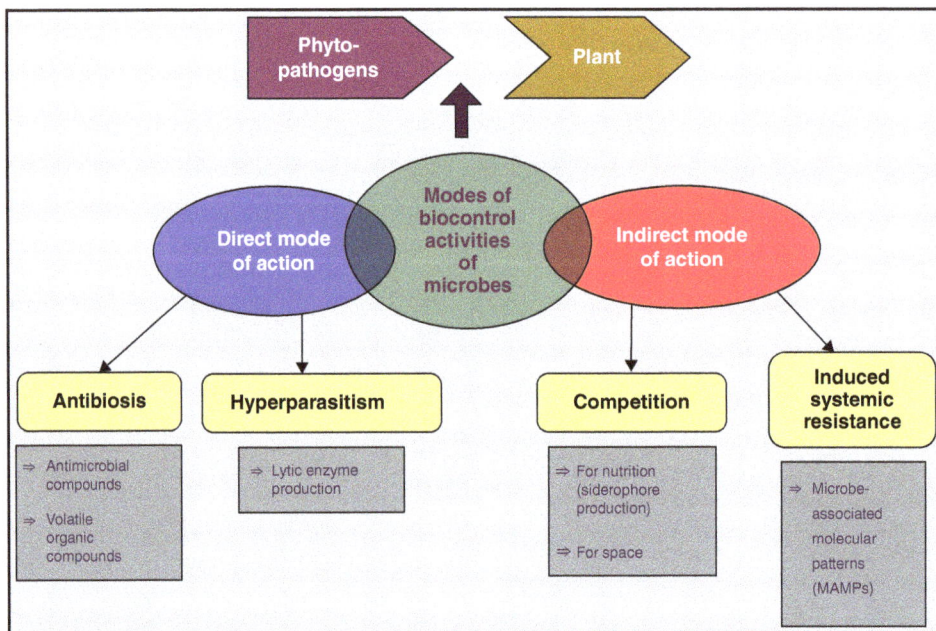

Fig. 16.3. Mode of action of microbial biocontrol agents.

against disease. The two frequently observed indirect biocontrol mechanisms are:

1. Competition. This refers to competition for essential nutrients and space, which inhibits the phytopathogen. Together with soil edaphic factors that hinder disease-causing bacteria and fungi, disease suppression by mixed communities of microbes or other organisms in the soil may also occur (e.g. through the effect of suppressive soils). Potential competitive microbes indirectly interact with such pathogens by rapidly consuming nutrient sources such as sugars, pollen and plant exudates on the plant surface, which are essential for pathogen infection. Thus, outcompeted pathogens cannot infect the host plant (Whipps, 2001; Bais *et al.*, 2006).

2. Eliciting defence pathways in the host by beneficial microbes. Plant-beneficial microbes activate a defence mechanism called induced systemic resistance (ISR), similar to pathogen-induced systemic acquired resistance (SAR). Triggering ISR in the host is done by conserved microbe-specific molecules called microbe-associated molecular patterns (MAMPs), which are recognized by the plant's pattern recognition receptors (PRRs). Microorganisms present MAMPs such as flagellin, elongation factor Tu, peptidoglycan, siderophores, lipopolysaccharides, β-glucans, fungal chitin and salicylic acid to the PRRs, thus triggering ISR in the host (Newman *et al.*, 2013).

16.3.3 Mode of action of microbial biocontrol agents

Antibiosis is the most common direct mechanism mentioned above. Different species of MBCAs within a genus (e.g., *Trichoderma*) may operate through various mechanisms, and, often, one MBCA may suppress the phytopathogen through two or more mechanisms. Understanding these modes of action is critical for successful disease management by helping select the optimal strains, thus substantially increasing their efficacy and consistency in the field. The method of action is critical for characterizing potential dangers to individuals or the environment and determining the risks of developing resistance against the MBCA. Plant growth-promoting bacteria (PGPB) may also indirectly counteract the deleterious

effects of phytopathogens. They may strengthen the host's resistance against the pathogen (Pieterse *et al.*, 2014; Conrath *et al.*, 2015). For example, the microbial enzyme 1-aminocyclopropane-1-carboxylate (ACC) deaminase can degrade the ethylene precursor ACC, thereby enhancing plant tolerance to stressors such as drought, flood, salinity or pathogen attack (Nascimento *et al.*, 2018; Saghafi *et al.*, 2020). Microbes can also influence the levels of phytohormones involved in plant development, for example auxins and cytokinins (Leach *et al.*, 2017).

A newer area of study is quorum sensing (QS). QS is a process dependent on cell density that regulates a wide range of bacterial behaviours (Fetzner, 2015). QS enables bacteria to change their behaviour once a certain concentration threshold of signalling molecules is reached. Non-pathogenic bacteria may also use the same signalling molecules as pathogens and can, thus, contribute to disease progression or inhibition, depending on how they interfere (Legein *et al.*, 2020). Several types of small peptides secreted by bacteria have dual roles. For example, the lantipeptides nisin and subtilin generated by *Lactococcus lactis* and *Bacillus subtilis* are implicated in QS and antibacterial attributes (Kleerebezem, 2004). Quorum quenching (QQ) is the chemical or enzymatic suppression of QS to reverse QS-regulated behaviours (Sikdar and Elias, 2020). QQ can also be implemented for specific bacterial phytopathogen inhibition by beneficial bacteria.

16.4 Microbial Agents Against Bacterial Phytopathogens

Plant-associated microorganisms can be beneficial (PGPBs) or harmful (phytopathogens). Bacteria dwell on all plant surfaces (epiphytes), while some reside inside plants (endophytes). Some are permanent inhabitants, while others are temporary visitors. The commonly occurring bacterial phytopathogens belong to one of five genera: *Ralstonia*, *Agrobacterium*, *Erwinia*, *Pseudomonas* and *Xanthomonas* (Michalak *et al.*, 2022). *Ralstonia* spp., particularly *R. solanacearum*, cause a devastating plant disease known as bacterial wilt in tomatoes and potatoes (Choudhary *et al.*, 2018). Crown gall disease is caused by *Agrobacterium* spp. in various dicotyledonous (broad-leaved) plants (Lai *et al.*, 2006). In apples and pears, fire blight disease is caused by *Erwinia* spp. (Hermann and Stenzel, 2019). *Pseudomonas* pathogens include *P. syringae*, a widespread bacterial pathogen causing several plant diseases, including bacterial speck in tomatoes (Paula Kuyat Mates *et al.*, 2019). *Xanthomonas* is a large genus of bacteria responsible for several severe plant diseases (An *et al.*, 2020). For example, *Xanthomonas* wilt in bananas causes considerable economic loss as this plant is a major cash and food crop (Timilsina *et al.*, 2020).

As bacterial phytopathogens cause significant economic loss in crops and ornamental plants, it is essential to control these pathogens to increase productivity (Mansfield *et al.*, 2012). Research on microbial agents for the biocontrol of various bacterial phytopathogens is increasing as sustainable agricultural practices prefer MBCAs as prospective biological control agents for managing pathogen-specific plant diseases. These agents include various microorganisms that can suppress or inhibit bacterial phytopathogens (Table 16.1).

16.4.1 Mechanisms of biocontrol agents against bacterial phytopathogens

- Viruses that infect and kill specific bacteria are called bacteriophages and can be used to control pathogenic bacteria. Bacteriophages are specific for a particular strain of bacteria. In a recent study, phytopathogenic bacteria *Xanthomonas campestris* pv. *campestris* was reportedly inhibited by a lytic bacteriophage, DSM 1706, of the *Tectiviridae* family. It was isolated from the Karun River water in Iran (Elikaei *et al.*, 2023). Bacterial spot (*X. euvesicatoria* pv. *euvesicatoria*) of pepper is suppressed by the *X euvesicatoria*-specific bacteriophage BsXeu269p/3 (Shopova *et al.*, 2023).

- Certain beneficial bacteria are known to outcompete pathogenic bacteria. Examples include *Bacillus* and *Pseudomonas* strains, which produce antibiotics and other compounds that inhibit phytopathogens (Bonaterra *et al.*, 2022). *Streptomyces* spp. are known to act as biocontrol agents to control various plant diseases. For example, *Streptomyces* sp. AN090126 showed broad-spectrum antagonistic activity against bacterial wilt in tomatoes (*Ralstonia solanacearum*) and leaf spot disease in red pepper (*X. euvesicatoria*). It was isolated from a soil sample in Korea (Le *et al.*, 2022).

- In some cases, a combination of beneficial microorganisms provides broad-spectrum protection against bacterial phytopathogens. These consortia may include antagonistic

Table 16.1. Common microbial biocontrol agents used against bacterial phytopathogens.

Bacterial phytopathogen	Microbial biocontrol agent	Mode of action	Reference
Agrobacterium spp.	*Agrobacterium radiobacter*	Competition	Abd-El-Aziz *et al.*, 2021
	Agrobacterium tumefaciens	Competition	Sharma *et al.*, 2017
	Trichoderma spp.	Mycoparasitism, competition	Tyśkiewicz *et al.*, 2022
Erwinia spp.	*Bacillus* spp.	Competition, antibiosis	Czajkowski *et al.*, 2012
	Bacteriophages	Lytic infection	Gayder *et al.*, 2023
Pseudomonas spp.	*Pseudomonas fluorescens*	Competition, antibiosis	Raaijmakers *et al.*, 2009
	Streptomyces spp.	Antibiosis	Yang *et al.*, 2019
Ralstonia spp.	*Bacillus velezensis*	Antibiosis, competition, induced systemic resistance (ISR)	Dong *et al.*, 2023
	Streptomyces spp.	Antibiosis	Ling *et al.*, 2020
	Trichoderma spp.	Mycoparasitism, ISR	Yendyo *et al.*, 2017
Xanthomonas spp.	Bacteriophages	Lytic infection	Stefani *et al.*, 2021
	Bacillus spp.	Antibiosis	Hernández-Huerta *et al.*, 2023
	Streptomyces spp.	Antibiosis	Gao *et al.*, 2023

bacteria, mycoparasitic fungi and other biocontrol agents. Fungi such as *Trichoderma* spp. are known for parasitizing and killing plant pathogenic bacteria. They can colonize plant roots and act as a protective shield against pathogens. *T. asperellum* foliar treatment was reported to control bacterial spots in tomatoes (*Xanthomonas* sp.) (Chien and Huang, 2020).

- Plant growth-promoting rhizobacteria (PGPR) establish symbiotic relationships with plants, protecting them against pathogens by ISR. They are also reported to defend the host through antibiosis or competition for nutrients and space. Under greenhouse and field conditions, rhizospheric bacterium *Pseudomonas fluorescens* producing 2,4-diacetyl phloroglucinol (DAPG) showed antagonistic activity against *Xanthomonas oryzae* pv. *oryzae*, rice bacterial leaf blight pathogen (Shivalingaiah and Umesha, 2013). Rhizosperic *Pseudomonas* sp. BH25 suppressed *R. solanacearum* and promoted tomato seed germination (Maji and Chakrabartty, 2014).
- Endophytic fungi belonging to the genus *Muscodor* produce volatile organic compounds (VOCs) with bioactivity against phytopathogens in common beans. Another study reported that *Muscodor* spp. suppressed tomato wilt disease (*R. solanacearum*) and enhanced tomato yields (Guimaraes *et al.*, 2023). Two endophytic bacteria, *Enterobacter tabaci* and *Bacillus cereus*, could significantly inhibit *R. solanacearum*, a bacterial pathogen of eggplant (Malek *et al.*, 2023).
- Lytic enzymes (lysins) of various microbes, especially bacteriophages, can degrade the cell walls of pathogenic bacteria. They can control specific bacterial phytopathogens by causing immediate lysis of the target bacterial cell (Fischetti, 2005). A study on the clinically relevant pathogen, *Staphylococcus aureus* and *B. cereus*, showed Lst (lysostaphin) and PlyPH (an enzyme) as promising selective bacteriolytic enzymes (Bhagwat *et al.*, 2019).
- Some bacteria also have QQ ability. They can interfere with the QS communication systems of pathogenic bacteria, disrupting their ability to coordinate virulence. For example, a recombinant N-acyl homoserine lactone lactonase from an Antarctic strain of *Planococcus versutus* blocked the QS signalling pathway of *Pectobacterium carotovorum* subsp. *carotovorum* and protected the crop model (cabbage) (See-Too *et al.*, 2018).
- A relatively new way of formulating microbial agents is using silver nanoparticles. Nanoparticles conjugated with microbes or their antimicrobial products demonstrate biocontrol activity. Silver chloride nanoparticles (AgCl-NPs) incorporating the bacterial strain IMA13 showed activity against *R. solanacearum*, which causes bacterial wilt in tomatoes and potatoes (Abd Alamer *et al.*, 2021).

There are fewer studies of biocontrol activity against bacterial phytopathogens than those against fungal pathogens. Pathogenic bacteria of the same genera may cause different plant diseases. MBCAs against phytopathogens may use various mechanisms, such as inducing resistance or priming of plants, competing for space and nutrients or other methods disrupting growth conditions (Prajapati *et al.*, 2020) (Table 16.2). The choice of an antibacterial agent depends on three crucial factors: the specific pathogen, the type of crop and the environmental conditions. Integrated pest management (IPM) strategies often incorporate microbial bactericidal agents to manage plant diseases effectively (Angon *et al.*, 2023).

16.5 Microbial Agents Against Fungal Phytopathogens

Plant-pathogenic fungi have caused some of the world's worst famines. Fungal phytopathogens account for over 70% of plant diseases (Liu *et al.*, 2017). Fungal phytopathogens predominantly belong to genera such as *Alternaria, Phytophthora, Aspergillus, Botrytis, Cladosporium, Verticillium, Pythium, Fusarium* and *Rhizoctonia* (Dean *et al.*, 2012; Tyśkiewicz *et al.*, 2022; Dobrzyński *et al.*, 2023). *Alternaria* spp. cause leaf spots, rots and blights on over 3800 plants (Misawa and Kurose, 2021). *Phytophthora* spp. include the infamous *P. infestans* that causes late blights in potatoes and tomatoes (Fry *et al.*, 2015).

Table 16.2. Mode of action of microbial biocontrol agents useds against bacterial plant diseases.

Biocontrol agent	Disease	Plant	Causative bacteria	Mode of action	Reference
Agrobacterium radiobacter	Crown gall	Various woody plants	*Agrobacterium tumefaciens*	Competition	Escobar and Dandekar, 2003
Bacillus amyloliquefaciens	Brown rot	Potato	*Ralstonia solanacearum*	Antibiosis, induced systemic resistance (ISR)	Sharma *et al.*, 2021
Bacillus spp.	Angular leaf spot	Cucumber	*Pseudomonas syringae* pv. *lachrymans*	Antibiosis, competition	Abd El-Ghafar, 2000
Bacillus subtilis	Soft rot	Various vegetables	*Pectobacterium carotovorum*	Antibiosis, competition	Abd-El-Khair *et la.*, 2021
Bacillus thuringiensis	Citrus canker	Citrus	*Xanthomonas citri* subsp. *citri*	Antibiosis	Islam *et al.*, 2019
Bacteriophages	Bacterial spot	Tomato, pepper	*Xanthomonas* spp.	Lytic infection	Jones *et al.*, 2007
Burkholderia anthina	Black rot	Cabbage, crucifers	*Xanthomonas campestris* pv. *campestris*	Antibiosis, competition	Ye *et al.*, 2020
Pantoea agglomerans	Stewart's wilt	Maize	*Pantoea stewartii* subsp. *stewartii*	Competition	Triplett *et al.*, 2008
Pantoea agglomerans (E325)	Fire blight	Apple, pear	*Erwinia amylovora*	Competition	Malnoy *et al.*, 2012
Pseudomonas fluorescens	Bacterial leaf blight	Rice	*Xanthomonas oryzae* pv. *oryzae*	Antibiosis, ISR	Vidhyasekaran *et al.*, 2001
Pseudomonas fluorescens	Halo blight	Bean	*Pseudomonas savastanoi* pv. *phaseolicola*	Antibiosis, competition	Dönmez and Aliyeva, 2023
Pseudomonas fluorescens	Bacterial canker	Tomato	*Clavibacter michiganensis* subsp. *michiganensis*	Antibiosis, competition	Boudyach *et al.*, 2001
Pseudomonas spp.	Bacterial speck	Tomato	*Pseudomonas syringae* pv. *tomato*	Antibiosis, competition	Elsharkawy *et al.*, 2023
Streptomyces spp.	Bacterial wilt	Tomato, potato	*Ralstonia solanacearum*	Antibiosis	Ling *et al.*, 2020
Streptomyces spp.	Bacterial ring rot	Potato	*Clavibacter michiganensis* subsp. *sepedonicus*	Antibiosis	Banetashvili, 2019

Aspergillus spp. contaminate bread and potatoes and can produce problematic mycotoxins (Pitt, 2000). *Botrytis cinerea* is the causative agent for grey mould disease affecting many fruits, vegetables and ornamentals (Elad *et al.*, 2007). *Cladosporium* is one of the most common genera of fungi. They are highly phytopathogenic and may develop even in refrigerated conditions. *Cladosporium* spp. may cause spoilage and discoloration (Bullerman, 2003). *C. cladosporioides* causes leaf blight on garden peas (Ragukula and Makandar, 2023). *Verticillium* spp. cause wilts in several plants, such as maple, olive, pepper, redbud, rose and tulip (Fradin and Thomma, 2006).

Pythium spp. are responsible for damping off, a disease that causes seedlings to collapse and die (Agrios, 2005). *Fusarium* spp. are major plant pathogens causing diseases including crown rot, head blight and scab in onions, cabbage, bananas, tomatoes, etc. (Nelson *et al.*, 1994; Arie, 2019). In various crops (alfalfa, corn, grains and tobacco), *Rhizoctonia* spp. cause diseases such as damping-off, root rot and wire stem (Agarwal *et al.*, 2011).

These pathogens cause considerable economic losses in agriculture and horticulture. Effective management strategies are warranted for the control of these pathogens. In agriculture, microbial fungicides can suppress a wide variety of fungal phytopathogens and include bacteria, fungi and viruses (Table 16.3).

16.5.1 Mechanisms of biocontrol agents against fungal phytopathogens

- *Trichoderma* spp. are well-known biocontrol agents against various fungal phytopathogens. They may suppress pathogenic fungi by producing enzymes such as chitinase. For example, increased chitinase production was linked to the antagonistic trait of *T. virens* against *Rhizoctonia solani* (Ghasemi *et al.*, 2020). An endophytic *Trichoderma* sp. showed antagonistic activity against *R. solani*

Table 16.3. Common microbial biocontrol agents used against fungal phytopathogens.

Fungal phytopathogen	Microbial biocontrol agent	Mode of action	Reference
Alternaria spp.	*Trichoderma harzianum*	Mycoparasitism, competition for resources	Metz and Hausladen, 2022
	Bacillus subtilis	Antibiosis	Ongena and Jacques, 2008
Aspergillus spp.	*Gliocladium virens*	Mycoparasitism, antibiosis	Agarwal *et al.*, 2011
	Trichoderma harzianum	Mycoparasitism, competition for nutrients	Kifle *et al.*, 2017
	Streptomyces griseoviridis	Antibiosis	Campos-Avelar *et al.*, 2021
	Bacillus amyloliquefaciens	Antibiosis	Santoso *et al.*, 2021
	Aureobasidium pullulans	Antibiosis, competition	Podgórska-Kryszczuk, 2023
Cladosporium spp.	*Streptomyces griseoviridis*	Antibiosis	Yu *et al.*, 2022
Fusarium spp.	*Baculoviruses*	Lytic infection	Hanlon, 2007
	Trichoderma spp.	Mycoparasitism, competition for nutrients	Harman, 2006
Phytophthora spp.	*Pseudomonas* spp.	Antibiosis, induced systemic resistance (ISR)	Caulier *et al.*, 2018
	Bacillus spp.	Antibiosis	Bhusal and Mmbaga, 2020
	Trichoderma spp.	Mycoparasitism, competition for nutrients	Osorio-Hernández *et al.*, 2011
Pythium spp.	*Pseudomonas fluorescens*	Antibiosis, ISR	Haas and Défago, 2005
Rhizoctonia spp.	*Coniothyrium minitans*	Mycoparasitism	Whipps *et al.*, 2008
	Bacillus spp.	Antibiosis	Lemańczyk *et al.*, 2023.
	Arthrobacter and *Blastobotrys* strains (FP15 and FP12)	Antibiosis, ISR	Aggeli *et al.*, 2020
Verticillium spp.	*Trichoderma atroviride*	Mycoparasitism, competition for nutrients	Shoresh *et al.*, 2010
	Coniothyrium minitans	Mycoparasitism	Whipps *et al.*, 2008

through niche competition. Under green-house conditions, it protected rice plants against sheath blight disease (Chaudhary *et al.*, 2020; Doni *et al.*, 2023).

- *Trichoderma harzianum* associated with the seed-harvester ant, *Trichomyrmex scabriceps*, exhibited plant growth-promoting traits, increased production of defence enzymes such as phenylalanine ammonia-lyase (PAL) and peroxidase (PO) and 93% biocontrol activity against *Sclerotium rolfsii* (Kumari and Rastogi, 2023).
- Endophytic *Bacillus* spp. exhibited antifungal activity against rice fungal pathogens by inhibiting mycelial growth (Khaskheli *et al.*, 2020). Another study reported that endophytic *Streptomyces albus* could suppress fungal pathogens (*Fusarium fujikuroi* and *Scopulariopsis gossypii*) by inhibiting spore germination (Quach *et al.*, 2023).
- *Bacillus amyloliquefaciens* has been shown to control several fungal pathogens by inhibiting mycelial growth and spore germination (Ji *et al.*, 2013).
- *Pseudomonas fluorescens* EPS62e and *P. pseudoalcaligenes* AVO110 can suppress infections by *Rosellinia necatrix* via higher growth potential and nutrient use efficiency than the target pathogen (Cabrefiga *et al.*, 2007; Pliego *et al.*, 2008).
- Some mycoviruses (viruses that infect fungi) can weaken or kill pathogenic fungi. A study on mycoviruses such as SsHADV-1 transformed a phytopathogen, *Sclerotinia sclerotiorum*, into a beneficial symbiont to manage white mould and other crop diseases. The study found that biopriming dry bean seeds with hypovirulent *S. sclerotiorum* strains enhanced host resistance to infection. SsHADV-1-infected fungal strains activated the expression of plant immunity pathway genes in wheat, pea and sunflower (Fu *et al.*, 2023).
- Some commercial biological fungicides contain a combination of beneficial microorganisms, providing a broader spectrum of biocontrol. A mixture of *P. fluorescens* and *T. harzianum* exhibited increased biocontrol against *R. solani* via enhanced PAL, PO and polyphenol oxidase (PPO) levels in the host accompanied by higher plant vigour index (Kabdwal *et al.*, 2023).

Trichoderma and *Bacillus* species have often been used as microbial fungicides to combat plant diseases (Patel and Saraf, 2017; Muhammad *et al.*, 2022). Several other beneficial microorganisms can help effectively manage commonly occurring fungal diseases in plants, reducing the need for chemical pesticides (Table 16.4). These antagonistic microbial strains can be employed individually or in combination in IPM programmes to reduce the impact of fungal pathogens on crops.

16.6   Commercially Used Microbial Agents

The US Environmental Protection Agency (EPA) approved the first bacterial BCA, *Agrobacterium radiobacter* strain K84, in 1979 to ward off crown gall disease. The first fungal BCA, *Trichoderma harzianum* ATCC 20476, was registered by the EPA in 1989. Many MBCAs have been commercialized as microbial biobactericides and microbial biofungicides. Most EPA-registered species (64%) were discovered in the early 21st century, with the remaining 36% discovered during the last few years (Table 16.5) (Junaid *et al.*, 2013; Gehlot and Singh, 2018; Lahlali *et al.*, 2022).

Agricultural and horticultural crops are predominantly affected by fungal rather than bacterial diseases. Therefore, there are fewer commercially available biobactericides in the market. Table 16.6 lists some common commercially available biobactericides.

There is a diversity of formulations of MBCAs in the market. However, it is advisable to determine their modes of action, quality and shelf-lives before they are commercialized. Understanding the applicability of MBCAs under the IPM strategy and the type and timing of application is the key to achieving better outcomes (Kulkarni, 2015). Commercial products advertised as BCAs often confuse farmers and professionals as they do not claim direct bactericide/fungicide effectiveness, do not report colony-forming units (CFUs) or guarantee microorganism viability. These so-called commercial BCAs are often ineffective and reflect poorly on other effective commercially available MBCAs. Stringent regulations are needed to resolve these issues (Pertot *et al.*, 2015). The Central Insecticides Board and Registration Committee

Table 16.4. Mode of action of microbial biocontrol agents used against fungal plant diseases.

Biocontrol agent	Disease	Plant	Causative fungi	Mode of action	Reference
Ampelomyces quisqualis, Bacillus subtilis	Powdery mildew	Various crops	Erysiphe spp.	Mycoparasitism, antibiosis	Kanipriya et al., 2019
Bacillus amyloliquefaciens/ subtilis, Aureobasidium pullulans	Brown rot	Stone fruits	Monilinia spp.	Antibiosis, competition for nutrients	Rungjindamai, et al., 2013
Bacillus spp., Pseudomonas spp.	Brown patch	Turfgrass	Rhizoctonia solani	Antibiosis	Weller et al., 2002
Bacillus spp., Streptomyces spp.	Fusarium wilt	Tomato, banana, others	Fusarium spp.	Antibiosis	Anusha et al., 2019
Bacillus subtilis, Pseudomonas spp.	Downy mildew	Grape, cucumber, others	Peronospora spp.	Antibiosis, induced systemic resistance (ISR)	Elsharkawy et al., 2014
Bacillus subtilis, Pseudomonas spp.	Root rot	Various crops	Pythium spp.	Antibiosis, ISR	Khabbaz et al., 2015
Beauveria bassiana, Metarhizium spp.	Sooty mould	Citrus, other crops	Capnodium spp.	Entomopathogenic activity, reducing insect vectors	Zimmermann, 2008
Beauveria bassiana, Trichoderma spp.	Dutch elm disease	Elm	Ophiostoma ulmi	Mycoparasitism, competition for nutrients	Jacobi, 2001
Coniothyrium minitans	Sclerotinia stem rot (white mould)	Soybean, sunflower, canola, lettuce	Sclerotinia sclerotiorum, S. minor	Mycoparasitism	Bolton et al., 2006
Streptomyces spp., Bacillus spp.	Leaf spot	Various crops	Alternaria spp.	Antibiosis	Singh et al., 2017
Trichoderma harzianum, Clonostachys spp.	Rust	Wheat, coffee, others	Puccinia spp.	Mycoparasitism, competition for nutrients	Druzhinina and Kubicek, 2005
Trichoderma harzianum, Pseudomonas spp.	Tomato blight	Tomato	Phytophthora infestans	Antibiosis, competition for nutrients	Elad and Stewart, 2007
Trichoderma harzianum, Streptomyces spp.	Clubroot	Cruciferous crops	Plasmodiophora brassicae	Mycoparasitism, antibiosis	Li et al., 2020
Trichoderma spp., Bacillus amyloliquefaciens	Grey mould	Various crops	Botrytis cinerea	Mycoparasitism, antibiosis	Batta, 1999
Trichoderma spp.	Rhizoctonia root rot	Soybean, wheat, barley, cotton	Rhizoctonia solani	Mycoparasitism, competition, ISR	Shoresh et al., 2010
Trichoderma spp., Bacillus spp.	Verticillium wilt	Tomato, potato, others	Verticillium spp.	Mycoparasitism, competition for nutrients	Shoresh et al., 2010
Trichoderma spp., Bacillus spp.	Apple scab	Apple	Venturia inaequalis	Antibiosis, competition for nutrients	Jimenez et al., 2018
Trichoderma spp., Gliocladium spp.	Cotton root rot	Cotton	Phymatotrichum omnivorum	Mycoparasitism, competition for nutrients	Guigón-López et al., 2015
Trichoderma spp., Pseudomonas spp.	Black sigatoka	Banana	Mycosphaerella fijiensis	Mycoparasitism, ISR	Marín et al., 2003
Trichoderma spp., Pseudomonas spp.	Wilt disease	Various trees	Ceratocystis spp.	Antibiosis, ISR	Raja et al., 2019
Trichoderma viride, Pseudomonas fluorescens	Late blight	Tomato, potato	Phytophthora infestans	Antibiosis, ISR	Zegeye et al., 2011

Table 16.5. List of selected commercially available microbial biofungicides.

Common/ trade name	Biocontrol agent	Producer company	Target pathogen/disease
Actino-Iron	*Streptomyces lydicus* strain WYEC 108	Monsanto	*Fusarium* spp., *Pythium* spp., *Rhizoctonia* spp., *Phytophthora* spp., *Monosprascus* spp., *Sclerotinia* spp., *Verticillium* spp.
Actinovate	*Streptomyces lydicus*	Monsanto BioAg	Powdery and downy mildew
Amplitude	*Bacillus Amyloliquefaciens* strain F727	ProFarm Group (formerly Marrone Bio Innovations, Inc.)	White mould and downy mildew
AQ10	*Ampelomyces quisqualis*	Intrachem International	Powdery mildew
Asperello	*Trichoderma asperellum* strain T34	Biobest Group	*Fusarium* spp.
Bactvipe	*Pseudomonas fluorescens*	IPL Biologicals	Root rot, stem rot, collar rot, wilt, blights, leaf spot, anthracnose, downy mildew
Bioharz	*Trichoderma harzianum*	IPL Biologicals	Soilborne diseases of paddy
Bio Jodi	*Pseudomonas fluorescence, Bacillus subtilis*	Agriplex (Multiplex)	Rice blast and sheath blight of paddy, root and stem rot of tomato, chili and potato
Bio-Tam	*Trichoderma asperellum, T. gamsii*	Isagro	Soilborne diseases of various fruits, nuts, vegetables
Bio-Trek	*Trichoderma harzianum*	Isagro USA; distributed by Bayer CropScience	Soilborne diseases
Botector	*Aureobasidium pullans* strains DSM 14940 and 14941	Nufarm	Botrytis disease, brown rot
BotryStop	*Urocladium oudemansii* (U3 strain)	BioWorks, Inc.	Botrytis diseases
Cease	*Bacillus subtilis* QST 713	BioWorks, Inc.	Powdery mildew
Companion	*Bacillus subtilis* strain GB03	Growth Products, Ltd.	*Rhizoctonia* spp., *Pythium* spp., *Fusarium* spp., *Phytophthora* spp.
LifeGard	*Bacillus mycoides* isolate J (BmJ)	Certis USA, LLC	Citrus canker, white mould, early blight
Mycostop	*Streptomyces griseoviridis* strain K61	Lallemand Plant Care	Damping off caused by *Alternaria* spp., *Rhizoctonia solani, Fusarium* spp. and *Phytophthora* spp.
Nisarga	*Trichoderma viride*	Agriplex (Multiplex)	Root rot, damping off, fungal wilts
PreFence	*Streptomyces* sp. strain K61	Evergreen Growers Supply	*Fusarium* spp., *Alternaria* spp.
Prestop	*Gliocladium catenulatum* strain J1446	Lallemand Plant Care	Soilborne and foliar diseases
Rhapsody	*Bacillus subtilis* QST 713	Bayer CropScience	Powdery mildew, botrytis disease, anthracnose, leaf spot
RootShield	*Trichoderma harzianum*	BioWorks, Inc.	*Sclerotinia* spp., *Fusarium* spp.
Serenade	*Bacillus subtilis* strain QRD 713	Bayer CropScience	Scab, grey mould, powdery mildew

Continued

Product	Active agent	Company	Target
SoilGard	Gliocladium virens	OHP, Inc.	Pythium spp., Rhizoctonia spp.
Sonata	Bacillus pumilus strain QST 2808	Bayer CropScience	Powdery mildew
Stargus	Bacillus amyloliquefaciens strain F727	ProFarm Group (formerly Marrone Bio Innovations, Inc.)	Fungal diseases of vegetables, small fruits and ornamental crops
Subtilex	Bacillus subtilis strain MBI 600	Seed World	Powdery mildew
Tenet WP	Trichoderma asperellum, T. gamsii	ARBICO Organics	Soilborne diseases of fruits, nuts, vegetables, field, ornamental crops and turf grass
WRC-AP-1	Gliocladium virens strain GL-21	CERTIS	Damping-off and root rot in ornamentals and vegetables
Zen-O-Spore	Ulocladium oudemansii (U3 strain)	BioGro	Botrytis diseases in fruits and nuts
Zio	Pseudomonas chlororaphis strain AFS009	AgBiome	Brown patch and anthracnose

Table 16.6. List of selected commercially available microbial biobactericides.

Common/trade name	Biocontrol agent	Producer company	Target pathogen/disease
Actinovate AG	*Streptomyces lydicus* strain WYEC 108	Novozymes	Soilborne diseases
Ballad Plus	*Bacillus pumilus* strain QST 2808	Bayer CropScience	Bacterial diseases of vegetables and field crops
BlightBan	*Psuedomonas fluoroscens* strain A506	Nufarm	*Erwinia amylovora*
Blossom Protect	*Aureobasidium pullulans*	Bioworks, Inc.	Fire blight
Cease	*Bacillus subtilis* QST 713	BioWorks, Inc.	*Pseudomonas* spp., *Erwinia* spp., *Xanthomonas* spp.
Galltrol	*Agrobacterium radiobacter* strain K84	AgBiochem, Inc.	Crown gall disease
Nogall	*Agrobacterium radiobacter* strain K84	Bayer CropScience	Crown gall disease
Serenade ASO	*Bacillus subtilis* strain QRD 713	Bayer CropScience	*Xanthomonas* spp., *Erwinia* spp.

(CIB&RC) oversees pesticide registration and evaluates product chemistry, bioefficacy and toxicity in India. India's pesticide legislation faces several issues due to loopholes and a lack of enforcement (Handford *et al.*, 2015). Though it is vital to ensure the health and safety of biopesticides, pre-market licensing remains a significant component that slows the innovation process (Ruiu, 2018). Changes in the regulatory framework can facilitate easier registration of commercial MBCAs. Ongoing cooperation between big firms and small agrobiotechnology research centres will enhance the potential of MBCAs to become a high-growth industry in the future.

16.7 Conclusion

The key markets for plant biopesticides include gardens, homes, greenhouses, parks and organic agriculture. MBs are gaining increasing attractiveness over synthetic pesticides. This has led to commercializing several MBCAs against major fungal and bacterial phytopathogens. Commercialization of biopesticides still faces several challenges. Product commercialization,

for example, necessitates the discovery and development of novel microbial strains, the capacity to scale up the production of microbial cells and the appropriate formulation of the product. Furthermore, microbial performance and standardization in the field are issues that must be addressed. Though MBCAs currently occupy only a niche market, there is an acceleration in the research targeting the development of such agencies. For the widespread adoption of MBCAs in agriculture, research aiming at improving efficacy, reliability and innovative delivery systems is crucial. Raising funds or grants for researchers, entrepreneurs, manufacturers and marketers will encourage biopesticide research and production. Regulatory frameworks must also be developed to assure product safety and correct usage. Food safety and better environmental health are consumer demands that impact the biopesticide market. Eco-friendly and improved and cheaper formulations are making MBs more widely used. Therefore, biopesticides may become a more rational alternative to existing choices for phytopathogen management, especially as an improved balance between cost and efficiency is achieved in the near future.

References

Abd Alamer, I.S., Tomah, A.A., Ahmed, T., Li, B. and Zhang, J. (2021) Biosynthesis of silver chloride nanoparticles by rhizospheric bacteria and their antibacterial activity against phytopathogenic bacterium *Ralstonia solanacearum*. *Molecules* 27, 224.

Abd-El-Aziz, R., Abd El-Rahman, A. and Hendi, D. (2021) Control of *Agrobacterium tumefaciens* with essential oils compared to antagonistic *Agrobacterium radiobacter* strain K84. *Egyptian Journal of Phytopathology* 49, 80–92.

Abd El-Ghafar, N.Y. (2000) Biocontrol of bacterial angular leaf spot disease of cucumber. *Annals of Agricultural Science (Cairo)* 4(Special), 1437–1450.

Abd-El-Khair, H., Abdel-Gaied, T.G., Mikhail, M.S., Abdel-Alim, A.I. and El-Nasr, H.I.S. (2021) Biological control of *Pectobacterium carotovorum* subsp. *carotovorum*, the causal agent of bacterial soft rot in vegetables, in vitro and in vivo tests. *Bulletin of the National Research Centre* 45, 37.

Agarwal, T., Malhotra, A., Trivedi, P.P.C. and Biyani, M. (2011) Biocontrol potential of *Gliocladium virens* against fungal pathogens isolated from chickpea, lentil and black gram seeds. *Journal of Agricultural Technology* 7, 1833–1839.

Aggeli, F., Ziogas, I., Gkizi, D., Fragkogeorgi, G.A. and Tjamos, S.E. (2020) Novel biocontrol agents against *Rhizoctonia solani* and *Sclerotinia sclerotiorum* in lettuce. *BioControl* 65, 763–773.

Agrios, G.N. (2005) *Plant Pathology*. Elsevier.

Aktar, W., Sengupta, D. and Chowdhury, A. (2009) Impact of pesticides use in agriculture: their benefits and hazards. *Interdisciplinary Toxicology* 2, 1–12.

An, S.Q., Potnis, N., Dow, M., Vorhölter, F.J., He, Y.Q. *et al.* (2020) Mechanistic insights into host adaptation, virulence and epidemiology of the *Phytopathogen xanthomonas*. *FEMS Microbiology Reviews* 44, 1–32.

Angon, P.B., Mondal, S., Jahan, I., Datto, M., Antu, U.B. *et al.* (2023) Integrated pest management (IPM) in agriculture and its role in maintaining ecological balance and biodiversity. *Advances in Agriculture* 2023, 1–19.

Anusha, B.G., Gopalakrishnan, S., Naik, M.K. and Sharma, M. (2019) Evaluation of *Streptomyces* spp. and *Bacillus* spp. for biocontrol of fusarium wilt in chickpea (*Cicer arietinum* L.). *Archives of Phytopathology and Plant Protection* 52, 417–442.

Arie, T. (2019) *Fusarium* diseases of cultivated plants, control, diagnosis, and molecular and genetic studies. *Journal of Pesticide Science* 44, 275–281.

Ashishie, P.B. and Ashishie, C.A. (2018) Biopesticide, their ecological and toxicological effects (review). *International Journal of Sciences* 4, 21–25.

Ayilara, M.S., Adeleke, B.S. and Babalola, O.O. (2023) Correction to: Bioprospecting and challenges of plant microbiome research for sustainable agriculture, a review on soybean endophytic bacteria. *Microbial Ecology* 86, 1454–1454.

Bais, H.P., Weir, T.L., Perry, L.G., Gilroy, S. and Vivanco, J.M. (2006) The role of root exudates in rhizosphere interactions with plants and other organisms. *Annual Review of Plant Biology* 57, 233–266.

Bajsa, N., Fabiano, E. and Rivas-Franco, F. (2023) Biological control of phytopathogens and insect pests in agriculture: an overview of 25 years of research in Uruguay. *Environmental Sustainability* 6, 121–133.

Banetashvili, I. (2019) The actinomycetes-antagonists and their use against some potato disease caused by phytopathogenic bacteria. MSc thesis, Ivane Javakhishvili Tbilisi State University, Tbilisi, Georgia.

Barratt, B.I.P., Moran, V.C., Bigler, F. and van Lenteren, J.C. (2018) The status of biological control and recommendations for improving uptake for the future. *BioControl* 63, 155–167.

Batta, Y. (1999) Biological effect of two strains of microorganisms antagonistic to Botrytis cinerea causal organism of gray mold on strawberry. *An-Najah University Journal for Research A Natural Sciences* 13, 67–83.

Bérdy, J. (2005) Bioactive microbial metabolites: a personal view. *Journal of Antibiotics* 58, 1–26.

Bhagwat, A., Collins, C.H. and Dordick, J.S. (2019) Selective antimicrobial activity of cell lytic enzymes in a bacterial consortium. *Applied Microbiology and Biotechnology* 103, 7041–7054.

Bhusal, B. and Mmbaga, M.T. (2020) Biological control of phytophthora blight and growth promotion in sweet pepper by bacillus species. *Biological Control* 150, 104373.

Blair, A., Ritz, B., Wesseling, C. and Freeman, L.B. (2015) Pesticides and human health. *Occupational and Environmental Medicine* 72, 81–82.

Bolton, M.D., Thomma, B.P.H.J. and Nelson, B.D. (2006) *Sclerotinia sclerotiorum* (lib.) de bary: Biology and molecular traits of a cosmopolitan pathogen. *Molecular Plant Pathology* 7, 1–16.

Bonaterra, A., Badosa, E., Daranas, N., Francés, J., Roselló, G. *et al.* (2022) Bacteria as biological control agents of plant diseases. *Microorganisms* 10, 1759.

Boudyach, E.H., Fatmi, M., Akhayat, O., Benizri, E. and Aoumar, A.A.B. (2001) Selection of antagonistic bacteria of *Clavibacter michiganensis* subsp. *michiganensis* and evaluation of their efficiency against bacterial canker of tomato. *Biocontrol Science and Technology* 11, 141–149.

Bullerman, L.B. (2003) SPOILAGE/fungi in food – an overview. *Encyclopedia of Food Sciences and Nutrition* 2, 5511–5522.

Cabrefiga, J., Bonaterra, A. and Montesinos, E. (2007) Mechanisms of antagonism of Pseudomonas fluorescens EPS62e against Erwinia amylovora, the causal agent of fire blight. *International Microbiology* 10, 123.

Campos-Avelar, I., Noue, A., Durand, N., Cazals, G., Martinez, V. *et al.* (2021) Aspergillus flavus growth inhibition and aflatoxin B1 decontamination by streptomyces isolates and their metabolites. *Toxins* 13, 340.

Caulier, S., Gillis, A., Colau, G., Licciardi, F., Liépin, M. *et al.* (2018) Versatile antagonistic activities of soil-borne *Bacillus* spp. and *Pseudomonas* spp. against *Phytophthora infestans* and other potato pathogen. *Frontiers in Microbiology* 9, 143.

Chakraborty, N., Mitra, R., Pal, S., Ganguly, R., Acharya, K. *et al.* (2023) Biopesticide consumption in India: insights into the current trends. *Agriculture* 13, 557.

Chaudhary, S., Sagar, S., Lal, M., Tomar, A., Kumar, V. *et al.* (2020) Biocontrol and growth enhancement potential of *Trichoderma* spp. against *Rhizoctonia solani* causing sheath blight disease in rice. *Journal of Environmental Biology* 41, 1034–1045.

Chien, Y.-C. and Huang, C.-H. (2020) Biocontrol of bacterial spot on tomato by foliar spray and growth medium application of *Bacillus amyloliquefaciens* and *Trichoderma asperellum. European Journal of Plant Pathology* 156, 995–1003.

Choudhary, D.K., Nabi, S.U., Dar, M.S. and Khan, K.A. (2018) *Ralstonia solanacearum*: a wide spread and global bacterial plant wilt pathogen. *Journal of Pharmacognosy and Phytochemistry* 7, 85–90.

Conrath, U., Beckers, G.J., Langenbach, C.J. and Jaskiewicz, M.R. (2015) Priming for enhanced defense. *Annual Review of Phytopathology* 53, 97–119.

Czajkowski, R., Boer, W.J., Veen, J.A. and Wolf, J.M. (2012) Characterization of bacterial isolates from rotting potato tuber tissue showing antagonism to *Dickeya* spp. biovar 3 in vitro and in planta: *Dickeya* spp. antagonists from rotting potato tissue. *Plant Pathology* 61, 169–182.

Damalas, C.A. and Eleftherohorinos, I.G. (2011) Pesticide exposure, safety issues, and risk assessment indicators. *International Journal of Environmental Research and Public Health* 8, 1402–1419.

De O. Gomes, H., Menezes, J.M.C., da Costa, J.G.M., Coutinho, H.D.M., Teixeira, R.N.P. *et al.* (2020) A socio-environmental perspective on pesticide use and food production. *Ecotoxicology and Environmental Safety* 197, 110627.

Dean, R., Van Kan, J.A.L., Pretorius, Z.A., Hammond-Kosack, K.E., Di Pietro, A. *et al.* (2012) The top 10 fungal pathogens in molecular plant pathology. *Molecular Plant Pathology* 13, 414–430.

Dobrzyński, J., Jakubowska, Z., Kulkova, I., Kowalczyk, P. and Kramkowski, K. (2023) Biocontrol of fungal phytopathogens by *Bacillus pumilus. Frontiers in Microbiology* 14, 1194606.

Dong, H., Gao, R., Dong, Y., Yao, Q. and Zhu, H. (2023) Bacillus velezensis RC116 inhibits the pathogens of bacterial wilt and Fusarium wilt in tomato with multiple biocontrol traits. *International Journal of Molecular Sciences* 24, 8527.

Doni, F., Isahak, A., Fathurrahman, F. and Yusoff, W.M.W. (2023) Rice plants' resistance to sheath blight infection is increased by the synergistic effects of Trichoderma inoculation with SRI management. *Agronomy* 13, 711.

Dönmez, M.F. and Aliyeva, Z. (2023) Biological control of bean halo blight disease (*Pseudomonas savastanoi* pv. *phaseolicola*) with antagonist bacterial strains. *Gesunde Pflanzen* 75, 815–824.

Druzhinina, I. and Kubicek, C.P. (2005) Species concepts and biodiversity in *Trichoderma* and *Hypocrea*: from aggregate species to species clusters. *Journal of Zhejiang University Science B* 6, 100–112.

El Gayar, A. (2021) Water systems strategy relation with horticultural crops. *International Journal of Agricultural and Applied Sciences* 1, 1–15.

El-Saadony, M.T., Saad, A.M., Soliman, S.M., Salem, H.M., Ahmed, A.I. *et al.* (2022) Plant growth-promoting microorganisms as biocontrol agents of plant diseases: mechanisms, challenges and future perspectives. *Frontiers in Plant Science* 13, 923880.

Elad, Y. and Stewart, A. (2007) Microbial control of *Botrytis* spp. In: Elad, Y., Williamson, B., Tudzynski, P. and Delen, N. (eds) *Botrytis: Biology, Pathology and Control*. Springer, Dordrecht, pp. 223–241.

Elad, Y., Williamson, B., Tudzynski, P. and Delen, N. (eds) (2007) *Botrytis: Biology, Pathology and Control.* Springer, Dordrecht, thte Netherlands.

Elikaei, A., Poshtiban, S. and Soudi, M.R. (2023) Isolation and identification of bacteriophage effective on *Xanthomonas campestris* strains for their biocontrol. *Journal of Microbial Biology* 12, 77–95.

Elsharkawy, M.M., Kamel, S. and El-Khateeb, N.M.M. (2014) Biological control of powdery and downy mildews of cucumber under greenhouse conditions. *Egyptian Journal of Biological Pest Control* 24, 407–414.

Elsharkawy, M.M., Khedr, A.A., Mehiar, F., El-Kady, E.M., Alwutayd, K.M. *et al.* (2023) Rhizobacterial colonization and management of bacterial speck pathogen in tomato by *pseudomonas* spp. *Microorganisms* 11, 1103.

Escobar, M.A. and Dandekar, A.M. (2003) *Agrobacterium tumefaciens* as an agent of disease. *Trends in Plant Science* 8, 380–386.

Fenibo, E.O., Ijoma, G.N. and Matambo, T. (2021) Biopesticides in sustainable agriculture: a critical sustainable development driver governed by green chemistry principles. *Frontiers in Sustainable Food Systems* 5.

Fenibo, E.O., Ijoma, G.N. and Matambo, T. (2022) Biopesticides in sustainable agriculture: current status and future prospects. In: Mandal, S.D., Ramkumar, G., Karthi, S., Jin, F. (eds) *New and Future Development in Biopesticide Research: Biotechnological Exploration.* Springer, Singapore, pp. 1–53.

Fetzner, S. (2015) Quorum quenching enzymes. *Journal of Biotechnology* 201, 2–14.

Fischetti, V.A. (2005) Bacteriophage lytic enzymes: novel anti-infectives. *Trends in Microbiology* 13, 491–496.

Fradin, E.F. and Thomma, B.P.P. (2006) Physiology and molecular aspects of verticillium wilt diseases caused by *V. dahliae and V. albo-atrum. Molecular Plant Pathology* 7, 71–86.

Fry, W.E., Birch, P.R.J., Judelson, H.S., Grünwald, N.J., Danies, G. *et al.* (2015) Five reasons to consider *Phytophthora infestans* a reemerging pathogen. *Phytopathology* 105, 966–981.

Fu, M., Qu, Z., Pierre-Pierre, N., Jiang, D., Souza, F.L. *et al.* (2023) Exploring the mycovirus sshadv-1 as a biocontrol agent of white mold caused by *Sclerotinia sclerotiorum. Plant Diseases* 108, 624–634.

Gao, L., Kumaravel, K., Xiong, Q., Liang, Y., Ju, Z. *et al.* (2023) Actinomycins produced by endophyte *Streptomyces* spp. GLL-9 from navel orange plant exhibit high antimicrobial effect against *Xanthomonas citri* susp. *citri* and *Penicillium italicum. Pest Management Science* 79, 4679–4693.

Gayder, S., Kammerecker, S. and Fieseler, L. (2023) Biological control of the fire blight pathogen Erwinia amylovora using bacteriophages. *Journal of Plant Pathology* 106, 853–869.

Gehlot, P. and Singh, J. (eds) (2018) *Fungi and their Role in Sustainable Development: Current Perspectives.* Springer, Singapore.

Ghasemi, S., Safaie, N., Shahbazi, S., Shams-Bakhsh, M. and Askari, H. (2020) The role of cell wall degrading enzymes in antagonistic traits of *Trichoderma virens* against *Rhizoctonia solani. Iranian Journal of Biotechnology* 18, e2333.

Glare, T.R. and O'Callaghan, M. (2019) Microbial biopesticides for control of invertebrates: progress from New Zealand. *Journal of Invertebrate Pathology* 165, 82–88.

Gomiero, T., Pimentel, D. and Paoletti, M.G. (2011) Environmental impact of different agricultural management practices: Conventional vs. organic agriculture. *Critical Reviews in Plant Sciences* 30(1–2), 95–124.

Gonçalves, C.R. and Delabona, P.P. da S. (2022) Strategies for bioremediation of pesticides: challenges and perspectives of the Brazilian scenario for global application – a review. *Environmental Advances* 8, 100220.

Guigón-López, C., Vargas-Albores, F., Guerrero-Prieto, V., Ruocco, M. and Lorito, M. (2015) Changes in *Trichoderma asperellum* enzyme expression during parasitism of the cotton root rot pathogen *Phymatotrichopsis omnivora. Fungal Biology* 119, 264–273.

Guimaraes, S.D.S.C., Santos, Í.A.F.M., Nunes, P.S. de O., Mengez, G.A.L., Monteiro, M.C.P. *et al.* (2023) *Muscodor* spp. controls tomato wilt disease by *Ralstonia solanacearum* and increases yield and total soluble solids content in tomatoes. *Research Square* doi: 10.21203/rs.3.rs-3097277/v1.

Haas, D. and Défago, G. (2005) Biological control of soil-borne pathogens by fluorescent pseudomonads. *Nature Reviews Microbiology* 3, 307–319.

Handford, C.E., Elliott, C.T. and Campbell, K. (2015) A review of the global pesticide legislation and the scale of challenge in reaching the global harmonization of food safety standards. *Integrated Environmental Assessment and Management* 11, 525–536.

Hanlon, G.W. (2007) Bacteriophages: an appraisal of their role in the treatment of bacterial infections. *International Journal of Antimicrobial Agents* 30, 118–128.

Harman, G.E. (2006) Overview of mechanisms and uses of *Trichoderma* spp. *Phytopathology* 96, 190–194.

Hashemi, M., Tabet, D., Sandroni, M., Benavent-Celma, C., Seematti, J. *et al.* (2022) The hunt for sustainable biocontrol of oomycete plant pathogens, a case study of *Phytophthora infestans*. *Fungal Biology Reviews* 40, 53–69.

Heimpel, G.E. and Mills, N.J. (2017) *Biological Control*. Cambridge University Press, Cambridge, UK.

Hermann, D. and Stenzel, K. (2019) FRAC mode-of-action classification and resistance risk of fungicides. In: Jeschke, P., Witschel, M, Krämer, W. and Schirmer, U. (eds) *Modern Crop Protection Compounds*. Wiley, pp. 589–608.

Hernández-Huerta, J., Tamez-Guerra, P., Gomez-Flores, R., Delgado-Gardea, M.C.E., Robles-Hernández, L. *et al.* (2023) Pepper growth promotion and biocontrol against *Xanthomonas euvesicatoria* by *Bacillus cereus* and *Bacillus thuringiensis* formulations. *PeerJ* 11, e14633.

Hill, C., Guarner, F., Reid, G., Gibson, G.R., Merenstein, D.J. *et al.* (2014) Expert consensus document. The International Scientific Association for Probiotics And Prebiotics consensus statement on the scope and appropriate use of the term probiotic. *Nature Reviews Gastroenterology and Hepatology* 11, 506–514.

Hutchings, M.I., Truman, A.W. and Wilkinson, B. (2019) Antibiotics: past, present and future. *Current Opinion in Microbiology* 51, 72–80.

Islam, M.N., Ali, M.S., Choi, S.J., Hyun, J.W. and Baek, K.H. (2019) Biocontrol of citrus canker disease caused by *Xanthomonas citri* subspp. *citri* using an endophytic *Bacillus thuringiensis*. *Plant Pathology Journal* 35, 486.

Jacobi, W.R. (2001) Controlling dutch elm disease in the 21st century. *Plant Disease* 85, 236–245.

Ji, S.H., Paul, N.C., Deng, J.X., Kim, Y.S., Yun, B.S. *et al.* (2013) Biocontrol activity of *Bacillus amyloliquefaciens* CNU114001 against fungal plant diseases. *Mycobiology* 41, 234–242.

Jimenez, M., Castillo, F., Alcal, E., Morales, G., Valdes, R. *et al.* (2018) Biological effectiveness of *Bacillus* spp. and *Trichoderma* spp. on apple scab (*Venturia inaequalis*) in vitro and under field conditions. *European Journal of Physical and Agricultural Sciences* 6, 11.

Jones, J.B., Jackson, L.E., Balogh, B., Obradovic, A., Iriarte, F.B. *et al.* (2007) Bacteriophages for plant disease control. *Annual Review of Phytopathology* 45, 245–262.

Junaid, J.M., Dar, N.A., Bhat, T.A., Bhat, A.H. and Bhat, M.A. (2013) Commercial biocontrol agents and their mechanism of action in the management of plant pathogens. *International Journal of Modern Plant and Animal Sciences* 1, 39–57.

Kabdwal, B.C., Sharma, R., Kumar, A., Kumar, S., Singh, K.P. *et al.* (2023) Efficacy of different combinations of microbial biocontrol agents against sheath blight of rice caused by *Rhizoctonia solani*. *Egyptian Journal of Biological Pest Control* 33, 29.

Kanipriya, R., Rajendran, L., Raguchander, T. and Karthikeyan, G. (2019) Characterization of Ampelomyces and its potentiality as an effective biocontrol agent against Erysiphe cichoracearum DC causing powdery mildew disease in Bhendi (Abelmoschus esculentus L.). *Madras Agricultural Journal* 106.

Khabbaz, S.E., Zhang, L., Cáceres, L.A., Sumarah, M., Wang, A. *et al.* (2015) Characterisation of antagonistic *Bacillus* and *Pseudomonas* strains for biocontrol potential and suppression of damping-off and root rot diseases. *Annals of Applied Biology* 166, 456–471.

Khaskheli, M.A., Wu, L., Chen, G., Chen, L., Hussain, S. *et al.* (2020) Isolation and characterization of root-associated bacterial endophytes and their biocontrol potential against major fungal phytopathogens of rice (*Oryza sativa* L.). *Pathogens* 9, 172.

Kifle, M.H., Yobo, K.S. and Laing, M.D. (2017) Biocontrol of Aspergillus flavus in groundnut using *Trichoderma harzianum* stain kd. *Journal of Plant Diseases and Protection* 124, 51–56.

Kleerebezem, M. (2004) Quorum sensing control of lantibiotic production; nisin and subtilin autoregulate their own biosynthesis. *Peptides* 25, 1405–1414.

Knobeloch, L., Salna, B., Hogan, A., Postle, J. and Anderson, H. (2000) Blue babies and nitrate-contaminated well water. *Environmental Health Perspectives* 108, 675–678.

Köhl, J., Kolnaar, R. and Ravensberg, W.J. (2019) Mode of action of microbial biological control agents against plant diseases: relevance beyond efficacy. *Frontiers in Plant Science* 10, 00845.

Kulkarni, S. (2015) Commercialisation of microbial biopesticides for the management of pests and diseases. In: Awasthi, L.P. (ed.) *Recent Advances in the Diagnosis and Management of Plant Diseases*. Springer, pp. 1–10.

Kumari, P. and Rastogi, N. (2023) The occurrence of plant growth-promoting fungus, *Trichoderma harzianum* in the nests of seed-harvester ant, *Trichomyrmex scabriceps*. *Proceedings of the National Academy of Sciences, India – Section B Biological Sciences* 94.

Kumari, S., Kumar, S.C., Jha, M.N., Kant, R., Upendra, S. *et al.* (2014) Microbial pesticide: a boom for sustainable agriculture. *International Journal of Scientific and Engineering Research* 5, 1394–1397.

Lahlali, R., Ezrari, S., Radouane, N., Kenfaoui, J., Esmaeel, Q. *et al.* (2022) Biological control of plant pathogens: a global perspective. *Microorganisms* 10, 596.

Lai, E.M., Shih, H.W., Wen, S.R., Cheng, M.W., Hwang, H.H. *et al.* (2006) Proteomic analysis of Agrobacterium tumefaciens response to the *vir* gene inducer acetosyringone. *Proteomics* 6, 4130–4136.

Le, K.D., Yu, N.H., Park, A.R., Park, D.J., Kim, C.J. *et al.* (2022) *Streptomyces* sp. AN090126 as a biocontrol agent against bacterial and fungal plant diseases. *Microorganisms* 10, 791.

Leach, J.E., Triplett, L.R., Argueso, C.T. and Trivedi, P. (2017) Communication in the phytobiome. *Cell* 169, 587–596.

Legein, M., Smets, W., Vandenheuvel, D., Eilers, T., Muyshondt, B. *et al.* (2020) Modes of action of microbial biocontrol in the phyllosphere. *Frontiers in Microbiology* 11, 1619.

Lemańczyk, G., Lisiecki, K. and Piesik, D. (2023) Binucleate rhizoctonia strain: a potential biocontrol agent in wheat production. *Agronomy* 13, 523.

Li, J., Philp, J., Li, J., Wei, Y., Li, H. *et al.* (2020) *Trichoderma harzianum* inoculation reduces the incidence of clubroot disease in Chinese cabbage by regulating the rhizosphere microbial community. *Microorganisms* 8, 1325.

Ling, L., Han, X., Li, X., Zhang, X., Wang, H. *et al.* (2020) A Streptomyces sp. NEAU-HV9: isolation, identification, and potential as a biocontrol agent against *Ralstonia solanacearum* of tomato plants. *Microorganisms* 8, 351.

Liu, K., Newman, M., McInroy, J.A., Hu, C.H. and Kloepper, J.W. (2017) Selection and assessment of plant growth-promoting rhizobacteria for biological control of multiple plant diseases. *Phytopathology* 107, 928–936.

Liu, X., Cao, A., Yan, D., Ouyang, C., Wang, Q. *et al.* (2021) Overview of mechanisms and uses of biopesticides. *International Journal of Pest Management* 67, 65–72.

Lopes-Ferreira, M., Maleski, A.L.A., Balan-Lima, L., Bernardo, J.T.G., Hipolito, L.M. *et al.* (2022) Impact of pesticides on human health in the last six years in Brazil. *International Journal of Environmental Research and Public Health* 19, 3198.

Maji, S. and Chakrabartty, P.P.K. (2014) Biocontrol of bacterial wilt of tomato caused by *Ralstonia solanacearum* by isolates of plant growth promoting rhizobacteria. *Australian Journal of Crop Science* 8, 208–214.

Malek, A.A., Ali, N.S., Kadir, J., Vadamalai, G. and Saud, H.M. (2023) *Enterobacter tabaci* and *Bacillus cereus* as biocontrol agents against pathogenic *Ralstonia solanacearum* of eggplant. *Asian Journal of Tropical Biotechnology* 20, 24–30.

Malnoy, M., Martens, S., Norelli, J.L., Barny, M.-A., Sundin, G.W. *et al.* (2012) Fire blight: applied genomic insights of the pathogen and host. *Annual Review of Phytopathology* 50, 475–494.

Mansfield, J., Genin, S., Magori, S., Citovsky, V., Sriariyanum, M. *et al.* (2012) Top 10 plant pathogenic bacteria in molecular plant pathology. *Molecular Plant Pathology* 13, 614–629.

Marín, D.H., Romero, R.A., Guzmán, M. and Sutton, T.B. (2003) Black sigatoka: an increasing threat to banana cultivation. *Plant Disease* 87, 208–222.

Marrone, P.G. (2014) The market and potential for biopesticides. In: Gross, A., Coats, J.R., Duke, S.O. and Seiber, J.N. (eds) *Biopesticides: State of the Art and Future Opportunities*. American Chemical Society, pp. 245–258.

Metz, N. and Hausladen, H. (2022) Trichoderma spp. as potential biological control agent against Alternaria solani in potato. *Biological Control: Theory and Applications in Pest Management* 166, 104820.

Michalak, I., Aliman, J., Hadžiabulić, A. and Komlen, V. (2022) Novel trends in crop bioprotection. In: Chojnacka, K. and Saeid, A. (eds) *Smart Agrochemicals for Sustainable Agriculture*. Elsevier, San Diego, California, pp. 185–224.

Misawa, T. and Kurose, D. (2021) First report of parsley basal petiole rot caused by *Alternaria petroselini* and comparison with parsley leaf blight pathogen in terms of morphology, phylogeny and pathogenicity. *Journal of General Plant Pathology* 87, 196–199.

Muhammad, M., Abdul Wahab, R., Huyop, F.Z., Rusli, M.H., Syed Yaacob, S.N. *et al.* (2022) An overview of the potential role of microbial metabolites as greener fungicides for future sustainable plant diseases management. *Journal of Crop Protection* 11, 1–27.

Nascimento, F.X., Tavares, M.J., Rossi, M.J. and Glick, B.R. (2018) The modulation of leguminous plant ethylene levels by symbiotic rhizobia played a role in the evolution of the nodulation process. *Heliyon* 4, e01068.

Nelson, P.E., Dignani, M.C. and Anaissie, E.J. (1994) Taxonomy, biology, and clinical aspects of fusarium species. *Clinical Microbiology Reviews* 7, 479–504.

Newman, M.A., Sundelin, T., Nielsen, J.T. and Erbs, G. (2013) MAMP (microbe-associated molecular pattern) triggered immunity in plants. *Frontiers in Plant Science* 4, 139.

Nicholls, C.I. and Altieri, M.A. (2013) Plant biodiversity enhances bees and other insect pollinators in agroecosystems. a review. *Agronomy for Sustainable Development* 33, 257–274.

Ongena, M. and Jacques, PP. (2008) Bacillus lipopeptides: versatile weapons for plant disease biocontrol. *Trends in Microbiology* 16, 115–125.

Ortiz, A. and Sansinenea, E. (2022) The role of beneficial microorganisms in soil quality and plant health. *Sustainability* 14, 5358.

Osorio-Hernández, E., Hernández-Castillo, F.D., Gallegos-Morales, G., Rodríguez-Herrera, R. and Castillo-Reyes, F. (2011) In-vitro behavior of *Trichoderma* spp. against *Phytophthora capsici leonian*. *African Journal of Agricultural Research* 6, 4594–4600.

Paret, M., Dufault, N., Momol, T., Marois, J. and Olson, S. (2015) Integrated disease management for vegetable crops in Florida. PP-193. Institute of Food and Agricultural Sciences, University of Florida. Available at: https://journals.flvc.org/edis/article/download/119988/118083 (last accessed November 2024).

Patel, S. and Saraf, M. (2017) Biocontrol efficacy oftrichodermaasperellum MSST against tomato wilting by Fusarium oxysporum f. sp. lycopersici. *Archives of Phytopathology and Plant Protection* 50, 228–238.

Pathak, V.M., Verma, V.K., Rawat, B.S., Kaur, B., Babu, N. *et al.* (2022) Current status of pesticide effects on environment, human health and it's eco-friendly management as bioremediation: a comprehensive review. *Frontiers in Microbiology* 13, 962619.

Pathma, J., Kennedy, R.K., Bhushan, L.S., Shankar, B.K. and Thakur, K. (2021) Microbial biofertilizers and biopesticides: nature's assets fostering sustainable agriculture. In: Prasad, R., Kumar, V., Singh, J., Upadhyaya, C.P. (eds) *Recent Developments in Microbial Technologies*. Springer, Singapore, pp. 39–69.

Paula Kuyat Mates, A., Carvalho Pontes, N. and Almeida Halfeld-Vieira, B. (2019) Bacillus velezensis GF267 as a multi-site antagonist for the control of tomato bacterial spot. *Biological Control: Theory and Applications in Pest Management* 137, 104013.

Pertot, I., Alabouvette, C., Esteve, E.H. and Franca, S., 2015. Mini-paper. The use of microbial biocontrol agents against soil-borne diseases. Eip-Agri Focus Group Soil-Borne Diseases, Brussels.

Pham, J.V., Yilma, M.A., Feliz, A., Majid, M.T., Maffetone, N. *et al.* (2019) A review of the microbial production of bioactive natural products and biologics. *Frontiers in Microbiology* 10, 1404.

Pieterse, C.M., Zamioudis, C., Berendsen, R.L., Weller, D.M., Wees, S.C. *et al.* (2014) Induced systemic resistance by beneficial microbes. *Annual Review of Phytopathology* 52, 347–375.

Pitt, J.I. (2000) Toxigenic fungi and mycotoxins. *British Medical Bulletin* 56, 184–192.

Pliego, C., Weert, S., Lamers, G., Vicente, A., Bloemberg, G. *et al.* (2008) Two similar enhanced root-colonizing pseudomonas strains differ largely in their colonization strategies of avocado roots and *Rosellinia necatrix* hyphae. *Environmental Microbiology* 10, 3295–3304.

Podgórska-Kryszczuk, I. (2023) Biological control of *Aspergillus flavus* by the yeast *Aureobasidium pullulans* in vitro and on tomato fruit. *Plants* 12, 236.

Prajapati, S., Kumar, N., Kumar, S. and Maurya, S. (2020) Biological control a sustainable approach for plant diseases management: a review. *Journal of Pharmacognosy and Phytochemistry* 9, 1514–1523.

Quach, N.T., Vu, T.H.N., Nguyen, T.T.A., Le, P.C., Do, H.G. *et al.* (2023) Metabolic and genomic analysis deciphering biocontrol potential of endophytic streptomyces albus RC2 against crop pathogenic fungi. *Brazilian Journal of Microbiology* 54, 2617–2626.

Raaijmakers, J.M., Paulitz, T.C., Steinberg, C., Alabouvette, C. and Moënne-Loccoz, Y. (2009) The rhizosphere: a playground and battlefield for soilborne pathogens and beneficial microorganisms. *Plant and Soil* 321, 341–361.

Ragukula, K. and Makandar, R. (2023) *Cladosporium cladosporioides* causes leaf blight on garden pea in Telangana, India. *Plant Disease* doi: 10.1094/PDIS-09-22-2175-PDN.

Raja, Sunkad, G. and Amaresh, Y.S. (2023) Bioagents induced resistance to Ceratocystis fimbriata in pomegranate. *International Journal of Environment and Climate Change* 13, 2277–2287.

Ram, R.M., Pandey, A.K. and Singh, H.B. (2021) Biocontrol research in India. In: Deshmukh, S.K., Deshpande, M.V. and Satyanarayana, T. (eds) *Progress in Mycology*. Springer, Singapore, pp. 371–395.

Ruiu, L. (2018) Microbial biopesticides in agroecosystems. *Agronomy* 8, 235.

Rungjindamai, N., Xu, X.M. and Jeffries, P. (2013) Identification and characterisation of new microbial antagonists for biocontrol of *Monilinia laxa*, the causal agent of brown rot on stone fruit. *Agronomy* 3, 685–703.

Saghafi, D., Asgari Lajayer, B. and Ghorbanpour, M. (2020) Engineering bacterial ACC deaminase for improving plant productivity under stressful conditions. In: Sharma, V., Salwan, R. and Al-Ani, L.K.T. (eds) *Molecular Aspects of Plant Beneficial Microbes in Agriculture*. Elsevier, San Diego, California, pp. 259–277.

Samada, L.H. and Tambunan, U.S.F. (2020) Biopesticides as promising alternatives to chemical pesticides: a review of their current and future status. *OnLine Journal of Biological Sciences* 20, 66–76.

Santoso, I., Fadhilah, Q.G. and Maryanto, A.E. (2021) Antagonist effect of *Bacillus* spp. against *Aspergillus niger* CP isolated from cocopeat powder. *IOP Conference Series* 846, 012001.

Seenivasagan, R. and Babalola, O.O. (2021) Utilization of microbial consortia as biofertilizers and biopesticides for the production of feasible agricultural product. *Biology* 10, 1111.

See-Too, W.S., Convey, P., Pearce, D.A. and Chan, K.G. (2018) Characterization of a novel N-acylhomoserine lactonase, AidP, from Antarctic Planococcus sp. *Microbial Cell Factories* 17, 179.

Sharma, A., Khosla, K., Mahajan, R., Bharti, J., Gupta, A.K. *et al.* (2017) Antagonistic potential of native agrocin-producing non-pathogenic *Agrobacterium tumefaciens* strain UHFBA-218 to control crown gall in peach. *Phytoprotection* 97, 1–11.

Sharma, K., Kreuze, J., Abdurahman, A., Parker, M., Nduwayezu, A. *et al.* (2021) Molecular diversity and pathogenicity of Ralstonia solanacearum species complex associated with bacterial wilt of potato in Rwanda. *Plant Disease* 105, 770–779.

Shivalingaiah, U.S. and Umesha, S. (2013) Pseudomonas fluorescens inhibits the Xanthomonas oryzae pv. oryzae, the bacterial leaf blight pathogen in rice. *Can. J. Plant Protect* 1, 147–153.

Shopova, E., Brankova, L., Ivanov, S., Urshev, Z., Dimitrova, L. *et al.* (2023) *Xanthomonas euvesicatoria*-specific bacteriophage BsXeu269p/3 reduces the spread of bacterial spot disease in pepper plants. *Plants* 12, 3348.

Shoresh, M., Harman, G.E. and Mastouri, F. (2010) Induced systemic resistance and plant responses to fungal biocontrol agents. *Annual Review of Phytopathology* 48, 21–43.

Sikdar, R. and Elias, M. (2020) Quorum quenching enzymes and their effects on virulence, biofilm, and microbiomes: a review of recent advances. *Expert Review of Anti-Infective Therapy* 18, 1221–1233.

Singh, S., Harwani, D. and Harwani, D. (2017) Role of bacillus as plant growth promoting rhizobacteria. In: Harwani, D. (ed.) *Rhizotrophs: Plant Growth Promotion to Bioremediation*. Springer, pp. 245–269.

Someya, N., Tsuchiya, K., Yoshida, T., Noguchi, M.T., Akutsu, K. *et al.* (2007) Fungal cell wall degrading enzyme-producing bacterium enhances the biocontrol efficacy of antibiotic-producing bacterium against cabbage yellows. *Journal of Plant Diseases and Protection* 114, 108–112.

Spurgeon, D., Lahive, E., Robinson, A., Short, S. and Kille, PP. (2020) Species sensitivity to toxic substances: evolution, ecology and applications. *Frontiers in Environmental Science* 8, 588380.

Stefani, E., Obradović, A., Gašić, K., Altin, I., Nagy, I.K. *et al.* (2021) Bacteriophage-mediated control of phytopathogenic xanthomonads: a promising green solution for the future. *Microorganisms* 9, 1056.

Stenberg, J.A., Sundh, I., Becher, P.G., Björkman, C., Dubey, M. *et al.* (2021) When is it biological control? A framework of definitions, mechanisms, and classifications. *Journal of Pest Science* 94, 665–676.

Sun, Y., Su, Y., Meng, Z., Zhang, J., Zheng, L. *et al.* (2023) Biocontrol of bacterial wilt disease in tomato using *Bacillus subtilis* strain R31. *Frontiers in Microbiology* 14, 1281381.

Tal, A. (2018) Making conventional agriculture environmentally friendly: moving beyond the glorification of organic agriculture and the demonization of conventional agriculture. *Sustainability* 10, 1078.

Teja, A.R., Leona, G., Prasanth, J., Yatung, T., Singh, S. *et al.* (2023) Role of plant growth-promoting rhizobacteria in sustainable agriculture. In: Gangola, S., Kumar, S., Joshi, S. and Bhatt, P. (eds) *Advanced Microbial Technology for Sustainable Agriculture and Environment*. Academic Press, pp. 175–197.

Thakur, N., Kaur, S., Tomar, P., Thakur, S. and Yadav, A.N. (2020) Microbial biopesticides: current status and advancement for sustainable agriculture and environment. In: Rastegari, A.A., Yadav, A.N. and Yadav, N. (eds) *New and Future Developments in Microbial Biotechnology and Bioengineering*. Elsevier, pp. 243–282.

Timilsina, S., Potnis, N., Newberry, E.A., Liyanapathiranage, P., Iruegas-Bocardo, F. *et al.* (2020) Xanthomonas diversity, virulence and plant–pathogen interactions. *Nature Reviews Microbiology* 18, 415–427.

Triplett, E.W., Kaeppler, S.M. and Chelius, M.K. (2008) Wisconsin Alumni Research Foundation. Klebsiella pneumoniae inoculants for enhancing plant growth. US Patent 7,393,678. Available at: https://patents.google.com/patent/US7393678B2/en (last accessed November 2024).

Tyśkiewicz, R., Nowak, A., Ozimek, E. and Jaroszuk-Ściseł, J. (2022) Trichoderma: the current status of its application in agriculture for the biocontrol of fungal phytopathogens and stimulation of plant growth. *International Journal of Molecular Sciences* 23, 2329.

Upmanyu, N. and Malviya, V.N. (2020) Antibiotics: mechanisms of action and modern challenges. In: Chowdhary, P., Raj, A., Verma, D. and Akhter, Y. (eds) *Microorganisms for Sustainable Environment and Health.* Elsevier, pp. 367–382.

Vero, S., Garmendia, G., Allori, E., Sanz, J.M., Gonda, M. *et al.* (2023) Microbial biopesticides: diversity, scope, and mechanisms involved in plant disease control. *Diversity* 15, 457.

Vidhyasekaran, P.P., Kamala, N., Ramanathan, A., Rajappan, K., Paranidharan, V. *et al.* (2001) Induction of systemic resistance by *Pseudomonas fluorescens* Pf1 against *Xanthomonas oryzae* pv. *oryzae* in rice leaves. *Phytoparasitica* 29, 155–166.

Weller, D.M., Raaijmakers, J.M., Gardener, B.B.M. and Thomashow, L.S. (2002) Microbial populations responsible for specific soil suppressiveness to plant pathogens. *Annual Review of Phytopathology* 40, 309–348.

Whipps, J.M. (2001) Microbial interactions and biocontrol in the rhizosphere. *Journal of Experimental Botany* 52 (Spec. Issue), 487–511.

Whipps, J.M., Sreenivasaprasad, S., Muthumeenakshi, S., Rogers, C.W. and Challen, M.P. (2008) Use of *Coniothyrium minitans* as a biocontrol agent and some molecular aspects of sclerotial mycoparasitism. *European Journal of Plant Pathology* 121, 323–330.

Yang, Y., Zhang, S.-W. and Li, K.-T. (2019) Antagonistic activity and mechanism of an isolated *Streptomyces corchorusii* stain AUH-1 against phytopathogenic fungi. *World Journal of Microbiology and Biotechnology* 35, 145.

Ye, T., Zhang, W., Feng, Z., Fan, X., Xu, X. *et al.* (2020) Characterization of a novel quorum-quenching bacterial strain, *Burkholderia anthina* HN-8, and its biocontrol potential against black rot disease caused by *Xanthomonas campestris* pv. *campestris. Microorganisms* 8, 1485.

Yendyo, S., Ramesh, G.C. and Pandey, B.R. (2017) Evaluation of *Trichoderma* spp., *Pseudomonas fluorescens* and *Bacillus subtilis* for biological control of *Ralstonia* wilt of tomato. *F1000Research* 6, 2028.

Yu, X., Song, W., Huang, X., Wang, J., Wang, H. *et al.* (2022) A Streptomyces sp. Neau-Y11: isolation, identification, and potential as a biocontrol agent against cucumber anthracnose caused by colletotrichum orbiculare. *SSRN Electronic Journal* doi: 10.2139/ssrn.4314481.

Zegeye, E.D., Santhanam, A., Gorfu, D., Tessera, M. and Kassa, B. (2011) Biocontrol activity of *Trichoderma viride* and *Pseudomonas fluorescens* against *Phytophthora infestans* under greenhouse conditions. *Journal of Agricultural Technology* 7, 1589–1602.

Zimmermann, G. (2008) The entomopathogenic fungus metarhizium anisopliae and its potential as a biopesticide. *Outlooks on Pest Management* 19, 15–19.

Index

Note:- Page numbers in bold type refer to figures
Page numbers in italic type refer to tables.